Phytoplasmas: Genomes, Plant Hosts and Vectors

Phytoplasmas: Genomes, Plant Hosts and Vectors

2nd Edition

Edited by

Wei Wei

Research Plant Pathologist
Molecular Plant Pathology Laboratory, ARS-USDA,
Beltsville, Maryland, USA

CABI

CABI is a trading name of CAB International

CABI
Nosworthy Way
Wallingford
Oxfordshire OX10 8DE
UK

Tel: +44 (0)1491 832111
E-mail: info@cabi.org
Website: www.cabi.org

CABI
200 Portland Street
Boston
MA 02114
USA

Tel: +1 (617)682-9015
E-mail: cabi-nao@cabi.org

A catalogue record for this book is available from the British Library, London, UK.

ISBN-13: 9781800627017 (hardback)
9781800627024 (ePDF)
9781800627031 (ePub)

DOI: 10.1079/9781800627031.0000

Commissioning Editor: David Hemming
Editorial Assistant: Theresa Regueira
Production Editor: Rosie Hayden

Typeset by Straive, Pondicherry, India
Printed and bound in the UK by CPI Group (UK) Ltd, Croydon, CR0 4YY

Contents

Contributors

Sabrina Araujo, *Department of Physics, Universidade Federal do Paraná, Paraná, Brazil. Email: araujosbl@gmail.com*

Assunta Bertaccini, *Alma Mater Studiorum, University of Bologna – Plant Pathology, Bologna, Italy. Email: assunta.bertaccini@unibo.it*

Yi-Ching Chiu, *Institute of Biochemistry, National Chung Hsing University, Taichung, Taiwan; PhD Program in Microbial Genomics, National Chung Hsing University and Academia Sinica, Taichung, Taiwan. Email: jean31019@gmail.com*

Marina Dermastia, *National Institute of Biology, Ljubljana, Slovenia. Email: marina.dermastia@nib.si*

Matthew Dickinson, *School of Biosciences, University of Nottingham, Loughborough, UK. Email: matthew.dickinson1@nottingham.ac.uk*

Christopher H. Dietrich, *Illinois Natural History Survey, Prairie Research Institute, University of Illinois at Urbana-Champaign, Champaign, Illinois, USA. Email: chdietri@illinois.edu*

Marina Drčelić, *Department of Biology, Faculty of Science, University of Zagreb, Zagreb, Croatia. Email: marina.drcelic@biol.pmf.hr*

Simone Egger, *Molecular Biology and Microbiology, Group of Breeding Genomics, Laimburg Research Centre, Pfatten (Vadena), South Tyrol, Italy. Email: simone.egger@laimburg.it*

Jiancan Feng, *Henan Agricultural University, Zhengzhou, China. Email: jcfeng@henau.edu.cn*

Luciana Galetto, *Istituto per la Protezione Sostenibile delle Piante, CNR, Torino, Italy. Email: luciana.galetto@cnr.it*

Jennifer Hodgetts, *Department of Epidemiological Sciences, Animal and Plant Health Agency, Addlestone, UK. Email: jennifer.hodgetts@apha.gov.uk*

Katrin Janik, *Molecular Biology and Microbiology, Group of Functional Genomics, Laimburg Research Centre, Pfatten (Vadena), Italy. Email: katrin.janik@laimburg.it*

Barbara Jarausch, *Julius Kuehn-Institute, Siebeldingen, Germany. Email: barbara.jarausch@julius-kuehn.de*

Wolfgang Jarausch, *RLP AgroScience, Neustadt an der Weintraße, Germany. Email: wolfgang.jarausch@agroscience.rlp.de*

Shigeyuki Kakizawa, *National Institute of Advanced Industrial Science and Technology (AIST), Ibaraki, Japan. Email: s.kakizawa@aist.go.jp*

Yugo Kitazawa, *Yamaguchi University, Yamaguchi, Japan. Email: ykitazawa@yamaguchi-u.ac.jp*

Chih-Horng Kuo, *Institute of Plant and Microbial Biology, Academia Sinica, Taipei, Taiwan; Institute of Biotechnology, National Chung Hsing University, Taichung, Taiwan; PhD Program in Microbial Genomics, National Chung Hsing University and Academia Sinica, Taichung, Taiwan. Email: chk@gate.sinica.edu.tw*

Jidong Li, *Henan Agricultural University, Zhengzhou, China. Email: lijidong@henau.edu.cn*
Helen Mae Mejia, *Institute of Biochemistry, National Chung Hsing University, Taichung, Taiwan. Email: helennavalmejia@gmail.com*
Kensaku Maejima, *The University of Tokyo, Tokyo, Japan. Email: amaejima@g.ecc.u-tokyo.ac.jp*
Christa Malfertheiner, *Molecular Biology and Microbiology, Group of Functional Genomics, Laimburg Research Centre, Pfatten (Vadena), Italy. Emai:christa.malfertheiner@laimburg.it*
Carmine Marcone, *Department of Pharmacy, University of Salerno, Fisciano (Salerno), Italy. Email: cmarcone@unisa.it*
Cristina Marzachì, *Istituto per la Protezione Sostenibile delle Piante, CNR, Torino, Italy. Email: cristina.marzachi@cnr.it*
Cecilia Mittelberger, *Molecular Biology and Microbiology, Group of Functional Genomics, Laimburg Research Centre, Pfatten (Vadena), Italy. Email: cecilia.mittelberger@laimburg.it*
Rita Musetti, *Department of Land, Environment, Agriculture and Forestry (TESAF), University of Padova, Legnaro, Italy. Email: rita.musetti@unipd.it*
Shigetou Namba, *University of Tokyo, Tokyo, Japan. Email: anamba@g.ecc.u-tokyo.ac.jp*
Kenro Oshima, *Hosei University, Tokyo, Japan. Email: kenro@hosei.ac.jp*
Sabrina Palmano, *Istituto per la Protezione Sostenibile delle Piante, CNR, Torino, Italy. Email: sabrina.palmano@cnr.it*
Cecilia Parise, *DISAFA, Università degli Studi di Torino, Torino, Italy; Istituto per la Protezione Sostenibile delle Piante, CNR, Torino, Italy;. Email: cecilia.parise@unito.it*
Martina Šeruga Musić, *Department of Biology, Faculty of Science, University of Zagreb, Zagreb, Croatia. Email: martina.seruga.music@biol.pmf.hr*
Jonathan Shao, *Beltsville Agricultural Research Center, Animal Parasitic Diseases Laboratory, ARS-USDA, Beltsville, Maryland, USA. Email: Jonathan.shao@usda.gov*
Mattia Tabarelli, *Molecular Biology and Microbiology, Group of Functional Genomics, Laimburg Research Centre, Pfatten (Vadena), Italy. Email: mattia.tabarelli@laimburg.it*
Valeria Trivellone, *Illinois Natural History Survey, Prairie Research Institute, University of Illinois at Urbana-Champaign, Champaign, Illinois, USA. Email: valeria3@illinois.edu*
Wei Wei, *Molecular Plant Pathology Laboratory, ARS-USDA, Beltsville, Maryland, USA. Email: wei.wei@usda.gov*
Jun-Yi Yang, *Institute of Biochemistry, National Chung Hsing University, Taichung, Taiwan; PhD Program in Microbial Genomics, National Chung Hsing University and Academia Sinica, Taichung, Taiwan; Institute of Biotechnology, National Chung Hsing University, Taichung, Taiwan; Advanced Plant Biotechnology Center, National Chung Hsing University, Taichung, Taiwan. Email: jyang@nchu.edu.tw*
Shao-Shuai Yu, *Chinese Academy of Tropical Agricultural Sciences, Wenchang, China. Email: yushaoshuai@catas.cn; hzuyss@163.com*
Yan Zhao, *Molecular Plant Pathology Laboratory, ARS-USDA, Beltsville, Maryland, USA. Email: yan.zhao@usda.gov; yzhao888999@yahoo.com*
Xianbo Zheng, *Henan Agricultural University, Zhengzhou, China. Email: xianboz@henau.edu.cn*

Preface

Phytoplasmas have long intrigued researchers and challenged agriculture due to their unusual biology, complex interactions with hosts and insect vectors, and the severe diseases they cause in vital crops. Since their initial identification in 1967 as "mycoplasma-like organisms," our understanding has advanced significantly. Over time, research has shifted from visual assessments to molecular biology, genomics, and precision agriculture, revealing the complexity and adaptability of these fields.

This second edition builds on earlier ones by incorporating recent discoveries and technological advances. It aims to provide a comprehensive and current resource that highlights the ongoing research questions driving the field forward. The chapters cover sensor-based monitoring, precision agriculture for disease management, and cutting-edge sequencing techniques for identifying and tracking phytoplasmas.

Topics include advances in taxonomy, phylogenetics, and the use of genetic markers, such as the 16S rRNA gene, as well as hypotheses regarding seed transmission. A key focus is on the molecular interactions between phytoplasmas and their hosts, including how membrane proteins facilitate infection and disrupt plant processes, and how plants respond with structural, biochemical, and epigenetic changes. Some chapters explore RNA interference as a tool to block the spread of pathogens via insects.

Given the insects' crucial role in transmission, discussions include psyllid vector biology, control strategies, and the evolution of hemipteran–phytoplasma relationships to predict outbreaks better. Additional topics cover phytoplasma mobility within plants and complex infections involving multiple pathogens. Recent omics research provides new insights into how plants respond to infection at the genetic and metabolic levels, informing breeding and management strategies. Contributions from scientists worldwide and across disciplines enrich this edition, reflecting ongoing efforts to unravel this complex host–pathogen–vector system.

I hope this new edition serves as a valuable resource for plant scientists, entomologists, molecular biologists, students, and stakeholders working to protect crops from phytoplasma diseases. By integrating diverse topics, we aim to promote learning, collaboration, and innovative solutions for future challenges.

Acknowledgments

I sincerely thank all the contributors whose expertise, dedication, and thoughtful scholarship have shaped this volume. Each chapter demonstrates your commitment, as you have generously shared your time and insights to deepen understanding in this fast-changing field. Your contributions lay the foundation of this work, integrating diverse perspectives that will serve researchers, practitioners, and students well.

I am also deeply grateful to the reviewers who provided critical and constructive feedback on the manuscripts. Your careful evaluation and insightful suggestions improved the clarity, accuracy, and depth of the content. The quality of this book owes much to your willingness to share your expertise and engage so thoroughly with the material.

A special thanks to the editorial and production team, whose professionalism and attention to detail ensured that the book progressed smoothly through each stage of development. Your support, from coordinating reviews to managing the final proof, has been essential, and your commitment to excellence is evident in the final publication.

Finally, I extend my gratitude to the broader scientific community, including educators and practitioners, who continue to inspire progress in this field. Your ongoing efforts to understand and address the challenges posed by plant pathogens provide the context and motivation for volumes such as this.

It is my hope that this book will serve as a valuable resource, stimulate new ideas, and support ongoing discovery and innovation.

1 Sensor-based Monitoring Systems for Phytoplasma Disease Detection

Wolfgang Jarausch[1]* and Barbara Jarausch[2]

[1]*RLP AgroScience, Neustadt an der Weinstraße, Germany;* [2]*Julius Kuehn-Institute, Siebeldingen, Germany*

Abstract

Phytoplasma-associated diseases are responsible for important economic losses in perennial crops like fruit trees and grapevines. Timely uprooting of infected plants is an efficient strategy to limit the spread of diseases, which requires large-scale monitoring. Phytoplasmas induce biophysical and biochemical modifications in their plant hosts, leading to detectable alterations in leaf spectral patterns. These characteristics make spectral imaging a promising approach for large-scale, non-invasive detection of phytoplasma infections. Multispectral and hyperspectral imaging techniques have been employed for proximal and remote sensing of phytoplasma diseases in grapevine and fruit trees, using handheld devices, vehicle-mounted systems, drones, and satellites.

Introduction

The preventive monitoring of quarantine pests is becoming an increasing requirement imposed by regulatory authorities. For certain quarantine diseases like flavescence dorée (FD), as well as in some fruit-growing regions, monitoring is mandatory. However, phytoplasma disease monitoring in the field by experts is labor-intensive, costly, and time-consuming. It is also prone to errors due to the subjective nature of individual assessments. The lack of resources further hampers local and regional disease management strategies. To address these challenges, new automated monitoring methods are essential for effective disease detection in the future. Digitalization offers promising solutions, such as automated sensor-based detection and data processing by machine learning methods or georeferencing with a geographic information system (GIS). In recent decades, optical sensors have been increasingly used for plant disease detection as they can detect physiological changes in plants caused by biotic stresses (Mahlein *et al.*, 2018). For example, these sensors have been applied for the detection of grapevine diseases, such as FD and bois noir (BN), as well as fruit tree phytoplasma diseases like apple proliferation and pear decline.

Sensor-based detection can be conducted with handheld devices or entirely laboratory-based systems. These devices can also be mounted on vehicles or tractors for short-distance sensing, commonly referred to as "proximal sensing." In contrast, "remote sensing" involves the use of aircraft or satellites. The increasing application of drones or unmanned aerial vehicles (UAVs) is a middle ground between these

*Corresponding author: wolfgang.jarausch@agroscience.rlp.de

DOI: 10.1079/9781800627031.0001

methods, and their classification remains a topic of debate. For consistency, this chapter assigns UAV-based approaches to the "remote sensing" category, following the classification proposed by Lasalle (2021).

In recent years, a wide range of optical sensor-based approaches, including red-green-blue (RGB), thermal imaging, chlorophyll fluorescence, and spectral sensors, have been examined, with spectral sensors emerging as the most promising for disease detection (Mahlein, 2016). Plant reflectance in the visible (VIS; 400–700 nm), near-infrared (NIR; 700–1000 nm), and short-wave infrared (SWIR; 1000–2500 nm) light ranges has been used for phytoplasma disease detection by optical sensors. Applications include laboratory-based systems for NIR, hyperspectral analysis of leaf samples, ground-based monitoring systems on quads or tractors, and remote sensing by UAVs, drones, or satellite image analyses. These sensors can be implemented in automated systems, significantly reducing the workload and enhancing efficiency.

Spectral sensors can be categorized into multispectral and hyperspectral applications based on the number of wavelength bands they record. Multispectral sensors detect a limited number of relatively broad spectral bands, typically in the RGB and NIR ranges. In contrast, hyperspectral sensors capture hundreds of contiguous bands with high spectral resolution (Mahlein, 2016). Hyperspectral imaging (HSI) sensors also capture spatial information, leading to hyperspectral data cubes that consist of two spatial dimensions and one spectral dimension (Bock *et al.*, 2020).

Spectral sensors capture the light reflected by different objects (Bock *et al.*, 2020). In a hyperspectral image, each pixel contains detailed information about the reflection intensity for every spectral band measured. This combination of data results in a spectral signature or spectral profile, which depends on specific leaf properties arising from the interaction of light with plant tissue. These interactions are characterized by absorption, transmission, and reflection (Carter and Knapp, 2001; Blackburn, 2007). In the VIS range, reflectance is relatively low due to the strong absorption properties of leaf pigments such as chlorophylls, carotenoids, and anthocyanins (Gitelson *et al.*, 2001). At the transition from VIS to NIR, reflectance is considerably increased, a feature referred to as "red-edge" (Horler *et al.*, 1983). In the NIR range, reflectance is primarily influenced by internal leaf and cell structures (Seelig *et al.*, 2008). Reflectance intensity decreases in the SWIR range, where absorption is dominated by leaf chemicals and water absorption bands (Seelig *et al.*, 2008).

The increasing size of hyperspectral data provided by the sensors presents challenges for data processing, leading to increased complexity and higher costs. To address this, advanced, automatable analytical methods and statistical approaches are necessary for managing high-dimensional data. Neural networks (Pantazi *et al.*, 2017) and machine learning models (Almoujahed *et al.*, 2022) have been used for data analysis and automated detection. Ali *et al.* (2015) reviewed the most used machine learning algorithms, highlighting their advantages and limitations. These methods include regression, decision trees, support vector machines (SVMs), artificial neural networks (ANNs), deep learning, and ensemble techniques. Among these, ensemble modeling, a machine learning process for creating classification and regression models, has proven to be highly efficient in tasks such as crop disease detection through the utilization of hyperspectral data. This method has been widely adopted by the scientific community for similar objectives (Jang *et al.*, 2024). Key algorithms in this domain are random forests and extreme gradient boosting (XGBoost). The analysis can often be reduced to 5–10 selected key wavelengths instead of using the full spectrum. A prediction model based on support vector regression (SVR) showed performance comparable with or better than full-spectrum analysis, offering a more efficient approach.

Hyperspectral measurements are affected by multiple factors, but often only specific wavelength bands are of primary interest. Therefore, vegetation indices (VIs) have been developed to highlight one factor and reduce the impacts of others (Mahlein *et al.*, 2018). VIs are commonly used to measure plant vitality, nutrition, and biotic and abiotic stresses. One well-established index, the normalized difference vegetation index (NDVI), uses the reflectance of red and NIR light and is correlated with biomass, leaf area index, and chlorophyll content. Candiago *et al.* (2015) compared multiple VIs, such as the NDVI, and demonstrated their utility in detecting plant diseases in crops such as tomatoes and grapevines. As phytoplasmas induce symptoms similar to other stresses like chlorosis and growth reduction, VIs can also be used for phytoplasma disease detection (Knauer *et al.*, 2024; Al Masri *et al.*, 2025). Thus, the specificity of sensor-based detection for phytoplasma diseases depends on how uniquely phytoplasma diseases alter the plant's biochemical and biophysical properties.

Sensor-based Monitoring of Grapevine Yellows Diseases

Symptoms of grapevine yellows

The term grapevine yellows (GY) comprises a number of grapevine diseases that are associated with different phytoplasmas (Angelini *et al.*, 2018). Worldwide, phytoplasmas from ten taxonomically different 16Sr groups have been related to GY (Angelini *et al.*, 2018). In Europe, the main coexisting GY diseases are FD and BN. While the FD phytoplasma belongs to the 16SrV elm yellows group, BN is caused by '*Candidatus* Phytoplasma solani,' a member of the 16SrXII-A stolbur group. In Germany, one case of FD was been detected in the past, which was probably transmitted from adjacent alders to grapevines (Jarausch *et al.*, 2021). The infected vine was uprooted immediately, and Germany was considered FD-free again. However, another phytoplasma of the 16SrV group, responsible for Palatinate grapevine yellows (PGY) disease, is regularly detected in German vineyards (Maixner *et al.*, 1995b). Among these three GY diseases, only FD is spread epidemically from vine to vine by the leafhopper *Scaphoideus titanus* (Malembic-Maher *et al.*, 2020). This mode of transmission makes FD a quarantine disease in Europe (EPPO, 2016).

Although FD, PGY, and BN are members of different groups and subgroups, the symptoms they induce on leaves, shoots, and bunches are indistinguishable (Bertaccini *et al.*, 2014). Symptoms can be restricted to only a few shoots or affect the whole vine (Eveillard *et al.*, 2016). Most prominent are discolorations and necroses of leaf veins that may expand across the whole leaf blade. Depending on the variety, leaves exhibit an irregular yellowing or reddening, may roll down, and become crispy and brittle as the season progresses. Black pustules can appear along infected shoots and canes lignify only partially, remain rubbery, and thus are more affected by the winter cold. Infected grapevines can show reduced growth, flower abortion, and berry shriveling and drying, leading to severe yield losses (Angelini *et al.*, 2018). Furthermore, phytoplasma infection negatively affects berry quality, which is expressed by reduced soluble solid contents, higher acidity, and bitterness, subsequently influencing wine quality (Ember *et al.*, 2018).

Phytoplasma infection in grapevines disrupts photosynthetic activities in the leaves, leading to a reduction in chlorophyll and carotenoid levels (Bertamini *et al.*, 2002). This results in visible symptoms, such as yellowing in white grape varieties and reddening in red varieties, which can be used for the spectral detection of GY diseases.

New EU legislation mandates regular monitoring of FD, even in regions that are currently FD-free. Sensor-based detection methods could greatly facilitate compliance with this task. However, BN is widespread in FD-free regions like Germany, where PGY occasionally occurs. So far, all three GY diseases (FD, BN, and PGY) can only be distinguished by molecular diagnostic means (Maixner *et al.*, 1995a; Arnaud *et al.*, 2007; Pelletier *et al.*, 2009; Malembic-Maher *et al.*, 2020).

Spectral detection of grapevine yellows

Grapevines are typically affected by various biotic and abiotic disorders, many of which result in leaf discoloration. Biotic disorders include fungal and viral diseases or insect damage, while abiotic disorders involve problems such as iron and magnesium deficiencies or physical damage like branch breakage. For spectral detection of GY diseases, it is crucial to first differentiate healthy vines from phytoplasma-infected ones in both white and red cultivars. Secondly, spectral detection must also distinguish phytoplasma infection from other disorders. The most challenging aspect is differentiating between BN and FD, as well as between FD and PGY.

Albetis *et al.* (2017, 2018) explored the potential of multispectral imaging in combination with UAVs for airborne detection of FD symptoms under field conditions. Al-Saddik *et al.* (2017, 2018, 2019) used a portable spectroradiometer (350–2500 nm) to collect hyperspectral reflectance data from healthy and symptomatic leaves.

While much of the research has focused on FD detection, relatively limited attention has been given to other GY diseases. Bendel *et al.* (2020) investigated the potential to distinguish BN from PGY using hyperspectral imaging in the 400–2500 nm range. Recently, Thielert *et al.* (2024) reported a spectral distinction between BN and FD through hyperspectral measurements combined with machine learning methods.

Proximal sensing of grapevine yellows

Al-Saddik *et al.* (2017) first developed a method for the spectral detection of FD using a non-imaging portable spectroradiometer with a wavelength range of 350-2500 nm. Their study focused on two white cultivars (Vermentino and Chardonnay) and two red cultivars (Marselan and Grenache), conducted directly in the field in southern France. The dataset comprised four diseased and four healthy grapevines per cultivar, with 2–4 measurements taken from 2–4 leaves per plant. No other diseases or stress factors were considered. To identify the most discriminative wavelengths between healthy and infected leaves, the researchers employed a genetic algorithm feature selection tool. Spectral disease indices (SDIs) were designed by testing all possible combinations of wavelengths. The best-weighted combination of a single wavelength and a normalized difference was chosen to create the optimal index. These SDIs were compared with commonly used spectral vegetation indices (SVIs) like the NDVI. The results demonstrated that SDIs specifically tailored for FD detection performed better than traditional SVIs in most cases, achieving a classification precision exceeding 90% across all four cultivars. In 2019, Al-Saddik *et al.* improved this approach by replacing the computationally intensive genetic algorithm with the successive projection algorithm. With this improvement, the identification of FD-indicative spectral bands became faster with a low computational load, and the overall accuracy of disease classification increased to more than 96%.

In 2018, Al-Saddik *et al.* extended their approach to the distinction of GY from ESCA, a common fungal grapevine trunk disease (GTD). In this work, the authors combined hyperspectral data from the spectroradiometer with RGB image data captured by a digital camera. These spectral and textural data were analyzed by neural networks, and an overall accuracy of disease detection of 99% was achieved.

Another group of authors applied hyperspectral imaging to the analysis of different biotic and abiotic stress factors affecting grapevine leaves in southern France (Ryckewaert *et al.*, 2023). Healthy and FD-symptomatic leaves of seven red and white varieties were compared with leaves showing symptoms of water stress, fungal wood diseases, insect damage by buffalo treehopper (*Stictocephala bisonia*) or green leafhopper (*Empoasca vitis*), senescence, and various deficiencies. Hyperspectral images in the VIS-NIR range (400–900 nm) were acquired from individual leaves with an IQ Specim hyperspectral camera. The datasets were available in the INRAE (Institut national de recherche pour l'agriculture, l'alimentation et l'environnement) data repository (Ryckewaert, 2023). The first analysis calculated average spectra for each variety and symptom and performed principal component analysis (PCA). In addition to red and white varieties, different stress factors could be distinguished in the spectra. However, further analyses were required to exploit this dataset. A subset of data regarding FD had already been analyzed by Mas Garcia *et al.* (2021), who applied a combination of multivariate curve resolution-alternating least squares (MCR-ALS) and factorial discriminant analysis (FDA) to distinguish FD-infected leaves from healthy ones. High levels of correct pixel discrimination rates (classification rate (CR) = 85.1%) were obtained.

Bendel *et al.* (2020) applied hyperspectral imaging to detect BN and PGY in a laboratory system. Two hyperspectral line scanning sensors were implemented: (i) Hyspex VNIR 1800 to record spectra in the visible and NIR range (VNIR; 400–1000 nm); and (ii) Hyspex SWIR 384 that captures the SWIR range (1000–2500 nm). In this study, BN and PGY infections were analyzed in the white cultivars Riesling and Scheurebe, either in greenhouse-grown test plants or in field samples. The hyperspectral data were analyzed with four different machine learning algorithms.

The radial basis function network with relevance (rRBF) performed best for greenhouse plants, where symptomatic and healthy plants were classified with up to 96% accuracy. Multilayer perceptron network (MLP) performed better with field samples, with correct classifications of up to 100%. The ten most important wavelengths were identified for each disease detection approach. Regarding field samples of PGY, the most important wavelengths are around 557, 639, and 672 nm, and within the range of 801–940 nm. Some of these wavelengths are also relevant for the identification of BN-infected shoots of white varieties, for example 553, 637, and 812–966 nm, with an additional important wavelength at 741 nm. Obvious differences were observed between symptomatic shoots of white and red varieties, with 528, 586–626, and 673 nm being the key wavelength ranges for detecting phytoplasma-infected field materials. In contrast, a clear pattern can be seen in SWIR, with wavelengths around 1585, 2135, and 2300 nm across all three differentiation tasks. There are small differences between PGY- and BN-infected shoots between 1880 and 2000 nm. Here, important wavelengths are slightly shifted. Further differences become apparent in the range of 1050–1490 nm, with 1072 nm being important for the identification of PGY and 1350 nm for BN. In conclusion, while the spectral detection of GY (BN and/or PGY) in German cultivars was feasible, a clear distinction between BN- and PGY-infected plants could not be achieved.

Distinguishing between FD- and BN-infected vines is crucial for spectral monitoring in zones with low or no FD incidence. Thielert *et al.* (2024) focused on a case study conducted in northern Italy, where both diseases are present in the same vineyards or areas. The study aimed to evaluate the potential of hyperspectral imaging to differentiate between FD- and BN-infected leaves. Measurements were taken in a mobile laboratory equipped with a twin-hyperspectral push broom camera system that included Hyspex VNIR-1600 and Hyspex SWIR-320m-e cameras. Three to four symptomatic or healthy leaves per vine from six different vineyards of Chardonnay and Pinot gris cultivars were analyzed, totaling 427 leaves. Molecular analyses confirmed that about half of the symptomatic leaves were either FD- or BN-infected. Spectral data were automatically processed using machine learning (ML) techniques. Healthy and GY-infected leaves could be distinguished based on differences in the green and red wavelength ranges and shifts in the red edge-NIR spectrum. Differences in the SWIR region were also observed between 1600–1800 and 2000–2400 nm. An AutoML approach achieved phytoplasma detection accuracies of 94% in the VNIR range and 97% in the SWIR range. FD- and BN-infected leaves could be separated with accuracies of 79% in both the VNIR and SWIR ranges.

Jarausch *et al.* (2023a) reported their work on distinguishing GY-infected vines (with BN and PGY) from other biotic (virus infection and insect damage) or abiotic (magnesium and iron deficiency) stress symptoms by laboratory-based VNIR (400–1000 nm) imaging with an FX10 (Specim) camera. The images were taken from 100 control and 100 symptomatic leaves, evenly divided between white and red cultivars.

Jarausch *et al.* (2023a) also reported ground-based imaging using a quad that was equipped with a five-channel multispectral imaging (MSI) camera system (FS-3200T-10GE-NNC). This system captures RGB channels and two channels in the VNIR range (400–1000 nm) and includes a time-of-flight camera (Helios 2+HTP003S-001). An inertial navigation system (SBG, Ellipse-D) was installed for localization. In combination with two antennas, the system can estimate the position and orientation of the vehicle with centimeter-level accuracy. This enables precise georeferencing of the acquired multispectral data and derived parameters. Ground based multispectral data were collected via a quad in 13 vineyards across ten varieties.

A different approach to proximal sensing was reported by Boulent *et al.* (2020), who used a standard RGB camera for proximal imaging of FD symptoms in south-west France. By applying deep learning methods such as convolutional neural networks (CNNs) and fully convolutional networks (FCNs) to images of FD-infected and healthy grapevines of the cultivar Chardonnay, a positive classification rate of 98% was reached. Although the developed algorithm was not transferable to other cultivars, the study demonstrated the potential of machine learning methods. Tardif *et al.* (2022) applied the same proximal imaging approach with a standard RGB camera to differentiate GY-infected vines (no distinction between FD and BN was made), ESCA-infected vines, and healthy vines. The study acknowledged the challenge posed by different biotic and abiotic stress factors that can cause similar symptoms on grapevine leaves. As GY also shows typical symptoms on shoots and bunches, a two-step automated detection method was developed. In the first step, two deep-learning algorithms detected symptomatic shoots, symptomatic bunches, and symptomatic leaves. In the second step, random forest classifiers were used to categorize symptomatic leaves into FD-symptomatic, ESCA-symptomatic, and confounding. Based on more than 1000 images from five different grapevine varieties, the precision of symptom detection ranged from 0.67 to 0.82. However, the highest precision values were achieved only when the same grapevine varieties as those in the training dataset were analyzed.

Remote sensing of grapevine yellows

Albetis *et al.* (2017) evaluated the feasibility of discriminating FD symptoms in red and white grape cultivars from healthy vines using UAV-based multispectral imagery. The study was conducted in four vineyards in south-west France, focusing on the red cultivars Gamas and Duras, and the white cultivars Sauvignon and Colombard. A UAV platform (long-range DT-18) equipped with a multispectral MicaSense RedEdge™ sensor was used for data collection. The spectral signatures of healthy and symptomatic plants were examined through a set of 20 computed variables derived from the UAV images, including five spectral bands, 11 vegetation indices, and four biophysical parameters. Both univariate and multivariate classification approaches were applied. The best results were achieved for red cultivars, using both univariate and multivariate methods. Two vegetation indices based on the green and red spectral bands demonstrated a very high level of separability. However, results for white cultivars were less satisfactory, with misclassification rates ranging from 10% to 20%. This outcome was not surprising, as the leaf reddening of red cultivars was much more visible in the field than the leaf yellowing of white cultivars (Albetis *et al.*, 2017). In 2018, Albetis *et al.* extended their work to distinguish FD from other grapevine diseases with similar leaf discoloration using UAV imagery. This expanded study included additional analyses of grapevine trunk diseases (GTDs), such as ESCA and black dead arm. The authors employed the multispectral MicaSense RedEdge™ sensor and analyzed 24 variables, including five spectral bands, 15 vegetation indices, and four biophysical parameters. Vegetation indices were selected for their potential to detect abnormal vegetation behavior associated with stress or disease. While the differentiation of diseased vines from asymptomatic plants showed promising results, as in the previous study, the separation of FD from GTD remained challenging due to high pixel misclassification rates. These findings highlight that the identification of FD through remote sensing techniques continues to be a complex and unsolved issue.

Zottele *et al.* (2022) explored the possibility of FD and BN detection in a Chardonnay vineyard in Valsugana, Trentino (Italy). The study used two different UAVs: a prosumer drone equipped with an RGB camera (DJI Mavic PRO) and a professional multispectral UAV (DJI Matrice 210 RTK V2). The objective was to achieve real-time detection of GY, enabling online streaming of results during flight without post-processing. While the RGB images produced poor results due to blurring effects, the multispectral data analysis indicated that a single wavelength within the non-visible range (red edge or NIR) could be sufficient for effective discrimination.

In Germany, remote sensing of BN-infected grapevines by hyperspectral imaging was assessed for ten vineyards in Württemberg and two vineyards in Palatinate (Jarausch *et al.*, 2023a). An integrated system consisting of the Headwall Co-Aligned VNIR-SWIR Sensor and the DJI M600 Pro drone was used. GPS and LiDAR data were processed to create a digital elevation model (DEM) to orthorectify the hyperspectral images. These images were used to train a machine learning algorithm using the perClass Mira software to identify vines showing phytoplasma symptoms in real time. Although BN-infected vines were successfully localized in the vineyard, the approach needs further improvements.

Support systems for sensor-based detection of grapevine yellows

The sensor-based detection of GY can accurately classify symptoms but cannot directly identify the pathogen responsible. In countries and regions where the quarantine disease FD is not present or has a low incidence, molecular confirmation is essential to validate sensor-based findings. Typically, this confirmation is performed in centralized laboratories, which can delay the transfer of results to the field where timely control measures are required. To address this, a concept for a mobile laboratory that integrates spectral detection with molecular identification has been conceived (Thielert *et al.*, 2024; Jarausch *et al.*, 2025c). This mobile laboratory is mounted on a specialized vehicle with a four-wheel drive, which allows autonomous laboratory work directly in the field. The concept of the PhenoTruck^{AI} is based on three key axes: (i) large-scale screening of vineyards using remote sensing by a multispectral drone; (ii) hyperspectral screening of symptomatic leaf samples for phytoplasma infections in the first mobile laboratory; and (iii) molecular identification of FD in the second mobile laboratory. Multi- and hyperspectral data are automatically processed using different machine learning methods. Rapid molecular identification of FD infections is achieved by loop-mediated isothermal amplification (LAMP) according to European and Mediterranean Plant Protection Organization (EPPO) standards (Kogovšek *et al.*, 2015, 2017). A case study with a prototype of the PhenoTruck^{AI} was successfully conducted in Trentino and South Tyrol in the summer of 2023, demonstrating the feasibility of this approach (Thielert *et al.*, 2024).

FD is efficiently transmitted by the leafhopper *Scaphoideus titanus*, which has been introduced from North America to south-west Europe. Over the past decades, *S. titanus* has spread further northwards, transmitting FD to new viticultural regions. Interestingly, the FD phytoplasma is of European origin and is widespread in alders (Arnaud *et al.*, 2007; Malembic-Maher *et al.*, 2020). Occasionally, polyphagous leafhoppers like *Orientus ishidae* transmit FD from alder to grapevine (Jarausch *et al.*, 2023b). These rare FD-infected vines can become sources of new FD outbreaks if *S. titanus* is present. Since the first appearance of *S. titanus* in the Alsace region of France, Germany has been at the borderline of its spread. In 2024, *S. titanus* was first reported in a viticultural area in southern Germany (Askani *et al.*, 2024; Jarausch *et al.*, 2025a). Geodata-based risk maps have been established for vineyards adjacent to alder stands in all viticultural regions of southern and western Germany (Jarausch *et al.*, 2018, 2025b). These maps are designed to focus FD monitoring efforts on vineyards with a higher risk of natural FD infections. The first detection of FD in Germany validated the efficiency of this targeted approach (Jarausch *et al.*, 2021).

Sensor-based Monitoring of Fruit Tree Phytoplasma Diseases

Phytoplasmas of the 16SrX group cause economically important diseases in apple, pear, and stone fruit crops in Europe and neighboring regions. Apple proliferation (AP) is associated with '*Candidatus* Phytoplasma mali,' pear decline (PD) is associated with '*Ca.* Phytoplasma pyri,' and European stone fruit yellows (ESFY) with '*Ca.* Phytoplasma prunorum' (Seemüller and Schneider, 2004). These phytoplasmas are transmitted by psyllid vectors (Jarausch *et al.*, 2019; see also Chapter 12). As infected trees cannot be cured, controlling disease spread by insect vectors is the most promising management strategy. However, insecticide treatments are not always feasible and are limited due to environmental concerns. Therefore, promptly uprooting infected trees limits the chances of phytoplasma acquisition by insect vectors and thus slows the disease spread. AP is considered a quarantine pest in many countries and is classified as a regulated non-quarantine pest (RNQP) under EU regulations (EU Regulation 2016/2031). In some heavily affected apple-growing regions in Italy, uprooting infected trees and chemical control of insect vectors are mandatory measures (Janik *et al.*, 2020). However, regional-scale monitoring of AP remains labor-intensive, time-consuming, and expensive. Therefore, sensor-based detection methods, including proximal and remote sensing technologies, have been developed to facilitate the monitoring process and improve efficiency.

Symptoms of fruit tree phytoplasma diseases

Apple proliferation induces specific symptoms such as witches'-brooms, enlarged stipules, or small, poor-tasting fruits (Seemüller, 1990; Seemüller *et al.*, 2011a). These typical primary symptoms are used by experts for visual inspection in the field. However, the expression of witches'-brooms varies depending on climatic conditions and can differ significantly from year to year (Jarausch *et al.*, 2024). Enlarged stipules are more regularly expressed but are sometimes very difficult to detect (Jarausch *et al.*, 2024). Premature leaf reddening, considered a secondary symptom of AP (Janik *et al.*, 2020), is particularly suited for remote sensing applications. To explore this potential, Jarausch *et al.* (2024) analyzed the correlation between this symptom and the disease. '*Ca.* Phytoplasma mali' was detected by PCR in all trees showing primary symptoms, and almost all trees with primary symptoms also expressed premature leaf reddening. Depending on the year of monitoring, '*Ca.* Phytoplasma mali' was also detected by PCR in 71–97% of trees that showed leaf reddening without other AP symptoms. Further experiments under controlled conditions conducted by Jarausch *et al.* (2024) showed that AP-correlated leaf reddening is induced by cold nights and warm day temperatures, which are typical at the beginning of fall. The authors concluded that leaf reddening observed in late September/early October is a reliable symptom for remote sensing of AP.

Premature foliar reddening in AP-infected trees is the consequence of phytoplasma-induced chlorophyll breakdown (Bertamini *et al.*, 2003; Mittelberger *et al.*, 2017). Zorer *et al.* (2008) identified the anthocyane cyanidin-3-glucoside as the pigment responsible for this leaf reddening.

PD, one of the most important diseases affecting pear trees, can cause either a sudden death of the entire tree within a few weeks (quick decline) or a gradual degradation leading to death over several years (slow decline). Unlike AP, PD does not have clearly distinguishable symptoms (Seemüller, 1989; Seemüller *et al.*, 2011b). However, as with AP, premature leaf reddening is a non-specific but highly indicative symptom of PD infection in pears. In a recent study, Jarausch *et al.* (2023c) examined the correlation between premature leaf reddening and molecular detection of '*Ca.* Phytoplasma pyri.' Over a 3-year study, more than 80% of trees exhibiting reddening were confirmed to be infected with PD. The strongest correlation between phytoplasma detection and leaf reddening was observed in September. The authors concluded that early leaf reddening in pear is a sufficiently reliable indicator of PD, making it a valuable symptom to be used in remote sensing approaches.

ESFY disease is a serious disease affecting apricots and peaches. The most characteristic symptoms observed in many infected *Prunus* species include off-season growth, premature leaf bud break before flowering, and leaf yellowing or reddening accompanied by leaf roll in summer (Marcone *et al.*, 2010). Ongoing studies are currently underway to investigate the potential of using leaf yellowing in apricots as a target for sensor-based detection methods (W. Jarausch and B. Thielert, unpublished data).

Spectral detection of fruit tree phytoplasma diseases

Studies on the spectral detection of fruit tree phytoplasma diseases started in 2004. Zorer *et al.* (2008) analyzed AP disease in the heavily affected Trentino region in northern Italy. Under experimental conditions, the authors first induced AP-related leaf reddening and analyzed symptomatic and healthy plants with hyperspectral imaging in the VNIR range. The study revealed clear distinctions between healthy, reddened, and senescent leaves in the green and red wavelength ranges. Based on these findings, they developed a reddening index (RI) that could be applied in digital RGB imaging for both proximal and remote sensing. However, the lack of affordable, customized UAVs at the time prevented further development and application of this approach.

In 2021, NIR spectroscopy (1100–2500 nm) was successfully applied to distinguish AP-infected apple leaves from healthy ones (Barthel *et al.*, 2021). However, this is a destructive approach as it uses dried and ground leaves. It is still laboratory-based and labor-intensive. The authors analyzed 20 infected and 20 non-infected trees from two different orchards in South Tyrol (Italy) at different time points from May to October. The best results for AP detection were achieved in fall. The wavelength regions that contributed to the differentiation of infected and non-infected trees were associated with reductions in carbohydrates and nitrogen-containing organic compounds.

Barthel *et al.* (2023) developed a sensor-based method to distinguish AP-infected from healthy leaves in the field with a portable spectroradiometer. By analyzing leaf hyperspectral signatures within the 350–2500 nm range, they successfully identified relevant wavelengths for AP infection. Knauer *et al.* (2024) detected AP in a multitude of apple varieties by laboratory-based hyperspectral imaging. The same approach was applied to distinguish PD-infected pear leaves from healthy ones (U. Knauer *et al.*, unpublished data).

Zorer *et al.* (2008) made the first attempts to use the specific spectra of leaf reddening for remote sensing of AP. In a simple approach, feasible in the mountainous region of Trentino, the authors photographed the apple orchards in the valley from a hill with a multispectral camera. They also explored identifying highly AP-infested zones in satellite images on the basis of their RI index.

The progress in customized low-cost UAVs has now made it feasible to pursue the approach described by Zorer *et al.* (2008). McLeod (2021) tested the detection of AP-infected apple trees in an apple orchard in South Tyrol using a UAV equipped with both a multispectral and a hyperspectral camera. Similarly, Thielert *et al.* (2023) demonstrated in the Palatinate region of south-west Germany that UAVs equipped either with a hyperspectral or with a multispectral camera system could detect AP-infected apple trees, as well as PD-infected pear trees.

The availability of high-resolution satellite imagery has further advanced remote sensing capabilities. Recent studies have shown that it is now possible to detect AP-infected apple trees and PD-infected pear trees using satellite data (Kohler *et al.*, 2021, 2023; Jarausch *et al.*, 2023d; Al Masri *et al.*, 2025).

Proximal sensing of fruit tree phytoplasma diseases

Proximal sensing of AP has been achieved by hyperspectral imaging of leaf samples either in the field using a portable spectroradiometer connected to a leaf clip assembly (Barthel *et al.*, 2023) or in the laboratory with a VNIR and SWIR camera system (Knauer *et al.*, 2024). Barthel *et al.* (2023) measured ten randomly selected leaves each of nine non-infected and 14 AP-infected cv. Golden Delicious apple trees in a single orchard in South Tyrol. Measurements were taken at different time points from April to October. Using multivariate statistical methods, the study was able to distinguish infected from non-infected trees and identified wavelengths relevant for differentiation. The best results for AP detection were obtained in October and November, achieving a classification accuracy of 0.870. The most indicative wavelengths in the visible light range (500 and 682 nm) were correlated with AP-induced leaf reddening. In October, other relevant wavelengths in the SWIR region (1194, 1432, and 2074 nm) were identified to be linked to water absorption. The authors concluded that an AP infection leads to water stress in the tree due to root dysfunction. Knauer *et al.* (2024) adopted a different approach, focusing on AP-related reddened leaves. The leaves were either obtained from AP-infected *ex vitro* apple plants after experimental induction or from field samples of AP-infected trees. A total of 1,160 leaves from 21 different varieties across 16 orchards were sampled in two consecutive years in September and October in the Palatinate region of south-west Germany. Fresh symptomatic and healthy leaves were analyzed in the laboratory by hyperspectral imaging with a dual VNIR and SWIR camera setup. The data were processed for segmentation of the leaf area, feature extraction, classification, and subsequent analysis of the relevance of spectral bands. Different machine learning algorithms (random forest, SVM, MLP neural networks, and ensemble methods) were applied. The most relevant wavelengths for AP detection were identified with the random forest classifier in the VNIR range. Classification models like rRBF achieved an accuracy of 0.971 with the homogenous dataset of the *ex vitro* plant samples. A combination of different models was needed to achieve a satisfying classification of the heterogeneous field samples. By including the spatial distribution of spectral data, an accuracy of 0.751 was obtained. Common VIs such as NDVI also achieved relatively good separation between infected and healthy samples. For practical application, a processing pipeline for automated hyperspectral imaging was established (Knauer *et al.* (2024).

The same laboratory-based hyperspectral imaging approach was used to distinguish between PD-infected and healthy pear leaves (Thielert *et al.*, 2023; U. Knauer *et al.*, unpublished data). Field samples from different pear varieties were collected in September from orchards in the Palatinate region. This study focused on the spectral separation of infected, reddened, and healthy leaves. All sampled trees had been characterized using PCR to confirm the presence of PD. As with AP, the most relevant wavelengths for PD detection were identified in the visible spectrum (red light) and NIR (red edge) range.

Remote sensing of fruit tree phytoplasma diseases

Airborne detection of AP and PD at the orchard level was tested with UAVs (drones) in South Tyrol (McLeod, 2021) and the Palatinate region (Thielert *et al.*, 2023). McLeod (2021) conducted a case study using an octocopter (Soleon) equipped with a multispectral MicaSense Red Edge™ camera and a hyperspectral Rikola camera. Thielert *et al.* (2023) used a self-made drone with a hyperspectral Cubert HD185 camera and a commercially available multispectral drone from DJI (P4). Single images captured by the drones were georeferenced, orthorectified, and processed into orthophotographs of the orchards. Spectral data analysis was performed using different multivariate statistical methods and machine learning techniques (Thielert *et al.*, 2023). McLeod's study (2021) faced challenges due to black hail nets covering the orchard, which hampered the detection of AP and rendered UAV-based results inconclusive. In contrast, Thielert *et al.* (2023) achieved a correct classification rate of 76% for AP and PD using hyperspectral data. Based on these results, a spectral index was identified and applied to multispectral drone data, enabling the creation of precision maps for AP-infected apple orchards or PD-infected pear orchards. However, further research is needed to increase the accuracy of UAV-based detection methods for fruit tree phytoplasma diseases.

Zorer *et al.* (2008) analyzed satellite images from Quickbird captured during different periods in 2005 and 2006 using their RI. However, the resolution per pixel (70 × 70 cm) was too low, and the georeferencing was insufficiently accurate to reliably detect AP-infected trees.

Building on work conducted by Jarausch *et al.* (2023c, 2024), Knauer *et al.* (2024), and Thielert *et al.* (2023), AP and PD detection in multispectral satellite images has been further assessed. Kohler *et al.* (2021) used very high spatial resolution satellite imagery from WorldView-3 (0.31 m in panchromatic and 1 m in multispectral bands) to investigate the AP pathosystem. Additionally, the potential of PlanetScope (PS) imagery (3 m spatial resolution) was assessed as an initial step of the digital disease-identification process for identifying apple and pear orchards in satellite images. Land-use classification of orchards and vineyards using PS data and random forest algorithms yielded a high total accuracy of 0.98 (Al Masri *et al.*, 2025). A comparison of infected and non-infected trees in the satellite data revealed distinct differences in reflectance properties. For PD detection, SkySat images with 0.50 m resolution were combined with drone data to georeference the trees in the orchard (Kohler *et al.*, 2023; Al Masri *et al.*, 2025). New high spatial resolution satellite images from WorldView-3 (0.15 m in multispectral bands) were used to identify PD-infected trees. These images provided data from visual and near-infrared spectral channels, enabling the differentiation of PD-infected and healthy trees. Common indices for growth parameters like NDVI, SIPI (structure intensive pigment index), or MCARI2 (modified chlorophyll absorption ratio index 2) proved effective in this process (Al Masri *et al.*, 2025).

Satellite images with a lower spatial resolution were also applicable for identifying symptomatic rows, zones, or orchards (Jarausch *et al.*, 2023d). The satellite data allowed for accurate localization of "hot spots" while maintaining high spectral quality. However, the accuracy of the remote sensing for AP and PD detection strongly relies on timing. For AP, the optimal period was the beginning of October when most AP-infected trees showed premature leaf reddening before leaf fall (Jarausch *et al.*, 2024). For PD, mid-September yielded the best results in south-west Germany when the red leaves of the pear trees were most strongly correlated with the molecular detection of '*Ca.* Phytoplasma pyri' (Jarausch *et al.*, 2023c).

Conclusion

Sensor-based detection of plant diseases identifies physiological changes induced by the pathogen but does not directly identify the pathogen itself. Therefore, the accuracy of disease classification relies on the correlation between spectral signatures and the presence of the pathogen within the plant. Phytoplasma infections weaken the plant and induce a multitude of physiological changes, making it challenging to differentiate phytoplasma-specific effects from general stress responses. Common vegetation indices like NDVI already achieve relatively good results, but there is a need for more disease-specific indices to enhance automated, low-cost, multispectral approaches. This is particularly critical in crops like grapevines, where phytoplasma-induced symptoms need to be distinguished from other biotic and abiotic stress factors. Further research is required to implement new machine learning algorithms and develop sophisticated classification models.

Experience with fruit tree phytoplasma diseases shows that sensors mainly detect relevant wavelengths in the range associated with visible symptoms, such as leaf coloration and plant stress. However, latent infections remain difficult to identify. The advantage of remote sensing for phytoplasma diseases lies in its ability to monitor large areas quickly and efficiently. Depending on the method, the costs of remote sensing are significantly lower than visual inspections by experts on the ground. Visual inspections of larger areas are time-consuming and cannot always be done at the perfect time point. Some specific symptoms, like enlarged stipules, can easily be overlooked. The accuracy of large-scale visual monitoring is therefore comparable to the potential shown by the studies on remote sensing.

Remote sensing by UAVs is often limited by drone flight restrictions, limiting the areas that can be monitored. Low battery capacity in commercially available drones further restricts flight time, which is a great concern for large scale monitoring. Specialized UAVs equipped with customized multispectral cameras are essential for monitoring larger areas effectively.

Satellite imagery offers the advantage of covering all orchards within a given region, theoretically enabling comprehensive monitoring. However, this is only feasible under clear skies when the images are captured. Although tasking of satellite images within a certain time frame (without too many clouds) is possible, unfavorable weather conditions in the fall can interfere with AP or PD detection at the optimal time. A series of satellite images may be required to mitigate these limitations.

In conclusion, while sensor-based detection of phytoplasma diseases is feasible, further technical advancements are needed to make it practical for routine applications.

References

Al Masri, A., Kohler, K., Höhn, J., Runne, M. and Jarausch, W. (2025) Satellite-based detection of apple proliferation and pear decline. *Phytopathogenic Mollicutes* 15, 35–36. DOI: 10.5958/2249-4677.2025.00022.0.

Albetis, J., Duthoit, S., Guttler, F., Jacquin, A., Goulard, M. *et al.* (2017) Detection of Flavescence dorée grapevine disease using unmanned aerial vehicle (UAV) multispectral imagery. *Remote Sensing* 9, 308.

Albetis, J., Jacquin, A., Goulard, M., Poilvé, H., Rousseau, J. *et al.* (2018) On the potentiality of UAV multispectral imagery to detect Flavescence dorée and grapevine trunk diseases. *Remote Sensing* 11, 23.

Ali, I., Greifeneder, F., Stamenkovic, J., Neumann, M. and Notarnicola, C. (2015) Review of machine learning approaches for biomass and soil moisture retrievals from remote sensing data. *Remote Sensing* 7, 16398–16421. DOI: 10.3390/rs71215841.

Almoujahed, M.B., Krishnaswamy Rangarajan, A., Whetton, R., Vincke, D., Eylenbosch, D., Vermeulen, P., and Mouazen, A. (2022) Detection of fusarium head blight in wheat under field conditions using a hyperspectral camera and machine learning. *Computers and Electronics in Agriculture* 203, 107456. DOI: 10.1016/j.compag.2022.

Al-Saddik, H., Simon, J.C., and Cointault, F. (2017) Development of spectral disease indices for 'Flavescence dorée' grapevine disease identification. *Sensors* 17, 2772.

Al-Saddik, H., Laybros, A., Billiot, B., and Cointault, F. (2018) Using image texture and spectral reflectance analysis to detect yellowness and Esca in grapevines at leaf-level. *Remote Sensing* 10, 618, DOI: 10.3390/rs10040618.

Al-Saddik, H., Simon, J.-C., and Cointault, F. (2019) Assessment of the optimal spectral bands for designing a sensor for vineyard disease detection: The case of "Flavescence dorée". *Precission Agriculture* 20, 398–422.

Angelini, E., Constable, F., Duduk, B., Fiore, N., Quaglino, F., and Bertaccini, A. (2018) Grapevine phytoplasmas. In: Rao, G., Bertaccini, A., Fiore, N. and Liefting, L. (eds) *Phytoplasmas: Plant Pathogenic Bacteria – I*. Springer, Singapore, pp. 123–151.

Arnaud, G., Malembic-Maher, S., Salar, P., Bonnet, P., Maixner, M. *et al.* (2007) Multilocus sequence typing confirms the close genetic interrelatedness of three distinct Flavescence dorée phytoplasma strain clusters and group 16SrV phytoplasmas infecting grapevine and alder in Europe. *Applied Environmental Microbiology* 73, 4001–4010.

Askani, L., Zimmermann, O., Zimmermann, C., Rinke, F., Jarausch, B. *et al.* (2024). First report of *Scaphoideus titanus* (Hemiptera: Cicadellidae) in Germany (Baden-Württemberg). *EPPO Bulletin* 54, 366–368. DOI: 10.1111/epp.13053.

Barthel, D., Cullinan, C., Mejia-Aguilar, A., Chuprikova, E., McLeod, B.A. *et al.* (2023) Identification of spectral ranges that contribute to phytoplasma detection in apple trees – A step towards an on-site method, *Spectrochimica Acta Part A: Molecular and Biomolecular Spectroscopy* 303, 123246, DOI: 10.1016/j.saa.2023.123246.

Barthel, D., Dordevic, N., Fischnaller, S., Kerschbamer, C., Messner, M. *et al.* (2021) Detection of apple proliferation disease in *Malus* x *domestica* by near infrared reflectance analysis of leaves. *Spectrochimica Acta Part A: Molecular and Biomolecular Spectroscopy* 263, 120178.

Bendel, N., Backhaus, A., Kicherer, A., Köckerling, J., Maixner, M. *et al.* (2020) Detection of two different grapevine yellows in *Vitis vinifera* using hyperspectral imaging. *Remote Sensing* 12, 4151. DOI: 10.3390/rs12244151.

Bertaccini, A., Duduk, B., Paltrinieri, S., and Contaldo, N. (2014) Phytoplasmas and phytoplasma diseases: A severe threat to agriculture. *American Journal of Plant Sciences* 5, 1763–1788. DOI: 10.4236/ajps.2014.512191.

Bertamini, M., Nedunchezhian, N., Tomasi, F., and Grando, M. (2002) Phytoplasma [Stolbursubgroup (Bois Noir-BN)] infection inhibits photosynthetic pigments, ribulose-1,5-bisphosphate carboxylase and photosynthetic activities in field grown grapevine (*Vitis vinifera* L. cv. Chardonnay) leaves. *Physiological and Molecular Plant Pathology* 61, 357–366. DOI: 10.1006/pmpp.2003.0449.

Bertamini, M., Grando, M.S., and Nedunchezhian, N. (2003) Effects of phytoplasma infection on pigments, chlorophyll-protein complex and photosynthetic activities in field grown apple leaves. *Biologia Plantarum* 47, 237–242.

Blackburn, G.A. (2007) Hyperspectral remote sensing of plant pigments. *Journal of Experimental Botany* 58, 855–867. DOI: 10.1093/jxb/erl123.

Bock, C.H., Barbedo, J.G., Del Ponte, E.M., Bohnenkamp, D., and Mahlein, A.-K. (2020) From visual estimates to fully automated sensor-based measurements of plant disease severity: Status and challenges for improving accuracy. *Phytopathology Research* 2, 9. DOI: 10.1186/s42483-020-00049-8.

Boulent, J., St-Charles, P.-L., Foucher, S., and Théau, J. (2020) Automatic detection of Flavescence dorée symptoms across white grapevine varieties using deep learning. *Frontiers in Artificial Intelligence* 3, 564878, DOI: 10.3389/frai.2020.564878.

Candiago, S., Remondino, F., De Giglio, M., Dubbini, M., and Gattelli, M. (2015) Evaluating multispectral images and vegetation indices for precision farming applications from UAV images. *Remote Sensing* 7, 4026–4047.

Carter, G.A. and Knapp, A.K. (2001) Leaf optical properties in higher plants: Linking spectral characteristics to stress and chlorophyll concentration. *American Journal of Botany* 88, 677–684.

Ember, I., Bodor-Pesti, P., Zsófi, Z., Pálfi, Z., Ladányi, M. *et al.* (2018) Bois noir affects the yield and wine quality of *Vitis vinifera* L. cv.'Chardonnay'. *European Journal of Plant Pathology* 152, 185–197. DOI: 10.1007/s10658-018-1462-3.

EPPO (European and Mediterranean Plant Protection Organization) (2016) PM 7/079 (2) Grapevine flavescence dorée phytoplasma. *Bulletin OEPP/EPPO Bulletin* 46, 78–93.

Eveillard, S., Jollard, C., Labroussaa, F., Khalil, D., Perrin, M. *et al.* (2016) Contrasting susceptibilities to Flavescence dorée in *Vitis vinifera*, rootstocks and wild *Vitis* species. *Frontiers in Plant Science* 7, 176. DOI: 10.3389/fpls.2016.01762.

Gitelson, A.A., Merzlyak, M.N., and Chivkunova, O.B. (2001) Optical properties and nondestructive estimation of anthocyanin content in plant leaves. *Photochemistry and Photobiology* 74, 38–45. DOI: 10.1562/0031-8655(2001)0740038OPANEO2.0.CO2.

Horler, D., Dockray, M., and Barber, J. (1983) The red edge of plant leaf reflectance. *International Journal of Remote Sensing* 4, 273–288. DOI: 10.1080/01431168308948546.

Jang, S., Han, J., Cho, J., Jung, J., Lee, S., Lee, D., and Kim, J. (2024) Estimation of apple leaf nitrogen concentration using hyperspectral imaging-based wavelength selection and machine learning. *Horticulture* 10, 35. DOI: 10.3390/horticulturae10010035

Janik, K., Barthel, D., Oppedisano, T., and Anfora, G. (eds) (2020) *Scopazzi del melo: Stato attuale della ricerca. Apfeltriebsucht: Aktueller Stand der Forschung*, Fondazione Edmund Mach, San Michele all'Adige/Centro di Sperimentazione, Laimburg Italy.

Jarausch, B., Tedeschi, R., Sauvion, N., Gross, J., and Jarausch, W. (2019) Psyllid vectors. In: Bertaccini, A., Weintraub, P.G., Rao, G.P., and Mori, N. (eds) *Phytoplasmas: Plant Pathogenic Bacteria – II. Transmission and Management of Phytoplasma – Associated Diseases.* Springer, Singapore, pp. 53–78. DOI: 10.1007/978-981-13-2832-9.

Jarausch, B., Biancu, S., Kugler, S., Wetzel, T., Baumann, M. *et al.* (2021) First report of flavescence doree-related phytoplasma in a productive vineyard in Germany. *Plant Disease* 105, 3285–3285. DOI: 10.1094/Pdis-02-21-0330-Pdn.

Jarausch, B., Alisaac, E., Schumacher, P., Gauweiler, P., Gruna, R. *et al.*. (2023a) Development of multilevel monitoring systems for the identification of phytoplasmoses in German viticultural areas. *Phytopathogenic Mollicutes* 13, 133–135. DOI: 10.5958/2249-4677.2023.00067.1.

Jarausch, B., Markheiser, A., Jarausch, W., Biancu, S., Kugler, S., Runne, M., and Maixner, M. (2023b) Risk assessment for the spread of Flavescence dorée-related phytoplasmas from alder to grapevine by alternative insect vectors in Germany. *Microorganisms* 11, 2766. DOI: 10.3390/microorganisms11112766.

Jarausch, B., Askani, L., Kugler, S., Jarausch, W., Fuchs, R. *et al.* (2025a) A new invader in Germany: *Scaphoideus titanus*, the vector of "flavescence dorée" in grapevine. *Phytopathogenic Mollicutes* 15, 43–44. DOI: 10.5958/2249-4677.2025.00026.6.

Jarausch, W., Bischoff, F., Runne, M., and Trapp, M. (2018) GIS-basierte Risikoanalyse zur Ausbreitung von Flavescence dorée-Phytoplasmen von Wildhabitaten in angrenzende Weinberge. *Julius-Kühn-Archiv* 461, 343–344.

Jarausch, W., Menz, P., Al Masri, A., Runne, R., Thielert, B. *et al.* (2023c) Digital phytoplasmology: Remote sensing of fruit tree phytoplasma diseases. *Phytopathogenic Mollicutes* 13, 135–136. DOI: 10.5958/2249-4677.2023.00068.3.

Jarausch, W., Runne, M., Schwind, N., and Jarausch, B. (2023d) Leaf reddening as suitable symptom of pear decline for remote sensing. *Phytopathogenic Mollicutes* 13, 137–138. DOI: 10.5958/2249-4677.2023.00069.5.

Jarausch W., Runne M., Schwind N., Jarausch B., and Knauer U. (2024) Phytoplasma-induced leaf reddening as a monitoring symptom of apple proliferation disease with regard to the development of remote sensing strategies. *Agronomy* 14, 376. DOI: 10.3390/agronomy14020376.

Jarausch, W., Runne, M., Schell, T., Guerniche, D., Jarausch, B., and Trapp M (2025b) Analysis of the spread of Flavescence doreée-related phytoplasmas in naturally infected alder in Germany based on molecular and geodata. *Frontiers in Agronomy* 7, 1569408. DOI: 10.3389/fagro.2025.1569408.

Jarausch, W., Thielert, B., Michel, M., Menz, P., Runne, M. *et al.* (2025c) PhenoTruck[AI]: A mobile laboratory for hyperspectral and molecular detection of phytoplasma quarantine diseases like "flavescence dorée". *Phytopathogenic Mollicutes* 15, 21–22. DOI: 10.5958/2249-4677.2025.00010.X.

Knauer, U., Warnemünde, S., Menz, P., Thielert, B., Klein, L. *et al.* (2024) Detection of apple proliferation disease by hyperspectral sensors and machine learning based image analysis. *Sensors* 24, 7774. DOI: 10.3390/s24237774.

Kogovšek, P., Hodgetts, J., Hall, J., Prezelj, N., Nikolić, P. *et al.* (2015) LAMP assay and rapid sample preparation method for on-site detection of flavescence dorée phytoplasma in grapevine. *Plant Pathology* 64, 286–296. DOI: 10.1111/ppa.12266.

Kogovšek, P., Mehle, N., Pugelj, A., Jakomin, T., Schroers, H-J., Ravnikar, M., and Dermastia, M. (2017) Rapid loop-mediated isothermal amplification assays for grapevine yellows phytoplasmas on crude leaf-vein homogenate has the same performance as qPCR. *European Journal of Plant Pathology* 148, 75–84. DOI: 10.1007/s10658-016-1070-z.

Kohler, K., Al Masri, A., Jarausch, W., Knauer, U., Höhn, J., Sahib, L., and Khan, A. (2021) Potenzial von Satellitenbildern mit hoher räumlicher Auflösung zur Erkennung von Obstbäumen mit Phytoplasma-Erkrankungen. *Julius-Kühn-Archiv* 467, 553.

Kohler, K., Al Masri, A., Khan, A., Höhn, J., Warnemünde, S. *et al.* (2023) Satellitenbild-basiertes Auffinden von Apfel- und Birnbäumen mit Phytoplasmosen für Anbauer, Pflanzenschutzdienste und Agrochemie. *Julius-Kühn-Archiv* 475, 574–575.

Lasalle, G. (2021) Monitoring natural and anthropogenic plant stressors by hyperspectral remote sensing: Recommendations and guidelines based on a meta-review. *Science of the Total Environment* 788, 147758. DOI: 10.1016/j.scitotenv.2021.147758.

Mahlein, A.K. (2016) Plant disease detection by imaging sensors – Parallels and specific demands for precision agriculture and plant phenotyping. *Plant Disease* 100, 241–251. DOI: 10.1094/PDIS-03-15-0340-FE.

Mahlein, A.-K., Kuska, M.T., Behmann, J., Polder, G., and Walter, A. (2018) Hyperspectral sensors and imaging technologies in phytopathology: State of the art. *Annual Review of Phytopathology* 56, 535–558, DOI: 10.1146/annurev-phyto-080417-050100.

Maixner, M., Ahrens, U., and Seemüller, E. (1995a) Detection of the German grapevine yellows (Vergilbungskrankheit) MLO in grapevine, alternative hosts and a vector by a specific PCR procedure. *European Journal of Plant Pathology* 101, 241–250.

Maixner, M., Rüdel, M., Daire, X., and Boudon-Padieu, E. (1995b) Diversity of grapevine yellows in Germany. *Vitis* 34, 235–236.

Malembic-Maher, S., Desqué, D., Khalil, D., Salar, P., Bergey, B. *et al.* (2020) When a Palearctic bacterium meets a Nearctic insect vector: Genetic and ecological insights into the emergence of the grapevine Flavescence dorée epidemics in Europe. *PLoS Pathogens* 16, e1007967. DOI: 10.1371/journal.ppat.1007967.

Marcone, C., Jarausch, B., and Jarausch, W. (2010) *Candidatus* Phytoplasma prunorum, the causal agent of European stone fruit yellows: an overview. *Journal of Plant Pathology* 92, 19–34.

Mas Garcia, S., Ryckewaert, M., Abdelghafour, F., Metz, M., Moura, D. *et al.* (2021) Combination of multivariate curve resolution with factorial discriminant analysis for the detection of grapevine diseases using hyperspectral imaging. A case study: Flavescence doree. *Analyst* 146, 7730–7739. DOI: 10.1039/D1AN01735G.

McLeod, A. (2021) Precision mapping of apple proliferation using multi- and hyperspectral data. MSc. thesis, University of Munich, Germany.

Mittelberger, C., Pichler, C., Yalcinkaya, H., Erhart, T., Gasser, J. *et al.* (2017) Pathogen-induced leaf chlorosis: products of chlorophyll breakdown found in degreened leaves of phytoplasma-infected apple (*Malus × domestica* Borkh.) and apricot (*Prunus armeniaca* L.) trees relate to the pheophorbide a oxygenase/phyllobilin pathway. *Journal of Agricultural and Food Chemistry* 65, 2651–2660.

Pantazi, X.E., Moshou, D., Oberti, R., West, J., Mouazen, A.M., and Bochtis, D. (2017) Detection of biotic and abiotic stresses in crops by using hierarchical self organizing classifiers. *Precision Agriculture* 18, 383–393.

Pelletier, C., Salar, P., Gillet, J., Cloquemin, G., Very, P. *et al.* (2009) Triplex real-time PCR assay for sensitive and simultaneous detection of grapevine phytoplasmas of the 16SrV and 16SrXII-A groups with an endogenous analytical control. *Vitis* 48, 87–95.

Ryckewaert, M. (2023) Hyperspectral images of grape leaves including healthy leaves and leaves with biotic and abiotic symptoms. *Recherche Data Gouv* V1. Available at: https://entrepot.recherche.data.gouv.fr/dataset.xhtml?persistentId=doi:10.57745/WW7TY7 (last accessed June 2025).

Ryckewaert, M., Héran, D., Trani, J.P., Mas-Garcia, S., Feilhes, C. *et al.* (2023) Hyperspectral images of grapevine leaves including healthy leaves and leaves with biotic and abiotic symptoms. *Scientific Data* 10, 743. DOI: 10.1038/s41597-023-02642-w.

Seelig, H.D., Hoehn, A., Stodieck, L., Klaus, D., Adams III, W., and Emery, W. (2008) The assessment of leaf water content using leaf reflectance ratios in the visible, near-, and short-wave-infrared. *International Journal of Remote Sensing* 29, 3701–3713. DOI: 10.1080/01431160701772500.

Seemüller, E. (1989) Pear decline. In: Fridlund, P.R. (ed.) *Virus and Viruslike Diseases of Pome Fruits and Simulating Noninfectious Disorders*. Cooperative Extension College of Agriculture and Home Economics, Washington State University, Pullmann, WA. pp. 188–201.

Seemüller, E. (1990) Apple proliferation. In: Jones, A.L. and Aldwinkle, H.S. (eds) *Compendium of Apple and Pear Diseases*. APS Press, St Paul, MN, pp. 67–68.

Seemüller, E. and Schneider, B. (2004) '*Candidatus* Phytoplasma mali', '*Candidatus* Phytoplasma pyri' and '*Candidatus* Phytoplasma prunorum', the causal agents of apple proliferation, pear decline and European stone fruit yellows, respectively. *International Journal of Systematic and Evolutionary Microbiology* 54, 1217–1226.

Seemüller, E., Carraro, L., Jarausch, W., and Schneider, B. (2011a) Apple proliferation phytoplasma. In: Hadidi, A., Barba, M., Candresse, T., and Jelkmann, W. (eds) *Virus and Virus-like Diseases of Pome and Stone Fruits*. APS Press, St Paul, MN, pp. 67–73.

Seemüller, E., Schneider, B., and Jarausch, B. (2011b) Pear decline phytoplasma. In: Hadidi, A., Barba, M., Candresse, T., and Jelkmann, W. (eds) *Virus and Virus-like Diseases of Pome and Stone Fruits*. APS Press, St Paul, MN, pp. 77–84.

Tardif, M., Amri, A., Keresztes, B., Deshayes, A., Martin, D., Greven, M., and Da Costa, J.-P. (2022) Two-stage automatic diagnosis of Flavescence dorée based on proximal imaging and artificial intelligence: A multi-year and multi-variety experimental study. *OENO One* 56, 371–384. DOI: 10.20870/oeno-one.2022.56.3.5460.

Thielert, B., Menz, P., Warnemünde, S., Holstein, K., Klein, L. *et al.* (2023) Drohnenbasiertes Verfahren zur Detektion geschädigter Obstbäume in Obstbaumplantagen. *Informatik* 2023, 1615–1622. DOI: 10.18420/inf2023_165.

Thielert, B., Menz, P., Götte, G., Runne, M., Michel, M. *et al.* (2024) PhenoTruckAI: On-site hyperspectral measurement for distinction of quarantine grapevine disease "flavescence dorée" and non-quarantine disease "bois noir" in a mobile laboratory. *Informatik* 2024, 1253–1260. DOI: 10.18420/inf2024_110.

Zorer, R., Bianchedi, P.L., Mattedi, L., and Branz, A. (2008) Arrossamento fogliare autunnale. Relazioni con la presenza del fitoplasma e possibili utilizzi nel monitoraggio prossimale o remoto. In: Ioriatti, C. and Jarausch, W. (eds) *Scopazzi del melo – Apple proliferation*. Fondazione Edmund Mach, San Michele all'Adige, Italy, pp. 51–62.

Zottele, F., Gelmetti, A., Mattè, B., and Crocetta, P. (2022) Do UAVs help in the detection of grapevine yellows in vineyards that are difficult to reach? In: Baptista, A. and Cepêda, C. (eds) Extreme viticulture: from a cultural landscape to an economic and environmental sustainability. In: Baptista, A. and Cepêda, C. (eds) *Proceedings of the VII International Congress of Mountain and Steep Slopes Viticulture,* Universidade de Trás-os-Montes e Alto Douro (UTAD), Vila Real, Portugal, pp. 17–22.

2 Integrating Precision Agriculture Principles and Techniques into Phytoplasma Research for Enhanced Disease Management

Wei Wei,[1]* Yan Zhao,[1] and Jonathan Shao[2]

[1]*Molecular Plant Pathology Laboratory, ARS-USDA, Beltsville, Maryland, USA;*
[2]*Beltsville Agricultural Research Center, Animal Parasitic Diseases Laboratory, ARS-USDA, Beltsville, Maryland, USA*

Abstract

Phytoplasmas are harmful plant pathogens that threaten global agriculture by reducing crop yields and quality. With no cures currently available to treat entire fields, effective disease management strategies focus on early detection, vector control, and the development of resistant crop varieties. Precision agriculture offers an innovative approach by integrating remote sensing, artificial intelligence, machine learning, and data-driven techniques to improve disease monitoring, optimize resource allocation, and enhance pest management. This chapter highlights the potential of precision agriculture-driven strategies to advance phytoplasma research, providing sustainable solutions that promote disease prevention, effective mitigation, and long-term resilience in agriculture.

Introduction

Phytoplasmas are bacterial pathogens that cause severe diseases in nearly 1000 plant species worldwide, leading to significant agricultural losses (Wei *et al.*, 2024). Since phytoplasmas cannot be cultured *in vitro*, their identification requires molecular diagnostic techniques, which, although effective, can be labor-intensive, costly, and impractical for large-scale surveillance (Narayanasamy, 2011; Wei *et al.*, 2025). The lack of curative treatments makes early detection and precise disease management essential to mitigate crop losses and prevent further spread (Buja *et al.*, 2021). Recent advances in precision agriculture have demonstrated significant potential for enhancing phytoplasma disease detection, real-time monitoring, and targeted intervention strategies (Narmilan *et al.*, 2022; Knauer *et al.*, 2024; Portela *et al.*, 2024; Thielert *et al.*, 2024; Wei *et al.*, 2025). Precision agriculture utilizes automated, data-driven solutions that enable rapid identification, spatial disease mapping, and precision-targeted management, improving the overall efficiency of disease monitoring (Abbas *et al.*, 2023; Guebsi *et al.*, 2024).

Remote sensing plays a crucial role by enabling non-invasive disease detection through hyperspectral imaging, unmanned aerial vehicle (UAV)-based multispectral surveys, and geographic information systems (GIS) mapping (Alsadik *et al.*, 2024; Joshi *et al.*, 2024). These tools

*Corresponding author: wei.wei@usda.gov

DOI: 10.1079/9781800627031.0002

capture plants' early physiological and biochemical stress indicators, allowing for disease detection before visible symptoms develop (Alsadik *et al.*, 2024; Joshi *et al.*, 2024). Artificial intelligence (AI) and big data analytics further enhance precision agriculture by processing large datasets from omics studies, environmental conditions, and historical disease outbreaks to improve disease forecasting and risk assessment (Woźniak and Ijaz, 2024).

This chapter explores how precision agriculture techniques can effectively integrate into phytoplasma research to advance early disease detection, predictive analytics, and sustainable disease management strategies. By bridging plant pathology with advanced agricultural technologies, this approach seeks to develop scalable, technology-driven solutions that improve resource efficiency, minimize crop losses, and support sustainable farming practices.

AI-driven Hyperspectral and UAV Technologies for Early Detection of Phytoplasma Diseases

Studies on AI-driven hyperspectral and UAV-based technologies for phytoplasma disease detection have been published in the past decade. Eight investigations are summarized in this section, focusing on key findings, methodologies, and practical implications (Table 2.1). The first two focus on hyperspectral imaging and autoencoder-based AI models for detecting flavescence dorée (FD) in grapevines, highlighting both high-resolution and low-cost spectrometry solutions. The next three examine AI-driven and drone-assisted approaches for cassava phytoplasma disease (CPD) detection, utilizing deep learning models, GIS mapping, and real-time video transmission for large-scale monitoring. Another analysis explores hyperspectral imaging and AI for detecting apple proliferation (AP) disease, emphasizing spectral indices for non-invasive diagnosis. Research on AI-based prediction of phytoplasma-induced biotic stress demonstrates how neural network models can assess biochemical responses in infected plants. Finally, a UAV-mounted multispectral imaging system combined with machine learning is studied that was introduced to detect white leaf disease (WLD) in sugarcane, demonstrating how aerial remote sensing enhances early detection and disease management strategies.

Hyperspectral imaging and autoencoder-based AI for early detection of FD in grapevines

The study conducted by Silva *et al.* (2022) explores the application of hyperspectral imaging and machine learning for the early detection of FD disease in grapevines. Traditional methods rely on expert visual inspection, which can be inaccurate and time-consuming. Hyperspectral imaging offers a more advanced solution by capturing spectral data beyond what the human eye can perceive, allowing for the identification of early-stage disease symptoms before visible damage occurs.

To improve detection accuracy and efficiency, the study employed autoencoder-based machine learning models to process hyperspectral images and reduce data complexity. Two approaches were tested: a full-image analysis method and a patched-image method, where images were divided into smaller segments for detailed analysis. The patched approach demonstrated superior performance, achieving an accuracy increase of 0.2 and an area under the curve (AUC) increase of 0.34 (an AUC score of 1.00 is considered perfect) compared with traditional convolutional neural network (CNN) models. The study highlights the advantages of autoencoders in reducing the computational burden of processing hyperspectral data while retaining critical spectral information. The encoder classifier model outperformed the baseline CNN model, achieving 83% accuracy, 80% recall, and 83% AUC, demonstrating the potential of machine learning for automated grapevine disease detection (Shiva *et al.*, 2022; Table 2.1).

In practical applications, this method could significantly improve vineyard management by preventing unnecessary vine uprooting, reducing false positives, and allowing early intervention strategies to be implemented effectively (Shiva *et al.*, 2022). Future work includes expanding the dataset, integrating real-time detection algorithms, and applying high-resolution hyperspectral data to enhance precision and scalability. This research contributes to the precision agriculture advancement by combining AI and hyperspectral imaging to enhance disease monitoring and management in viticulture.

Table 2.1. AI and machine learning applications in phytoplasma disease detection.

Study(reference)	Host	Technique	Model	Device	Accuracy	Key findings
Hyperspectral imaging and autoencoder AI for FD (Silva *et al.*, 2022)	Grapevine	Hyperspectral imaging, autoencoder ML	Auto encoder classifier	Hyperspectral camera	83% (accuracy), 80% (recall), 83% (AUC)	Autoencoder improves spectral data processing; the patched approach outperforms CNN
Low-cost handheld spectrometry for FD (Imran *et al.*, 2023)	Grapevine	Handheld spectrometry, ML	Logistic regression, SVM, XGBoost, RF, cubist	Hamamatsu spectrometer (340–850 nm), NIRScan (900–1700 nm)	96% (GA + LR), 85% (SVM)	Handheld spectrometry effectively detects FD; the Hamamatsu sensor is more reliable than the NIRScan
Deep learning for CPD detection (Plata *et al.*, 2021)	Cassava	Deep learning (CNNs), NDVI-based processing	Faster R-CNN, SSD MobileNet v2, YOLOv4	Near-infrared imaging, automated annotation system	70–82.5% (Faster R-CNN), 51.25–67.5% (YOLOv4), 37.5–47.5% (SSD MobileNet v2)	Faster R-CNN is recommended for real-time CPD detection; NDVI visualizes symptoms
Real-time CPD detection with deep learning (Plata *et al.*, 2022a)	Cassava	CNN-based object detection, drone imaging	Faster R-CNN, SSD MobileNet v2, YOLOv4	Drone with HD camera, wireless video transmission	95% (Faster R-CNN), 85% (YOLOv4), 73% (SSD MobileNet v2)	Faster R-CNN is the most accurate but slow; YOLOv4 is the fastest; wireless video monitoring is effective
Drone-based GIS mapping for CPD (Plata *et al.*, 2022b)	Cassava	Drone-based GIS mapping, image analysis	Pixel and color intensity analysis, GIS heat map	Drone-based GIS system, GPS-based mapping	High precision in detecting infected plants via heat maps	GIS heat maps improve CPD detection; drone-based approach reduces manual labor
Hyperspectral imaging and AI for AP (Knauer *et al.*, 2024)	Apple	Hyperspectral Imaging, ML	rRBF classifier, NDVI, PRI indices	Dual-camera hyperspectral imaging system (400–2500 nm)	97.1% (laboratory), 73.1% (field)	Hyperspectral imaging detects AP via spectral differences; the rRBF classifier enhances classification
AI-based prediction of '*Ca.* P. solani'-induced biotic stress (Gatou *et al.*, 2024)	Peony, mint, and dill	ANN, PCA	MLP 8-9-13 ANN	Biochemical analysis tools, spectrometry	r^2 = 1.000 (high prediction accuracy)	ANN predicts biochemical stress responses; PCA differentiates symptomatic and asymptomatic plants
UAV and ML for white leaf phytoplasma disease in sugarcane (Narmilan *et al.*, 2022)	Sugarcane	UAV multispectral imaging, ML	XGBoost, random forest, decision tree, KNN	UAV multispectral camera	92% and 94% (XGB, RF, KNN), 91% and 93% (DT) for model validation and field test, respectively	XGB is the most reliable; red band is the most effective; UAV-based approach improves WLD monitoring

AI, artificial intelligence; ANN, artificial neural network; AP, apple proliferation; AUC, area under the curve; CNN, convolutional neural network; CPD, cassava phytoplasma disease; DT, decision tree; FD, flavescence dorée; GA, genetic algorithm; GIS, geographic information system; HD, high definition; KNN, K-nearest neighbors; LR, logistic regression; ML, machine learning; MLP, multilayer perceptron; NDVI, normalized difference vegetation index; PCA, principal component analysis; PRI, photochemical reflectance index; R-CNN, region-based convolutional neural network; RF, random forest; rRBF, random radial basis function; SVM, support vector machine; SSD, single shot detector; UAV, unmanned aerial vehicle; WLD, white leaf disease; XGB, XGBoost; YOLOv4, You Only Look Once v4.

Early detection of FD disease in vineyards using low-cost handheld spectrometry

In the study performed by Imran *et al.* (2023), the researchers explored the potential of low-cost handheld hyperspectral sensors for the early detection of FD in vineyards before symptoms become visible. Two portable spectrometers, Hamamatsu (340–850 nm) and NIRScan (900–1700 nm), were used to collect spectral data from 180 leaf samples across five vegetative growth stages. The research aimed to determine whether spectral differences could effectively distinguish between healthy and infected vines.

Analysis of the spectral data revealed that the Hamamatsu sensor detected significant differences between healthy and diseased leaves, particularly in the visible and near infrared (NIR) spectral region (400–850 nm). In contrast, the NIRScan sensor showed minimal differences, suggesting it was less effective in identifying early signs of FD infection. Changes in leaf reflectance patterns were attributed to alterations in pigment concentration and metabolism caused by the disease (Imran *et al.*, 2023).

To enhance classification accuracy, the study employed machine learning algorithms, including logistic regression (LR), support vector machine (SVM), XGBoost, random forest (RF), and cubist regression. Feature selection techniques, such as ensemble methods and genetic algorithms (GAs), improved model performance. The highest accuracy of 96% was achieved using GA-selected features from the Hamamatsu sensor with the LR classifier on test samples. When employing leave-one-out (LOO) cross-validation, the best accuracy of 85% was obtained using the SVM classifier (Table 2.1).

The study also identified the most effective spectral bands for FD detection. The Hamamatsu sensor provided optimal wavelengths at 374, 392, 395, 416, 479, 534, 622, 654, 683, 692, 713, 756, and 798 nm, while the NIRScan sensor identified some informative bands at 901, 1108, 1367, 1440, 1456, and 1462 nm. These results suggest that feature selection can significantly improve FD classification accuracy, particularly in the visible and NIR spectral range (Imran *et al.*, 2023). Despite its promising results, the study was conducted on a small dataset using a single grape variety (Garganega), limiting its generalizability. Future research should validate these findings across multiple grape varieties and vineyard locations. Integrating handheld hyperspectral sensors with drone-based or ground-based monitoring systems could provide a more scalable approach for large-scale FD detection.

AI and drones for CPD detection

The study conducted by Plata *et al.* (2021) presents a real-time detection method for CPD disease using deep learning-based object detection models. Given the challenges in traditional disease detection, the researchers integrated wireless video transmission, AI-based recognition, and drone technology to enhance precision agriculture applications. The study specifically explored the effectiveness of transfer learning neural networks in detecting CPD in actual field environments. To achieve this, three CNN models were custom-trained and evaluated: Faster R-CNN Inception v2, Single Shot Detector (SSD) MobileNet v2, and You Only Look Once (YOLO) v4. The Faster R-CNN model achieved the highest accuracy (*c.*72–82.5%) but was computationally slower, making it ideal for cases where precision is the top priority. YOLOv4, on the other hand, provided the fastest image processing rate (up to 25 frames per second (FPS)) but had slightly lower accuracy, making it more suitable for real-time applications requiring rapid analysis. The SSD MobileNet v2 model demonstrated moderate performance but was less accurate than the other models (Table 2.1).

A significant aspect of the study was the development of a wireless video transmission system that enabled remote monitoring of cassava fields. Using 2.4 and 5 GHz wireless networks, the system facilitated long-range real-time video transmission. A drone-mounted high definition (HD) camera was used to capture and analyze plant conditions, and a GPS-based data-logging system geotagged infected plants for further study. This integration of AI and drone-based imaging allowed for an automated and scalable approach to plant disease monitoring, reducing dependency on manual inspections. The field tests demonstrated that AI-based detection could successfully identify CPD in cassava plants, supporting early intervention strategies. The Faster R-CNN model was recommended for real-world deployment due to its superior detection accuracy (Plata *et al.*, 2021). The study highlights the practical applications of AI in precision agriculture, showing how machine learning, drones, and wireless technology can improve disease surveillance and farm management.

This research provides a cost-effective and automated alternative to traditional plant disease diagnosis. By combining AI-driven analysis, real-time video transmission, and remote sensing, the study introduces a new approach to early disease detection and resource-efficient crop management. Future work includes expanding training datasets, optimizing hyperparameters, and refining real-time processing speeds to improve detection efficiency further. The findings demonstrate the potential of AI-powered precision agriculture in minimizing crop losses and enhancing food security.

Deep learning for CPD disease classification

In 2022, Plata *et al.* (2022a) investigated the use of deep learning-based CNNs for real-time detection of CPD. Traditional manual identification methods are slow and inefficient, particularly on large farms. To address this, the research compared the performance of three CNN models, including Faster R-CNN Inception v2, SSD MobileNet v2, and YOLOv4, to determine the most effective approach for CPD detection between healthy and infected plants. The findings showed that Faster R-CNN Inception v2 achieved the highest accuracy at 95%, making it the most reliable model for detecting CPD-infected plants. YOLOv4, while slightly less accurate at 85%, demonstrated superior image processing speed, handling 25 FPS, making it ideal for real-time applications. SSD MobileNet v2, the least accurate at 73%, still performed efficiently with a speed of 20 FPS. Although Faster R-CNN Inception v2 processed images at only 11 FPS, its accuracy made it the recommended model for CPD detection (Table 2.1).

To support model training, the study collected over 50,000 raw NIR images of CPD-infected cassava from various plantations. The images were processed using normalized difference vegetation index (NDVI) calculations, enhancing the visibility of disease symptoms. An automated annotation system was employed to label and prepare images for training. The research also developed a semi-unsupervised deep learning approach, incorporating a rolling learning rate search cross-validation (RLRSC-V) algorithm to optimize training performance (Plata *et al.*, 2022a). In terms of training time, YOLOv4 reached optimal accuracy in just 17 hours, whereas Faster R-CNN Inception v2 took 53 hours, and SSD MobileNet v2 required 56 hours. Model optimization efforts included score thresholding and computational efficiency improvements, particularly for the Faster R-CNN model. The study concluded that YOLOv4 is the best choice for real-time processing. Faster R-CNN Inception v2 remains the most accurate and reliable model for CPD detection.

Although the study produced promising findings, the authors recognized certain limitations. Wireless video transmission for real-time monitoring was effective only within a 2 km range, and external factors (background noise) such as lighting conditions, weather, and signal interference impacted detection accuracy. Future research aims to enhance spectral sensitivity, integrate a non-invasive 2D cassava scanner, and extend the model's application to detect other diseases and pests in cassava fields.

Drone-based GIS mapping for CPD

Plata *et al.* (2022b) developed a drone-based GIS mapping system to detect and monitor CPD using the ArcGIS JavaScript API to generate real-time heat maps of infected areas. The system was designed using the rapid application development (RAD) methodology, ensuring an iterative and efficient approach. By integrating drone-captured data with GIS, the system provided spatial insights into disease spread, allowing for more informed decision-making in disease management.

To detect CPD, the study employed several image analysis techniques, including pixel intensity and color intensity analysis to classify disease severity, as well as ratio and coordinate plotting to map affected areas. Raw data collected by drones were processed using GPS-based spatial analysis, enabling the precise localization of infected plants. These techniques significantly improved the accuracy and efficiency of disease detection compared with traditional methods (Table 2.1). The heat map visualization proved to be an effective tool for understanding the extent of CPD spread. The system assigned color intensities based on the severity of infections: yellow for low infection rates, orange for moderate infection rates, and red for severe infections. This visual representation enabled farm managers to identify disease hotspots and implement targeted control measures, reducing the risk of further spread (Plata *et al.*, 2022b).

Drone-based data collection played a crucial role in the system's effectiveness. High-resolution images and GPS coordinates of cassava fields were captured and processed automatically to generate CPD-infected area maps. In a test conducted over a 500 m^2 cassava farm, the system successfully detected four infected stalks among 5000 plants. This level of precision highlights the potential of drone-assisted GIS mapping in large-scale agricultural disease monitoring (Plata *et al.*, 2022b). The GIS-based system significantly enhanced efficiency and accuracy in CPD detection. Compared with traditional manual inspections, the web-based system provided real-time updates, allowing faster intervention in controlling disease outbreaks. The system reduced the time and labor required for disease monitoring by offering automated data processing and mapping.

Despite its success, the study identified several areas for improvement. Future enhancements could include customizable heat map features, such as adjustable color schemes and scaling options to improve visualization. Refining the system's graphical user interface (GUI) would also enhance usability, making it more accessible for farm managers. Further research is also needed to expand the system's capabilities to detect other plant diseases and to increase detection sensitivity through improved image-processing techniques.

Hyperspectral imaging and AI for AP detection

Apple proliferation is a major disease affecting European fruit production, caused by '*Candidatus* Phytoplasma mali.' The study by Knauer *et al.* (2024) investigated the potential of hyperspectral imaging combined with machine learning techniques to detect AP in apple trees efficiently. Traditional detection methods rely on visual symptoms and PCR analysis, which can be time-consuming and labor-intensive. This research aimed to develop a non-invasive detection system based on the spectral signatures of infected leaves.

The study collected 1160 leaf samples over two growing seasons (2019–2020) and used a dual-camera hyperspectral imaging system covering 400–2500 nm to capture spectral data. PCR testing was conducted on all samples to validate the imaging results. The research found that leaf reddening in early fall (a symptom of AP) correlates strongly with infection and is detectable via hyperspectral imaging. The spectral differences between healthy and infected leaves were most prominent in the visible and NIR bands (500–700 nm).

Machine learning algorithms were applied to classify infected and healthy trees. The random radial basis function (rRBF) classifier achieved the highest accuracy of 97.1% in a controlled environment and 73.1% when applied to field samples across multiple apple varieties. The integration of spatial–spectral data improved classification accuracy to 75.1%, demonstrating that considering multiple leaves per tree enhances detection reliability (Table 2.1). The study also tested spectral indices such as the NDVI and photochemical reflectance index (PRI) to simplify data processing. These indices provided robust indicators of AP infection, allowing for potential integration into low-cost multispectral sensors for large-scale monitoring.

AI-based prediction of phytoplasma-induced biotic stress in plants

Gatou *et al.* (2024) explored the effects of biotic stress induced by '*Ca.* P. solani' on different plant species, including peony (*Paeonia tenuifolia*), mint (*Mentha* × *piperita*), and dill (*Anethum graveolens*). The pathogen significantly affects plant health, leading to stunting, yellowing, leaf curling, and reduced productivity, which can result in economic losses. The research aimed to identify biochemical changes caused by the infection and use artificial neural network (ANN) modeling to predict the impact of the disease.

The phylogenetic analysis of Serbian '*Ca.* P. solani' strains, based on 16S rRNA and *plsC* gene sequences, indicated genetic homogeneity among the strains, regardless of their host plants. This suggests a stable genetic structure of the pathogen across different plant species. The study confirmed that the pathogen is closely related to other '*Ca.* P. solani' strains that have been reported globally. Significant biochemical changes were observed in infected plants compared with asymptomatic ones. Total sugar content was nearly doubled in symptomatic leaves, while lipid peroxidation intensity significantly increased, indicating higher oxidative stress. The study also found that reduced glutathione levels were slightly but consistently elevated in all tested species, suggesting an adaptive antioxidant response to the infection. Phenylalanine ammonia-lyase (PAL) activity was 1.25–3.5 times higher in symptomatic leaves, particularly in mint and dill, which correlated with increased tannin and proanthocyanidin content. Photosynthetic pigments were severely affected, with chlorophyll a and b levels reduced by 2.45–5.08 times in infected leaves, leading to reduced photosynthetic efficiency. Interestingly, antioxidant capacity varies across species. Mint and dill had enhanced superoxide anion scavenging activity (above 80% neutralization), while peony exhibited a strong 2,2-diphenyl-1-picrylhydrazyl (DPPH) radical neutralization ability (over 90%), indicating a species-specific response to oxidative stress (Gatou *et al.*, 2024).

Principal component analysis (PCA) was performed to analyze biochemical patterns. This statistical approach effectively differentiated plant responses to phytoplasma infection, revealing clear separations between asymptomatic and symptomatic samples based on chemical parameters such as total polyphenolics, tannins, flavonoids, and antioxidant activity. An ANN model was developed to predict the effects of biotic stress. The multilayer perceptron (MLP) 8-9-13 architecture provided the best performance, achieving a coefficient of determination (r^2) of 1.000 during training, indicating high prediction accuracy (Table 2.1). The ANN model successfully estimated disease impact based on biochemical markers, demonstrating the potential of AI-driven early warning systems for plant disease detection.

UAV and machine learning for early detection of WLD in sugarcane

In the study performed by Narmilan *et al.* (2022), an advanced approach was introduced for detecting WLD in sugarcane fields by integrating UAV-mounted multispectral imaging and machine learning models. Traditional detection methods, which rely on visual inspection and manual scouting, are labor-intensive and inefficient. This research aims to provide a cost-effective, automated, and scalable solution for early WLD detection to improve disease management and minimize economic losses in sugarcane production. To classify sugarcane health, the study tested four supervised machine learning algorithms, including XGBoost (XGB), RF, decision tree (DT), and K-nearest neighbors (KNN), on UAV-acquired multispectral images. The images were segmented into five categories: ground, shadow, healthy plants, early symptoms, and severe symptoms of WLD. Among these models, XGB, RF, and KNN achieved the highest accuracy of 94%, while DT performed slightly lower at 93%. The XGB model was identified as the most reliable due to its balance of precision, recall, and training efficiency (Table 2.1).

The study also evaluated the effectiveness of vegetation indices (VIs) in detecting WLD. The most significant VIs for classification were the modified soil-adjusted vegetation index (MSAVI), NDVI, and excess green (ExG). Among the spectral bands, the red band proved most effective for XGB and RF, whereas the green band was more relevant for DT (Narmilan *et al.*, 2022). These findings suggest that combining spectral indices with machine learning models enhances disease detection accuracy. To ensure the robustness of the approach, the models were validated in a different field within the Gal-Oya Plantation, Sri Lanka. The XGB, RF, and KNN models maintained 92% accuracy in validation, demonstrating their reliability for practical deployment. While slightly less accurate, the DT model still achieved 91% accuracy, reinforcing the effectiveness of the machine learning pipeline (Table 2.1).

The research highlights the advantages of UAV-based detection systems over traditional methods. The proposed system offers a fast, cost-effective, and scalable solution for monitoring large sugarcane fields. Early detection of WLD allows farmers to implement timely intervention strategies, reducing disease spread and improving crop yields. Future improvements include incorporating hyperspectral imaging, deep learning models, and disease-specific VIs to enhance detection accuracy further. Expanding the dataset across different sugarcane varieties and growth stages will also improve model generalization and scalability.

AI, hyperspectral imaging, and UAV technology are revolutionizing plant disease detection by enabling early, accurate, and scalable monitoring (For more details about remote sensing-based detection of phytoplasma diseases, see Chapter 1). These advancements enhance precision agriculture, reducing reliance on manual inspections and improving disease management. Future efforts should focus on real-time integration and broader applicability to optimize crop protection strategies further.

Databases

Databases are essential in plant pathology, providing repositories for pathogen classification, disease tracking, and host–pathogen interactions. They support molecular identification, predictive modeling, and risk assessment, aiding in accurate diagnostics and disease management. In phytoplasma research, specialized databases facilitate strain classification, vector identification, and epidemiological analysis, enhancing early detection and surveillance. By integrating genetic and taxonomic data, these resources contribute to precision agriculture, regulatory policies, and plant health security.

*i*PhyClassifier

*i*PhyClassifier is a bioinformatics tool for classifying and identifying phytoplasma strains based on 16S rRNA gene sequences (Zhao *et al.*, 2009) (Table 2.2). The tool analyzes 16S rRNA gene sequences, comparing them with a reference database of known phytoplasma strains. It assigns the sample to a specific 16Sr group and subgroup, using similarity coefficients and restriction fragment length polymorphism (RFLP) patterns. The tool provides an automated approach to phytoplasma classification, supporting research on these pathogens' epidemiology and geographic distribution. It is used in plant pathology to distinguish between different phytoplasma strains, facilitating the study of their genetic relationships and evolutionary patterns. The classification results assist in tracking disease spread and assessing potential impacts on affected crops. *i*PhyClassifier is commonly used in combination with other molecular diagnostic techniques, including PCR and RFLP analysis. These methods generate sequence data, which can then be processed through *i*PhyClassifier to determine the strain classification. The tool is applied in laboratory settings where plant samples undergo DNA extraction and sequencing before analysis. The use of *i*PhyClassifier requires molecular biology expertise and sequencing capabilities. It is employed in research institutions and agricultural laboratories for phytoplasma detection and classification. The tool contributes to studies on phytoplasma diversity, genetic variation, and plant disease management, providing sequence-based classification for further analysis.

CpnClassiPhyR

CpnClassiPhyR is a bioinformatics tool used to classify and identify phytoplasma strains based on the *chaperonin-60* (*cpn60*) gene (Muirhead *et al.*, 2019) (Table 2.2). This gene provides an alternative to the 16S rRNA-based classification, offering a different perspective on genetic variation among phytoplasma strains. The tool compares the *cpn60* gene sequences from phytoplasma samples with a reference database to determine classification groups. The *cpn60* gene has greater sequence variability than 16S rRNA, allowing for the identification of genetic differences between closely related strains. CpnClassiPhyR is used in molecular plant pathology research to study phytoplasma diversity and distribution. It can support epidemiological studies by identifying phytoplasma strains present in different plant species and geographic regions.

Table 2.2. Comparison of phytoplasma databases: classification, diagnostics, and epidemiological resources.

Database (reference)	Focus	Primary use	Data type	Application	Citation and usage
*i*PhyClassifier: https://plantpathology.ba.ars.usda.gov/cgi-bin/resource/iphyclassifier.cgi (Zhao *et al.*, 2009)	Phytoplasma classification based on the 16S rRNA gene	Classifies phytoplasma strains, assigns 16Sr groups and subgroups	16S rRNA gene sequences, RFLP patterns	Research institutions, phytoplasma classification, and epidemiology studies	Extensively cited, widely used in phytoplasma classification research
CpnClassiPhyR: http://cpnclassiphyr.ca/ (Muirhead *et al.*, 2019)	Phytoplasma classification based on the *cpn60* gene	Provides higher-resolution classification of closely related strains	*cpn60* gene sequences	Molecular plant pathology research, strain differentiation	Used in specialized phytoplasma genetic studies
*i*PhyDSDB: https://plantpathology.ba.ars.usda.gov/iphydsdb/iphydsdb.html (Wei *et al.*, 2024)	Phytoplasma disease symptoms and host plant database	Documents phytoplasma-associated disease symptoms and host plants	Symptom descriptions, host plants, and molecular data	Disease recognition, epidemiology, and agricultural studies	Referenced in symptom-based disease studies
Hemiptera–Phytoplasma–Plant Biological Interaction Database: http://trivellone.speciesfile.org/ (Trivellone, 2019)	Phytoplasma transmission via Hemiptera vectors and plant interactions	Tracks phytoplasma transmission through insect vectors and host plants	Insect vector species, phytoplasma strains, and host plants	Vector ecology, phytoplasma disease monitoring, epidemiology	Used in niche phytoplasma transmission research
EPPO-Q-bank-phytoplasma section: https://qbank.eppo.int/phytoplasmas/ (Bertaccini *et al.*, 2024)	Phytoplasma identification, molecular diagnostics, and host–vector records	Provides genetic sequences, PCR primers, and diagnostic protocols for phytoplasma detection and classification	Genetic sequences, PCR primers, diagnostic protocols, host and vector records	Plant health diagnostics, regulatory plant protection, phytoplasma detection, and monitoring	Used by plant health authorities, cited in regulatory and phytosanitary research

All websites last accessed June 2025.
PCR, polymerase chain reaction; RFLP, restriction fragment length polymorphism.

The tool is applied in laboratory settings, where DNA is extracted from plant samples and sequenced before analysis. This tool is often used alongside other molecular techniques, such as PCR and sequencing-based phylogenetic analysis. The classification results contribute to research on phytoplasma genetics and variation, supporting studies on plant diseases and their causes.

*i*PhyDSDB

*i*PhyDSDB (Interactive Phytoplasma Disease Symptom Database) is a database designed to collect, organize, and provide information on phytoplasma-associated diseases and their symptoms. It serves as a resource for researchers studying phytoplasma infections by compiling data on plant hosts, disease symptoms, and associated 16S rRNA-based classifications (Wei *et al.*, 2024) (Table 2.2). The database contains records of phytoplasma-infected plant species, detailing symptoms such as yellowing, leaf deformation, stunting, phyllody, and witches'-broom. It also includes links between symptoms and specific 16Sr groups and subgroups, helping researchers analyze correlations between phytoplasma strains and the diseases they cause. *i*PhyDSDB is used in plant pathology and phytoplasma research to support disease identification and classification. Integrating molecular data and symptom descriptions provides a reference for comparing newly observed phytoplasma infections with documented cases. The database can assist in epidemiological studies, helping track the distribution and impact of phytoplasma diseases across different plant species and regions. This resource is typically accessed through an online platform, where users can search for specific phytoplasma strains, disease symptoms, or affected plant species. It is primarily used in research institutions, agricultural studies, and diagnostic laboratories to study phytoplasma-related plant diseases and their characteristics.

Hemiptera–Phytoplasma–Plant Biological Interaction Database

The Hemiptera–Phytoplasma–Plant Biological Interaction Database is a resource that compiles information on the relationships between phytoplasmas, their Hemiptera insect vectors, and host plants. It is designed to support research on phytoplasma transmission, vector ecology, and plant disease dynamics by integrating data on these biological interactions (Trivellone, 2019) (Table 2.2). The database includes records of Hemiptera insect species that serve as vectors for phytoplasma transmission, detailing their host range, feeding behavior, and transmission efficiency. It also documents phytoplasma strains associated with these vectors and the plant species they infect. Information on disease symptoms, geographic distribution, and molecular classification (e.g. 16Sr groups and subgroups) is also included. This database is used in entomology, plant pathology, and epidemiology to study phytoplasma transmission pathways and understand how insect vectors contribute to the spread of plant diseases. Researchers can use it to analyze patterns of host–vector interactions, identify potential reservoir hosts, and track the movement of phytoplasma-associated diseases across different regions. The database provides a reference for scientists investigating vector-borne phytoplasma diseases. It is typically accessed in research institutions and agricultural studies, where it supports the development of strategies for managing phytoplasma-related plant diseases by targeting insect vectors and understanding their role in disease spread.

EPPO-Q-bank, phytoplasma section

The phytoplasma section of the EPPO-Q-bank is a specialized component of the database that provides reference data for the identification and classification of phytoplasmas. It contains validated molecular and taxonomic information that supports diagnostic laboratories, plant health professionals, and researchers in detecting and monitoring phytoplasma-associated diseases (Bertaccini *et al.*, 2024) (Table 2.2). This section includes genetic sequences, PCR primers, and diagnostic protocols for phytoplasma identification. It provides 16S rRNA gene sequences, which are commonly used for phytoplasma classification, as well as additional molecular markers that aid in differentiating closely related strains. The database also links phytoplasma strains to their host plants and insect vectors, helping track transmission pathways and epidemiological trends. EPPO-Q bank's phytoplasma section is designed to support regulatory plant protection measures and phytosanitary diagnostics. It serves as a reference for national plant protection organizations and research institutions, ensuring accurate identification and classification of phytoplasma species. The database is regularly updated in collaboration with experts and laboratories specializing in plant pathogens, making it a valuable tool for plant health surveillance and risk assessment.

Enhancing Plant Health with Data

As plant diseases, including phytoplasma disease, become more complex and geographically widespread, the role of curated, accessible, and up-to-date biological databases becomes increasingly critical. The databases, tools, and resources described in this chapter form the foundation for data-driven disease management strategies, enhancing the ability of precision agriculture to adapt to emerging phytoplasma disease challenges. These resources also contribute to broader goals in plant health security and food system resilience.

Future Perspectives of Precision Agriculture in Phytoplasma Research

The integration of AI, remote sensing, and specialized databases will shape the future of precision agriculture in phytoplasma research. AI can revolutionize disease monitoring by processing vast amounts of environmental, visual, and sensor data to identify early indicators of crop stress that may suggest phytoplasma infection. When combined with high-resolution images from drones and satellites, remote sensing facilitates non-invasive observation of large agricultural areas, enabling quick identification of regions that require further scrutiny. While AI cannot replace molecular diagnostics or directly classify phytoplasma, it enhances decision-making by helping prioritize samples for laboratory analysis and increasing the efficiency of surveillance operations. Curated databases, which store verified genetic, taxonomic, and host–vector interaction data, are crucial to this system for confirming diagnoses and understanding disease dynamics. Nonetheless, challenges persist, such as the scarcity of labeled training data for AI models, variability in symptom presentation among plant species, and the need for standardization and interoperability among databases. Furthermore, successful implementation demands strong collaboration among data scientists, pathologists, and farmers, along with continued investment in digital infrastructure. Despite these challenges, the fusion of AI, remote sensing, and robust databases presents a hopeful avenue for more proactive, precise, and scalable management of plant diseases related to phytoplasma.

References

Abbas, A., Zhang, Z., Zheng, H., Alami, M.M, Al Refaei, A.F. *et al.* (2023) Drones in plant disease assessment, efficient monitoring, and detection: A way forward to smart agriculture. *Agronomy* 13, 1524. DOI: 10.3390/agronomy13061524.

Alsadik, B., Ellsäßer, F.J., Awawdeh, M., Al-Rawabdeh, A., Almahasneh, L. *et al.* (2024) Remote sensing technologies using UAVs for pest and disease monitoring: A review centered on date palm trees. *Remote Sensing* 16, 4371. DOI: 10.3390/rs16234371.

Bertaccini, A., van de Bilt, J.L.J., Contaldo, N., Cottyn, B., Damm, U. *et al.* (2024) EPPO-Q-bank: a curated database to support plant pest diagnostic activities. *EPPO Bulletin* 54, 361–365. DOI: 10.1111/epp.13063.

Buja, I., Sabella, E., Monteduro, A.G., Chiriacò, M.S., De Bellis, L., Luvisi, A., and Maruccio, G. (2021) Advances in plant disease detection and monitoring: From traditional assays to in-field diagnostics. *Sensors* 21, 2129. DOI: 10.3390/s21062129.

Gatou, P., Tsiara, X., Spitalas, A., Sioutas, S., and Vonitsanos, G. (2024) Artificial intelligence techniques in grapevine research: A comparative study with an extensive review of datasets, diseases, and techniques evaluation. *Sensors* 24, 6211. DOI: 10.3390/s24196211.

Guebsi, R., Mami, S., and Chokmani, K. (2024). Drones in precision agriculture: A comprehensive review of applications, technologies, and challenges. *Drones* 8, 686. DOI: 10.3390/drones8110686.

Imran, H.A., Zeggada, A., Ianniello, I., Melgani, F., Polverari, A. *et al.* (2023) Low-cost handheld spectrometry for detecting flavescence dorée in vineyards. *Applied Sciences* 13, 2388. DOI: 10.3390/app13042388.

Joshi, P., Sandhu, K.S., Dhillon, G.S., Chen, J., and Bohara, K. (2024) Detection and monitoring wheat diseases using unmanned aerial vehicles (UAVs). *Computers and Electronics in Agriculture* 224, 109158. DOI: 10.1016/j.compag.2024.109158.

Knauer, U., Warnemünde, S., Menz, P., Thielert, B., Klein, L. *et al.* (2024) Detection of apple proliferation disease using hyperspectral imaging and machine learning techniques. *Sensors* 24, 7774. DOI: 10.3390/s24237774.

Muirhead, K., Pérez-López, E., Bahder, B.W., Hill, J.E., and Dumonceaux, T. (2019) The CpnClassiPhyR is a resource for cpn60 universal target-based classification of phytoplasmas. *Plant Diseases* 103, 2494–2497. DOI: 10.1094/PDIS-03-19-0454-A.

Narayanasamy, P. (2011) Detection of bacterial and phytoplasmal pathogens. In: *Microbial Plant Pathogens – Detection and Disease Diagnosis*, Vol. 2. Springer, Dordrecht, pp. 5-169. DOI: 10.1007/978-90-481-9769-9_5.

Narmilan, A., Gonzalez, F., Salgadoe, A.S.A., and Powell, K. (2022) Detection of white leaf disease in sugarcane using machine learning techniques over UAV multispectral images. *Drones* 6, 230. DOI: 10.3390/drones6090230.

Plata, I.T., Panganiban, E.B., Alado, D.B., Taracatac, A.C., Bartolome, B.B., and Labuanan, F.R.E. (2021) A recognition method for cassava phytoplasma disease (CPD) real-time detection based on transfer learning neural networks. *International Journal of Advanced Computer Science and Applications* 12. DOI: 10.14569/IJACSA.2021.0121234, 12(12).

Plata, I.T., Panganiban, E.B., Alado, D.B., Taracatac, A.C., Bartolome, B.B., and Labuanan, F.R.E. (2022a) Comparative performance analysis of real-time methods for cassava phytoplasma disease (CPD) detection based on deep learning neural networks. *International Journal of Emerging Technology and Advanced Engineering* 12, 55–64. DOI: 10.46338/ijetae0222_07.

Plata, I.T., Panganiban, E.B., Alado, D.B., Taracatac, A.C., Bartolome, B.B., and Labuanan, F.R.E. (2022b) Drone-based geographical information system (GIS) mapping of cassava phytoplasma disease (CPD) for precision agriculture. *International Journal of Emerging Technology and Advanced Engineering* 12, 1–9. DOI: 10.46338/ijetae0222_01.

Portela, F., Sousa, J.J., Araújo-Paredes, C., Peres, E., Morais, R., and Pádua, L. (2024) A systematic review on the advancements in remote sensing and proximity tools for grapevine disease detection. *Sensors* 24, 8172. DOI: 10.3390/s24248172.

Silva, D.M., Bernardin, T., Fanton, K., Nepaul, R., Pádua, L., Sousa, J.J., and Cunha, A. (2022). Automatic detection of Flavescence dorée grapevine disease in hyperspectral images using machine learning. *Procedia Computer Science* 196, 125–132. DOI: 10.1016/j.procs.2021.12.019.

Thielert, B., Menz, P., Götte, G., Runne, M., Michel, M. *et al.* (2024) PhenoTruckAI: On-site hyperspectral measurement for distinction of quarantine grapevine disease "flavescence dorée" and non-quarantine disease "bois noir" in a mobile laboratory. *Informatik* 2024, 1253–1260.

Trivellone, V. (2019) An online global database of Hemiptera-Phytoplasma-Plant biological interactions. *Biodiversity Data Journal* 7), e34859. DOI: 10.3897/BDJ.7.e34859.

Wei, W., Shao, J., Zhao, Y., Inaba, J., Ivanauskas, A. *et al.* (2024) *i*PhyDSDB: Phytoplasma Disease and Symptom Database. *Biology* 13, 657. DOI: 10.3390/biology13090657.

Wei, W., Shao, J., and Zhao, Y. (2025) Leveraging artificial intelligence and big data to advance phytoplasma disease detection and crop health management. *Phytopathogenic Mollicutes* 15, 15–16. DOI: 10.5958/2249-4677.2025.00007.6.

Woźniak, M. and Ijaz, M.F. (2024) Editorial: Recent advances in big data, machine, and deep learning for precision agriculture. *Frontiers in Plant Science* 15, 1367538. DOI: 10.3389/fpls.2024.1367538.

Zhao, Y., Wei, W., Lee, I.M., Shao, J., Suo, X., and Davis, R.E. (2009) Construction of an interactive online phytoplasma classification tool, iPhyClassifier, and its application in analysis of the peach X-disease phytoplasma group (16SrIII). *International Journal of Systematic and Evolutionary Microbiology* 59, 2582–2593. DOI: 10.1099/ijs.0.009688-0.

3 Exploring Phytoplasma Genomes through Advanced Sequencing and Enrichment Techniques

Helen Mae Mejia,[1] Yi-Ching Chiu,[1] Chih-Horng Kuo,[2*] and Jun-Yi Yang[1*]
[1]*Institute of Biochemistry, National Chung Hsing University, Taichung, Taiwan;*
[2]*Institute of Plant and Microbial Biology, Academia Sinica, Taipei, Taiwan*

Abstract
In the past decade, affordable next-generation sequencing (NGS) has increased access to phytoplasma genomes, offering new insights into their virulence, evolution, and taxonomy. However, challenges remain due to the lack of axenic culture, low titer in hosts, and the genomes' repetitive, biased composition, leading to fragmented assemblies with NGS short reads. This chapter reviews advanced methods to address these issues, including DNA enrichment techniques and the integration of third-generation sequencing to improve assembly quality and completeness.

Introduction

Phytoplasmas are specialized bacteria that lack cell walls and are known to infect a wide range of plant species, causing significant agricultural losses (Hogenhout *et al.*, 2008). Understanding the molecular interactions and protein composition of phytoplasmas is crucial for comprehending their pathogenic mechanisms and devising effective control strategies (Huang *et al.*, 2021). As obligate intracellular parasites often rely on host cells for growth and replication, isolating and purifying phytoplasmas poses a significant challenge (Hogenhout and Segura Music, 2009). Cultivating phytoplasmas in cell-free media has also been challenging, making most traditional research methods in microbiology impractical.

To circumvent the challenges of obtaining an axenic culture of phytoplasmas for microbiological studies, genome sequencing and analysis have emerged as a critical approach over the past two decades (Hogenhout and Segura Music, 2009; Huang *et al.*, 2023). The rapid advancement in phytoplasma genomics has been largely driven by the affordability of next-generation sequencing (NGS) technologies such as the Illumina platforms (Huang *et al.*, 2023). Nevertheless, the difficulties associated with obtaining phytoplasma DNA of sufficient quality and quantity remain a significant technical hurdle. Furthermore, the short-read length of NGS technologies often hinders the generation of high-quality genome assemblies, thereby limiting the utility of genome analyses. Fortunately, several technological advancements have been made in recent years, including the development of phytoplasma DNA enrichment protocols and third-generation sequencing technologies for producing long reads (Table 3.1).

*Corresponding author: chk@gate.sinica.edu.tw and jyang@nchu.edu.tw

DOI: 10.1079/9781800627031.0003

Table 3.1. Comparison of key features between immunoprecipitation-based enrichment, iodixanol gradient centrifugation-based enrichment, and methyl-CpG-binding domain (MBD)-mediated enrichment of phytoplasma DNA.

Feature	Immunoprecipitation-based enrichment	Iodixanol gradient centrifugation-based enrichment	Methyl-CpG-binding domain-mediated enrichment
Principle	Targets immunodominant membrane proteins (IDPs) with specific antibodies to isolate phytoplasma	Separates phytoplasma based on density differences between phytoplasma and host cell using iodixanol gradients	The MBD protein selectively interacts with the CpG-methylated sites in the host DNA, capturing and separating it from phytoplasma DNA
Effectiveness	Captures up to 36% phytoplasma reads, reducing host DNA contamination, but still allows some co-purification	Achieves up to 14-fold increase in phytoplasma reads with a 1.7–2-fold reduction of host DNA	Increases the phytoplasma DNA by 10–30-fold
Limitations	Requires specific antibodies for each phytoplasma strain; prone to co-purification of plant material	Sensitive to gradient concentration and density; requires multiple centrifugation steps	Requires ensuring the removal of other microorganisms prior to DNA extraction
Advantages	Enables the study of host–pathogen interactions through proteomics	Non-toxic, cost-effective, and practical; improves genome assembly completeness	Inexpensive and time-efficient, requiring fewer steps prior to DNA sequencing
Applicability	Useful for in-depth studies of host–phytoplasma interactions	Ideal for generating high-quality phytoplasma genomes for comparative genomics	Useful for whole-genome sequencing of unculturable, intracellular, and plant-pathogenic bacteria
Key equipment	Specific antibodies, protein A/G beads, and auxiliary agents for elution	Centrifuge and iodixanol-based gradient solutions (Optiprep™)	NEBNext Microbiome Enrichment Kit and magnetic bead separator
Assembly improvement	Enables complete genome coverage when combined with long-read sequencing	Increases genome size by 15,000–137,000 bp, depending on the strain	Achieves 90–96% completeness in genome assembly when coupled with next-generation sequencing
Time and cost efficiency	Time-consuming and strain-specific antibody generation may increase the cost	Faster and more economical compared with CsCl gradients; safer for researchers	Faster and more economical compared with density gradient centrifugation method

This chapter reviews these advancements and highlights the biological insights obtained from analyzing the high-quality genome assemblies made possible by these advanced methodologies. These approaches have enabled more detailed investigations into phytoplasma genome structure and function, providing an important foundation for studying the pathogenic mechanisms of phytoplasmas and developing prevention and control strategies.

Immunoprecipitation-based Phytoplasma Enrichment

The spread of phytoplasmas relies on interactions between phytoplasma immunodominant membrane proteins (IDPs) and host proteins (Konnerth *et al.*, 2016). IDPs, which are abundant in phytoplasma cell membranes, are typically categorized into three groups: immunodominant membrane protein (Imp), immunodominant membrane protein A (IdpA), and antigenic membrane protein (Amp) (Kakizawa *et al.*, 2006). These proteins play a crucial role in phytoplasmas' physiological functions and interactions with host plants. Therefore, utilizing IDPs as targets for immunoprecipitation can facilitate the purification of phytoplasma in infected plants.

Immunoprecipitation selectively captures IDPs using specific antibodies, enabling researchers to discern specific interactions between phytoplasma and their hosts. This technique involves binding the target IDP with antibodies, followed by capturing the antibody–protein complex using auxiliary agents like protein A/G beads. Subsequent wash steps remove non-specifically bound proteins, and appropriate methods (e.g. low pH buffer or competitive elution) are employed to separate and purify phytoplasma-containing complexes. Readings obtained through antibody enrichment procedures show up to 36% mapping to the phytoplasma genome (Tan *et al.*, 2021), effectively isolating and purifying phytoplasma from plant tissues, overcoming limitations posed by complex host–pathogen interactions. However, reads originating from plant hosts still constitute a significant portion of the total reads. This result may stem from interactions between phytoplasma and plant cells, leading to the co-purification of plant cells during the process.

To enhance enrichment outcomes, further purification of antibodies used for immunoprecipitation can be considered. However, generating antibodies specific to a particular phytoplasma might not be suitable for other phytoplasma strains, as Imp exhibits substantial variations even within the same species, with low similarity between phytoplasma groups (Morton *et al.*, 2003; Kakizawa *et al.*, 2006). Therefore, to enrich multiple phytoplasma strains through immunoprecipitation, antibodies targeting individual IDPs need to be generated. Additionally, selecting infected plant samples with high phytoplasma titers may aid in overcoming challenges associated with varying phytoplasma cell counts among different plants propagated from the same infected plant (Christensen *et al.*, 2004).

A study in 2021 demonstrated the use of immunoprecipitation-based enrichment on '*Candidatus* Phytoplasma aurantifolia' strain NCHU2014, resulting in the first complete genome available for the 16SrII phytoplasma using Oxford Nanopore Technologies' (ONT) MinION long-read DNA sequencing (Tan *et al.*, 2021). In this study, 22.5–36.0% raw reads were obtained from phytoplasma after the enrichment protocol, in which all segments of the phytoplasma genome are covered in the final assembly. In comparison, samples processed without enrichment yielded only 8.0–23.2% phytoplasma reads, leaving approximately 37 kb of the genome uncovered.

Immunoprecipitation-based enrichment is a powerful tool for studying host–phytoplasma interactions. By targeting IDPs, this approach facilitates the purification of phytoplasma from infected plant tissues, providing valuable insights into phytoplasma pathogenesis and host manipulation. However, further refinement of antibody specificity and selection of appropriate plant samples are necessary to enhance enrichment outcomes. Additionally, downstream analyses of enriched phytoplasma proteins via proteomic approaches can contribute to a deeper understanding of phytoplasma biology and may lead to the development of novel control strategies against phytoplasma-related diseases.

Density Gradient using Iodixanol as a Medium for Phytoplasma Enrichment

Centrifugation has long been employed in scientific research to separate specific cellular components from the rest of the cell matrix based on size and density gradients. Phytoplasma DNA and host DNA have distinct physical characteristics and biological makeup, making it possible to separate them, however technically challenging (Oyola *et al.*, 2012; Tran-Nguyen and Schneider, 2013). In density gradient centrifugation, the separation of cell components is based on their buoyant densities using certain reagents with specific densities. The use of post-DNA extraction cesium chloride (CsCl)–bisbenzimide gradient centrifugation followed by pulsed-field gel electrophoresis (PFGE) has been reported to be effective in enriching the DNA separation of phytoplasma from the host plant (Tran-Nguyen and Schneider, 2013). However, on top of the health hazard of CsCl, this approach is neither economical nor time-efficient; it requires specialized equipment, an extended amount of time for processing, and expensive reagents.

In response to these concerns, the use of iodixanol density gradient – a non-toxic, non-ionic, and iso-osmotic density gradient medium – was developed as an alternative approach to enrich the phytoplasma prior to DNA extraction (Rodrigues Jardim *et al.*, 2022). The study provided significant evidence of increased phytoplasma DNA sequences and reduced host DNA sequences for potato and capsicum infected with '*Ca.* P. australasia'-related strains. After homogenization and differential centrifugation of phytoplasma-infected samples, density gradient centrifugation was done using iodixanol (Optiprep™), diluted in three different concentrations (14%, 22%, and 30%), followed by the separation of cells based on their densities. The potential of the method was evaluated using quantitative PCR (qPCR) and reference mapping of the metagenomic reads. Results showed that the Illumina sequencing reads of '*Ca.* P. australasia'-related strains were increased by approximately 14-fold in potato and threefold in capsicum after iodixanol density gradient enrichment. Moreover, host DNA reads were reduced by approximately 1.7-fold and twofold in the potato and capsicum samples, respectively. Phytoplasma genome assemblies after the enrichment were found to be approximately 15,000 bp larger in the potato-associated strain and approximately 137,000 bp larger in the capsicum-associated strain, compared with assemblies from DNA extracted from infected samples without enrichment.

The iodixanol density gradient is well suited for generating complete and accurate metagenomic assembled genomes for uncultivated phytopathogenic bacteria such as phytoplasmas. This approach is non-toxic and cost- and time-efficient, thus safe and practical for researchers to execute. The improvements in the obtained genome assemblies can facilitate comparative genomics and taxonomic inference.

Methyl-CpG-binding Domain-mediated Phytoplasma DNA Enrichment Method

While the density gradient method for phytoplasma DNA enrichment has been proven effective, there are still some limitations for some researchers. As a result, efforts have continued to develop alternative approaches that offer comparable or improved efficiency while further reducing time and expense. In a 2021 study conducted by Nijo *et al.*, researchers demonstrated the effectiveness of employing methyl-CpG-binding domain (MBD) proteins to enrich phytoplasma DNA, utilizing the principle of DNA methylation.

DNA methylation is an important epigenetic mechanism that significantly regulates gene expression and cell differentiation. In eukaryotic organisms, this process involves the addition of a methyl group (-CH_3) to the carbon 5 position of cytosine bases within CpG sequences. In contrast, prokaryotic DNA methylation primarily occurs at the exocyclic amino group on the carbon 6 position of adenine residues. The distinction in DNA methylation mechanisms enables an effective separation of prokaryotic and eukaryotic DNA. After extracting the total DNA from phytoplasma-infected plants, the NEBNext Microbiome Enrichment Kit was used to enhance the separation of phytoplasma DNA and host DNA. This kit includes a genetically fused MBD2 protein and an Fc tail (MBD2-Fc), which is linked to magnetic beads that are covalently bonded to a truncated form of recombinant protein A. The total DNA of the samples was incubated with these beads, during which the MBD2-Fc protein selectively interacts with the CpG-methylated sites in the host DNA, effectively capturing and separating it from phytoplasma DNA. It was then subjected to magnetic separation, allowing the beads to accumulate on the walls of the tubes while the supernatant, which contained phytoplasma-enriched DNA, was collected. Qualitative PCR analysis revealed that the MBD-mediated enrichment method increased the phytoplasma DNA levels by 10–30 times compared with those of the total DNA without enrichment. This enrichment method was coupled with NGS and was able to provide the first draft genome sequence of the '*Ca.* P. asteris' hydrangea phyllody strain, with a total contig length of 597,762 bp including 521 coding sequences, one rRNA operon, and 30 tRNA genes (Nijo *et al.*, 2021).

Other studies by Ranebennur *et al.* (2022) and Debonville *et al.* (2022) utilized the MBD-mediated enrichment method, specifically the NEBNext Microbiome Enrichment Kit, to enrich the phytoplasma DNA extracted from '*Ca.* Phytoplasma' species strain-infected sesame plants and from *Euscelidius variegatus* fed with flavescence dorée phytoplasma-infected broad beans, respectively. Qualitative PCR revealed a 30-fold enrichment of the phytoplasma DNA from *E. variegatus* following the method, which consequently generated a highly complete circular chromosome of 654,223 bp containing 506 protein-coding genes. While the extent of DNA enrichment achieved from the infected sesame plants was not reported, the process led to the generation of a draft genome comprising 60 contigs, totaling 553,228 bp in length, with over 400 times depth coverage and 95.21% estimated completeness.

These studies have demonstrated the effectiveness of MBD proteins in enriching phytoplasma DNA, thereby aiding whole-genome sequencing, as they offer a more concentrated and representative sample of the target DNA. This also marks a significant advancement in the field of genomics, providing researchers with powerful tools to explore and characterize phytoplasma genomes.

Illumina Sequencing

Illumina sequencing is one of several platforms of the NGS method and is by far the most widely used and essential technique for many biotechnology applications. So far, the draft genomes of many phytoplasmas have been assembled *de novo* following an Illumina sequencing-by-synthesis workflow, which has four basic steps: (i) sample preparation; (ii) cluster generation; (iii) sequencing; and (iv) data analysis (Huang *et al.*, 2023). Sample preparation involves cutting isolated DNA molecules into smaller fragments, followed by the ligation of appropriate adaptors. After that is the generation of clusters by isothermal amplification of each DNA fragment on a solid phase (a glass slide with lanes) composed of two types of oligonucleotide sequences. The ligated adapter of the fragment strand is complementary to one of the oligonucleotide sequences. Hybridization of these complementary sequences initiates synthesis, which is simultaneously repeated multiple times through "bridge amplification," approximately 1000 copies of each to create a clonal cluster. The sequencing begins by washing off the reverse strand, blocking the 3′ end of the remaining oligonucleotide to avoid unnecessary priming, and then adding a fluorescently tagged nucleotide to the remaining forward strand. The fluorescent tag is unique for each of the different bases (A, T, C, G) that are incorporated into the chain and will later be excited by the light source to give a distinct signal for detection. All identical strands are read simultaneously, and the process is repeated multiple times depending on the desired read length. The emission wavelength, as well as its intensity, determines the base call. Since DNA molecules are cut into smaller fragments during library construction, the sequence read is also short (ranging from 50 to 300 bp) but with multiple overlaps, so they can be assembled to generate a longer sequence (Slatko *et al.*, 2018; Cheng *et al.*, 2023). Since the phytoplasma genome contains repeat-rich sequences such as the potential mobile units (PMUs) (Bai *et al.*, 2006) or sequence-variable mosaics (SVMs) (Jomantiene *et al.*, 2007), Illumina sequencing may not be ideal for generating complete genome assemblies. Fortunately, the development of third-generation sequencing technologies can complement such NGS technologies.

Third-generation Sequencing Technologies: PacBio and Oxford Nanopore Technologies

The most recent addition to genome sequencing technologies is third-generation single-molecule sequencing. Unlike the short reads (up to 300 bp) generated by NGS, third-generation sequencing can generate over 10,000 bp reads, which resolves the longstanding problem of assembling a genome with high stretches of repeat sequence regions. Pacific Biosciences (PacBio) and ONT are the leading platforms for this advancement, both widely used in gene isoform identification, *de novo* transcriptomic analysis, and metagenomics because of their long sequence readability, an area where NGS cannot compete. Both PacBio and ONT are superior when it comes to reading length and throughput. However, each has its limitations that prevent it from completely replacing short-read sequencing in transcriptomics (Slatko *et al.*, 2018; Udaondo *et al.*,

2021). PacBio sequencing or single molecule real-time (SMRT) sequencing can sequence up to 30–50 kb or longer. The method involves the isolation of DNA or RNA samples and then the generation of the "SMRTbell" library by ligating adaptors to the double-stranded sample to create a circular template. Primers and bioengineered DNA polymerase are also added to the library, which is then placed on the instrument's SMRT flow cell, consisting of millions of small chambers called a zero-mode waveguide (ZMW). A DNA molecule is immobilized inside one of these wells. The ligated adaptor and the primer are complementary; therefore, synthesis is initiated. The nucleotides are uniquely tagged with phospho-linked fluorophores, emitting light each time they are added to the growing chain. The reaction is imaged and recorded in real time, monitoring the time lapse between each base incorporation, and occurs simultaneously in millions of other wells. PacBio SMRT sequencing offers a relatively straightforward assembly of a complete genome sequence using only a few SMRT cells and also determines methylation patterns. However, limitations include lower yield, higher error rates, and a higher cost per base (Slatko *et al.*, 2018; Udaondo *et al.*, 2021).

Nanopore-based DNA sequencing has been commercialized by ONT. The machine used is small, handheld, with a small chamber of a few nanometers in diameter, as a DNA or RNA inlet. The technology relies on a nanopore, basically a very small hole that serves as a biosensor and is embedded in an electrically resistant polymer membrane. Long dsDNA molecules are first bound to DNA polymerase, and when an electric field is applied, they start moving to the other side of the chamber through the nanopore. During this process, small voltages are used across the pore, and the nucleotide provides a distinct electronic signal, giving rise to a current that can be easily measured in real time. The magnitude of the current depends on the base composition of the DNA sample; hence, just by measuring the magnitude of the electric current across the nanopore, the base composition of the polynucleotide chain can be obtained (Slatko *et al.*, 2018; Udaondo *et al.*, 2021). The limitation of ONT is a high error rate of up to 15%, but it is tolerable when large amounts of the same sequence are sequenced.

Several studies have obtained complete phytoplasma genome assemblies using these third-generation sequencing technologies. For example, the complete genome assembly of jujube witches'-broom (JWB) phytoplasma was generated using the PacBio RS II platform; the circular chromosome of 750,803 bp contains 694 protein-encoding genes and four PMUs (Wang *et al.*, 2018). In comparison, the ONT platforms have been more popular. In 2021, the complete genome of '*Ca.* P. aurantifolia' strain NCHU2014 was assembled using ONT MinION. The assembly consists of a circular chromosome with 635,584 bp, 471 protein-coding genes containing 33 transporter genes, and 27 putative effector genes (Tan *et al.*, 2021). The complete genome assembly of flavescence dorée phytoplasma was generated in 2022, which is a circular chromosome of 654,223 bp containing 506 protein-coding genes (Debonneville *et al.*, 2022). Another complete genome assembly was generated from '*Ca.* P. luffae' strain NCHU2019, a circular chromosome with 769 kb, 75 kb repeats, and at least 13 PMUs (*c.*25% of its chromosome) (Huang *et al.*, 2022). Lastly, the complete genome assembly generated from '*Ca.* P. ziziphi' strain Hebei-2018, using ONT GridION X5, is reported to have a circular chromosome of 764,108 bp, including 701 protein-coding genes with assigned functions (Xue *et al.*, 2023).

Biological Insights from Complete Assemblies

Obtaining a complete, gap-free genome assembly of any organism is essential for understanding its nature and genetic diversity. Phytoplasmas are bacterial pathogens that infect plants through phloem-feeding insect vectors. Therefore, the availability of complete genome sequences provides a valuable resource in understanding their chromosomal organization, gene content, metabolic pathways, possible molecular interactions with the hosts, and virulence mechanisms. Phytoplasma symptoms such as phyllody and witches'-broom are induced by the secreted protein effectors, encoded mostly by genes associated with PMUs (Sugio and Hogenhout, 2012). PMUs account for up to 25% of the genome (Huang *et al.*, 2022). These PMUs contain symptom-determinant genes and can be horizontally transferred from one phytoplasma strain to another (Chung *et al.*, 2013). Therefore, comprehensive identification of PMUs by obtaining complete genome assemblies can infer synteny and phylogenetic relationships with other phytoplasma species and strains (Huang *et al.*, 2022; Tokuda *et al.*, 2023). The first complete phytoplasma sequence assembled was that of the OY strain ('*Ca.* P. asteris'), line OY-M (Oshima *et al.*, 2004). The OY-M genome was found to lack several essential genes for sustaining cellular life. Notably, it lacks the gene that encodes ATP-synthase subunits, suggesting that phytoplasmas rely on their host plants for energy (Oshima *et al.*, 2004). The confirmation of gene absence is only possible when the complete genome assemblies are available. Later on, the absence of these genes was confirmed in other phytoplasma strains, as well as other obligate symbiotic bacteria (Namba, 2019).

Another important example is a study by Huang *et al.* (2022) that reported the complete genome sequence of '*Ca.* P. luffae' strain NCHU2019 and provided information regarding the importance of phytoplasma PMUs in the evolution of phytoplasma genome organization and effector gene content. Aside from horizontal gene transfer to another phytoplasma, PMUs also contribute to the duplication of effector genes in the genome, providing redundant elements for subfunctionalization or neofunctionalization. Nearly 80% of the variance in phytoplasma genome sizes can be explained by the PMU abundance, which accounts for the genetic diversity and virulence (Huang *et al.*, 2022). Interestingly, PMUs are not universally present in all phytoplasmas. In a rare exception, the complete genome sequence of flavescence dorée phytoplasma was found to lack any PMU-like region (Debonneville *et al.*, 2022). This discovery emphasizes the need for further studies into the roles of PMU in the evolution of diverse phytoplasmas.

References

Bai, X., Zhang, J., Ewing, A., Miller, S.A., Jancso Radek, A. *et al.* (2006) Living with genome instability: The adaptation of phytoplasmas to diverse environments of their insect and plant hosts. *Journal of Bacteriology* 188, 3682–3696. DOI: 10.1128/JB.188.10.3682–3696.2006.

Cheng, C., Fei, Z., and Xiao, P. (2023) Methods to improve the accuracy of next-generation sequencing. *Frontiers in Bioengineering and Biotechnology* 11, 982111. DOI: 10.3389/fbioe.2023.982111.

Christensen, N.M., Nicolaisen, M., Hansen, M., and Schulz, A. (2004) Distribution of phytoplasmas in infected plants as revealed by real-time PCR and bioimaging. *Molecular Plant-Microbe Interactions* 17, 1175–1184. DOI: 10.1094/MPMI.2004.17.11.1175.

Chung, W.-C., Chen, L.-L., Lo, W.-S., Lin, C.-P., and Kuo, C.-H. (2013) Comparative analysis of the peanut witches'-broom phytoplasma genome reveals horizontal transfer of potential mobile units and effectors. *PLoS One* 8, e62770. DOI: 10.1371/journal.pone.0062770.

Debonneville, C., Mandelli, L., Brodard, J., Groux, R., Roquis, D. *et al.* (2022) The complete genome of the "Flavescence dorée" phytoplasma reveals characteristics of low genome plasticity. *Biology* 11, 953. DOI: 10.3390/biology11070953.

Hogenhout, S.A. and Segura Music, M. (2009) Phytoplasma genomics, from sequencing to comparative and functional genomics – What have we learnt? In: Weintraub, P.G. and Jones, P. (eds) *Phytoplasmas: Genomes, Plant Hosts and Vectors*. CAB International, Wallingford, UK, pp. 19–36.

Hogenhout, S.A., Oshima, K., Ammar, E.D., Kakizawa, S., Kingdom, H.N. *et al.* (2008) Phytoplasmas: bacteria that manipulate plants and insects. *Molecular Plant Pathology* 9, 403–423. DOI: 10.1111/j.1364-3703.2008.00472.x.

Huang, C.-T., Cho, S.-T., Lin, Y.-C., Tan, C.-M., Chiu, Y.-C. *et al.* (2022) Comparative genome analysis of '*Candidatus* Phytoplasma luffae' reveals the influential roles of potential mobile units in phytoplasma evolution. *Frontiers in Microbiology* 13, 773608. DOI: 10.3389/fmicb.2022.773608.

Huang, C.-T., Pei, S.-C., and Kuo, C.-H. (2023) Genomic studies on Asian phytoplasmas. In: Tiwari, A.K., Oshima, K., Yadav, A., Esmaeilzadeh-Hosseini, S.A., Hanboonsong, Y. and Lakhanpaul, S. (eds) *Phytoplasma diseases in Asian countries: Characterization, Epidemiology, and Management*. Academic Press, Cambridge, MA, pp. 67–83.

Huang, W., Maclean, A.M., Sugio, A., Maqbool, A., Busscher, M. *et al.* (2021) Parasitic modulation of host development by ubiquitin-independent protein degradation. *Cell* 184, 5201–5214, e12. DOI: 10.1016/j.cell.2021.08.029.

Jomantiene, R., Zhao, Y., and Davis, R.E. (2007) Sequence-variable mosaics: Composites of recurrent transposition characterizing the genomes of phylogenetically diverse phytoplasmas. *DNA and Cell Biology* 26, 557–564. DOI: 10.1089/dna.2007.0610.

Kakizawa, S., Oshima, K., and Namba, S. (2006) Diversity and functional importance of phytoplasma membrane proteins. *Trends in Microbiology* 14, 254–256. DOI: 10.1016/j.tim.2006.04.008.

Konnerth, A., Krczal, G., and Boonrod, K. (2016) Immunodominant membrane proteins of phytoplasmas. *Microbiology* 162, 1267–1273. DOI: 10.1099/mic.0.000331.

Morton, A., Davies, D.L., Blomquist, C.L., and Barbara, D.J. (2003) Characterization of homologues of the apple proliferation immunodominant membrane protein gene from three related phytoplasmas. *Molecular Plant Pathology* 4, 109–114. DOI: 10.1046/j.1364-3703.2003.00155.x.

Namba, S. (2019) Molecular and biological properties of phytoplasmas. *Proceedings of the Japan Academy, Series B* 95, 401–418. DOI: 10.2183/pjab.95.028.

Nijo, T., Iwabuchi, N., Tokuda, R., Suzuki, T., Matsumoto, O. *et al*. (2021) Enrichment of phytoplasma genome DNA through a methyl-CpG binding domain-mediated method for efficient genome sequencing. *Journal of General Plant Pathology* 87, 154–163. DOI: 10.1007/s10327-021-00993-z.

Oshima, K., Kakizawa, S., Nishigawa, H., Jung, H.-Y., Wei, W. *et al*. (2004) Reductive evolution suggested from the complete genome sequence of a plant-pathogenic phytoplasma. *Nature Genetics* 36, 27–29. DOI: 10.1038/ng1277.

Oyola, S.O., Otto, T.D., Gu, Y., Maslen, G., Manske, M. *et al*. (2012) Optimizing Illumina next-generation sequencing library preparation for extremely AT-biased genomes. *BMC Genomics* 13, 1–12. DOI: 10.1186/1471-2164-13-1.

Ranebennur, H., Kirdat, K., Tiwarekar, B., Rawat, K., Chalam, V.C. *et al*. (2022). Draft genome sequence of '*Candidatus* Phytoplasma australasia', strain SS02 associated with sesame phyllody disease. *3 Biotech* 12, 107. DOI: 10.1007/s13205-022-03163-w.

Rodrigues Jardim, B., Tran-Nguyen, L.T.T., Gambley, C., Rodoni, B., and Constable, F.E. (2022) Iodixanol density gradients as an effective phytoplasma enrichment approach to improve genome sequencing. *Frontiers in Microbiology* 13, 937648. DOI: 10.3389/fmicb.2022.937648.

Slatko, B.E., Gardner, A.F., and Ausubel, F.M. (2018) Overview of next-generation sequencing technologies. *Current Protocols in Molecular Biology* 122, e59. DOI: 10.1002/cpmb.59.

Sugio, A. and Hogenhout, S.A. (2012) The genome biology of phytoplasma: modulators of plants and insects. *Current Opinion in Microbiology* 15, 247–254. DOI: 10.1016/j.mib.2012.04.002.

Tan, C.-M., Lin, Y.-C., Li, J.-R., Chien, Y.-Y., Wang, C.-J. *et al*. (2021) Accelerating complete phytoplasma genome assembly by immunoprecipitation-based enrichment and MinION-based DNA sequencing for comparative analyses. *Frontiers in Microbiology* 12, 766221. DOI: 10.3389/fmicb.2021.766221.

Tokuda, R., Iwabuchi, N., Kitazawa, Y., Nijo, T., Suzuki, M. *et al*. (2023) Potential mobile units drive the horizontal transfer of phytoplasma effector phyllogen genes. *Frontiers in Genetics* 14, 1132432. DOI: 10.3389/fgene.2023.1132432.

Tran-Nguyen, L.T.T. and Schneider, B. (2013) Cesium chloride-bisbenzimide gradients for separation of phytoplasma and plant DNA. In: Dickinson, M. and Hodgetts, J. (eds) *Phytoplasma: Methods in Molecular Biology*. Humana Press, Totowa, NJ, pp. 381–393.

Udaondo, Z., Sittikankaew, K., Uengwetwanit, T., Wongsurawat, T., Sonthirod, C. *et al*. (2021) Comparative analysis of PacBio and Oxford Nanopore sequencing technologies for transcriptomic landscape identification of Penaeus monodon. *Life* 11, 862. DOI: 10.3390/life11080862.

Wang, J., Song, L., Jiao, Q., Yang, S., Gao, R. *et al*. (2018) Comparative genome analysis of jujube witches'-broom Phytoplasma, an obligate pathogen that causes jujube witches'-broom disease. *BMC Genomics* 19, 1–12. DOI: 10.1186/s12864-018-5075-1.

Xue, C., Zhang, Y., Li, H., Liu, Z., Gao, W. *et al*. (2023) The genome of *Candidatus* phytoplasma ziziphi provides insights into their biological characteristics. *BMC Plant Biology* 23, 251. DOI: 10.1186/s12870-023-04243-6.

4 Advances and Challenges in Phytoplasma Taxonomy

Yan Zhao* **and Wei Wei**
Molecular Plant Pathology Laboratory, ARS-USDA, Beltsville, Maryland, USA

Abstract
Phytoplasmas are non-helical, mycoplasma-like, wall-less bacteria that significantly impact economies and the environment by causing diseases in nearly 1000 plant species. Their genetic and ecological diversity necessitates an informative taxonomic system to understand their range and uniqueness. However, their limited metabolic capacity and high host dependency prevent *in vitro* cultivation, therefore excluding them from polyphasic characterization. Molecular analyses of conserved genes and whole-genome sequencing have become the primary methods for their characterization. This chapter reviews the history and current status of phytoplasma taxonomy, covering nomenclature, classification, and identification. Topics include the 2004 original and 2022 revised guidelines for '*Candidatus* Phytoplasma species' delineation; opinions on their taxonomic rank; the ongoing need for a pragmatic group/subgroup classification scheme; criteria for recognizing new groups and subgroups; and common and advanced molecular identification practices. Despite recent advances that address key issues, challenges remain in enhancing phytoplasma taxonomy in the post-genomic era.

Introduction

Phytoplasmas, previously termed mycoplasma-like organisms (MLOs), are non-helical, cell wall-less bacteria that infect vascular plants. Parasitizing phloem sieve elements and being transmitted by phloem-feeding insects, phytoplasmas have immense economic and environmental impacts as they are known to affect nearly 1000 plant species and are responsible for numerous plant diseases worldwide (Wei and Zhao, 2022). Phytoplasmas are diverse in their genetic variability, plant host range, insect vector specificity, and geographic distribution. To understand the breadth of phytoplasma diversity and the uniqueness of each species and related strains, it is crucial to have an informative taxonomic system that allows for accurate identification, classification, and nomenclature of diverse phytoplasmas. However, despite their economic and environmental impacts, phytoplasmas are still among the least studied plant pathogens (Wei *et al.*, 2013). Decades of efforts to isolate phytoplasmas and establish a pure axenic phytoplasma culture remain unsuccessful, partly due to, as hinted by available genome data, phytoplasmas' limited metabolic capacities and therefore high host dependency (Oshima *et al.*, 2004; Kube *et al.*, 2012; Zhao *et al.*, 2014). The unavailability of many required phenotypic characters has

*Corresponding author: yan.zhao@usda.gov

 Phytoplasmas: Genomes, Plant Hosts and Vectors, 2nd Edition. (Ed. W. Wei)
DOI: 10.1079/9781800627031.0004

excluded phytoplasmas from polyphasic microbial species delineation. As a result, phytoplasmas still do not have a formal taxonomic rank or placement yet (Zhao *et al.*, 2015). Presently, phytoplasmas are accommodated in a provisional genus, '*Candidatus* Phytoplasma' (IRPCM, 2004; Bertaccini *et al.*, 2022). Nevertheless, the history of phytoplasma taxonomy has witnessed gradual yet profound transformations over time. This chapter provides an overview of the history and current status of the phytoplasma taxonomy, with a primary focus on the advances achieved since the last edition of the book and the lingering challenges in this field.

Brief History of Phytoplasma Taxonomy

It has been known for over a century that many plants are susceptible to a specific type of disease that is collectively referred to as "yellows diseases." The epidemiology of one such disease, aster yellows, has been the subject of numerous studies since the 1920s (Kunkel, 1924, 1925, 1926). However, for decades, the causal agents of yellows diseases were mistakenly believed to be viruses (Severin, 1942; Black, 1953; Chiykowski, 1962; Freitag, 1964), although no such presumed virus was ever isolated. In 1967, through electron microscopic studies, Doi and colleagues discovered that cell wall-less bacteria, rather than viruses, were responsible for aster yellows as well as other yellows diseases. They found MLOs in the phloem elements of plants affected by various yellows diseases, including aster yellows, mulberry dwarf, potato witches'-broom, and paulownia witches'-broom diseases (Doi *et al.*, 1967). The term MLO was adopted because of its morphological resemblance to human- and animal-infecting mycoplasmas. Later, the term MLO was replaced by a new trivial name, phytoplasma (ICSB Subcommittee on the Taxonomy of Mollicutes, 1993).

The biodiversity of phytoplasmas was recognized even at the time when the etiological agents of phytoplasmal diseases were mistakenly presumed to be viruses. From the 1960s to most of the 1980s, the differentiation of phytoplasmas (including previously termed MLOs and wrongly presumed "yellows viruses") was made based on the geographic origins of the diseases, the identities of affected plant hosts, the suspected insect vectors, and the observed disease symptoms (Granados and Chapman, 1968; Chiykowskiand Sinha, 1989; McCoy *et al.*, 1989). However, such a superficial descriptive approach could not provide a reliable framework to classify phytoplasmas because: (i) different phytoplasma strains may cause diseases with similar symptoms or share a common vector; (ii) a single phytoplasma strain may induce different symptoms in different hosts; and (iii) the same phytoplasma species may cause different diseases in different geographic areas (Zhao *et al.*, 2010). Early knowledge of the genetic relationships of phytoplasma strains came from serological and nucleic acid hybridization studies in the late 1980s and the 1990s.Such studies differentiated phytoplasmas into mutually distinct serological groups based on an individual strain's response to specific antibodies, and identified an array of genomic strain clusters based on DNA hybridization characteristics of any given strain to specific DNA probes (Lin and Chen, 1985; Lee and Davis, 1988; McCoy *et al.*, 1989; Kuske *et al.*, 1991; Harrison *et al.*, 1992; Lee *et al.*, 1992a, 1992b, 1993a; Kison *et al.*, 1994; Schneider and Seemüller, 1994; Marcone *et al.*, 1999). In the 1990s and the ensuing decade, the expanded application of PCR, restriction fragment length polymorphism (RFLP) analysis, and advanced DNA sequencing technologies in phytoplasma research made amplification and comparative analyses of evolutionarily conserved genes, especially 16S rRNA gene sequences, possible. Phylogenetic studies of 16S rRNA genes and other conserved core housekeeping genes indicated that extant phytoplasmas shared a common evolutionary root and are descended from low GC Gram-positive bacteria in the *Bacillus-Clostridium* group (Woese *et al.*, 1980; Weisburg *et al.*, 1989; Gundersen *et al.*, 1994; Sears and Kirkpatrick, 1994; Zhao *et al.*, 2005; Wei *et al.*, 2007b). After evolutionary divergence from an acholeplasma-like last common ancestor, phytoplasmas emerged as a discrete clade, and an array of widely divergent phytoplasma lineages have evolved in adaptation to a broad range of bio- and geoecological niches (Gundersen *et al.*, 1994; Lee *et al.*, 2000; Davis *et al.*, 2005; Wei *et al.*, 2008a). Phylogenetic analysis of the 16S rRNA gene sequences also revealed subclades within the phytoplasma clade and served as the basis for delineation of mutually distinct lineages (Kuske and Kirkpatrick, 1992;

Namba *et al.*, 1993; Gundersen *et al.*, 1994; Schneider *et al.*, 1995; Seemüller*et al.*, 1998). In addition to phylogenetic analysis, comparative RFLP profiling of a 1.25 kb PCR-amplified 16S rRNA gene fragment (termed F2nR2 fragment) of diverse phytoplasma strains led to the establishment of a set of reference RFLP profiles, which formed the basis for the16S rRNA gene-based group/subgroup classification scheme (Lee *et al.* 1993b, 1998). In the late 2000s, the introduction of computer-simulated RFLP analysis transformed the traditional laboratory enzymatic/gel electrophoretic-based RFLP analysis to DNA sequence-based virtual RFLP. Such transformation significantly accelerated the pace of new phytoplasma group/subgroup delineation and new phytoplasma discovery (Wei *et al.*, 2007a, 2007b, 2008b; Cai *et al.*, 2008; Zhao *et al.*, 2009b). Figure 4.1 highlights some of the key events that shaped the taxonomy of phytoplasmas.

While the research on genotypic-based identification of phytoplasmas progressed rapidly, phenotypic characterization stalled over the same period. For decades, countless attempts were made to achieve isolation and axenic cultivation of phytoplasmas. While several reports claimed success (Hampton *et al.*, 1969; Lin *et al.*,

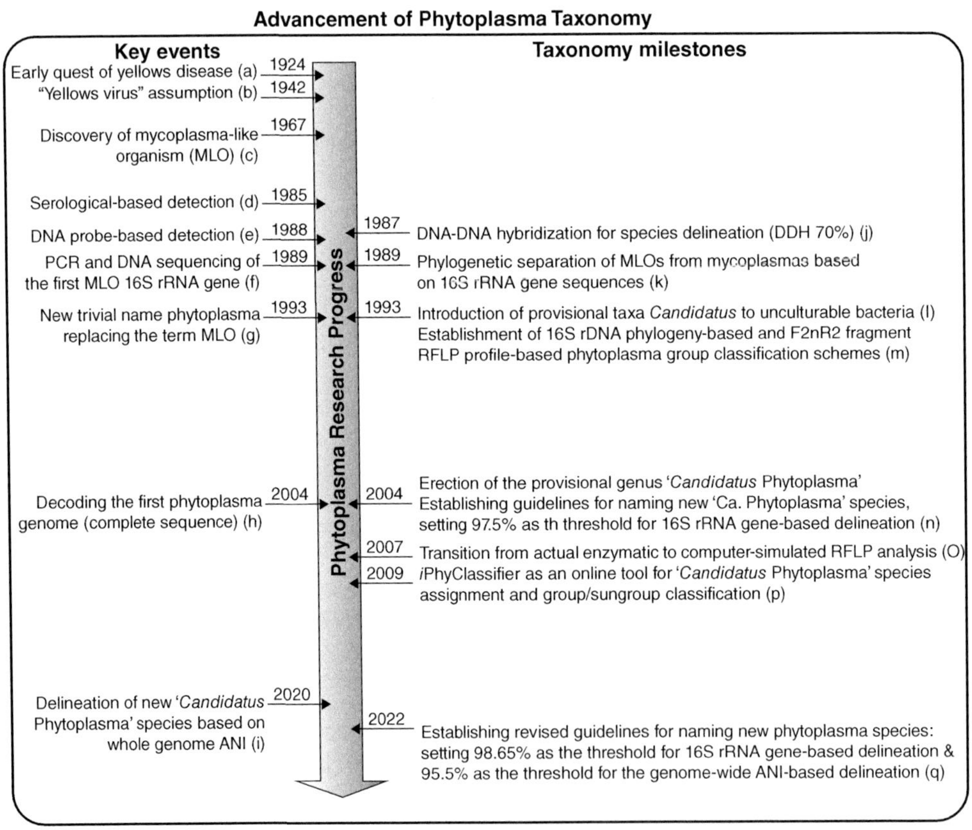

Fig. 4.1. Key events of phytoplasma research and milestones of phytoplasma taxonomy. (a–q) Citations of the original publications describing the key research events and taxonomic milestones: a, Kunkel, 1924; b, Severin, 1942; c, Doi *et al.*, 1967; d, Lin and Chen, 1985; e, Lee and Davis, 1988; f, Lim and Sears, 1989; g, ICSB Subcommittee on the Taxonomy of Mollicutes, 1993; h, Oshima *et al.*, 2004; i, Kirdat *et al.*, 2021; Zhao *et al.*, 2021; j, Kirkpatrick *et al.*, 1987; k, Lim and Sears, 1989; l, Murray and Schleifer, 1994; m, Kuske and Kirkpatrick, 1992; Lee *et al.*, 1993b; n, IRPCM, 2004; o, Wei *et al.*, 2007a; p, Zhao *et al.*, 2009b; q, Bertaccini *et al.*, 2022. (Adapted from Wei and Zhao (2022), Open Access Journal, with modifications.)

1970; Lombardo and Pignattelli, 1970; Ghosh *et al.*, 1971), none were substantiated by independent confirmation. More recently, renewed efforts have been rewarded with limited growth of phytoplasma colonies in complex media (Contaldo *et al.*, 2016, 2019). Nevertheless, sustained pure culture in cell-free media has not yet been demonstrated for any phytoplasma. Thus, for phytoplasmas, many of the basic phenotypic characters required for polyphasic species delineation remain inaccessible. Completion of genome sequencing of the first few representative phytoplasma strains in the mid-2000s (Oshima *et al.*, 2004; Bai *et al.*, 2006; Kube *et al.*, 2008; Tran-Nguyen *et al.*, 2008) revealed that phytoplasmas lack many genes known to be essential to a free-living bacterium with the minimum gene complement (Fraser *et al.*, 1995; Glass *et al.*, 2006). Such missing essential genes included those encoding enzymes/transporters required for carbohydrate intake and metabolism, energy generation, nucleotide metabolism, cell membrane genesis, and redox homeostasis (Zhao *et al.*, 2014). Conceivably, the absence of those essential genes and the consequent lack of corresponding critical metabolic capabilities and cellular functions have diminished the ability of phytoplasmas to survive and multiply in cell-free media. Therefore, it is likely that, in the foreseeable future, the advancement of phytoplasma taxonomy will continue to rely on a culture-independent genotypic approach based on heredity (genome) information.

Nomenclature and Delineation of '*Candidatus* Phytoplasma' Species

In the bacterial taxonomic system, taxa are arranged hierarchically at seven different levels. From the highest (most general) to the lowest (most specific), the seven ranks are kingdom, phylum, class, order, family, genus, and species. While the means and ways by which bacteria are characterized evolve constantly in line with conceptual innovations and technological advances, for species delineation, the traditional polyphasic approach remains the gold standard. The approach integrates the results from phenotypic and genotypic characterizations and reflects the ecological nature of the bacteria (Kämpfer and Glaeser, 2012). Since phenotypic characters such as morphological, physiological, and chemotaxonomic traits are heavily culture-dependent, they are largely inaccessible for phytoplasmas. To date, no validated species name or type genus name has been given to any phytoplasma. Currently, all phytoplasma strains are accommodated within a provisional genus, '*Ca.* Phytoplasma' (IRPCM, 2004). The term *Candidatus* was first introduced into the bacteria taxonomy in 1994, granting non-culturable bacteria the appropriate status of potential taxa based on their 16S rRNA gene sequences (Murray and Schleifer, 1994). *Candidatus* is not a rank, nor is it governed by the International Code of Nomenclature of Prokaryotes (Oren, 2017). As with other unculturable bacteria, the higher taxa of phytoplasmas (Bacteria/Mycoplasmatota (formerly Tenericutes)/Mollicutes/Acholeplasmatales/Incertae sedis – Family II) were named in the absence of a type genus and species, with the *Candidatus* status being used to reserve the putative lower rank taxa – the genus and species (Gasparich *et al.*, 2020).

The original 2004 guidelines for '*Candidatus* Phytoplasma' species delineation

In 2004, based on consensus among phytoplasmologists and in agreement with the International Committee of Systematic Bacteriology (ICSB) Subcommittee for the Taxonomy of Mollicutes, the International Research Programme on Comparative Mycoplasmology (IRPCM) Phytoplasma/Spiroplasma Working Team–Phytoplasma Taxonomy Group adopted a taxonomic rule that had been established for recording properties of uncultured organisms (Murray and Schleifer, 1994; Murray and Stackebrandt, 1995) and erected a genus-level provisional taxon '*Ca.* Phytoplasma' to accommodate plant-pathogenic nonhelical mollicutes (IRPCM, 2004). In order to prevent the nomenclatural confusion that may arise from the description of poorly differentiated taxa, the Working Team established seven "rules" for the description of new taxa within the provisional genus. Among the seven rules, the following four are most relevant to the delineation of new '*Ca.* Phytoplasma' species (referred to as the 2004 guidelines or the original guidelines hereafter in this chapter).

1. A novel '*Ca.* Phytoplasma' species description should refer to a single, unique 16S rRNA gene sequence greater than 1200 bp. The phytoplasma strain whose 16S rRNA gene is used to describe a '*Ca.* Phytoplasma' species is called the "reference strain." Strains whose 16S rRNA genes have even minimal difference from the corresponding reference strain are referred to as "related strains."
2. The 16S rRNA gene sequence shares less than 97.5% sequence similarity (identity) with that of any previously described '*Ca.* Phytoplasma' species.
3. In cases where the phytoplasma's 16S rRNA sequence shares >97.5% sequence identity, the phytoplasma under consideration should clearly represent an ecologically separated population. To justify that a phytoplasma whose 16S rRNA gene shares more than 97.5% sequence identity with a previously named '*Ca.* Phytoplasma' species represents an ecologically separated population (thus a new '*Ca.* Phytoplasma' species), it is required to pass all three critical tests: (i) the two phytoplasmas have different natural plant hosts or at least their behaviors are significantly different in the same plant host; (ii) they are transmitted by different insect vectors; and (iii) there is evidence of significant molecular diversity as demonstrated either by DNA-DNA hybridization, serological reaction, or PCR-based assay.
4. The rank of subspecies should not be used.

In addition, the 16S rRNA gene of any novel '*Ca.* Phytoplasma' species should possess at least one unique sequence region in addition to the signature sequence that is characteristic of phytoplasmas: 5′-CAAGAYBATKATGTKTAGCYGGDCT-3′. This phytoplasma signature sequence corresponds to the annealing site of phytoplasmal universal forward primer R16F2n, which, together with reverse primer R16R2, has served as a basic tool for phytoplasma-specific 16S rDNA amplification (Gundersen and Lee, 1996). Unique identifiers other than the 16S rRNA gene and specific host information should also be included in the new species description. A novel '*Ca.* Phytoplasma' species can be named after the place or the plant host where it was discovered. Below are two examples of descriptions of '*Ca.* Phytoplasma' species (Zhao *et al.*, 2009a).

'*Candidatus* Phytoplasma americanum' (a.mer.i.can'um. N.L. neut. adj. *americanum* pertaining to America):

The reference strain is APPTW12-NE. Related phytoplasma strains include APPTW1-TX, APPTW2-TX, APPTW9-NE, APPTW10-NE, APPTW13-NE, and APPTW1833 #6-TX, which are associated with potatoes exhibiting purple top syndrome in Texas and Nebraska, USA.

[(*Mollicutes*) NC; NA; O, wall-less; NAS (GenBank accession number DQ-174122), oligonucleotide sequences of unique regions of the 16S rRNA gene are 5′-GTTTCTTCGGAAA-3′ (68–80), 5′-GTTAGAAATGACT-3′ (142–153), and 5′-GCTGGTGGCTT-3′ (1438–1448); P (*Solanum tuberosum*, phloem); M] (Lee *et al.*, 2006).

'*Candidatus* Phytoplasma tamaricis' (N.L. gen. n. *tamaricis* of *tamarix*, the scientific name of salt cedar; epithet referring to the plant host).

The reference strain is SCWB1^R.

[(*Mollicutes*) NC; NA; O, wall-less; NAS (GenBank accession number FJ432664), oligonucleotide sequences of unique regions of the 16S rRNA gene are: 5′-ATTAGGCATCTAGTAACTTTG-3′, 5′-TGCTCAACATTGTTGC-3′, 5′-AGCTTTGCAAAGTTG-3′, and 5′-TAACAGAGGTTATCAGAGTT-3′; P (*Tamarix chinensis*, phloem); M].

Under the 2004 guidelines, 47 '*Ca.* Phytoplasma' species were formally described (Table 4.1, subsection I.a, and references therein). The reference strains of these *Candidatus* species were discovered from 43 plant species in 24 countries. While most of the 47 '*Ca.* Phytoplasma' species were delineated based on the threshold 16S rRNA gene sequence identity score (<97.5%); nine were differentiated and named based on their distinctive biological/ecological properties. For example, in the 16SrX cluster, '*Ca.* P. mali,' '*Ca.* P. pyri,' and '*Ca.* P. prunorum' are etiological agents of economically devastating fruit tree diseases, including apple proliferation (AP), pear decline (PD), and European stone fruit yellows (ESFY), respectively. While the pairwise 16S RNA gene sequence identity scores among the reference strains of these three phytoplasmas range from 98.6% to 99.0% (not qualifying for separate species per sequence identity score), the three phytoplasmas clearly have different

Table 4.1. '*Candidatus* Phytoplasma' species and their reference strains.

Reference strain	Host	Country	16Sr group classification	GenBank accession no.	Reference
I. Taxa delineated under 2004 guidelines					
I.a. Formally described '*Ca.* Phytoplasma' species					
'*Ca.* P. asteris' OAY	Evening primrose	USA	16SrI-A	M30790	Lee *et al.*, 2004a
'*Ca.* P. lycopersici' Santa Cruz	Tomato	Bolivia	16SrI-Y	EF199549	Arocha *et al.*, 2007
'*Ca.* P. tritici' WBD *rrnA*	Wheat	China (mainland)	16SrI-C	NZ_AVAO01000003	Zhao *et al.*, 2021
rrnB				NZ_AVAO01000003	
'*Ca.* P. aurantifolia' WBDL*	Lime	Oman	16SrII-B	U15442	Zreik *et al.*, 1995
'*Ca.* P. australasia' PpYC†	Papaya	Australia	16SrII-D	Y10097	White *et al.*, 1998
'*Ca.* P. pruni' PX11CT1 *rrnA*	Peach	USA	16SrIII-A	JQ044393	Davis *et al.*, 2013
rrnB				JQ044392	
'*Ca.* P. ulmi' EY1	Elm	USA	16SrV-A	AY197655	Lee *et al.*, 2004b
'*Ca.* P. ziziphi' JWB-G1	Jujube	Japan	16SrV-B	AB052876	Jung *et al.*, 2003a
Ca. P. rubi' RuS	*Rubus*	Italy	16SrV-E	AY197648	Malembic-Maher *et al.*, 2011
'*Ca.* P. balanitae' BltWB	Balanites	Myanmar	16SrV-new	AB689678	Win *et al.*, 2013
'*Ca.* P. trifolii' CP	Alsike clover	Canada	16SrVI-A	AY390261	Hiruki & Wang, 2004
'*Ca.* P. sudamericanum' PassWB-Br3	Passion fruit	Brazil	16SrVI-I	GU292081	Davis *et al.*, 2012
'*Ca.* P. fraxini' AshY1	Ash	USA	16SrVII-A	AF092209	Griffiths *et al.*, 1999
'*Ca.* P. luffae' LfWB *rrnA*	Luffy	China (Taiwan)	16SrVIII-A	AF248956	Davis *et al.*, 2017
rrnB				AF353090	
'*Ca.* P. phoenicium' AlmWB	Almond	Lebanon	16SrIX-D	AF515636	Verdin *et al.*, 2003
'*Ca.* P. mali' AP15	Apple	Italy	16SrX-A	AJ542541	Seemüller & Schneider, 2004
'*Ca.* P. pyri' PD1	Pear	Germany	16SrX-C	AJ542543	Seemüller & Schneider, 2004
'*Ca.* P. spartii' SpaWB	Spanish broom	Italy	16SrX-D	X92869	Marcone *et al.*, 2004a
'*Ca.* P. prunorum' ESFY-G1	Peach	Germany	16SrX-F	AJ542544	Seemüller & Schneider, 2004
'*Ca.* P. oryzae' RYD-J	Rice	Japan	16SrXI-A	AB052876	Jung *et al.*, 2003b
'*Ca.* P.sacchari' SCGS	Sugarcane	India	16SrXI-B	MN889545	Kirdat *et al.*, 2021
'*Ca.* P. cirsii' CirYS	Creeping thistle	Czech	16SrXI-E	KR869146	Šafářová *et al.*, 2016
'*Ca.* P. solani' STOL11	*Capsicum annuum*	Serbia	16SrXII-A	AF248959	Quaglino *et al.*, 2013
'*Ca.* P. australiense' AUSGY	Grapevine	Australia	16SrXII-?	L76865	Davis *et al.*, 1997
'*Ca.* P. japonicum' JPH	Hortensia	Japan	16SrXII-D	AB010425	Sawayanagi *et al.*, 1999

Continued

Table 4.1. Continued

Reference strain	Host	Country	16Sr group classification	GenBank accession no.	Reference
'*Ca.* P. fragariae' StrawY	Strawberry	Lithuania	16SrXII-E	DQ086423	Valiunas *et al.*, 2006
'*Ca.* P. convolvuli' BY-S57/11	*Convolvulus arvensis*	Serbia	16SrXII-H	JN833705	Martini *et al.*, 2012
'*Ca.* P. hispanicum' MPV	Periwinkle	Mexico	16SrXIII-A	AF248960	Davis *et al.*, 2016
'*Ca.* P. meliae' ChTY-Mo3	Chinaberry	Argentina	16SrXIII-G	KU850940	Fernández *et al.*, 2016
'*Ca.* P. cynodontis' BGWL-C1	Bermuda grass	Italy	16SrXIV-A	AJ550984	Marcone *et al.*, 2004b
'*Ca.* P. brasiliense' HibWB26	Hibiscus	Brazil	16SrXV-A	AF147708	Montano *et al.*, 2001
'*Ca.* P. graminis' SCYLP	Sugarcane	Cuba	16SrXVI-A	AY725228	Arocha *et al.*, 2005
'*Ca.* P. caricae' PAY	Papaya	Cuba	16SrXVII-A	AY725234	Arocha *et al.*, 2005
'*Ca.* P. americanum' APPTW12-NE	Potato	USA	16SrXVIII-A	DQ174122	Lee *et al.*, 2006
'*Ca.* P. castaneae' CnWB	Chestnut	Korea	16SrXIX-A	AB054986	Jung *et al.*, 2002
'*Ca.* P. rhamni' BWB	Buckthorn	Germany	16SrXX-A	X76431	Marcone *et al.*, 2004a
'*Ca.* P. pini' Pin127S	Pine	Spain	16SrXXI-A	AJ632155	Schneider *et al.*, 2005
'*Ca.* P. palmicola' LYDM 178	Coconut palm	Mozambique	16SrXXII-A	KF751387	Harrison *et al.*, 2014
'*Ca.* P. omanense' IM-1	*Cassia italica*	Oman	16SrXXIX-A	EF666051	Al-Saady *et al.*, 2008
'*Ca.* P. tamaricis' SCWB1	Salt cedar	China (Mainland)	16SrXXX-A	FJ432664	Zhao *et al.*, 2009a
'*Ca.* P. costaricanum' SoyST1c1	Soybean	Costa Rica	16SrXXXI-A	HQ225630	Lee *et al.*, 2011
'*Ca.* P. malaysianum' MaPV	Periwinkle	Malasia	16SrXXXII-A	EU371934	Nejat *et al.*, 2013
'*Ca.* P. allocasuarinae' AlloY	Allocasuarina	Australia	16SrXXXIII	AY135523	Marcone *et al.*, 2004a
'*Ca.* P. wodyetiae' FPYD Bangi-2	Foxtail palm	Malasia	16SrXXXVI-A	KC844879	Naderali *et al.*, 2017
'*Ca.* P. stylosanthis' StLL	Stylosanthes	Australia	16SrXXXVII-A	MT431550	Rodrigues Jardim *et al.*, 2021
'*Ca.* P. noviguineense' BCS-Bo	Coconut palm	New Guinea	N/D[‡]	LC228755	Miyazaki *et al.*, 2018
'*Ca.* P. dypsidis RID7692	Palm (*Dypsis poivriana*)	Australia	N/D[‡]	MT536195	Jones *et al.*, 2021
I.b. Incidentally cited taxa					
'*Ca.* P. palmae' LY[§]	Coconut palm	USA	16SrIV-A	U18747	IRPCM, 2004; Bertaccini *et al.*, 2022
'*Ca.* P. cocostanzaniae' LD[§]	Coconut palm	Tanzania	16SrIV-C	X80117	IRPCM, 2004; Bertaccini *et al.*, 2022
'*Ca.* P. vitis' FD70	Grapevine	France	16SrV-C	AF176319	IRPCM, 2004

I.c. Potentially new '*Ca.* Phytoplasma' species					
Buckland Valley grapevine yellows phytoplasma	Grapevine	Australia	16SrXXIII-A	AY083605	Wei *et al.*, 2007a
Sorghum bunchy shoot phytoplasma	Sorghum	Australia	16SrXXIV-A	AF509322	Wei *et al.*, 2007a
Weeping tea witches'-broom phytoplasma	Weeping tea tree	Australia	16SrXXV-A	AF521672	Wei *et al.*, 2007a
Sugarcane phytoplasma Mauritius-1	Sugarcane	Mauritius	16SrXXVI-A	AJ539179	Wei *et al.*, 2007a
Sugarcane phytoplasma Mauritius-2	Sugarcane	Mauritius	16SrXXVII-A	AJ539180	Wei *et al.*, 2007a
Derbid phytoplasma	Derbid	Cuba	16SrXXVIII-A	AY744945	Wei *et al.*, 2007a
II. Taxa delineated under 2022 guidelines					
'*Ca.* P. citri' WBDL	Lime	Oman	16SrII	JAOSJF000000000	Rodrigues Jardim *et al.*, 2023
'*Ca.* P. australasiaticum'					Rodrigues Jardim *et al.*, 2023
Subspecies australasiaticumPR08	*Parthenium* sp.	India	16SrII	GCA_015239935	Rodrigues Jardim *et al.*, 2023
Subspecies taiwanenseNCHU2014	*Echinacea purpurea*	China (Taiwan)	16SrII	GCA_001307505	Rodrigues Jardim *et al.*, 2023
Subspecies ipomoeaeo7C	*Ipomoea*	Australia	16SrII	GCA_024425275	Rodrigues Jardim *et al.*, 2023
'*Ca.* P. asiaticum' PR34	*Parthenium hysterophorus*	India	16SrII	GCA_015100165	Rodrigues Jardim *et al.*, 2023
'*Ca.* P. gossypii' BAWM-BFA-CoWB	*Gossypium*	Burkina Faso	16SrII	JAOSIT000000000	Rodrigues Jardim *et al.*, 2023
'*Ca.* P. crotalariae' BAWM-OMN-P53	*Crotalaria*	Oman	16SrII	JAOSIR000000000	Rodrigues Jardim *et al.*, 2023
'*Ca.* P. bonamiae' BAWM-225	*Bonamia*	Australia	16SrII	JAOSIQ000000000	Rodrigues Jardim *et al.*, 2023
'*Ca.* P. planchoniae' BAWM-156b	*Planchonia*	Australia	16SrII	JAOSIP000000000	Rodrigues Jardim *et al.*, 2023
'*Ca.* P. fabacearum' BAWM-027	*Fabaceae*	Australia	16SrII	JAOSIM000000000	Rodrigues Jardim *et al.*, 2023
'*Ca.* P. melaleucae' BAWM-155c	*Melaleuca*	Australia	16SrXXV-A	JAOSID000000000	Rodrigues Jardim *et al.*, 2023

*'*Ca.* P. aurantifolia' has been renamed to '*Ca.* P. citri' (Rodrigues Jardim *et al.*, 2023).
†'*Ca.* P. aurantifolia' was once abolished (IRPCM, 2004) but has recently been reinstated and renamed as '*Ca.* P. australasiaticum' (Rodrigues Jardim *et al.*, 2023).
‡The 16Sr group classifications of the reference strains of '*Ca.* P. noviguineense' and '*Ca.* P. dypsidis' have not been determined. According to the result from a virtual RFLP analysis using *i*PhyClassifier in this study, the reference strains of '*Ca.* P. noviguineense' and '*Ca.* P. dypsidis' may represent two new 16Sr groups.
§Two once incidentally cited species '*Ca.* P. palmae' and '*Ca.* P. cocostanzaniae' were formally described during 2022 guideline revision (Bertaccini *et al.*, 2022).

natural plant host ranges, and each induces a different set of characteristic symptoms in infected plants. The three phytoplasmas are transmitted preferentially by different vectors: AP is transmitted by *Cacopsylla picta* and *C. melanoneura*; PD is transmitted by *C. pyricola* and *C. pyri*; and ESFY is transmitted by *C. pruni*. In addition, the three phytoplasmas differ in serological properties as well as in several key molecular markers, including the 16S–23S rDNA spacer region and randomly cloned protein-encoding genes (Seemüller and Schneider, 2004). Such mutually distinctive biological properties of the three phytoplasmas clearly demonstrate that they are ecologically separated populations.

Similarly, in the 16SrV cluster, '*Ca*. P. ulmi' and '*Ca*. P. ziziphi' share 99.0% 16S rRNA gene sequence identity but have been recognized as two distinct taxa because they occupy different ecological niches and exhibit strikingly different geographic distributions. Whereas '*Ca*. P. ziziphi' and related strains are found exclusively in Asia, '*Ca*. P. ulmi' and related strains are present predominantly in North America. The ecological differences between the two phytoplasmas include their plant hosts, insect vectors (*Scaphoideus luteolus* vs *Hishimonus sellatus*), and unique signature sequences (Jung *et al*., 2003a; Lee *et al*., 2004b). Another taxon in the 16SrV cluster, '*Ca*. P. rubi' was differentiated from other closely related '*Ca*. Phytoplasma' taxa based on their host (*Rubus*) and vector (*Macropsis fuscula*) specificities and combined sequences of the *tuf*, *rplV-rpsC*, *rplF-rplR*, *map*, and *uvrB-degV* genetic loci (Malembic-Maher *et al*., 2011).

In the 16SrXII cluster, '*Ca*. P. solani' was distinguished from the other taxa, especially from '*Ca*. P. australiense,' by its unique gene pool, vectorship (*Hyalesthes obsoletus*), and host range (*Capsicum annuum*, potato, and cultivated grapevines) (Quaglino *et al*., 2013). In the description of '*Ca*. P. solani' a new term, "distinguishing sequence blocks" (DSBs), was introduced to designate its signature sequences. By using a combination of two specific sequence regions, DSBs can distinguish the signature sequences of closely related '*Ca*. Phytoplasma' species more effectively.

More recently, the etiological agent of wheat blue dwarf (WBD) disease, previously known as a '*Ca*. P. asteris'-related strain and a member of the subgroup 16SrI-C, was characterized as a distinctive ecological lineage and that it deserved to be recognized as a separate species, '*Ca*. P. tritici' (Zhao *et al*., 2021). The WBD phytoplasma, the reference strain of '*Ca*. P. tritici,' not only has its unique transmitting vector (*Psammotettix striatus*) but also elicits a distinctive symptom in its predominant plant host (wheat). In addition, the WBD phytoplasma possesses molecular characteristics that further demonstrate its striking divergence from '*Ca*. P. asteris.' Such molecular characteristics include lineage-specific antigenic membrane proteins and a lower than 95% genome-wide average nucleotide identity (ANI) score with genomes of all sequenced '*Ca*. P. asteris' strains (Zhao *et al*., 2021). While genome-wide ANI is not a criterion or requirement for '*Ca*. Phytoplasma' species delineation under the original IRPCM 2004 guidelines, the comparative genomic analysis result certainly strengthens the reasoning that '*Ca*. Phytoplasma tritici' is distinctly different from '*Ca*. P. asteris' both ecologically and genetically.

Based on the 2004 guidelines, one previously described taxon, '*Ca*. P. australasia' (White *et al*., 1998), was abolished because its 16S rRNA gene sequence shares 99.5% identity with that of '*Ca*. P. aurantifolia.' There is no evidence that the two phytoplasmas are transmitted by different vectors nor the presence of significant molecular diversity (IRPCM, 2004). The taxon was later reinstated and renamed after the 2022 revision of the '*Ca*. Phytoplasma' species delineation guidelines. It is worth noting that the reference strains of two formally described taxa, '*Ca*. P. hispanicum' MPV (Davis *et al*., 2016) and '*Ca*. P. meliae' ChTY-Mo3 (Fernández *et al*., 2016), share 98.9% sequence identity in their 16S rRNA genes (which is higher than 97.5%, the threshold value specified in the 2004 guidelines), and no evidence is established that the two phytoplasmas represent two ecologically separated populations. Since the description of '*Ca*. P. hispanicum' was published in early September 2016 in the *International Journal of Systematic and Evolutionary Microbiology* (IJSEM), merely 3 months earlier than that of '*Ca*. P. meliae,' it is quite possible that the delineation of '*Ca*. P. meliae' was incomplete as the authors had little opportunity to examine and include the information regarding '*Ca*. Phytoplasma hispanicum.'

There are three economically important phytoplasmas that have been incidentally cited in the literature even before the establishment of the 2004 guidelines. '*Ca.* P. palmae' and '*Ca.* P. cocostanzaniae' are associated with lethal yellowing and lethal decline diseases in coconut palms, respectively. '*Ca.* P. vitis' is responsible for flavescence dorée disease in cultivated grapevines (Table 4.1, subsection I.b). These incidentally cited taxa may be described based on their unique biological properties under the 2004 guidelines, but the task remained unaccomplished until 2022 when '*Ca.* P. palmae' and '*Ca.* P. cocostanzaniae' were described along with the revised guidelines (Bertaccini *et al.*, 2022). '*Ca.* P. vitis' still awaits formal description. In addition, six potentially new '*Ca.* Phytoplasma' species were identified based on their lower than 97.5% sequence identity, the threshold score under the 2004 guidelines (Table 4.1 subsection I.c) (Wei *et al.*, 2007b). These phytoplasmas affect weeping tea tree, grapevine, sorghum, sugarcane, and derbid. The weeping tea tree phytoplasma was later described as '*Ca.* P. melaleucae' (Rodrigues Jardim *et al.*, 2023); the other five putative taxa remain to be formally described.

Revised 2022 guidelines and issues of their implementation

In 2022, with the approval of the International Committee on Systematics of Prokaryotes (ICSP) Subcommittee on the Taxonomy of Mollicutes, Bertaccini *et al.* (2022) proposed revised guidelines for '*Ca.* Phytoplasma' species delineation (referred to as the 2022 guidelines or revised guidelines hereafter in this chapter). The major revisions include:

1. The minimum length of the 16S rRNA fragment required for '*Ca.* Phytoplasma' species delineation was extended from 1200 to 1500 bp, which covers approximately 95% of the entire16S rRNA gene.
2. The threshold 16S rRNA gene sequence identity score for recognizing new '*Ca.* Phytoplasma' species was changed from 97.5% to 98.65%, and the term "member strains" was introduced.
3. Genome-wide ANI was introduced as a new criterion for delineating '*Ca.* Phytoplasma' species, and the threshold ANI score was set to 95–96%.
4. If a phytoplasma shares >98.65% 16S rRNA gene sequence identity and >95% genome ANI with that of a previously established species, a multilocus sequence analysis (MLSA) approach can be used to demarcate a new species. Five housekeeping genes (*tufB*, *secY*, *secA*, *rplV-rpsC*, and *groEL*) were recommended for such a purpose.

In recent years, with the rapid advancement of genome sequencing technology and the ever-increasing availability of genome sequence data, bacteria taxonomists explored taxonomic coherence between the 16S rRNA gene sequence identity and the whole-genome information for species demarcation of prokaryotes (Kim *et al.*, 2014; Parks *et al.*, 2020; Hugenholtz*et al.*, 2021). While the 16S rRNA gene still serves as the backbone for bacterial taxonomy, in many bacterial genera, the 16S rRNA gene alone is not sufficient to differentiate species, and in some cases, comparative sequence analyses of multiple housekeeping genes are needed to determine phylogenetic relationships (Fox *et al.*, 1992, Dingle *et al.*, 2001; Maiden, 2006). The available genome sequence data facilitated whole-genome sequence-based genotypic characterization. In particular, the whole-genome ANI calculation has emerged as a robust approach for assessing species boundaries and estimating the genetic relatedness between two genomes. Ample data have shown that whole-genome ANI is correlated with the traditional microbiological concept of DNA–DNA hybridization relatedness for defining species (Richter and Rosselló-Móra, 2009; Jain *et al.*, 2018). In culturable bacteria, an ANI value of 95–96% has been generally accepted for circumscribing species (Richter and Rosselló-Móra, 2009; Kim *et al.*, 2014). In 2019, *Bergey's Manual* suggested the beginning of the transition from 16S rRNA gene sequence-based to whole-genome-based taxonomy in bacteria (Hugenholtz *et al.*, 2021). On the other hand, results from pairwise comparison of the 16S rRNA gene sequence identity score and the corresponding whole-genome ANI score indicated that, in many bacteria, an ANI value of 95–96% corresponds to a 16S rRNA gene sequence identity score of 98.65%. Based on such data, bacterial taxonomists twice revised the 16S rRNA gene sequence identity threshold value for delineating new bacterial species, changing from 97% to 98.7% and then to the current 98.65% (Stackebrandt 2006; Kim *et al.*, 2014; Chun *et al.*, 2018).

In theory, the 2022 revised guidelines embraced recent developments in bacterial taxonomy and adopted nearly identical criteria (threshold scores) for both 16S rRNA gene sequence identity-based and whole-genome ANI-based '*Ca.* Phytoplasma' species delineation. While it represents a milestone in the advancement of phytoplasma taxonomy, some provisions in the 2022 guidelines lack clarity and precision, which may hinder the delineation of new '*Ca.* Phytoplasma' species, and therefore require clarification and amendment. The shortcomings of the revised guidelines have been reviewed (Wei and Zhao, 2022). The following is a recap of some key issues.

The 2022 revised guidelines introduced a new term, "member strain," through the following statement:

> Strains sharing >98.65% sequence identity when compared with the reference strain are considered members of the respective '*Ca.* Phytoplasma' species. Strains showing identity <98.65% to the reference strain, but >98.65% with other strains of the same '*Ca.* Phytoplasma' species should be considered as related to this '*Ca.* Phytoplasma' species.

This statement drew a sharp contrast to the definition of a '*Ca.* Phytoplasma' species that the 2004 original guidelines spelled out:

> The strain from which this sequence was obtained should be named the "reference strain" and not the "type strain." Strains in which even minimal differences in the 16S rRNA gene sequence from the reference strain are detected do not "belong" to the *Candidatus* species but are "related" to it.

For culturable bacteria, the type strain of a validated species is well characterized through the traditional polyphasic approach, and other strains with similarly defined characteristics may be assigned to the same species as member strains. On the other hand, for any '*Ca.* Phytoplasma' species, there is no well-defined type strain. The reference strain is designated based on limited DNA sequence information only, and such limited information is not sufficient for defining member strains. Therefore, preserving the term "related strain" and its definition given by the 2004 original guidelines is appropriate.

Practically, determining a so-called "member strain" based on 16S rRNA gene sequence identity scores specified in the 2022 revised guidelines could lead to erroneous assignment of a single given strain to more than one species. For example, the 16S rRNA gene of the alder yellows phytoplasma strain ALY (AY197646) shares 98.98%, 99.35%, and 99.7% sequence identity with that of '*Ca.* P. ziziphi' reference strain JWB-G1 (AB052876), '*Ca.* P. rubi' reference strain RuS (AY197648), and '*Ca.* P. ulmi' reference strain EY1 (AY197655), respectively. According to the 2022 guidelines, ALY would be a member strain of '*Ca.* P. ziziphi,' '*Ca.* P. rubi,' and '*Ca.* P. ulmi' simultaneously. Likewise, the 16S rRNA gene of the plum leptonecrosis phytoplasma strain LNp (JQ868450) shares 99.93%, 98.86%, and 98.66% sequence identity with that of '*Ca.* P. prunorum' reference strain ES-FY-G1 (AJ542544), '*Ca.* P. pyri' reference strain PD1 (AJ542543), and '*Ca.* P. mali' reference strain AP15 (AJ542541), respectively. Therefore, according to the 2022 guidelines, LNp would be a member strain of '*Ca.* P. prunorum,' '*Ca.* P. pyri,' and '*Ca.* P. mali' at the same time. Both conceptually and practically, a strain can be related to more than one species but cannot be a member of more than one species simultaneously. For this reason, the term "member strain" should be abolished, and the term "related strain," as coined in the 2004 original guidelines, should be restored.

The 2022 revised guidelines allowed naming new '*Ca.* Phytoplasma' species based on the sequence identity scores of two out of five designated housekeeping genes (*tufA*, *secY*, *secA*, *rplV-rpsC*, and *groEL*) if a phytoplasma under study shares >98.65% identity in 16S rRNA gene sequence and >95% whole-genome ANI with previously established species. Thus far, about 20,000 phytoplasma-derived nucleotide sequences have been deposited in the nucleotide database at the National Center for Biotechnology Information (NCBI). Most of the accessions are sequences of housekeeping genes, including approximately 8000 16S rRNA gene sequences, 1300 ribosomal protein-encoding gene sequences (*rplV-rplO-rpsC* locus), 880 *secY* gene sequences, and 570 *tufA* gene sequences. The remaining entries are relatively less-conserved genes often used for differentiation of closely related phytoplasma strains, such as *vmp*, *cpn60*, *amp*, *map*, and *SecA*. To date, there is no single gene other than the 16S rRNA gene that has universal primers capable of amplifying the target sequences from strains of all phytoplasma groups. Even widely used generic *rp* and *secY* primers can only amplify phytoplasmas that belong to certain 16Sr groups. Without universal primers and sufficient sequence data for a thorough comparative analysis of these five designated house keeping genes, it is still difficult to establish objective criteria for MLSA-based species delineation. In fact, at the present time, the most valuable application of MLSA in new species' delineation is to provide supporting evidence in verifying the molecular divergence of ecological populations: if a phytoplasma under study clearly represents an ecologically separated population, MLSA could be used to demonstrate significant molecular diversity in addition to fulfilling the unique vectorship and host specificity requirement. In such cases, the MLSA assay should not be limited to just the five genes listed in the 2022 guidelines. Evidence of significant molecular diversity from other housekeeping genes, or even better, genes encoding membrane proteins, should be accepted as well.

Under the 2022 revised guidelines, nine new '*Ca.* Phytoplasma' species were described (Table 4.1, section II) (Rodrigues Jardim *et al.* 2023), all of which were delineated based on the whole-genome ANI criteria. Eight of the nine newly described species belong to the 16SrII cluster. Among them, '*Ca.* P. citri' is a recommended new nomenclature, replacing previously described '*Ca.* P. aurantifolia' (Zreik *et al.*, 1995); '*Ca.* P. australasiaticum' is a new nomenclature and a reinstatement of a previously abolished taxon '*Ca.* P. australasia' (White *et al.*, 1998; IRPCM, 2004). The other six new taxa in this cluster are spinoffs from previously named '*Ca.* P. aurantifolia,' as their pairwise whole-genome ANI scores fall below the threshold value (95–96%). Interestingly, the pairwise 16S rRNA gene sequence identity scores among the eight freshly described '*Ca.* Phytoplasma' species within the 16SrII cluster range from 98.1% to 99.7%, with most of them being above 98.7%. This appears to be different to the observation from culturable bacteria, where a whole-genome ANI value of 95–96% corresponds to a 16S rRNA gene sequence identity score of 98.65%. Does this indicate phytoplasma genome sequences are more variable than those of culturable bacteria? If this is true, it would be intriguing to learn what ANI value would correspond to a 16S rRNA gene sequence identity score of 98.65% in phytoplasma.

While lowering the bar for 16S rRNA gene sequence identity-based species delineation and introducing MLSA- and ANI-based new criteria, the 2022 guidelines ignored one important aspect – recognition of potential ecological species – to which the 2004 original guidelines paid special attention, as it reflects the ecological nature of bacteria. The rule (c) of the original guidelines provided a means for recognizing such an "ecological population." In the genomic era, certain details for delineation of such species may well warrant modification; an ecological traits-based delineation provision should be retained in the revised guidelines.

Could phytoplasmas be a higher than genus-level taxon?

Thus far, the evolutionarily conserved 16S rRNA gene remains the most powerful phylogenetic marker for taxonomic resolution of prokaryotes. The 16S rRNA gene-derived phylogenetic tree has been considered a valid representation of organismal genealogy (Woese, 2000), and it continues to serve as the backbone of prokaryotic systematics (Ludwig and Klenk, 2010). Over the past decades, while many more phytoplasmas have been discovered, the topology of the phytoplasma phylogenetic tree (as inferred from the phytoplasma 16S rRNA gene) remains essentially the same as in the last edition of this chapter, written 15 years ago (Zhao *et al.*, 2010, 2015). Phytoplasmas comprise a phylogenetically coherent clade distinct from all other mollicutes, with the acholeplasma clade and the anaeroplasma clade being the closest neighboring clades. Within the monophyletic phytoplasma clade, individual *Candidatus* species are divided into three distinct subclades (Fig. 4.1). An alignment of 16S rRNA gene sequences from the reference strains of the 63 taxa (including formally described, incidentally cited, and potentially new '*Ca.* Phytoplasma' species listed in Table 4.1) revealed that pairwise sequence identity scores ranged from 79.4% to 99.7%. The scores above 97.5% were contributed by the nine '*Ca.* Phytoplasma' species that were differentiated based on their distinctive biological properties under the 2004 original guidelines and by the eight new taxa delineated based on the whole-genome ANI criterion (Rodrigues Jardim *et al.*, 2023) under the 2022 revised guidelines. Interestingly, the same set of phytoplasma 16S rRNA gene sequences shares 84.5–91.0% sequence identity with that of *Acholeplasma palmae* (L33734), the closest known relative of phytoplasmas. Such scores indicate that phytoplasma lineages have evolved divergently from a common ancestor after the emergence of the phytoplasma clade and leave phytoplasma researchers/taxonomists with a dilemma: either all extant phytoplasma species do not belong to a single genus, or *A. palmae* is a member of the phytoplasma genus. Since the 16S rRNA gene-derived phylogenetic tree has been considered a valid representation of organismal genealogy (Woese, 2000), the topology of the phytoplasma clade provides a clue to this dilemma (Zhao *et al.*, 2010). As shown in Fig. 4.2, the phytoplasma clade is divided into three distinct subclades. Within each subclade, phytoplasmas are grouped coherently with a much narrower 16S rRNA gene sequence similarity score range, and with higher sequence identity scores with each other than with *A. palmae*. Studies on dinucleotide relative abundance (DRA) of 16S rDNA sequences also indicate that DNA physical signatures of the phytoplasmas in the three subclades are noticeably different (Zhao *et al.*, 2010). These results point to the question of whether the three phytoplasma subclades should be ranked at the genus level, and consequently, '*Candidatus* Phytoplasma' should be promoted to a family-level taxon.

Over the last decade, another question raised regarding the taxonomic placement of phytoplasmas was whether '*Ca.* Phytoplasma' should be retained in the order Acholeplasmatales. Presently, the class Mollicutes holds four orders, Mycoplasmatales, Entomoplasmatales, Acholeplasmatales, and Anaeroplasmatales; and phytoplasmas are placed under the order Acholeplasmatales. It has been argued that although phytoplasmas likely descended from a Gram-positive, low GC walled bacterium (Gundersen *et al.*, 1994) and are most closely related to acholeplasmas among the mollicutes (Davis *et al.*, 1997), the ecological, nutritional, biochemical, physiological, genome architectural, and phylogenetic properties clearly distinguish phytoplasmas from acholeplasmas and all other taxa in each of the four existing orders of class Mollicutes (Zhao *et al.*, 2015):

Habitat specificity

Microbial habitat is an important property that separates the four existing orders in the class Mollicutes (Razin, 1992). Phytoplasmas inhabit a highly specialized dual niche where they are restricted to phloem sieve elements of infected plants and to the bodies of specialized phloem-feeding insects. Thus, phytoplasmas are transkingdom parasites of plants and insect vectors. Their entire life cycle is obligate parasitism, having no free-living stage. The lifestyle of phytoplasmas clearly distinguishes them from saprophytic, free-living acholeplasmas (Acholeplasmatales), from human and animal parasitic mycoplasmas and ureaplasmas (Mycoplasmatales), and from obligately anaerobic bovine/ovine luminal anaeroplasmas and asteroplasmas (Anaeroplasmatales). Habitat clearly obviates the inclusion of phytoplasmas within any of these three orders. The order Entomoplasmatales accommodates two anthropod-associated families, Entomoplasmataceae and Spiroplasmataceae. Members of the former exclusively inhabit insects, and the latter inhabit crustaceans or insects, and do not infect plants except for three species (*Spiroplasma kunkelii*, *S. citri*, and *S. phoenicium*). The three spiroplasmas are distinctly different from phytoplasmas in morphology (Davis *et al.*, 1972) and phylogenetic position (Zhao *et al.*, 2005; Gasparich, 2010), as well as physiological and genomic properties. Since helical morphology is a distinguishing feature of the family Spiroplasmataceae (Skripal, 1983), phytoplasmas cannot be appropriately accommodated within Spiroplasmataceae.

Distinguishing nutritional, biochemical, and physiological features inferred from genome data

The nutritional requirements and other physiological properties of the mollicutes are largely determined by their habitat. While phytoplasmas

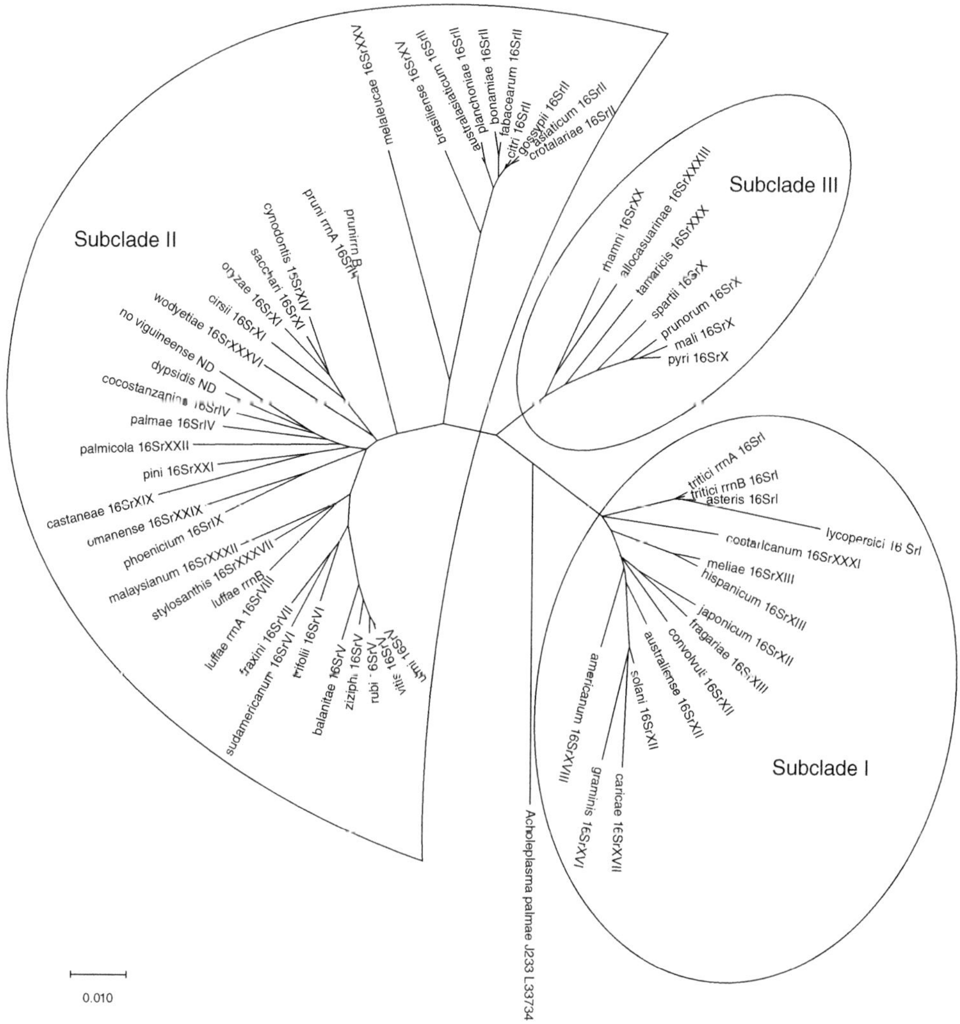

Fig. 4.2. Phylogenetic tree inferred from 16S rRNA gene sequences analysis. Minimum evolution analysis was conducted using the close neighbor interchange (CNI) algorithm implemented in the Molecular Evolutionary Genetics Analysis program (MEGA X; Kumar *et al.*, 2018). The initial tree for the CNI search was obtained by the neighbor-joining method. The reliability of the analysis was subjected to a bootstrap test with 1000 replicates. The taxa used in the phylogenetic tree reconstruction are reference strains of 57 '*Candidatus* Phytoplasma' species (see Table 4.1). Acholeplasmapalmae served as an out-group during the phylogenetic tree reconstruction. There were a total of 1599 positions in the final dataset. Evolutionary analyses were conducted in MEGA X. The scale bar represents the number of nucleotide substitutions per site.

cannot be cultured, such properties can be inferred from genomics and other omics data. Fundamental distinctions in metabolic pathways and physiological processes set phytoplasmas apart from other mollicutes. Those distinctive features include: (i) sterol requirement; (ii) lack of both phosphoenolpyruvate (PEP)-dependent sugar phosphotransferase system (PTS) and hexokinase; and (iii) absence of genes encoding enzymes essential for regulating cellular redox potentials.

Historically, the requirement versus nonrequirement for sterol in axenic culture media has been an important criterion to distinguish between taxonomic categories within the class Mollicutes. While this criterion of exogenous sterol dependence was initially used to differentiate taxa at the generic level (Acholeplasma vs Mycoplasma; Edward and Freundt, 1970), it eventually became a determining factor for separating mollicutes at the ordinal level (Acholeplasmatales vs Mycoplasmatales and Entomoplasamas; Freundt *et al.*, 1984; Tully *et al.*, 1993). The order Acholeplasmatales encompasses only species that can grow in sterol-free media. Biochemical and omics studies have revealed that acholeplasmas are able to grow in sterol-free media because they possess a comprehensive set of genes encoding enzymes for biosynthesis of fatty acids, isoprenoids, and carotenoids (Lazarev *et al.*, 2011; Kube *et al.*, 2012). In contrast, none of the completely sequenced phytoplasma genomes possesses such genes (Oshima *et al.*, 2004; Bai *et al.*, 2006; Kube *et al.*, 2008; Tran-Nguyen *et al.*, 2008; Andersen *et al.*, 2013).

Utilizing energy stored in PEP, the PTS powers cross-membrane transport of glucose, mannose, fructose, and other sugar molecules; therefore, it is essential to the life of any free-living bacterium (Glass *et al.* 2006). Hexokinase catalyzes phosphorylation of hexose, mainly glucose, yielding glucose 6-phosphate, the entry point of glycolysis and other glucose utilization pathways (Nelson and Cox, 2008). The absence of both PTS and hexokinase genes in phytoplasma genomes indicates that other forms of carbohydrate must be imported, converted, and metabolized by phytoplasma cells (Kube *et al.*, 2012). On the other hand, some phytoplasma genomes possess genes encoding malate/citrate symporters and enzymes that convert malate to pyruvate, indicating malate may be a carbon source for phytoplasmas (Kube *et al.*, 2012). In sharp contrast, the genomes of all studied mollicutes, including acholeplasmas, mycoplasmas, ureaplasmas, spiroplasmas, and mesoplasmas, all have genes encoding PTS and hexokinases, but do not have genes encoding malate/citrate symporters (Zhao *et al.*, 2014, 2015). Such distinction clearly sets phytoplasmas apart from acholeplasmas and other mollicutes.

Nicotinamide adenine dinucleotide (NAD^+/NADH) is a coenzyme pair that participates in numerous metabolic redox reactions (Belenky *et al.* 2007; Pollak *et al.* 2007). Additionally, some NAD^+ is converted to NAD phosphate ($NADP^+$), performing the electron transfer function in living cells. The ratios of NAD^+/NADH and $NADP^+$/NADPH are vital factors in maintaining cellular redox homeostasis (Schafer and Buettner, 2001). While NH_3/glutamine-dependent NAD^+ synthetase (NADS), nicotinate-nucleotide adenylyltransferase (NadD), and nicotinate phosphoribosyltransferase (PncB) are among the key enzymes essential for NAD^+ synthesis (Olland *et al.*, 2002; Magni *et al.*, 2009; Bi *et al.*, 2011), inorganic polyphosphate/ATP-NAD kinase (PPNK) plays a key role in converting NAD^+ to $NADP^+$ and balancing NADH/NAD^+ and NADPH/$NADP^+$ ratios (Kawai and Murata, 2008). However, the genomes of phytoplasmas lack genes encoding these important proteins (Zhao *et al.*, 2014). Thioredoxin is a ubiquitous redox protein found in all studied eukaryotic and prokaryotic cells (Buchanan *et al.*, 2012). It exists in both oxidized (Trx-S2) and reduced [Trx(SH)2] forms. The enzyme thioredoxin reductase catalyzes the conversion of Trx-S2 to [Trx(SH)2], and the latter is a powerful hydrogen/electron donor that can reduce the disulfide in various substrate proteins. One of the important functions of thioredoxin is maintaining cellular redox homeostasis (Hanschmann *et al.*, 2013). While phytoplasmas possess a gene encoding thioredoxin (*trxA*), they do not have a gene encoding thioredoxin reductase (*trxB*). Since both thioredoxin reductase and NADPH are required to maintain the reducing potential of thioredoxin, the lack of both a thioredoxin reductase gene and an ATP-NAD kinase gene (and therefore NADPH) in the genomes of phytoplasmas raises a question as to how oxidized thioredoxin is being reduced in phytoplasma cells, and ultimately, how redox homeostasis is achieved in phytoplasma cells. Further compounding the situation, phytoplasma genomes also lack genes encoding glutaredoxin and glutathione reductase. It was hypothesized that oxidized thioredoxin molecules in phytoplasma cells may have to be exported to host cells, where they become reduced by host thioredoxin reductase and NADPH, and "recharged" thioredoxin molecules would then be shuttled back into phytoplasma cells (Zhao *et al.*, 2014). No other known mollicutes lack such genes essential to cellular redox self-regulation.

Distinctive genomic architecture

Phytoplasmas possess a genome architecture that is unique among all members of the class Mollicutes. While mobile genetic elements play a significant role in the evolution of all studied organisms, it is especially remarkable in the case of phytoplasma emergence and evolution (Wei *et al.*, 2008a). Phytoplasma genomes are uniquely featured in possessing densely clustered repetitive sequences termed sequence variable mosaics (SVMs; Jomantiene and Davis, 2006), which were likely formed by recurrent and targeted attacks by ancient phages at the root of phytoplasma evolution (Wei *et al.*, 2008a). It has been hypothesized that the phytoplasma genomes evolved through two mutually complementary genome-sculpting mechanisms: gene loss and horizontal acquisition (Zhao *et al.*, 2014). Thus, the evolutionary shrinkage of phytoplasma genomes was partially countered by repeated and targeted chromosomal integration of phage genomes, and by further gene acquisition through targeted insertion of mobile gene cassette-like elements (Jomantiene *et al.*, 2007; Wei *et al.*, 2008a). No SVM-like structures were found in any other mollicutes.

Phylogenetic separation from other mollicutes

The evolutionarily conserved 16S rRNA gene is a powerful phylogenetic marker for taxonomic resolution of prokaryotes (Stackebrandt, 2009; Ludwig and Klenk, 2010). In the 1990s, based on comparative and phylogenetic analyses of then available phytoplasma 16S rRNA gene sequences, a conclusion was reached that phytoplasmas comprise a phylogenetically coherent group that is distinct from other prokaryotes, including mycoplasmas and achloleplasmas (Gundersen *et al.*, 1994). As more and more phytoplasmas were being discovered in the ensuing years, several updated analyses were conducted, and the results showed that the monophyletic status of phytoplasmas remained concrete. In the tree, phytoplasmas form a monophyletic clade, discrete from all other mollicutes, with the acholeplasma clade and the anaeroplasma clade being the closest neighboring clades. The separation between the phytoplasma and acholeplasma clades is as compelling as the separation between the anaeroplasma and acholeplasma clades (Zhao *et al.*, 2015). In addition to phylogenetic relationships, the average pairwise sequence distance (APSD) scores also support the separation of the phytoplasma clade from the acholeplasma clade and the anaeroplasma clade.

Based on this evidence and reasoning, it was argued that phytoplasmas possess distinctive ecological, nutritional, biochemical, physiological, genome architectural, and phylogenetic properties that are not referable to the descriptions of the order Acholeplasmatales or any other existing order of the class Mollicutes. Therefore, a future prospect of establishing a new ordinal level taxon to host phytoplasmas was postulated (Zhao *et al.*, 2015). While the current International Code of Nomenclature of Prokaryotes (ICNP) lacks a means to incorporate *Candidatus* organisms into the formal hierarchical classification system, a consensus has emerged within the scientific community that "given the genomic era we are in, there is a greater need than ever to determine a means for the ICNP to embrace yet-to-be cultured species" (personal communication with IJSEM Editor Professor Richard Birtles as cited in Zhao *et al.*, 2015).

16S rRNA Gene-based Group/ Subgroup Classification

In parallel with *Candidatus* species assignment, phytoplasmas are classified into groups and subgroups based on their genetic relatedness. Unlike the taxonomic nomenclature system, the group/subgroup classification scheme is designed to meet practical needs. It is clear that the number of known phytoplasmas, along with the new ones discovered in the field on a regular basis, far exceeds the number of known '*Ca.* Phytoplasma' species. Furthermore, the term "related strains" cannot sufficiently distinguish or describe multiple strains that represent distinct lineages. A convenient group and subgroup classification system is needed to facilitate proper study of and easy reference to diverse phytoplasma strains.

Comparison of the phylogeny-based and RFLP-based classification schemes

Two major phytoplasma group classification schemes have been developed and extensively cited in the literature; both schemes are based on genetic information coded in 16S rRNA genes. One classification scheme is based on phylogenetic analysis of 16S rRNA gene sequences (Kuske and Kirkpatrick, 1992; Namba *et al.*, 1993; Gundersen *et al.*, 1994; Schneider *et al.*, 1995; Seemüller *et al.*, 1998) and the other is based on RFLP analysis of PCR-amplified 16S rRNA gene fragments (Lee *et al.*, 1993b, 1998, 2000; Schneider *et al.*, 1993). The way phytoplasmas are grouped under the phylogeny-based classification scheme is in excellent alignment with '*Ca.* Phytoplasma' species delineations (Jung *et al.*, 2002; Firrao *et al.*, 2005). However, this phylogeny-based scheme lacks a mechanism for dealing with subgroup-level classification; therefore, it does not address the needs of differentiating distinct phytoplasma lineages within a phylogenetic group or strain cluster. On the other hand, the RFLP profiling-based classification scheme offers a mechanism that, by distinguishing subtle pattern differences, is able to identify and differentiate distinct subgroup lineages among phytoplasmas within individual groups. Since the RFLP profiling-based scheme exploits an adequate subset of characters (recognition sites of a defined set of 17 restriction enzymes) present in the 16S rRNA genes (the F2nR2 fragment of *c.*1250 bp), the phytoplasma groups delineated using this scheme are consistent with 16S rRNA gene phylogeny. Because of its practicality, this RFLP profiling-based scheme gained popularity among phytoplasmatologists and has become a useful tool in phytoplasma identification.

Some researchers might think that the RFLP approach is obsolete, but the truth is that RFLP analysis still plays an important role in the classification and differentiation of many unculturable and fastidious bacteria and fungi (Silvester *et al.*, 2017; Edel, 2020; Claussen and Schmidt, 2021). Examples include classifications of the genera *Basidiobolus* and *Vibrio*. According to Google Scholar, from 2017 to 2022, more than 15,000 research articles have been published on the classification and differentiation of bacteria and fungi using the RFLP analysis approach, including nearly 1600 articles on phytoplasmas. Computer-simulated virtual RFLP analysis undoubtedly enhanced the effectiveness and value of the RFLP analysis approach. One example of the usefulness of the group/subgroup classification system is that hundreds of phytoplasma strains around the globe have been assigned as '*Ca.* P. asteris'-related strains, which grossly masks the differences among the strains. The classification scheme can differentiate the strains into more than two dozen subgroups in the aster yellows (AY) phytoplasma group (16SrI), with each subgroup having its own unique RFLP profile. Some subgroups are only (or predominantly) present in certain geological regions and are associated with specific ecological niches (Lee *et al.*, 2000, 2004a). Virtual RFLP analysis directly reveals "visible markers" that other sequence-based analyses, such as pairwise sequence comparisons and phylogenetic analysis, cannot provide.

Furthermore, in certain cases, the current '*Ca.* Phytoplasma' species assignment system may even have difficulty in determining which '*Ca.* Phytoplasma' species a given strain should be affiliated with. For example, the 16S rRNA gene sequence of a *Bischofiapolycarpa* witches'-broom phytoplasma strain (GenBank accession KJ452548) of the elm yellows phytoplasma group shares 99.1–99.3% identity with '*Ca.* P. ulmi'- and '*Ca.* P. ziziphi'-related strains. So, what species should this strain be affiliated with, '*Ca.* P. ulmi' or '*Ca.* P. ziziphi'? The RFLP-based group/subgroup classification system can at least provide distinguishing RFLP markers to separate them and classify the strain into a new subgroup other than 16SrV-A and 16SrV-B. This example clearly demonstrates that the group/subgroup classification system provides more strain-specific information than the term '*Candidatus* Phytoplasma' species-related strain, and the RFLP profile can serve as a molecular marker to assist diagnosticians and regulatory agencies in distinguishing closely related phytoplasma strains more effectively (Wei and Zhao, 2022).

Advantages of computer-simulated RFLP analysis

Since the inception of the RFLP-based classification scheme, 37 groups and more than 150 subgroups have been delineated (Wei and Zhao, 2022). The classification scheme's rapid expansion was achieved through computer-simulated RFLP analysis (Wei *et al.*, 2007a, 2008b; Zhao *et al.*, 2009b). By mimicking laboratory restriction enzyme digestions and subsequent gel electrophoresis, computer-simulated 16S rDNA analysis produces virtual RFLP patterns, allowing high-throughput differentiation and identification of phytoplasma strains. Virtual RFLP analysis eliminates the need for visual pattern comparison and manual similarity coefficient calculation. More importantly, it circumvents the obstacle that the conventional RFLP analysis faces: as the number of known phytoplasmas grows rapidly, no single research group has a complete or near-complete collection of phytoplasma strains as sources of DNA for comparative analysis. Virtual RFLP analysis is operated on DNA sequences but retains the principles and criteria of the original classification scheme. Using accurate DNA sequence data, the virtual gel patterns generated by computer-simulated RFLP analysis can faithfully duplicate the classic and authoritative patterns established by conventional enzymatic and gel electrophoresis-based RFLP analysis. Novel pattern types generated from virtual RFLP analysis using DNA sequences from newly discovered phytoplasma strains have also been swiftly confirmed by actual enzymatic gel electrophoresis (Cai *et al.*, 2008; Quaglino *et al.*, 2009; Zhao *et al.*, 2009a).

Criteria and an important issue related to 16Sr group and subgroup delineation

As new phytoplasmas are discovered at an ever-increasing pace in emerging and re-emerging plant diseases around the globe, the scheme for classification of phytoplasmas is experiencing an ongoing rapid expansion, and more phytoplasma groups and subgroups are being recognized. Delineation of new groups and subgroups should follow certain criteria. The criterion for the delineation of a new subgroup is fairly simple. A new 16Sr subgroup is recognized if the collective F2nR2 RFLP profile of a phytoplasma has a 0.97 or lower pattern similarity coefficient with that of each previously established subgroup within a given group (Wei *et al.*, 2008b). For the establishment of a new 16Sr group, however, two criteria must be met. First, the F2nR2 RFLP profile of the representative strain must have a similarity coefficient of 0.85 or less with each of the subgroup patterns of all previously recognized 16Sr groups (Zhao *et al.*, 2009b). Second, a new group must host at least one '*Ca.* Phytoplasma' species (Lee *et al.*, 1998; Zhao and Davis, 2016).

An issue related to the phytoplasma group/subgroup classification is the lack of a platform for registration of phytoplasma groups and subgroups. For most 16Sr groups, since their delineation is often coupled with the naming or descriptions of novel '*Ca.* Phytoplasma' species, they are coincidentally "archived" in the IJSEM, as the latter is required to be published in this journal. However, such practice does not prevent new 16Sr groups from being published in other journals if the groups host previously described '*Ca.* Phytoplasma' species. For example, the erection of the *Allocasuarina* phytoplasma group (16SrXXXIII) was published elsewhere (Bertaccini *et al.*, 2014). Due to the unavailability of a tracking system, the same group number was inadvertently assigned to another new group, the Chilean grapevine phytoplasma (Pérez-López *et al.*, 2015). The subgroups are even more difficult to track as new subgroup designations may be published in various plant pathology, crop protection, and general plant biology journals, and the number of 16Sr subgroups is far greater. Establishing a platform, preferably an online portal, for registering newly established 16Sr groups and subgroups has become urgent in maintaining the integrity of the phytoplasma classification scheme. It has been suggested that *i*PhyClassifier could be an ideal place to host such a portal as it is a widely used online tool for phytoplasma classification and taxonomic assignment (Zhao and Davis, 2016).

Identification and Fine Differentiation of Diverse Phytoplasma Strains

Early diagnosis of phytoplasmal diseases and accurate identification of phytoplasmas are vital for devising effective disease management strategies. This will not only limit direct economic losses from plant death/damage, but also impede the spread of disease and protect international trade. Plants infected by phytoplasmas often exhibit visual symptoms. Characteristic symptoms include virescence, phyllody, cauliflower-like inflorescence, and witches'-broom growth (MacLean *et al.*, 2011; Wei *et al.*, 2013, 2019). Additionally, phytoplasma infection can induce more general symptoms that are indistinguishable from other plant diseases, even nutritional deficiencies. Such general symptoms include leaf discoloration (such as leaf yellowing), general decline, stem fasciation, and stunting (MacLean *et al.*, 2011; Wu *et al.*, 2012; Wei *et al.*, 2013, 2022). In some cases, phytoplasma-infected plants may stay symptomless (Zwolińska *et al.*, 2019). Since phytoplasmas cannot be cultured *in vitro*, the culture-dependent metrics and characteristics for bacterial identification cannot be applied. Phytoplasma detection and characterization heavily rely on molecular diagnostic techniques. With the rapid development of molecular diagnostic techniques, a variety of fast, sensitive, and cost-effective phytoplasma detection methods have been developed, ranging from PCR, nested PCR, real-time PCR, droplet digital PCR (ddPCR), and LAMP (loop-mediated isothermal amplification) to CRISPR (clustered regularly interspaced short palindromic repeats)-based assays (Wei and Zhao, 2022, and reference therein). Some of these diagnostic assays are devised based on conserved DNA sequence blocks of phytoplasma genes such as the 16S rRNA gene, *rp* gene, *secY* gene, and *tuf* gene, etc., for universal or 16Sr group-specific detection, while the others are designed to target unique sequences for strain-, lineage-, or species-specific detection.

Currently, for phytoplasmal disease diagnosis and phytoplasma identification, a widely adopted procedure is: (i) PCR or nested PCR amplification of phytoplasma DNA using universal primers targeting the 16S rRNA gene; (ii) determining the nucleotide sequence of the resulting PCR amplicon; (iii) determining the '*Ca.* Phytoplasma' species affiliation and group/subgroup classification status of the phytoplasma using *i*PhyClassifier; and (iv) finer differentiation of the strains using other molecular markers. Oligonucleotide primers commonly used for PCR-based phytoplasma detection and identification are listed in Table 4.2 along with corresponding references. Among the primer pairs, P1/P7 is useful for general phytoplasma detection. It can generate amplicons of appropriately 1.7 kb from diverse phytoplasma DNA templates. The P1/P7 amplicon covers a nearly full-length 16S rRNA gene, the 16S–23S RNA intergenic spacer region, and a partial 23S rRNA gene. However, primer pair P1/P7 may occasionally produce false-positive results by amplifying DNA templates from closely related non-phytoplasma bacteria. Primer pair P1A/16S-SR offers a higher specificity without a noticeable loss of sensitivity. It produces an amplicon of approximately 1.5 kb (covering a nearly full-length 16S rRNA gene) and is recommended. Primer pair R16F2n/R16R2 has the highest specificity; it generates an approximately 1.25 kb amplicon termed the 16S rDNA F2nR2 fragment. The F2nR2 fragment fulfills the minimum requirement for assigning the '*Ca.* Phytoplasma' species affiliation and serves as the basis for 16Sr group/subgroup classification of phytoplasmas. In addition, Table 4.2 also lists cocktail primers for a universal phytoplasma detection (DNA barcoding) method based on the conserved sequence blocks in the elongation factor Tu (*tuf*) gene (Makarova *et al.*, 2012).

Once the 16S rRNA gene sequence becomes available, the *i*PhyClassifier (Zhao *et al.*, 2009b; https://plantpathology.ba.ars.usda.gov/cgi-bin/resource/iphyclassifier.cgi; last accessed June 2025) is a handy tool for the identification of the phytoplasma strain under investigation. The *i*PhyClassifier can operate on a single or multiple query sequences. An ideal query sequence is a full or nearly full-length 16S rDNA. The minimum requirement is the entire F2nR2 region of the 16S rDNA. The *i*PhyClassifier aligns the query sequences with that of the reference strains of all previously described '*Candidatus* Phytoplasma' species, calculates pairwise sequence identity scores, and assigns each of the phytoplasmas under study to the most appropriate '*Ca.* Phytoplasma' species as its related strain according to the guidelines from the Phytoplasma Taxonomy Group of the International Research Programme on Comparative Mycoplasmology. The *i*PhyClassifier also simulates laboratory restriction enzyme digestions with a defined set of 17 enzymes (*AluI*, *BamHI*, *BfaI*, *BstUI* (*ThaI*), *DraI*, *EcoRI*, *HaeIII*, *HhaI*, *HinfI*, *HpaI*, *HpaII*, *KpnI*, *Sau3AI* (*MboI*), *MseI*, *RsaI*, *SspI*, and *TaqI*) and mimics subsequent gel electrophoresis.

Table 4.2. Oligonucleotide primers for phytoplasma detection and identification.

Primer pair	Target locus	Primer/probe sequence (5′-3′)	Amplicon size (bp)	Specificity	Type of PCR assay	Reference
P1/P7	*rrn*	F: AAGAGTTTGATCCTGGCTCAGGATT R: CGTCCTTCATCGGCTCTT	1800	Universal	End-point	Deng & Hiruki. 1991 Schneider *et al.*, 1995
P1A/16S-SR	*rrn*	F: AACGCTGGCGGCGCGCCTAATAC R: GGTCTGTCAAAACTGAAGATG	1500	Universal	End-point	Lee *et al.*, 2004b
R16F2n/R16R2	*rrn*	F: GAAACGACTGCTAAGACTGG R: TGACGGGCGGTGTGTACAAACCCCG	1250	Universal	End-point	Gundersen & Lee, 1996 Lee *et al.*, 1993b
P1/Tint	*rrn*	F: AAGAGTTTGATCCTGGCTCAGGATT R: TCAGGCGTGTGCTCTAACCAGC	1600	Universal	End-point	Smart *et al.*, 1996
Christensen Q-PCR	16S rDNA	F: CGTACGCAAGTATGAAACTTAAAGGA P: TGACGGGACTCCGCACAAGCG-TAMRA R: TCTTCGAATTAAACAACATGATCCA	75	Universal	TaqMan Q-PCR	Christensen *et al.*, 2004
JH-F1/JH-R (probe JH-P uni)	23S rDNA	F: GGTCTCCGAATGGGAAAACC P: FAM-AACTGAAATATCTAAGTAAC R: CTCGTCACTACTACCRGAATCGTTATTAC	152	Universal	TaqMan Q-PCR	Hodgetts *et al.*, 2009
JH-Fall/JH-R (probe JH-P uni)	23S rDNA	F: ATTTCCGAATGGGGCAACC P: FAM-AACTGAAATATCTAAGTAAC R: CTCGTCACTACTACCRGAATCGTTATTAC	151	Universal	TaqMan Q-PCR	Hodgetts *et al.*, 2009
P1/AYint	*rrn*	F: AAGAGTTTGATCCTGGCTCAGGATT R: TACAATTTGCAAGCAAGTTAC	1500	16SrI	End-point	Smart *et al.*, 1996
P1/WXint	*rrn*	F: AAGAGTTTGATCCTGGCTCAGGATT R: GACAGTGCTTATAACTTTTA	1600	16SrIII	End-point	Smart *et al.*, 1996
fB1/rULWS1	*rrn*	F: GACCCTTCAAAAGGTCTTAG R: CGTCTTTTATATAAGAGAAACA	1500	16SrV	End-point	Smart *et al.*, 1996
P1/BLTVAint	*rrn*	F: AAGAGTTTGATCCTGGCTCAGGATT R: GATGATTTTAGTATATATAGTCC	1450	16SrVI	End-point	Smart *et al.*, 1996
fB1/rASHYS	*rrn*	F: GACCCTTCAAAAGGTCTTAG R: GCAGGACCGTTTATATTAATC	1500	16SrVII	End-point	Smart *et al.*, 1996
P1/PYLRint	*rrn*	F: AAGAGTTTGATCCTGGCTCAGGATT R: CCCGGCCATTATTAATTTTTATC	1550	16SrX-B, C	End-point	Smart *et al.*, 1996
rpF1/rpR1	*rpsS-rplV*	F: GGACATAAGTTAGGTGAATTT R: ACGATATTTAGTTCTTTTTGG	1245–1389	16SrI, III, IV, V, VII, VIII, IX, XIII	End-point	Lim & Sears, 1992

Continued

Table 4.2. Continued

Primer pair	Target locus	Primer/probe sequence (5´-3´)	Amplicon size (bp)	Specificity	Type of PCR assay	Reference
rpL2F3/rp(I)R1A	*rplB-rplP*	F: WCCTTGGGGYAAAAAAGCTC R: GTTCTTTTTGGCATTAACAT	1600	16SrI, III, IV, V, VI, VII, IX, X, XII, XIII, XVIII	End-point	Martini *et al.*, 2007
rpF1C/rp(I)R1A	*rplP-rplV*	F: ATGGTDGGDCAYAARTTAGG R: GTTCTTTTTGGCATTAACAT	1212–1386	16SrI, II, III, IV, V, VI, VII, IX, X, XII, XIII, XVIII	End-point	Martini *et al.*, 2007
rp(I)F1A/rp(I)R1A	*rpsC-rplP*	F: TTTTCCCCTACACGTACTTA R: GTTCTTTTTGGCATTAACAT	1200	16SrI	End-point	Lee *et al.*, 2004a
rp(II)F1/rp(II)R1	*rpsS-rpsC*	F: GCTCTTACTCGTAAAYATGTAGT R: TTACTTGATTTTCTGGTTTTGA	1200	16SrII	End-point	Martini *et al.*, 2007
rp(III)F1/rp(III)R1	*rpsS-rpsC*	F: TTAGAGAAGGCATTAAAC R: CTCTTTCCCCATCTAGGACG	1200	16SrIII	End-point	Martini *et al.*, 2007
rp(V)F1/rpR1	*rpsS-rplP*	F: AGGCGATAAAAAAGTTTCAAAA R: GGCATTAACATAATATATTATG	1200	16SrV	End-point	Lee *et al.*, 1998 Lim & Sears, 1992
rp(VI)F2/rp(VI)R2	*rplV-rpsC*	F: GGTTGTTGATTTAATTCGTGGTC R: CCAGATATTCGTCTAGTATCAGAA	1000	16SrVI	End-point	Martini *et al.*, 2007
rp(VIII)F2/rp(VIII) R2	*rplV-rpsC*	F: AGTTGTCGATTTAATTCGTGGCA R: CAGCAGATATTTGTCTAGTATCTGCG	1000	16SrVII, VIII	End-point	Martini *et al.*, 2007
rp(IX)F2/rp(IX)R2	*rplV-rpsC*	F: GCACAAGCTATTTTAATGTTTACACCC R: CAAAGGGACTAAACCTAAAG	800	16SrIX	End-point	Martini *et al.*, 2007
rpStolF/rpStolR	*rplB-rpsC*	F: CGTACAAAATAATCGGGAGA R: CGAAACAAAAGGTTTACGAG	1372	16SrXII	End-point	Martini *et al.*, 2007
SecAfor1/ SecArev3	*secA*	F: GARATGAAAACTGGRGAAGG R: GTTTTRGCAGTTCCTGTCATNCC	840	16SrI, II, III, IV, V, VI, VII, IX, X, XI, XII, XIII	End-point	Hodgetts *et al.*, 2008
SecAfor2/ SecArev3	*secA*	F: GAYGARGSWAGAACKCCT R: GTTTTRGCAGTTCCTGTCATNCC	480	16SrI, II, III, IV, V, VI, VII, IX, X, XI, XII, XIII	End-point	Hodgetts *et al.*, 2008
L15F1/MapR1	*rplO-secY-(adk)-map*	F: CCTGGTAGTGGYAMTGGWAAAAC R: ATTARRAATATARGGYTCTTCRTG	2200–2800	16SrI, II, III, IV, V, VI, VII, VIII, X, XII, XIII, XVIII	End-point	Lee *et al.*, 2010

L15F1/MapR1	*rplO-secY-(adk)-map*	F: CCTGGTAGTGGYAMTGGWAAAAC R: ATTARRAATATARGGYTCTTCRTG	2200–2800	16SrI, II, III, IV, V, VI, VII, VIII, X, XII, XIII, XVIII	End-point	Lee *et al.*, 2010
L15F1A-a/MapR1A-a	*rplO-secY-adk-map*	F: TGGWAAAACTKCBGGWAARGG R: AAGMTKYACCRATDCCATG	2800	16SrI, XII, XIII	End-point	Lee *et al.*, 2010
L15F1A-b/MapR1A-b	*rplO-secY-map*	F: GGWAAAACYTSHGGYMRVGGHCATAAAGG R: CCWATMCCRTGWCCDGWAAAA	2200	16SrII, III,V, VI	End-point	Lee *et al.*, 2010
L15F1A(I)/MapR1A (I)	*rplO-secY-adk-map*	F: CTTCTGGTAAAGGACATAAAGG R: GTTCTTCGTGCAAAGATGTACC	2800	16SrI	End-point	Lee *et al.*, 2010
L15F1A(II)/MapR1A (II)	*rplO-secY-map*	F: CTTGCGGTCGCGGCCATAAAGG R: GGTTCTTTCGTGTAAAGATTTACC	2200	16SrII	End-point	Lee *et al.*, 2010
L15F1A(III)/MapR1A (III)	*rplO-secY-map*	F: CTTCTGGTAAAGGACATAAAGG R: GGTTCTTTCGTGCAATTGCAAACC	2200	16SrIII	End-point	Lee *et al.*, 2010
L15F1A(VI)/MapR1A (VI)	*rplO-secY-map*	F: CTTCAGGYAARGGTCATAAAGG R: GGTTTCTTTCATCAAGTCTAGTACC	2200	16SrVI	End-point	Lee *et al.*, 2010
SecYF1(II)/SecYR1 (II)	*rplO-secY*	F: CGCGTATAGGTTTTGAAGGTG R: CCTGCCATTTTCATTATAGCG	1600	16SrII	End-point	Lee *et al.*, 2010
SecYF1(III)/SecYR1 (III)	*rplO-secY*	F: CTAGACCAGGTTTTGAAGG R: GACCTGCTTTTCTCATTATAGC	1600	16SrIII	End-point	Lee *et al.*, 2010
SecYF1(VI)/SecYR1 (VI)	*rplO-secY*	F: CTAGATTAGGATTYGAGGG R: GACCRCCAAAACCTTGATAATC	1600	16SrVI	End-point	Lee *et al.*, 2010
SecYF1(X)/SecYR1 (X)	*rplO-secY*	F: GGTGGTTGTTAGACCAGGTTT R: GGAATACCYTGAACAACTAC	1600	16SrX	End-point	Lee *et al.*, 2010
SecYF1a/SecYR1 (XII)	*rplO-secY*	F: GGACAATTAGCWCGTTCAGG R: CAGGAACTAACTTCCCTTGA	1,600	16SrXII	End-point	Lee *et al.*, 2010
fTufil/rTufil	*tuf*	F: CAAGTTGGTGTTCCAAA R: GTTGTCACCTGCTTGAGC	1000	16SrI, III, XII	End-point	Schneider *et al.*, 1997
fTufAy/rTufAy	*tuf*	F: GCTAAAAGTAGAGCTTATGA R: CGTTGTCACCTGGCATTACC	940	16SrI, XII	End-point	Schneider *et al.*, 1997
Tuf 340/Tuf890 cocktail	*tuf*	F(340A): GCTCCTGAAGAAARAGAACGTGG F(340B): ACTAAAGAAGAAAAAGAACGTGG R(890RA): ACTTGDCCTCTTTCKACTCTACCAGT R(890RB): ATTTGTCCTCTTTCWACACGTCCTGT R(890RC): ACCATTCCTCTTTCAACACGTCCAGT	550	16SrI, II, III, IV, V, VI, VII, IX, X, XI, XII, XV, XX	End-point	Makarova *et al.*, 2012

Continued

Table 4.2. Continued

Primer pair	Target locus	Primer/probe sequence (5´-3´)	Amplicon size (bp)	Specificity	Type of PCR assay	Reference
Tuf400/Tuf835 cocktail*	*tuf*	F(400A): GTAAAACGACGGCCAGTGAAACA GAAAAACGTCAYTATGCTCA F(400B): GTAAAACGACGGCCAGTGA AACTTCTAAAAGACATTACGCTCA F(400C): GTAAAACGACGGCCAGT GAAACATCAAAAAGACAYTATGCTCA F(400D): GTAAAACGACGGCCAGT GAAACAGAAAAAAGACAYTATGCTCA F(400E): GTAAAACGACGGCCAGT CAAACAGCTAAAAGACATTATYCTCA R(835RA): TAATACGACTCACTATAGGGAA CATCTTCWACHGGCATTAAGAAAGG R(835RB): TAATACGACTCACTATAGGG AACACCTTCAATAGGCATTAAAAAWGG R(835RC): TAATACGACTCACTATAGGG AACATCTTCTATAGGTAATAAAAAAGG	440	16SrI, II, III, IV, V, VI, VII, IX, X, XI, XII, XV, XX	End-point	Makarova *et al.*, 2012

F, forward primer; P, probe; PCR, polymerase chain reaction; R, reverse primer.
*For Tuf400/Tuf835 cocktail primers, the underlined sequences GTAAAACGACGGCCAGT and TAATACGACTCACTATAGGG are M13 and T7 tags, respectively.

The operation generates virtual gel images that exhibit RFLP patterns. Following pairwise comparison of RFLP profiles, the *i*PhyClassifier automatically calculates pattern similarity coefficient scores. Based on the score, the *i*PhyClassifier gives instant suggestions on the group and subgroup classification status of the phytoplasma strains under study. Since the 16Sr group/subgroup classification is based on the F2nR2 fragment, if the query sequence is longer than the F2nR2 fragment, the *i*PhyClassifier will automatically trim the query sequence to the full-length F2nR2 fragment. Additional functions of the *i*PhyClassifier include delineation of potentially new phytoplasma groups and subgroups, as well as new '*Ca.* Phytoplasma' species.

The operation of the *i*PhyClassifier is solely dependent on the query sequence provided by the users. Any error in query sequences that misrepresents the phytoplasma strains under study could result in erroneous '*Candidatus* Phytoplasma' species assignment and group/subgroup classification. It is the user's responsibility to ensure the accuracy of their query sequence before using *i*PhyClassifier. Another important issue that requires attention in conducting virtual RFLP analysis using *i*PhyClassifier is the *rrn* interoperon sequence heterogeneity. The genomes of the vast majority of known phytoplasma strains harbor two ribosomal RNA operons, *rrnA* and *rrnB*. In some phytoplasma strains with two heterogeneous *rrn* operons, if the sequence variations between the two operons fall into restriction sites within the 16S rRNA gene F2nR2 region, then two different virtual 16Sr RFLP pattern types will result from virtual RFLP analysis. This could cause erroneous assignment of the same phytoplasma to two different 16S rRNA subgroups. The same issue has also been raised in conventional RFLP analysis (Zhao *et al.*, 2010). To help resolve this issue, a three-letter subgroup designation was proposed (Wei *et al.*, 2008b). For example, paulownia witches'-broom phytoplasma – a member of the previously delineated subgroup 16SrI-D – possesses two sequence-heterogeneous rRNA operons displaying two different 16Sr RFLP patterns, 16SrI-B and 16SrI-D. According to the three-letter designation proposal, the subgroup classification of PaWB was suggested to be 16SrI-[B/D]D. Another example is the reference strain of '*Ca.* Phytoplasma tritici,' wheat blue dwarf phytoplasma WBD, whose subgroup classification of WBD is 16SrI-[R/S]C (Zhao *et al.*, 2021). In such subgroup designations, the first and second letters in square parentheses denote the RFLP pattern types of *rrn*A and *rrn*B, respectively, and the third letter designates the 16Sr subgroup. In conventional RFLP analysis, a composite banding pattern may arise from two sequence-heterogeneous 16S rRNA genes. A composite pattern is suspected when the sum of the sizes of DNA fragments is greater than the expected size of the F2nR2 region (1.25 kb). In virtual RFLP analysis, a composite pattern can be reconstructed by superimposing two individual virtual gel images derived from the analysis of two sequence-heterogeneous *rrn* operons.

References

Al-Saady, N.A., Khan, A.J., Calari, A., Al-Subhi, A.M., and Bertaccini, A. (2008) '*Candidatus* Phytoplasma omanense', associated with witches'-broom of *Cassia italica* (Mill.) Spreng. in Oman. *International Journal of Systematic and Evolutionary Microbiology* 58, 461–466. DOI: 10.1099/ijs.0.65425-0.

Andersen, M.T., Liefting, L.W., Havukkala, I., and Beever, R.E. (2013) Comparison of the complete genome sequence of two closely related isolates of '*Candidatus* Phytoplasma australiense' reveals genome plasticity. *BMC Genomics* 14, 529. DOI: 10.1186/1471-2164-14-529.

Arocha, Y., López, M., Piñol, B., Fernández, M., Picornell, B. *et al.* (2005) '*Candidatus* Phytoplasma graminis' and '*Candidatus* Phytoplasma caricae', two novel phytoplasmas associated with diseases of sugarcane, weeds and papaya in Cuba. *International Journal of Systematic and Evolutionary Microbiology* 55, 2451–2463. DOI: 10.1099/ijs.0.63797-0.

Arocha, Y., Antesana, O., Montellano, E., Franco, P., Plata, G., and Jones, P. (2007) '*Candidatus* Phytoplasma lycopersici', a phytoplasma associated with 'hoja de perejil' disease in Bolivia. *International Journal of Systematic and Evolutionary Microbiology* 57, 1704–1710. DOI: 10.1099/ijs.0.64851-0.

Bai, X.D., Zhang, J.H., Ewing, A., Miller, S.A., Radek, A.J. *et al.* (2006) Living with genome instability: The adaptation of phytoplasmas to diverse environments of their insect and plant hosts. *Journal of Bacteriology* 188, 3682–3696. DOI: 10.1128/JB.188.10.3682-3696.2006.

Belenky, P., Bogan, K.L., and Brenner, C. (2007) NAD+ metabolism in health and disease. *Trends in Biochemical Sciences* 32, 12–19. DOI: 10.1016/j.tibs.2006.11.006.

Bertaccini, A., Duduk, B., Paltrinieri, S., and Contaldo, N. (2014) Phytoplasmas and phytoplasma diseases: A severe threat to agriculture. *American Journal of Plant Science* 5, 1763–1788. DOI: 10.4236/ajps.2014.512191.

Bertaccini, A., Arocha-Rosete, Y., Contaldo, N., Duduk, B., Fiore, N. *et al.* (2022) Revision of the '*Candidatus* Phytoplasma' species description guidelines. *International Journal of Systematic and Evolutionary Microbiology* 72, 005353. DOI: 10.1099/ijsem.0.005353.

Bi, J., Wang, H., and Xie, J. (2011) Comparative genomics of NAD(P) biosynthesis and novel antibiotic drug targets. *Journal of Cellular Physiology* 226, 331–340. DOI: 10.1002/jcp.22419.

Black, L.M. (1953) Transmission of plant viruses by cicadellids. In: Smith, K.M. and Lauffer, M.A. (eds) *Advances in Virus Research*. Academic Press, New York, 69–89. DOI: 10.1016/s0065-3527(08)60462-5.

Buchanan, B.B., Holmgren, A., Jacquot, J.P., and Scheibe, R. (2012) Fifty years in the thioredoxin field and a bountiful harvest. *Biochimica et Biophysica Acta* 1820, 1822–1829. DOI: 10.1016/j.bbagen.2012.07.006.

Cai, H., Wei, W., Davis, R.E., Chen, H., and Zhao, Y. (2008) Genetic diversity among phytoplasmas infecting *Opuntia* spp.: virtual RFLP analysis identifies new subgroups in the peanut witches'-broom phytoplasma group. *International Journal of Systematic and Evolutionary Microbiology* 58, 1448–1457. DOI: 10.1099/ijs.0.65615-0.

Chiykowski, L.N. (1962) Clover phyllody virus in Canada and its transmission. *Canadian Journal of Botany* 40, 397–404.

Chiykowski, L.N. and Sinha, R.C. (1989) Differentiation of MLO disease by means of symptomatology and vector transmission. *Zentralblatt für Bakteriologie* Suppl. 20, 280–287.

Christensen, N.M., Nicolaisen, M., Hansen, M., and Schulz, A. (2004) Distribution of phytoplasmas in infected plants as revealed by real-time PCR and bioimaging. *Molecular Plant-Microbe Interactions* 17, 1175–1184. DOI: 10.1094/MPMI.2004.17.11.1175.

Chun, J., Oren, A., Ventosa, A., Christensen, H., Arahal, D.R. *et al.* (2018) Proposed minimal standards for the use of genome data for the taxonomy of prokaryotes. *International Journal of Systematic and Evolutionary Microbiology* 68, 461–466. DOI: 10.1099/ijsem.0.002516.

Claussen, M. and Schmidt, S. (2021) Differentiation of *Basidiobolus* spp. isolates: RFLP of a diagnostic PCR amplicon matches sequence-based classification and growth temperature preferences. *Journal of Fungi* 7, 110. DOI: 10.1099/ijsem.0.002516.

Contaldo, N., Satta, E., Zambon, Y., Paltrinieri, S., and Bertaccini, A. (2016) Development and evaluation of different complex media for phytoplasma isolation and growth. *Journal of Microbiological Methods* 127, 105–110. DOI: 10.1016/j.mimet.2016.05.031.

Contaldo, N., D'Amico, G., Paltrinieri, S., Diallo, H.A., Bertaccini, A., and Arocha-Rosete, Y. (2019) Molecular and biological characterization of phytoplasmas from coconut palms affected by the lethal yellowing disease in Africa. *Microbiological Research* 223–225, 51–57. DOI: 10.1016/j.micres.2019.03.011.

Davis, R.E., Worley, J.F., Whitcomb, R.F., Ishijima, R., and Steere, R.L. (1972) Helical filaments produced by a mycoplasmalike organism associated with corn stunt disease. *Science* 176, 521–523. DOI: 10.1126/science.176.4036.521.

Davis, R.E., Dally, E.L., Gundersen, D.E., Lee, I.M., and Habili, N. (1997) '*Candidatus* Phytoplasma australiense', a new phytoplasma taxon associated with Australian grapevine yellows. *International Journal of Systematic Bacteriology* 47, 262–269. DOI: 10.1099/00207713-47-2-262.

Davis, R.E., Jomantiene, R., and Zhao, Y. (2005) Lineage-specific decay of folate biosynthesis genes suggests ongoing host adaptation in phytoplasmas. *DNA and Cell Biology* 24, 832–840. DOI: 10.1089/dna.2005.24.832.

Davis, R.E., Zhao, Y., Dally, E.L., Jomantiene, R., Lee, I.M., Wei, W., and Kitajima, E.W. (2012) '*Candidatus* Phytoplasma sudamericanum', a novel taxon, and strain PassWB-Br4, a new subgroup 16SrIII-V phytoplasma, from diseased passion fruit (*Passiflora edulis* f. *flavicarpa* Deg.). *International Journal of Systematic and Evolutionary Microbiology* 62, 984–989. DOI: 10.1099/ijs.0.030528-0.

Davis, R.E., Zhao, Y., Dally, E.L., Lee, I.M., Jomantiene, R., and Douglas, S.M. (2013) '*Candidatus* Phytoplasma pruni', a novel taxon associated with X-disease of stone fruits, *Prunus* spp.: multilocus characterization based on 16S rRNA, *secY*, and ribosomal protein genes. *International Journal of Systematic and Evolutionary Microbiology* 63, 766–776. DOI: 10.1099/ijs.0.041202-0.

Davis, R.E., Harrison, N.A., Zhao, Y., Wei, W., and Dally, E.L. (2016) '*Candidatus* Phytoplasma hispanicum', a novel taxon associated with Mexican periwinkle virescence disease of *Catharanthus roseus*. *International Journal of Systematic and Evolutionary Microbiology* 66, 3463–3467. DOI: 10.1099/ijsem.0.001264.

Davis, R.E., Zhao, Y., Wei, W., Dally, E.L., and Lee, I.M. (2017) '*Candidatus* Phytoplasma luffae', a novel taxon associated with witches' broom disease of loofah, *Luffa aegyptica* Mill. *International Journal of Systematic and Evolutionary Microbiology* 67, 3127–3133. DOI: 10.1099/ijsem.0.002079.

Deng, S. and Hiruki, C. (1991) Amplification of 16S rRNA genes from culturable and nonculturable Mollicutes. *Journal of Microbiological Methods* 14, 53–61. DOI: 10.1016/0167-7012(91)90007-D.

Dingle, K.E., Colles, F.M., Wareing, D.R.A., Ure, R., Fox, A.J. *et al.* (2001) Multilocus sequence typing system for *Campylobacter jejuni*. *Journal of Clinical Microbiology* 39, 14–23. DOI: 10.1128/JCM.39.1.14-23.2001.

Doi, Y., Teranaka, M., Yora, K., and Asuyama, H. (1967) Mycoplasma or PLT-group-like microorganisms found in the phloem elements of plants infected with mulberry dwarf, potato witches'-broom, aster yellows, or paulownia witches'-broom. *Annals of the Phytopathological Society of Japan* 33, 259–266. DOI: 10.3186/jjphytopath.33.259.

Edel, V. (2020) Use of PCR and RFLP in fungal systematics. In: Frisvard, J.C.,Bridge, P.D., and Arora, D.K. (eds) *Chemical Fungal Taxonomy*. CRC Press, Boca Raton, FL, pp. 51–76.

Edward, D.G.Ff. and Freundt, E.A. (1970) Amended nomenclature for strains related to *Mycoplasma laidlawii*. *Journal of General Microbiology* 62, 1–2. DOI: 10.1099/00221287-62-1-1.

Fernández, F.D., Galdeano, E., Kornowski, M.V., Arneodo, J.D., and Conci, L.R. (2016) Description of '*Candidatus* Phytoplasma meliae', a phytoplasma associated with Chinaberry (*Melia azedarach* L.) yellowing in South America. *International Journal of Systematic and Evolutionary Microbiology* 66, 5244–5251. DOI: 10.1099/ijsem.0.001475.

Firrao, G., Gibb, K., and Streten, C. (2005) Short taxonomic guide to the genus '*Candidatus* phytoplasma'. *Journal of Plant Pathology* 87, 249–263.

Fox, G.E., Wisotzkey, J.D., and Jurtshuk Jr., P. (1992) How close is close: 16S rRNA sequence identity may not be sufficient to guarantee species identity. *International Journal of Systematic and Evolutionary Microbiology* 42, 166–170. DOI: 10.1099/00207713-42-1-166.

Fraser, C.M., Gocayne, J.D., White, O., Adams, M.D., Clayton, R.A. *et al.* (1995) The minimal gene complement of *Mycoplasma genitalium*. *Science* 270, 397–403. DOI: 10.1126/science.270.5235.397.

Freitag, J.H. (1964) Interaction and mutual suppression among three strains of aster yellows virus. *Virology* 24, 401–413. DOI: 10.1016/0042-6822(64)90350-7.

Freundt, E.A., Whitcomb, R.F., Barile, M.F., Razin, S., and Tully, J.G. (1984) Proposal for elevation of the family Acholeplasmataceae to ordinal rank: Acholeplasmatales. *International Journal of Systematic Bacteriology* 34, 346–349. DOI: 10.1099/00207713-34-3-346.

Gasparich, G.E. (2010) Spiroplasmas and phytoplasmas: microbes associated with plant hosts. *Biologicals* 38, 193–203. DOI: 10.1016/j.biologicals.2009.11.007.

Gasparich, G.E., Bertaccini, A., and Zhao, Y. (2020) *Candidatus* Phytoplasma. In: Trujillo, M.E., Dedysh, S., DeVos, P., Hedlund, B., Kämpfer, P., Rainey, F.A. and Whitman, W.B. (eds) *Bergey's Manual of Systematics of Archaea and Bacteria*. Wiley, Hoboken, NJ in association with Bergey's Manual Trust. DOI: 10.1002/9781118960608.gbm01259.pub3.

Ghosh, S.K., Raychaudhuri, S.P., Varma, A., and Nariani, T.K. (1971) Isolation and culture of mycoplasma associated with citrus greening disease. *Current Science* 40, 299–300.

Glass, J.I., Assad-Garcia, N., Alperovich, N., Yooseph, S., Lewis, M.R. *et al.* (2006) Essential genes of a minimal bacterium. *Proceedings of the National Academy of Sciences of the USA* 103, 425–430. DOI: 10.1073/pnas.0510013103.

Granados, R.R. and Chapman, R.K. (1968) Identification of some new aster yellows virus strains and their transmission by the aster leafhopper *Macrosteles fascifrons*. *Phytopathology* 58, 1685–1692.

Griffiths, H.M., Sinclair, W.A., Smart, C.D., and Davis, R.E. (1999) The phytoplasma associated with ash yellows and lilac witches'-broom: '*Candidatus* Phytoplasma fraxini'. *International Journal of Systematic Bacteriology* 49, 1605–1614. DOI: 10.1099/00207713-49-4-1605.

Gundersen, D.E. and Lee, I.-M. (1996) Ultrasensitive detection of phytoplasmas by nested-PCR assays using two universal primer pairs. *Phytopathologia Mediterranea* 35, 144–151.

Gundersen, D.E., Lee, I.-M., Rehner, S.A., Davis, R.E., and Kingsbury, D.T. (1994) Phylogeny of mycoplasmalike organisms (phytoplasmas): a basis for their classification. *Journal of Bacteriology* 176, 5244–5254. DOI: 10.1128/jb.176.17.5244-5254.1994.

Hampton, R.O., Stevens, J.G., and Allen, T.C. (1969) Mechanically transmissible mycoplasma from naturally infected peas. *Plant Disease Reporter* 53, 499–503.

Hanschmann, E.M., Godoy, J.R., Berndt, C., Hudemann, C., and Lillig, C.H. (2013) Thioredoxins, glutaredoxins, and peroxiredoxins – molecular mechanisms and health significance: From cofactors to antioxidants to redox signaling. *Antioxidants and Redox Signaling* 19, 1539–1605. DOI: 10.1089/ars.2012.4599.

Harrison, N.A., Bourne, C.M., Cox, R.L., Tsai, J.H., and Richardson, P.A. (1992) DNA probes for detection of mycoplasmalike organisms associated with lethal yellowing disease of palms in Florida. *Phytopathology* 82, 216–224. DOI: 10.1094/Phyto-82-216.

Harrison, N.A., Davis, R.E., Oropeza, C., Helmick, E.E., Narváez, M. *et al.* (2014) '*Candidatus* Phytoplasma palmicola', associated with a lethal yellowing-type disease of coconut (*Cocos nucifera* L.) in Mozambique. *International Journal of Systematic and Evolutionary Microbiology* 64, 1890–1899. DOI: 10.1099/ijs.0.060053-0.

Hiruki, C. and Wang, K.R. (2004) Clover proliferation phytoplasma: '*Candidatus* Phytoplasma trifolii'. *International Journal of Systematic and Evolutionary Microbiology* 54, 1349–1353. DOI: 10.1099/ijs.0.02837-0.

Hodgetts, J., Boonham, N., Mumford, R., Harrison, N., and Dickinson, M. (2008) Phytoplasma phylogenetics based on analysis of *secA* and 23S rRNA gene sequences for improved resolution of candidate species of '*Candidatus* Phytoplasma'. *International Journal of Systematic and Evolutionary Microbiology* 58, 1826–1837. DOI: 10.1099/ijs.0.65668-0.

Hodgetts, J., Boonham, N., Mumford, R., and Dickinson, M. (2009) Panel of 23S rRNA gene-based real-time PCR assays for improved universal and group-specific detection of phytoplasmas. *Applied and Environmental Microbiology* 75, 2945–2950. DOI: 10.1128/AEM.02610-08.

Hugenholtz, P., Chuvochina, M., Oren, A., Parks, D.H., and Soo, R.M. (2021) Prokaryotic taxonomy and nomenclature in the age of big sequence data. *ISME Journal* 15, 1879–1892. DOI: 10.1038/s41396-021-00941-x.

ICSB Subcommittee on the Taxonomy of Mollicutes (1993) Minutes of the interim meetings, 1 and 2 August, 1992, Ames, Iowa. *International Journal of Systematic Bacteriology* 43, 394–397. DOI: 10.1099/00207713-43-2-394.

IRPCM (International Research Programme on Comparative Mycoplasmology) Phytoplasma/Spiroplasma Working Team–Phytoplasma Taxonomy Group (2004) '*Candidatus* Phytoplasma', a taxon for the wall-less, non-helical prokaryotes that colonize plant phloem and insects. *International Journal of Systematic and Evolutionary Microbiology* 54, 1243–1255. DOI: 10.1099/ijs.0.02854-0.

Jain, C., Rodriguez-R, L.M., Phillippy, A.M., Konstantinidis, K.T. and Aluru, S. (2018) High throughput ANI analysis of 90K prokaryotic genomes reveals clear species boundaries. *Nature Communications* 9, 5114. DOI: 10.1038/s41467-018-07641-9.

Jomantiene, R. and Davis, R.E. (2006) Clusters of diverse genes existing as multiple, sequence-variable mosaics in a phytoplasma genome. *FEMS Microbiology Letters* 255, 59–65. DOI: 10.1111/j.1574-6968.2005.00061.x.

Jomantiene, R., Zhao, Y., and Davis, R.E. (2007) Sequence-variable mosaics: composites of recurrent transposition characterizing the genomes of phylogenetically diverse phytoplasmas. *DNA and Cell Biology* 26, 557–564. DOI: 10.1089/dna.2007.0594.

Jones, L.M., Pease, B., Perkins, S.L., Constable, F.E., Kinoti, W.M. *et al.* (2021) '*Candidatus* Phytoplasma dypsidis', a novel taxon associated with a lethal wilt disease of palms in Australia. *International Journal of Systematic and Evolutionary Microbiology* 71. DOI: 10.1099/ijsem.0.004818.

Jung, H.-Y., Sawayanagi, T., Kakizawa, S., Nishigawa, H., Miyata, S. *et al.* (2002) '*Candidatus* Phytoplasma castaneae', a novel phytoplasma taxon associated with chestnut witches' broom disease. *International Journal of Systematic and Evolutionary Microbiology* 52, 1543–1549. DOI: 10.1099/00207713-52-5-1543.

Jung, H.Y., Sawayanagi, T., Kakizawa, S., Nishigawa, H., Wei, W. *et al.* (2003a) '*Candidatus* Phytoplasma ziziphi', a novel phytoplasma taxon associated with jujube witches'-broom disease. *International Journal of Systematic and Evolutionary Microbiology* 53, 1037–1041. DOI: 10.1099/ijs.0.02393-0.

Jung, H.Y., Sawayanagi, T., Wongkaew, P., Kakizawa, S., Nishigawa, H. *et al.* (2003b) '*Candidatus* Phytoplasma oryzae', a novel phytoplasma taxon associated with rice yellow dwarf disease. *International Journal of Systematic and Evolutionary Microbiology* 53, 1925–1929. DOI: 10.1099/ijs.0.02631-0.

Kämpfer, P. and Glaeser, S.P. (2012) Prokaryotic taxonomy in the sequencing era – The polyphasic approach revisited. *Environmental Microbiology* 14, 291–317. DOI: 10.1111/j.1462-2920.2011.02614.x.

Kawai, S. and Murata, K. (2008) Structure and function of NAD kinase and NADP phosphatase: Key enzymes that regulate the intracellular balance of NAD(H) and NADP(H). *Bioscience, Biotechnology, and Biochemistry* 72, 919–930. DOI: 10.1271/bbb.70738.

Kim, M., Oh, H.S., Park, S.C., and Chun, J. (2014) Towards a taxonomic coherence between average nucleotide identity and 16S rRNA gene sequence similarity for species demarcation of prokaryotes. *International Journal of Systematic and Evolutionary Microbiology* 64, 346–351. DOI: 10.1099/ijs.0.059774-0.

Kirdat, K., Tiwarekar, B., Thorat, V., Sathe, S., Shouche, Y., and Yadav, A. (2021) '*Candidatus* Phytoplasma sacchari', a novel taxon associated with Sugarcane Grassy Shoot (SCGS) disease. *International Journal of Systematic and Evolutionary Microbiology* 71. DOI: 10.1099/ijsem.0.004591.

Kirkpatrick, B.C., Stenger, D.C., Morris, T.J., and Purcell, A.H. (1987) Cloning and detection of DNA from a nonculturable plant pathogenic mycoplasma-like organism. *Science* 238, 197–200. DOI: 10.1126/science.238.4824.197.

Kison, H., Schneider, B., and Seemüller, E. (1994) Restriction fragment length polymorphism within the apple proliferation mycoplasmalike organism. *Journal of Phytopathology* 141, 395–401. DOI: 10.1111/j.1439-0434.1994.tb01480.x.

Kube, M., Schneider, B., Kuhl, H., Dandekar, T., Heitmann, K. *et al.* (2008) The linear chromosome of the plant-pathogenic mycoplasma *'Candidatus* Phytoplasma mali'. *BMC Genomics* 9, 306. DOI: 10.1186/1471-2164-9-306.

Kube, M., Mitrovic, J., Duduk, B., Rabus, R., and Seemüller, E. (2012) Current view on phytoplasma genomes and encoded metabolism. *Scientific World Journal* 2012, 185942. DOI: 10.1100/2012/185942.

Kumar, S., Stecher, G., Li, M., Knyaz, C., and Tamura, K. (2018) MEGA X: Molecular Evolutionary Genetics Analysis across computing platforms. *Molecular Biology and Evolution* 35, 1547–1549. DOI: 10.1093/molbev/msy096.

Kunkel, L.O. (1924) Insect transmission of aster yellows. *Phytopathology* 14, 54.

Kunkel, L.O. (1925) Insect transmission and host range of aster yellows. *Science* 62, 524. DOI: 10.1126/science.62.1601.524.

Kunkel, L.O. (1926) Incubation period of aster yellows in its insect host. *Phytopathology* 16, 67.

Kuske, C.R. and Kirkpatrick, B.C. (1992) Phylogenetic relationships between the western aster yellows mycoplasmalike organism and other prokaryotes established by 16S rRNA gene sequence. *International Journal of Systematic Bacteriology* 42, 226–233. DOI: 10.1099/00207713-42-2-226.

Kuske, C.R., Kirkpatrick, B.C., and Seemüller, E. (1991) Differentiation of virescence MLOs using western aster yellows mycoplasma-like organism chromosomal DNA probes and restriction fragment length polymorphism analysis. *Journal of General Microbiology* 137, 153–159. DOI: 10.1099/00221287-137-7-153.

Lazarev, V.N., Levitskii, S.A., Basovskii, Y.I., Chukin, M.M., Akopian, T.A. *et al.* (2011) Complete genome and proteome of *Acholeplasma laidlawii*. *Journal of Bacteriology* 193, 4943–4953. DOI: 10.1128/JB.05059-11.

Lee, I.-M. and Davis, R.E. (1988) Detection and investigation of genetic relatedness among aster yellows and other mycoplasmalike organisms by using cloned DNA and RNA probes. *Molecular Plant-Microbe Interactions* 1, 303-310. DOI: 10.1094/MPMI-1-303.

Lee, I.-M., Davis, R.E., Chen, T.-A., Chiykowski, L.N., Fletcher, J., Hiruki, C., and Schaff, D.A. (1992a) A genotype-based system for identification and classification of mycoplasmalike organisms (MLOs) in the aster yellows MLO strain cluster. *Phytopathology* 82, 977–986. DOI: 10.1094/Phyto-82-977.

Lee, I.-M., Gundersen, D.E., Davis, R.E., and Chiykowski, L.N. (1992b) Identification and analysis of a genomic strain cluster of mycoplasmalike organisms associated with Canadian peach (eastern) X-disease, Western X-disease, and clover yellow edge. *Journal of Bacteriology* 174, 6694–6698. DOI: 10.1128/jb.174.21.6694-6698.1992.

Lee, I.-M., Davis, R.E., and Hsu, H.-T. (1993a) Differentiation of strains in the aster yellows mycoplasmalike organism strain cluster by serological assays with monoclonal antibodies. *Plant Disease* 77, 815–-817. DOI: 10.1094/PD-77-0815.

Lee, I.-M., Hammond, R.W., Davis, R.E., and Gundersen, D.E. (1993b) Universal amplification and analysis of pathogen 16S rDNA for classification and identification of mycoplasmalike organisms. *Phytopathology* 83,834-842. DOI: 10.1094/Phyto-83-834.

Lee, I.-M., Gundersen-Rindal, D.E., Davis, R.E., and Bartoszyk, I.-M. (1998) Revised classification scheme of phytoplasmas based on RFLP analyses of 16S rRNA and ribosomal protein gene sequences. *International Journal of Systematic Bacteriology* 48, 1153–1169. DOI: 10.1099/00207713-48-4-1153.

Lee, I.-M., Davis, R.E., and Gundersen-Rindal, D.E. (2000) Phytoplasma: phytopathogenic mollicutes. *Annual Review of Microbiology* 54, 221–255. DOI: 10.1146/annurev.micro.54.1.221.

Lee, I.-M., Gundersen-Rindal, D.E., Davis, R.E., Bottner, K.D., Marcone, C., and Seemüller, E. (2004a) '*Candidatus* Phytoplasma asteris', a novel phytoplasma taxon associated with aster yellows and related diseases. *International Journal of Systematic and Evolutionary Microbiology* 54, 1037–1048. DOI: 10.1099/ijs.0.02837-0.

Lee, I.-M., Martini, M., Marcone, C., and Zhu, S.F. (2004b) Classification of phytoplasma strains in the elm yellows group (16SrV) and proposal of '*Candidatus* phytoplasma ulmi' for the phytoplasma associated with elm yellows. *International Journal of Systematic and Evolutionary Microbiology* 54, 337–347. DOI: 10.1099/ijs.0.02697-0.

Lee, I.-M., Bottner, K.D., Secor, G., and Rivera-Varas, V. (2006) *'Candidatus* Phytoplasma americanum', a phytoplasma associated with a potato purple top wilt disease complex. *International Journal of Systematic and Evolutionary Microbiology* 56, 1593–1597. DOI: 10.1099/ijs.0.64251-0.

Lee, I.-M., Bottner, K.D., Zhao, Y., Davis, R.E., and Harrison, N. (2010) Phylogenetic analysis and delineation of phytoplasmas based on the *secY* gene. *International Journal of Systematic and Evolutionary Microbiology* 60, 2887–2897. DOI: 10.1099/ijs.0.019695-0.

Lee, I.-M., Bottner-Parker, K.D., Zhao, Y., Villalobos, W., and Moreira, L. (2011) *'Candidatus* Phytoplasma costaricanum' a novel phytoplasma associated with an emerging disease in soybean (*Glycine max*). *International Journal of Systematic and Evolutionary Microbiology* 61, 2822–2826. DOI: 10.1099/ijs.0.028597-0.

Lim, P.O. and Sears, B.B. (1989) 16S rRNA sequence indicates that plant-pathogenic mycoplasmalike organisms are evolutionarily distinct from animal mycoplasmas. *Journal of Bacteriology* 171, 5901–5906. DOI: 10.1128/jb.171.11.5901-5906.1989.

Lim, P.O. and Sears, B.B. (1992) Evolutionary relationships of a plant pathogenic mycoplasmalike organism and *Acholeplasma laidlawii* deduced from two ribosomal protein gene sequences. *Journal of Bacteriology* 174, 2606–2611. DOI: 10.1128/jb.174.8.2606-2611.1992.

Lin, C.-P. and Chen, T.-A. (1985) Monoclonal antibodies against the aster yellows agent. *Science* 227, 1233–1235. DOI: 10.1126/science.227.4691.1233.

Lin, S.C., Lee, C.S., and Chin, R.J. (1970) Isolation and cultivation of, and inoculation with a mycoplasma causing white leaf disease of sugarcane. *Phytopathology* 60, 795–797. DOI: 10.1094/Phyto-60-795.

Lombardo, G. and Pignattelli, P. (1970) Cultivation in a cell-free medium of a mycoplasma-like organism from *Vinca rosea* with phyllody symptoms of the flowers. *Annals of Microbiology* 20, 83–88.

Ludwig, W. and Klenk, H.-P. (2010) Overview: A phylogenetic backbone and taxonomic framework for procaryotic systematics. In: Krieg, N.R., Staley, J.T., Brown, D.R., and Hedlund, B. (eds) *Bergey's Manual of Systematic Bacteriology*, Vol. 4. Springer, New York, pp. 49–65. DOI: 10.1007/978-0-387-68572-4_3.

MacLean, A.M., Sugio, A., Makarova, O.V., Findlay, K.C., Grieve, V.M. *et al.* (2011) Phytoplasma effector SAP54 induces indeterminate leaf-like flower development in *Arabidopsis* plants. *Plant Physiology* 157, 831–841. DOI: 10.1104/pp.111.181586.

Magni, G., Di Stefano, M., Orsomando, G., Raffaelli, N., and Ruggieri, S. (2009) NAD(P) biosynthesis enzymes as potential targets for selective drug design. *Current Medicinal Chemistry* 16, 1372–1390. DOI: 10.2174/092986709787846511.

Maiden, M.C. (2006) Multilocus sequence typing of bacteria. *Annual Review of Microbiology* 60, 561–588. DOI: 10.1146/annurev.micro.59.030804.121325.

Makarova, O., Contaldo, N., Paltrinieri, S., Kawube, G., Bertaccini, A., and Nicolaisen, M. (2012) DNA barcoding for identification of *'Candidatus* Phytoplasmas' using a fragment of the elongation factor Tu gene. *PLoS One* 7, e52092. DOI: 10.1371/journal.pone.0052092.

Malembic-Maher, S., Salar, P., Filippin, L., Carle, P., Angelini, E., and Foissac, X. (2011) Genetic diversity of European phytoplasmas of the 16SrV taxonomic group and proposal of '*Candidatus* Phytoplasma rubi'. *International Journal of Systematic and Evolutionary Microbiology* 61, 2129–2134. DOI: 10.1099/ijs.0.028175-0.

Marcone, C., Neimark, H., Ragozzino, A., Lauer, U., and Seemüller, E. (1999) Chromosome sizes of phytoplasmas composing major phylogenetic groups and subgroups. *Phytopathology* 89, 805–810. DOI: 10.1094/PHYTO.1999.89.9.805.

Marcone, C., Gibb, K.S., Streten, C., and Schneider, B. (2004a) '*Candidatus* Phytoplasma spartii', *'Candidatus* Phytoplasma rhamni' and *'Candidatus* Phytoplasma allocasuarinae', respectively associated with spartium witches'-broom, buckthorn witches'-broom and allocasuarina yellows diseases. *International Journal of Systematic and Evolutionary Microbiology* 54, 1025–1029. DOI: 10.1099/ijs.0.02837-0.

Marcone, C., Schneider, B., and Seemüller, E. (2004b) '*Candidatus* Phytoplasma cynodontis', the phytoplasma associated with Bermuda grass white leaf disease. *International Journal of Systematic and Evolutionary Microbiology* 54, 1077–1082. DOI: 10.1099/ijs.0.02837-0.

Martini, M., Lee, I.-M., Bottner, K.D., Zhao, Y., Botti, S. *et al.* (2007) Ribosomal protein gene-based phylogeny for finer differentiation and classification of phytoplasmas. *International Journal of Systematic and Evolutionary Microbiology* 57, 2037–2051. DOI: 10.1099/ijs.0.65013-0.

Martini, M., Marcone, C., Mitrović, J., Maixner, M., Delić, D. *et al.* (2012) *'Candidatus* Phytoplasma convolvuli', a new phytoplasma taxon associated with bindweed yellows in four European countries. *International Journal of Systematic and Evolutionary Microbiology* 62, 2910–2915. DOI: 10.1099/ijs.0.038612-0.

McCoy, R.E., Caudwell, A., Chang, C.J., Chen, T.A., Chiykowski, L.N. *et al.* (1989) Plant diseases associated with mycoplasmalike organisms. In: Whitcomb, R.F. and Tully, J.G. (eds) *The Mycoplasmas*, Vol. 5. Edited by. Academic Press, New York, pp. 545–560.

Miyazaki, A., Shigaki, T., Koinuma, H., Iwabuchi, N., Rauka, G.B. *et al.* (2018) '*Candidatus* Phytoplasma noviguineense', a novel taxon associated with Bogia coconut syndrome and banana wilt disease on the island of New Guinea. *International Journal of Systematic and Evolutionary Microbiology* 68, 170–175. DOI: 10.1099/ijsem.0.002490.

Montano, H.G., Davis, R.E., Dally, E.L., Hogenhout, S., Pimentel, J.P., and Brioso, P.S.T. (2001) '*Candidatus* Phytoplasma brasiliense', a new phytoplasma taxon associated with hibiscus witches' broom disease. *International Journal of Systematic and Evolutionary Microbiology* 51, 1109–1118. DOI: 10.1099/00207713-51-3-1109.

Murray, R.G.E. and Schleifer, K.H. (1994) Taxonomic notes: a proposal for recording the properties of putative taxa of procaryotes. *International Journal of Systematic Bacteriology* 44, 174–176. DOI: 10.1099/00207713-44-1-174.

Murray, R.G. and Stackebrandt, E. (1995) Taxonomic note: implementation of the provisional status *Candidatus* for incompletely described procaryotes. *International Journal of Systematic Bacteriology* 45, 186–187. DOI: 10.1099/00207713-45-1-186.

Naderali, N., Nejat, N., Vadamalai, G., Davis, R.E., Wei, W. *et al.* (2017) '*Candidatus* Phytoplasma wodyetiae', a new taxon associated with yellow decline disease of foxtail palm (*Wodyetia bifurcata*) in Malaysia. *International Journal of Systematic and Evolutionary Microbiology* 67, 3765–3772. DOI: 10.1099/ijsem.0.002187.

Namba, S., Oyaizu, H., Kato, S., Iwanami, S., and Tsuchizaki, T. (1993) Phylogenetic diversity of phytopathogenic mycoplasmalike organisms. *International Journal of Systematic Bacteriology* 43, 461–467. DOI: 10.1099/00207713-43-3-461.

Nejat, N., Vadamalai, G., Davis, R.E., Harrison, N.A., Sijam, K. *et al.* (2013) '*Candidatus* Phytoplasma malaysianum', a novel taxon associated with virescence and phyllody of Madagascar periwinkle (*Catharanthus roseus*). *International Journal of Systematic and Evolutionary Microbiology* 63, 540–548. DOI: 10.1099/ijs.0.041467-0.

Nelson, D.L. and Cox, M.M. (2008) Glycolysis, gluconeogenesis, and the pentose phosphate pathway. In: Ahr, K. (ed.) *Lehninger Principles of biochemistry*, 5th edn. W.H. Freeman and Company, New York, pp. 527–568.

Olland, A.M., Underwood, K.W., Czerwinski, R.M., Lo, M.C., Aulabaugh, A. *et al.* (2002) Identification, characterization, and crystal structure of *Bacillus subtilis* nicotinic acid mononucleotide adenylyltransferase. *Journal of Biological Chemistry* 277, 3698–3707. DOI: 10.1074/jbc.M109933200.

Oren, A. (2017) A plea for linguistic accuracy – also for *Candidatus* taxa. *International Journal of Systematic and Evolutionary Microbiology* 67, 1085–1094. DOI: 10.1099/ijsem.0.001715.

Oshima, K., Kakizawa, S., Nishigawa, H., Jung, H.-Y., Wei, W. *et al.* (2004) Reductive evolution suggested from the complete genome sequence of a plant-pathogenic phytoplasma. *Nature Genetics* 36, 27–29. DOI: 10.1038/ng1277.

Parks, D.H., Chuvochina, M., Chaumeil, P.A., Rinke, C., Mussig, A.J., and Hugenholtz, P. (2020) A complete domain-to-species taxonomy for Bacteria and Archaea. *Nature Biotechnology* 38, 1079–1086. DOI: 10.1038/s41587-020-0501-8.

Pérez-López, E., Luna-Rodríguez, M., Olivier, C.Y., and Dumonceaux, T.J. (2015) The underestimated diversity of phytoplasmas in Latin America. *International Journal of Systematic and Evolutionary Microbiology* 65, 492–513. DOI: 10.1099/ijsem.0.000726.

Pollak, N., Dölle, C., and Ziegler, M. (2007) The power to reduce: pyridine nucleotides – Small molecules with a multitude of functions. *Biochemical Journal* 402, 205–218. DOI: 10.1042/BJ20061638.

Quaglino, F., Zhao, Y., Bianco, P., Wei, W., Casati, P., Durante, G., and Davis, R.E. (2009) New 16Sr subgroups and distinct SNP lineages among grapevine Bois noir phytoplasma populations. *Annals of Applied Biology* 154, 279-289. DOI: 10.1111/j.1744-7348.2008.00291.x.

Quaglino, F., Zhao, Y., Casati, P., Bulgari, D., Bianco, P.A., Wei, W., and Davis, R.E. (2013) '*Candidatus* Phytoplasma solani', a novel taxon associated with stolbur and bois noir related diseases of plants. *International Journal of Systematic and Evolutionary Microbiology* 63, 2879–2894. DOI: 10.1099/ijs.0.044750-0.

Razin, S. (1992) Mycoplasma taxonomy and ecology. In: Maniloff, J., McElhaney, R.N., Finch, L.R., and Baseman, J.B. (es) *Mycoplasmas: Molecular Biology and Pathogenesis*.American Society for Microbiology, Washington, DC, pp. 3–22.

Richter, M. and Rosselló-Móra, R. (2009) Shifting the genomic gold standard for the prokaryotic species definition. *Proceedings of the National Academy of Sciences of the USA*, 106, 19126–19131. DOI: 10.1073/pnas.0906412106.

Rodrigues Jardim, B., Kinoti, W.M., Tran-Nguyen, L.T.T., Gambley, C., Rodoni, B., and Constable, F.E. (2021) '*Candidatus* Phytoplasma stylosanthis', a novel taxon with a diverse host range in Australia characterised using multilocus sequence analysis of 16S rRNA, secA, tuf, and rp genes. *International Journal of Systematic and Evolutionary Microbiology* 71. DOI: 10.1099/ijsem.0.004589.

Rodrigues Jardim, B., Tran-Nguyen, L.T.T., Gambley, C., Al-Sadi, A.M., Al-Subhi, A.M. *et al.* (2023) The observation of taxonomic boundaries for the 16SrII and 16SrXXV phytoplasmas using genome-based delimitation. *International Journal of Systematic and Evolutionary Microbiology* 73. DOI: 10.1099/ijsem.0.005977.

Šafářová, D., Zemánek, T., Válová, P., and Navrátil, M. (2016) '*Candidatus* Phytoplasma cirsii', a novel taxon from creeping thistle [*Cirsium arvense* (L.) Scop]. *International Journal of Systematic and Evolutionary Microbiology* 66, 1745–1753. DOI: 10.1099/ijsem.0.000943.

Sawayanagi, T., Horikoshi, N., Kanehira, T., Shinohara, M., Bertaccini, A. *et al.* (1999) '*Candidatus* Phytoplasma japonicum', a new phytoplasma taxon associated with Japanese Hydrangea phyllody. *International Journal of Systematic Bacteriology* 49, 1275–1285. DOI: 10.1099/00207713-49-3-1275.

Schafer, F.Q. and Buettner, G.R. (2001) Redox environment of the cell as viewed through the redox state of the glutathione disulfide/glutathione couple. *Free Radical Biology and Medicine* 30, 1191–1212. DOI: 10.1016/s0891-5849(01)00480-4.

Schneider, B. and Seemüller, E. (1994a) Studies on the taxonomic relationships of mycoplasmalike organisms by Southern blot analysis. *Journal of Phytopathology* 141, 173–185. DOI: 10.1111/j.1439-0434.1994.tb01458.x.

Schneider, B., Ahrens, U., Kirkpatrick, B.C., and Seemüller, E. (1993) Classification of plant-pathogenic mycoplasma-like organisms using restriction-site analysis of PCR-amplified 16S rDNA. *Journal of General Microbiology* 139, 519–527. DOI: 10.1099/00221287-139-3-519.

Schneider, B., Seemüller, E., Smart, C.D., and Kirkpatrick, B.C. (1995) Phylogenetic classification of plant pathogenic mycoplasmalike organisms or phytoplasmas. In: Razin, R. and Tully, J.G. (eds) *Molecular and Diagnostic Procedures in Mycoplasmology*. Academic Press, San Diego, pp. 369–380. DOI: 10.1016/B978-012583805-4/50040-6.

Schneider, B., Gibb, K.S. and Seemüller, E. (1997) Sequence and RFLP analysis of the elongation factor Tu gene used in differentiation and classification of phytoplasmas. *Microbiology* 143 3381–3389. DOI: 10.1099/00221287-143-10-3381.

Schneider, B., Torres, E., Martín, M.P., Schroder, M., Behnke, H.D., and Seemüller, E. (2005) '*Candidatus* Phytoplasma pini', a novel taxon from *Pinus silvestris* and *Pinus halepensis*. *International Journal of Systematic and Evolutionary Microbiology* 55 303–307. DOI: 10.1099/ijs.0.63334-0.

Sears, B.B. and Kirkpatrick, B.C. (1994) Unveiling the evolutionary relationships of plant pathogenic mycoplasmalike organisms. *ASM News* 60 307–312.

Seemüller, E. and Schneider, B. (2004) '*Candidatus* Phytoplasma mali', '*Candidatus* Phytoplasma pyri' and '*Candidatus* Phytoplasma prunorum', the causal agents of apple proliferation, pear decline and European stone fruit yellows, respectively. *International Journal of Systematic and Evolutionary Microbiology* 54, 1217–1226. DOI: 10.1099/ijs.0.02823-0.

Seemüller, E., Marcone, C., Lauer, U., Ragozzino, A., and Göschl, M. (1998) Current status of molecular classification of the phytoplasmas. *Journal of Plant Pathology* 80, 3–26.

Severin, H.H. (1942) Infection of perennial delphiniums by California-aster-yellows virus. *Hilgardia*, 14, 411–440. DOI: 10.3733/hilg.v14n08p411.

Silvester, R., Alexander, D., Antony, A.C., and Hatha, M. (2017) GroEL PCR-RFLP – An efficient tool to discriminate closely related pathogenic Vibrio species. *Microbial Pathogenesis* 105, 196–200. DOI: 10.1016/j.micpath.2017.02.028.

Skripal, I.G. (1983) Revival of the name Spiroplasmataceae fam. nov., nom. rev., omitted from the 1980 approved lists of bacterial names. *International Journal of Systematic Bacteriology* 33, 408–409. DOI: 10.1099/00207713-33-2-408.

Smart, C.D., Schneider, B., Blomquist, C.L., Guerra, L.J., Harrison, N.A. *et al.* (1996) Phytoplasma-specific PCR primers based on sequences of the 16S-23S rRNA spacer region. *Applied and Environmental Microbiology* 62, 2988–2993. DOI: 10.1128/aem.62.8.2988-2993.1996.

Stackebrandt, E. (2006) Taxonomic parameters revisited: tarnished gold standards. *Microbiology Today* 33, 152–155.

Stackebrandt, E. (2009) Phylogeny based on 16S rRNA/DNA. In: *Encyclopedia of Life Sciences (ELS)*. Wiley, Chichester. DOI: 10.1002/9780470015902.a0000462.pub2

Tran-Nguyen, L.T., Kube, M., Schneider, B., Reinhardt, R., and Gibb, K.S. (2008) Comparative genome analysis of '*Candidatus* Phytoplasma australiense' (subgroup tuf-Australia I; rp-A) and '*Ca.* Phytoplasma asteris' strains OY-M and AY-WB. *Journal of Bacteriology* 190, 3979–3991. DOI: 10.1128/JB.01301-07.

Tully, J.G., Bove, J.M., Laigret, F., and Whitcomb, R.F. (1993) Revised taxonomy of the class Mollicutes: Proposed elevation of a monophyletic cluster of arthropod-associated mollicutes to ordinal rank (Entomoplasmatales ord. nov.), with provision for familial rank to separate species with nonhelical morphology (Entomoplasmataceae fam. nov.) from helical species (Spiroplasmataceae), and emended descriptions of the order Mycoplasmatales, family Mycoplasmataceae. *International Journal of Systematic Bacteriology* 43, 378–385. DOI: 10.1099/00207713-43-2-378.

Valiunas, D., Staniulis, J., and Davis, R.E. (2006) '*Candidatus* Phytoplasma fragariae', a novel phytoplasma taxon discovered in yellows diseased strawberry, *Fragaria* × *ananassa*. *International Journal of Systematic and Evolutionary Microbiology* 56, 277–281. DOI: 10.1099/ijs.0.63935-0.

Verdin, E., Salar, P., Danet, J.L., Choueiri, E., Jreijiri, F. *et al.* (2003) '*Candidatus* Phytoplasma phoenicium' sp. nov., a novel phytoplasma associated with an emerging lethal disease of almond trees in Lebanon and Iran. *International Journal of Systematic and Evolutionary Microbiology* 53, 833–838. DOI: 10.1099/ijs.0.02453-0.

Wei, W. and Zhao, Y. (2022) Phytoplasma taxonomy: Nomenclature, classification and identification. *Biology* 11, 1119. DOI: 10.3390/ biology11081119.

Wei, W., Davis, R.E., Lee, I.-M., and Zhao, Y. (2007a) Computer-simulated RFLP analysis of 16S rRNA genes: Identification of ten new phytoplasma groups. *International Journal of Systematic and Evolutionary Microbiology* 57, 1855–1867. DOI: 10.1099/ijs.0.65000-0.

Wei, W., Lee, I.-M., Davis, R.E., Suo, X., and Zhao, Y. (2007b) Virtual RFLP analysis of 16S rDNA sequences identifies new subgroups in the clover proliferation phytoplasma group. *Bulletin of Insectology* 60, 349–350.

Wei, W., Davis, R.E., Jomantiene, R., and Zhao, Y. (2008a) Ancient, recurrent phage attacks and recombination shaped dynamic sequence-variable mosaics at the root of phytoplasma genome evolution. *Proceedings of the National Academy of Sciences of the USA* 105, 11827–11832. DOI: 10.1073/pnas.0805237105.

Wei, W., Lee, I.-M., Davis, R.E., Suo, X., and Zhao, Y. (2008b) Automated RFLP pattern comparison and similarity coefficient calculation for rapid delineation of new and distinct phytoplasma 16S rDNA subgroup lineages. *International Journal of Systematic and Evolutionary Microbiology* 58, 2368–2377. DOI: 10.1099/ijs.0.65868-0.

Wei, W., Davis, R.E., Nuss, D.L., and Zhao, Y. (2013) Phytoplasmal infection derails genetically preprogrammed meristem fate and alters plant architecture. *Proceedings of the National Academy of Sciences of the USA* 110, 19149–19154. DOI: 10.1073/pnas.1318489110.

Wei, W., Davis, R.E., Bauchan, G., and Zhao, Y. (2019) New symptoms identified in phytoplasma-infected plants reveal extra stages of pathogen-induced meristem fate-derailment. *Molecular Plant-Microbe Interactions* 32, 1314–1323. DOI: 10.1094/MPMI-03-19-0080-R.

Wei, W., Inaba, J., Zhao, Y., Mowery, J.D., and Hammond, R. (2022) Phytoplasma infection blocks starch breakdown and triggers chloroplast degradation, leading to premature leaf senescence, sucrose reallocation, and spatiotemporal redistribution of phytohormones. *International Journal of Molecular Sciences* 23, 1810. DOI: 10.3390/ijms23031810.

Weisburg, W.G., Tully, J.G., Rose, D.L., Petzel, J.P., Oyaizu, H. *et al.* (1989) A phylogenetic analysis of the mycoplasmas: basis for their classification. *Journal of Bacteriology* 171, 6455–6467. DOI: 10.1128/jb.171.12.6455-6467.1989.

White, D.T., Blackall, L.L., Scott, P.T., and Walsh, K.B. (1998) Phylogenetic positions of phytoplasmas associated with dieback, yellow crinkle and mosaic diseases of papaya, and their proposed inclusion in '*Candidatus* Phytoplasma australiense' and a new taxon, '*Candidatus* Phytoplasma australasia'. *International Journal of Systematic Bacteriology* 48, 941–951. DOI: 10.1099/00207713-48-3-941.

Win, N.K., Lee, S.Y., Bertaccini, A., Namba, S., and Jung, H.Y. (2013) '*Candidatus* Phytoplasma balanitae' associated with witches'-broom disease of *Balanites triflora*. *International Journal of Systematic and Evolutionary Microbiology* 63, 636-640. DOI: 10.1099/ijs.0.041467-0.

Woese, C.R. (2000) Interpreting the universal phylogenetic tree. *Proceedings of the National Academy of Sciences of the USA* 97, 8392–8396. DOI: 10.1073/pnas.97.15.8392.

Woese, C.R., Maniloff, J., and Zablen, L.B. (1980) Phylogenetic analysis of the mycoplasmas. *Proceedings of the National Academy of Sciences of the USA* 77, 494–498. DOI: 10.1073/pnas.77.1.494.

Wu, W., Cai, H., Wei, W., Davis, R.E., Lee, I.M., Chen, H., and Zhao, Y. (2012) Identification of two new phylogenetically distant phytoplasmas from *Senna surattensis* plants exhibiting stem fasciation and shoot proliferation symptoms. *Annals of Applied Biology* 160, 25–34. DOI: 10.1111/j.1744-7348.2011.00519.x.

Zhao, Y. and Davis, R.E. (2016) Criteria for phytoplasma 16Sr group/subgroup delineation and the need of a platform for proper registration of new groups and subgroups. *International Journal of Systematic and Evolutionary Microbiology* 66, 2121–2123. DOI: 10.1099/ijsem.0.000999.

Zhao, Y., Davis, R.E., and Lee, I.-M. (2005) Phylogenetic positions of '*Candidatus* Phytoplasma asteris' and *Spiroplasma kunkelii* as inferred from multiple sets of concatenated core house-keeping proteins. *International Journal of Systematic and Evolutionary Microbiology* 55, 2131–2141. DOI: 10.1099/ijs.0.63655-0.

Zhao, Y., Sun, Q., Wei, W., Davis, R.E., Wu, W., and Liu, Q. (2009a) '*Candidatus* Phytoplasma tamaricis', a novel taxon discovered in witches'-broom diseased salt cedar (*Tamarix chinensis* Lour.). *International Journal of Systematic and Evolutionary Microbiology* 59, 2496–2504. DOI: 10.1099/ijs.0.010256-0.

Zhao, Y., Wei, W., Lee, I.-M., Shao, J., Suo, X., and Davis, R.E. (2009b) Construction of an interactive online phytoplasma classification tool, *i*PhyClassifier, and its application in analysis of the peach X-disease phytoplasma group (16SrIII). *International Journal of Systematic and Evolutionary Microbiology* 59, 2582–2593. DOI: 10.1099/ijs.0.010249-0.

Zhao, Y., Wei, W., Davis, R.E., and Lee, I.-M. (2010) Recent advances in 16S rRNA gene-based phytoplasma differentiation, classification and taxonomy. In: Weintraub, P. and Jones, P. (eds) *Phytoplasmas: Genomes, Plant Hosts and Vectors*. CABI International, Wallingford, UK, pp. 64–92. DOI: 10.1079/9781845935306.0064.

Zhao, Y., Davis, R.E., Wei, W., Shao, J., and Jomantiene, R. (2014) Phytoplasma genomes: Evolution through mutually complementary mechanisms, gene loss and horizontal acquisition. In: Gross, D., Lichens-Park, A. and Kole, C. (eds) *Genomics of Plant-Associated Bacteria*.: Springer, Heidelberg, pp. 234–271. DOI: 10.1007/978-3-642-55378-3_10.

Zhao, Y., Davis, R.E., Wei, W., and Lee, I.-M. (2015) Should '*Candidatus* Phytoplasma' be retained within the order Acholeplasmatales? *International Journal of Systematic and Evolutionary Microbiology* 65, 1075–1082. DOI: 10.1099/ijs.0.000060.

Zhao, Y., Wei, W., Davis, R.E., Lee, I.-M., and Bottner-Parker, K.D. (2021) The agent associated with blue dwarf disease in wheat represents a new phytoplasma taxon, '*Candidatus* Phytoplasma tritici'. *International Journal of Systematic and Evolutionary Microbiology* 71. DOI: 10.1099/ijsem.0.004604.

Zreik, L., Carle, P., Bové, J.M., and Garnier, M. (1995) Characterization of the mycoplasmalike organism associated with witches'-broom disease of lime and proposition of a *Candidatus* taxon for the organism, '*Candidatus* Phytoplasma aurantifolia'. *International Journal of Systematic Bacteriology* 45, 449–453. DOI: 10.1099/00207713-45-3-449.

Zwolińska, A., Krawczyk, K., Borodynko-Filas, N., and Pospieszny, H. (2019) Non-crop sources of rapeseed phyllody phytoplasma ('*Candidatus* Phytoplasma asteris': 16SrI-B and 16SrI-(B/L)L), and closely related strains. *Crop Protection* 119, 59–68. DOI: 10.1016/j.cropro.2019.01.010.

5 Phytoplasma Phylogeny and Detection: The 16S rRNA and Alternate Genes

Jennifer Hodgetts[1*] **and Matthew Dickinson**[2]
[1]*Department of Epidemiological Sciences, Animal and Plant Health Agency, Addlestone, UK;* [2]*School of Biosciences, University of Nottingham, Loughborough, UK*

Abstract
Classification of phytoplasma species has been based upon sequence analysis of the 16S rRNA gene since the first use of DNA sequencing. Whilst whole-genome sequencing has become commonplace for a wide range of taxa, intrinsic features of phytoplasmas mean the widescale and routine application of this method remains elusive. However, precise taxonomic identification based solely on the highly conserved 16S rRNA has limitations. As a result, numerous other genes have been studied to aid in the discrimination of species and strains/isolates. This chapter reviews the most commonly used genes, describing and discussing their applications, and the range of diagnostic approaches that have been developed, both using the 16S rRNA and a range of other markers.

Introduction

Phytoplasma diagnostics and phylogenetics have historically been based on the 16S rRNA gene and the 16S–23S rRNA spacer region because of the availability of universal primers for this region. However, it is well known that phylogenetics based on a single, highly conserved gene has limitations, particularly when it comes to defining subgroups and strain differences within subgroups. The approach that is now being used for most other organisms is to rebuild trees by combining sequence data from a range of different genes (multilocus sequence analysis (MLSA)) and increasingly, phylogenomics, utilizing whole-genome sequences for phylogenetic analysis. However, the cost and technical challenges of generating whole-genome sequences for phytoplasmas have meant that phylogenomics is not routinely used. Furthermore, finding universal primers that can be used for additional genes in the phytoplasmas has proven difficult because of the sequence divergence between strains within the genus. Primers for genes such as *tuf* and the *rp* operon have been designed for use within specific 16Sr groups and have been useful for subgroup discrimination. Semi-universal primers have been designed for the *rp* operon, *tuf*, *cpn60*, and *rpoB* genes, and universal primers for the *secA* gene. Results from these genes confirm the phylogenetic groupings based on the 16S rRNA gene but provide better resolution between isolates and confirm the extensive variation in both DNA and amino acid sequences between different phytoplasmas. These new sequences are also being developed into alternative phytoplasma diagnostic systems which are aimed at providing rapid methods for both identifying the

*Corresponding author: jennifer.hodgetts@apha.gov.uk

 Phytoplasmas: Genomes, Plant Hosts and Vectors, 2nd Edition. (Ed. W. Wei)
DOI: 10.1079/9781800627031.0005

presence of phytoplasmas and assigning them to specific phylogenetic groups.

Phylogenetics and Diagnostics Based on the rRNA Operon

16S rRNA gene

The rRNA operon in bacteria consists of the 16S rRNA gene followed by an internal transcribed spacer (ITS) and the 23S rRNA gene. In the case of the phytoplasmas, there are two copies of this operon, and the ITS typically contains a single tRNA within it. Since the application of molecular biology, numerous PCR primer combinations have been devised to amplify parts of this operon for identification, diagnostics, and phylogenetics (Fig. 5.1).

Some of these, such as P1 and P7, are universal primers that work on DNA from all phytoplasma phylogenetic groups, whilst others are group-specific (Smart *et al.*, 1996; Firrao *et al.*, 2005). However, diagnostics based on these primers can be problematic, with occasional false positives, particularly amplifying any *Bacillus* spp. that might be present in a plant sample (Harrison *et al.*, 2002). In addition, it is important to guard against false negatives since there is no internal control built into the diagnostic test to confirm that a negative result is due to a lack of phytoplasma and not PCR inhibition. The presence of false negatives due to PCR inhibitors can be tested through "spiking" controls, in which DNA preparations are tested combined with DNA that is known to amplify. Still, this approach does not rule out the possibility that the DNA extraction might have been unsuccessful and/or the DNA degraded. Therefore, separate assays using primers such as those for the plant cytochrome C oxidase (COX assay) should be used. However, both of these approaches are expensive and labour-intensive, especially if many samples are to be tested, for example, in screening assays to identify insect vectors. More recently, real-time PCR assays have been developed for both generic and specific phytoplasma detection based on the 16S rRNA gene (amongst others). Real-time PCR assays have the advantage of being better automated and less labour-intensive than conventional PCR (Rubio *et al.*, 2017), such that appropriate controls can

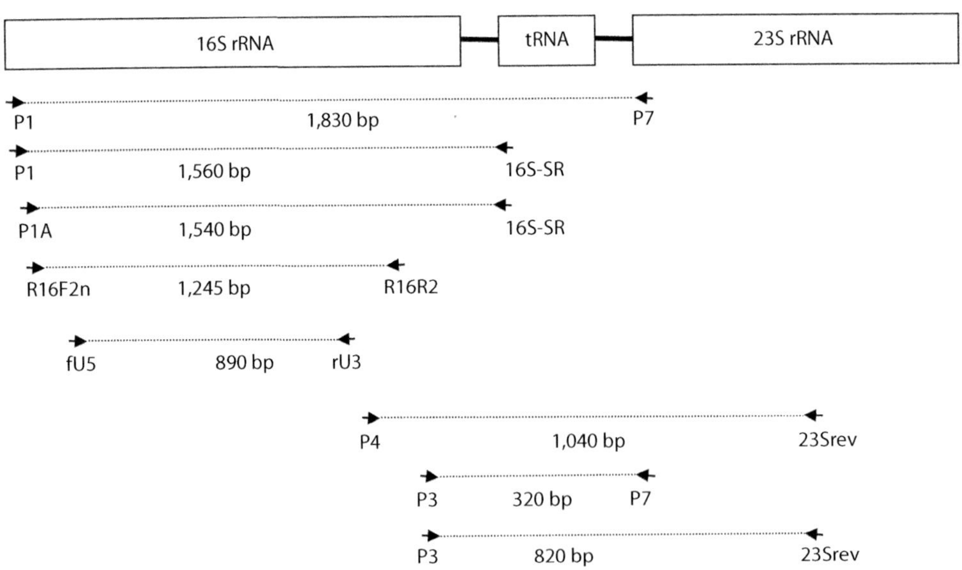

Fig. 5.1. Diagrammatic representation of the 16S–23S rRNA operon, demonstrating the position of some of the various universal primers that have been developed for PCR amplification of this region from phytoplasmas. Primer names are given under the arrows, and the sizes of the expected amplicons are shown between the dotted lines. Not drawn to scale.

be conducted more easily. In addition, a 16S rRNA gene-based oligonucleotide array system consisting of 21–33 nt long oligonucleotides has been developed to identify samples from different subgroups (Nicolaisen and Bertaccini, 2007).

Over the years during which phytoplasmas have been studied by molecular means, two parallel systems for classification have been developed, the 16Sr group classification system (e.g. 16SrI) and the '*Candidatus* Phytoplasma' taxon system (e.g. '*Ca.* P. asteris'), reflecting the evolution and mainstream deployment of different molecular methods. Both systems are based upon analysis of the 16S rRNA gene and have been used in parallel for many years, leading to some taxonomic confusion. A recent suggestion is that the systems serve subtly different purposes: the '*Ca.* Phytoplasma' taxon system for nomenclature (i.e. species delineation and description), and the 16Sr group classification system for classification (i.e. groups and subgroups) (Wei and Zhao, 2022). Unfortunately, the complexity of these systems often leads to publications describing strains by only one system, which creates further uncertainty around precise strain identification.

Initially, the most widely used approach was the 16Sr group classification system, which is based on restriction enzyme digest profiles of the 16S rRNA PCR products. In this system, universal primers are used to amplify a specified region of the 16S rRNA gene, and the PCR product is digested with numerous (17) specific restriction enzymes and the profiles analyzed, normally using polyacrylamide gel electrophoresis to achieve good resolution of the digest products. The restriction fragment length polymorphism (RFLP) profiles are then compared, and specific patterns are used to classify phytoplasmas into groups (with roman numerals) and subgroups (with letters) (Lee *et al.*, 1998; Wei *et al.*, 2007). However, problems with maintaining reference strains have made it difficult to undertake conventional RFLP analysis and compare patterns by gel electrophoresis for all the different groups and subgroups. This also makes it problematic to incorporate new strains and subgroups into the system as they are described. Furthermore, obtaining the required resolution for accurate RFLP profile size determination can be technically demanding, and without direct comparison with all existing strains, interpretation is challenging. Therefore, performing *in vitro* RFLP is operationally inconvenient, and its use is declining (Wei and Zhao, 2022).

As a result, Wei *et al.* (2007, 2008) devised a system based on sequences deposited at the National Center for Biotechnology Information's (NCBI) nucleotide sequence database, in which the restriction digestion profiles are simulated for a broad range of enzymes using *in silico* methods such as the AcaClone pDRAW32 system (http://acaclone.com; last accessed August 2025). Similarity coefficients for the restriction fragments are then calculated by defined formulae (Nei and Li, 1979; Lee *et al.*, 1998; Wei *et al.*, 2008) to reflect the number of shared and distinct fragments between any given strains. When completed, this analysis assigned ten new 16Sr groups, classified as 16SrI to 16SrXXVIII (Wei *et al.*, 2007). The so-called "virtual RFLP" approach was widely adopted by laboratories with access to DNA sequencing and later led to the evolution of this method into the *i*PhyClassifier (Zhao *et al.*, 2013), an online interactive tool for classification and taxonomic assignment. The *in silico* system has the potential to work well for sequences of new phytoplasmas in that they can be readily assigned to appropriate groups or subgroups. However, it does require the complete and accurate sequencing of a 1.25 kb region of the 16S rRNA gene to ensure that the RFLP analysis is correct. While this is possible in some phytoplasma diagnostics laboratories, it is not appropriate as a routine diagnostic technique in laboratories in many countries where equipment and facilities are limited. There is also a real danger that slight misreads during sequencing will result in sequences being used for the *in silico* analysis that are not accurate, resulting, in turn, in inaccurate restriction profiles and phylogenetic classifications. Unfortunately, there are already a number of phytoplasma 16S rRNA gene sequences in the NCBI databases that clearly contain errors and/or are incorrectly annotated, so care needs to be taken in using this approach. There is a case to be made that phytoplasma sequences should be independently confirmed before they are submitted to databases, particularly if there is an indication that they may result in a new group or subgroup. A further complication with basing a system on RFLP analysis can come from the presence of heterogeneity between the two copies of the 16S rRNA gene, as was shown for *Phormium* yellow leaf phytoplasma (Liefting *et al.*, 1996).

Developed in parallel with the 16Sr classification system is the '*Candidatus* Phytoplasma' taxon system, in which there are currently around 50 formally described species. This approach is also based primarily on the 16S rRNA gene sequence and was first proposed in 2004 (IRPCM, 2004), more recently updated (Bertaccini *et al.*, 2022), and is the system that should be used for formal taxonomic species description. Strains within a species should share at least 98.65% sequence identity within the 16S rRNA gene (previously, this was set at 97.5%) (IRPCM, 2004; Firrao *et al.*, 2005; Bertaccini *et al.*, 2022). Phylogenetic analysis of the entire length of the 16S rRNA gene to assign strains to species has the advantage of including all sequence variation within the gene fragment, compared with *in silico* RFLP where only sequence differences present at the 17 restriction enzyme sites are analyzed, making this a more powerful tool. However, this system also recognizes that 16S rRNA sequences alone are not sufficient for defining species, and other criteria such as host ranges and/or vectors are used to separate species; for example, '*Ca.* P. mali' and '*Ca.* P. pyri,' which would be within a single species based on the 16S rRNA sequences alone. Indeed, the recent revision to the guidelines also highlights the importance of using non-16S rRNA genes, with recommendations to use other housekeeping genes in instances where strains share >98.65% similarity in 16S rRNA gene sequences (Bertaccini *et al.*, 2022). This mirrors the approach taken with other bacterial genera, where it is acknowledged that housekeeping genes are useful for differentiating between lineages due to their more rapid evolution (Ogier *et al.*, 2019). Clearly, nomenclature and classification systems in the future cannot be based on the 16S rRNA gene alone, and this chapter will now detail some of the other genes that are being used to help provide a more robust approach to phytoplasma phylogenetics and classifications.

Other sequences derived from the rRNA operon

While the 16S rRNA gene-based systems have provided a solid basis for phytoplasma diagnostics and phylogenetics, there have been ongoing attempts to devise and develop alternative primers based on other phytoplasma genes, which might be better for distinguishing between different isolates and provide further definition of the groups and subgroups. Initially these studies were based on sequences adjacent to the 16S rRNA gene, in particular the 16S–23S ITS region, with a universal primer P7 developed that could be used as a reverse primer from the 23S rRNA (Smart *et al.*, 1996) (Fig. 5.1). Sequencing of the spacer region clearly showed that there was much more sequence variation in this region than in the 16S rRNA gene, presumably because there are fewer evolutionary constraints on the spacer. However, certain anomalies and inconsistencies also became apparent. Phylogenetic trees based on the ITS region alone are poorly supported by bootstrap values and place certain isolates in the wrong groups (Hodgetts *et al.*, 2008), and there is also evidence that variations occur in this region between the two copies of the rRNA operon within particular phytoplasmas. For example, one of the two rRNA operons in *Stylosanthes* little leaf has been shown to lack a tRNAIle in the 16S–23S ITS region (De La Rue *et al.*, 2001; Rodrigues Jardim *et al.*, 2021).

Universal primers were developed to extend the sequencing of the phytoplasma rRNA operon further into the 23S rRNA gene (Hodgetts *et al.*, 2007). The 23Srev primer, which was developed for universal amplification of bacterial 23S rRNA genes (Anthony *et al.*, 2000), has been shown to work well on phytoplasmas in combination with phytoplasma-specific forward primers and has provided an additional 500 bp of 23S rRNA gene-sequence information for a number of different phytoplasmas beyond the P7 primer (Fig. 5.1). The phylogenetic trees based on these partial 23S rRNA gene sequences are similar to those obtained for 16S rRNA gene sequences. However, there are a few minor anomalies (Hodgetts *et al.*, 2008), probably due to the relatively short nature of this 23S rRNA sequence (500 bp) when compared with the 16S rRNA gene (approximately 1500 bp). The study of this gene region allowed the development of a universal real-time PCR assay which is useful for high-throughput laboratory diagnostics, avoiding some of the issues with 16S rRNA-based detection assays (cross-reaction with non-target species), and allows the leveraging of advantages of real-time PCR compared with conventional PCR (Hodgetts *et al.*, 2009). Universal primers for the 23S rRNA gene have recently been developed for rapid, in-field diagnostics using loop-mediated isothermal amplification (LAMP) (Akahori *et al.*, 2024), detecting 31 '*Ca.* Phytoplasma' species. However, whilst useful for diagnostics, due to the short amplicon size these primers are not suitable for phylogenetic analysis or classification. The entire 23S rRNA gene is approximately 2850 bp, and it may ultimately be possible to design universal primers that amplify this entire sequence for all phytoplasmas and can then be exploited for phylogenetic analysis.

Alternative Genes for Group and Subgroup Identification

Primers based on the 16S rRNA gene, in combination with RFLP analysis, have proved very useful for the identification of groups and subgroups of phytoplasmas and have been the backbone of initial identification work since phytoplasmas were first discovered. However, DNA sequencing has shown that there are differences in the sequences between isolates within subgroups that cannot be identified by digestion with the conventional restriction enzymes and gel systems commonly used for RFLP analysis. There are also issues when defining subgroups, particularly if these are based on a single, well-conserved gene, and by setting the similarity coefficient at different levels, it is possible to change group and subgroup structures. The '*Ca.* Phytoplasma' definition system tries to take account of these issues by basing species on more than just gene sequences, using criteria such as host range and vector as well.

The approach that is now being used in other organisms, such as, for example, to define the "tree of eukaryotes," is to rebuild trees that were originally based on single genes by combining sequence data for a number of different genes (Keeling *et al.*, 2005). This can be achieved by concatenating the sequence data from a number of genes into a single sequence and building the phylogeny based on this or, alternatively, the trees derived for the different genes can be combined using appropriate software into larger "supertrees." For example, James *et al.* (2006) used a dataset in which six genes are combined (nu-SSU, nu-LSU, 5.8S rRNA, *rpb1*, *rpb2*, and *tef1*) to construct a comprehensive phylogenetic analysis of the fungi.

In the phytoplasmas, initial attempts to add alternative genes into phylogenetic studies were performed using genes such as the ribosomal protein genes *rpl22* and *rps3* (Lim and Sears, 1992; Gundersen *et al.*, 1994; Toth *et al.*, 1994) and the *tuf* gene (Schneider *et al.*, 1997). The primers used for these studies were generally designed based on the sequences that were available at the time for culturable mollicutes such as *Acholeplasma laidlawii*, *Spiroplasma citri*, and *Mycoplasma mycoides* and were found to work well for a number of phytoplasma taxonomic groups but not for all, as is discussed in more detail here.

A study by Wang *et al.* (2003) used molecular evolution analysis of the 16S rRNA and 16/23S spacer region to hypothesize that phytoplasmas evolved independently into two branches from a common ancestor. They named these the "AS branch" containing the aster yellows (16SrI) and stolbur (16SrXII) isolates, and the "WB branch" containing all other isolates. Subsequently, as complete genome sequences have become available for different phytoplasmas, it has been possible to try and design primers for other genes based on sequence alignments. However, the first publicly available phytoplasma genome sequences – from two 16SrI aster yellows phytoplasmas (onion yellows (Oshima *et al.*, 2004) and aster yellows witches'-broom (Bai *et al.*, 2006)), 16SrXII '*Ca.* P. australiense' (Tran-Nguyen *et al.*, 2008), and 16SrX apple proliferation (Kube *et al.*, 2008) – demonstrated that challenges would continue despite the additional genomic data. The first three isolates are in the AS branch of phytoplasmas, which generally have genomes that are larger than the other phytoplasmas, on average by 230 kb. Sequencing and comparative genomics have shown that these three genomes are circular and contain large numbers of potential mobile units (PMUs), which comprise up to 23% of these phytoplasma genomes and exhibit features of bacterial pathogenicity islands. Furthermore, whilst roughly 30% of the genome shows good synteny between the "closely related" 16SrI onion yellows and aster yellows witches'-broom genomes, less than 10% of the genome shows good synteny between these isolates and that of the 16SrXII '*Ca.* P. australiense' genome. The apple proliferation phytoplasma, however, is in the WB branch of phytoplasmas, which generally have smaller genomes with less repetitive DNA and are believed to have evolved independently. This genome is linear, has only one partial PMU sequence, and shows little synteny with the AS branch phytoplasmas. In phylogenetic trees, apple proliferation and the other 16SrX phytoplasmas also form a quite distinct phylogenetic grouping closer to the AS group than the other WB phytoplasmas.

Technological improvements and decreasing costs make the rapid production of whole-genome sequences for many other bacterial plant pathogens commonplace (Harrison *et al.*, 2023). However, despite substantive efforts to generate complete, closed phytoplasma genomes, technical challenges due to the inherent nature of phytoplasmas (Musić *et al.*, 2019) unfortunately mean there is still only partial sequence information available for many more diverse phytoplasma groups. Between the first genome publication in 2004 and 2021, only seven complete and around 30 draft genomes were published (Tan *et al.*, 2021). Furthermore, partial genomes limit the comparative genomics analysis that can be conducted (Ricker *et al.*, 2012) and mean these resources cannot be fully exploited. Nonetheless, evidence mounts to suggest that these other phytoplasmas have gene sequences that are very poorly conserved in comparison with the published whole-genome sequences. For example, a comparison of the complete genome sequence of two closely related isolates of '*Ca.* P. australiense' (from cottonbush in Australia and strawberry in New Zealand) revealed high genome plasticity, with an 80 kbp size difference largely due to the number and size of PMUs (Andersen *et al.*, 2013). Another analysis of inferred phylogeny using a set of 84 core genes conserved among the published genome sequences provided better resolution for resolving close relationships. However, this also demonstrated that sequence divergence rates were extremely high among phytoplasmas, with an average nucleotide identity (ANI) and average amino acid identity (AAI) analysis suggesting some isolates are too divergent to be considered as belonging to the same genus (Musić *et al.*, 2019). Interesting, despite the two isolates of '*Ca.* P. australiense' showing large overall differences, the variation between housekeeping genes was generally less than 1% at a nucleotide level (Andersen *et al.*, 2013). The consequence of this is that in attempts that have made to align sequences for known genes from published phytoplasma sequences and to design primers based on these that are universally applicable to all phytoplasmas, it has proven very difficult to find any well-conserved sequences to design such universal primers from. It is even more difficult to design the two sets of universal primers required for the nested PCR assays that are generally required for detecting phytoplasmas at low titre. This section will now discuss in more detail the primers that have been designed and used to date for genes other than the rRNA operon.

Ribosomal protein (*rp*) operon genes

The initial cloning of *rp* genes from a phytoplasma was achieved by using a heterologous probe derived from the *Mycoplasma capricolum* ribosomal protein gene operon to isolate a DNA fragment encoding the proteins rpL2 and rpS19 from a phytoplasma infecting *Oenothera* (Lim and Sears, 1991). Subsequently, the same approach was used to clone a second DNA fragment, encoding proteins rpL22 and rpS3 (Lim and Sears, 1992). The *rp* genes are part of a single, large operon, the *rp* operon, which contains at least 21 genes in phytoplasmas sequenced to date. In the literature, these genes and the ribosomal proteins they encode are sometimes given alternative names. Still, the general order that they are found in the phytoplasma genome is: *rpsJ* (encoding ribosomal protein S10), *rplC* (L3), *rplD* (L4), *rplW* (L23), *rplB* (L2), *rpsS* (S19), *rplV* (L22), *rpsC* (S3), *rplP* (L16), *rpmC* (L29), *rpsQ* (S17), *rplN* (L14), *rplX* (L24), *rplE* (L5), *rpsN* (S14), *rpsH* (S8), *rplF* (L6), *rplR* (L18), *rpsE* (S5), *rpmD* (L30), and *rplO* (L15).

Based on this initial *Oenothera* phytoplasma sequencing, primers rpF1 (5′-GGACATAAGTTAGGTGAATTT-3′) and rpR1 (5′-ACGATATTTAGTTCTTTTTGG-3′) were designed, which were initially shown to amplify the *rplV* (L22), *rpsC* (S3), and part of the *rplP* (L16) gene from

Acholeplasma laidlawii. Gundersen *et al.* (1994) extended these studies and used these primers along with a second set of primers, rpF2 (5′-TCTCGTACTTTTCGTGG-3′) and rpR2 (5′-ACCTTTAGCTCTTGGAA-3′), to amplify a 1245–1389 bp region encompassing the same genes from 16SrI, III, V, VII, VIII, IX, and X group phytoplasmas. However, there were some phylogenetic groups, notably the 16SrII group, for which these primers failed to amplify PCR products. Additional primers have subsequently been designed such that, by using the right combinations of primers, it is now possible to amplify these same genes from phytoplasmas belonging to all the phylogenetic groups, and these primers are referred to as semi-universal (Martini *et al.*, 2007). In particular, the degenerate primer pair rpF1C (5′-ATGGTDGGDCAYAARTTAGG-3′) and rp(I)R1A (5′-GTTCTTTTTGGCATTAACAT-3′) have been shown to amplify a 1212–1386 bp product for these genes from groups 16SrI, II, III, IV, V, VI, VII, IX, X, XII, XIII, and XVIII. Primers for these genes have also been designed that are group-specific. In addition, Martini *et al.* (2007) have designed a forward primer rpL2F3 (5′-WCCTTGGGGYAAAAAAGCTC-3′) from the *rplB* (L2) gene that was originally identified by Lim and Sears (1991). This primer works in combination with primer rp(I)R1A to amplify a 1600 bp product encompassing the 3′ end of the *rplB* (L2) gene, the *rpsS* (S19), *rplV* (L22), *rpsC* (S3), and *rplP* (L16) genes, and the 5′ end of the *rpmC* (L29) gene. These primers have been shown to work on groups 16SrI, III, IV, V, VI, VII, IX, X, XII, XIII, and XVIII phytoplasmas but not on isolates from groups 16SrII or XI.

The PCR products amplified using these *rp* operon primers have been used in RFLP analysis in combination with 16S rRNA gene RFLP analysis to assign phytoplasmas to 16Sr-rp subgroups (Lee *et al.*, 1998). Once sequencing became more accessible, researchers also used them to produce phylogenetic trees that are nearly congruent with those derived from 16S rRNA sequencing (Martini *et al.*, 2007). Furthermore, the added resolution provided by genes encoding proteins has enabled the subdividing of the groups into more distinct subclades than the 16S rRNA gene. For example, in separate studies, the 16SrII, IV, V, IX, XII, and XVIII groups have all been resolved into further distinct subclades by *rp* gene analysis. As more strains are studied with alternate genes, the ability of these to allow the identification of distinct genetic lineages that could not be readily resolved by analysis of the 16S rRNA gene sequence has been repeatedly demonstrated. For example, in a multigene study of the pigeon pea witches'-broom group (16SrIX), the sequence similarities between strains within subgroup 16SrIX-C were found to be 99.6% (based on the 16S rRNA gene) and 94.5% (based on the *rp* gene), and 99.5% (16S rRNA gene) compared with 98.2% (*rp* gene) within 16SrIX-E strains (Lee *et al.*, 2012). It is possible that other *rp* gene universal primers could be designed in the future based on the phytoplasma genome sequences, but as yet, no such primer sequences have been published, and many studies effectively utilize the existing primers.

Elongation factor Tu (*tuf*) gene

The *tuf* gene, encoding the elongation factor Tu (EF-Tu), is a well-conserved gene with a central role in translation (Schneider *et al.*, 1997), and there is a single copy of this gene in the phytoplasma genome. This gene has often been used in phylogenetic studies for other bacteria. Two primers for the amplification of the *tuf* gene, fTuf1 (5′-CACATTGACCACGGTAAAAC-3′) and rTuf1 (5′-CCACCTTCACGAATAGAGAAC-3′), were initially designed for phytoplasmas based on alignment of the *Mycoplasma pneumoniae*, *M. genitalium*, *M. gallisepticum*, *M. hominis*, and *Ureaplasma urealyticum* sequences. These primers amplified products of the expected size (1000 bp) for the 16SrI aster yellows, 16SrIII green valley X and vaccinium witches'-broom, and 16SrXII stolbur groups but failed to amplify from 16SrII faba bean phyllody and 16SrX apple proliferation (Schneider *et al.*, 1997). It has been subsequently found that these primers also fail to amplify from the 16SrIV coconut lethal yellowing and 16SrXXII coconut lethal decline phytoplasmas.

Additional primers based on these early phytoplasma sequences were subsequently designed, but none have been found that can act as universal primers. The main use of *tuf* gene primers was subsequently, and remains today, to establish subgroups within the 16Sr groups particularly within the 16SrI aster yellows group (Marcone *et al.*, 2000) and the 16SrXII '*Ca.* P. australiense' group (Streten and Gibb, 2005). In a study on the aster yellows group, the AY-specific primers fTufAy (5′-GCTAAAAGTAGAGCTTATGA-3′) and rTufAy (5′-CGTTGTCACCTGGCATTACC-3′) (Schneider *et al.*, 1997), which amplify a 940 bp product, were used on 70 phytoplasma isolates in conjunction with the 16S rRNA gene primers (Marcone *et al.*, 2000). RFLP analyses of the 16S rRNA gene PCR products divided the isolates into ten 16Sr subgroups (-A, -B, -C, -D, -E, -F, -K, -L, -M, and -N), whilst the *tuf* gene RFLP profiles only resulted in seven subgroups (-A, -B, -C, -D, -E, -F, and -G). In particular, the 16Sr subgroups -B, -D, -L, and -M all gave the same *tuf* gene profiles. On the basis of these profiles, and in conjunction with pathological aspects, the authors concluded that the 16SrI-A, -B, -C, -D, -E, -F, -K, and -N subgroups were substantially supported as distinct subgroups. In a similar study on the '*Ca.* P. australiense' phytoplasmas, Streten and Gibb (2005) examined 11 strains that had previously been indistinguishable using 16S rRNA gene analysis, using the same aster yellows *tuf* gene primers (fTufAY and rTufAY) in conjunction with primers rpF1 and rpR1 for the *rplV* and *rpsC* genes. RFLP and sequence analysis were able to separate the isolates into three *tuf* gene subgroups (*tuf* Australia I, *tuf* New Zealand I, and *tuf* New Zealand II) and three *rp* groups (*rp*A, *rp*B, and *rp*C). By combining the data, four distinct subgroups were established.

Whilst these studies demonstrated the resolving power of the *tuf* gene, the long size of amplicons can mean they are impractical for routine sequencing-based identification (e.g. in diagnostic, rather than research, laboratories). With the increasing popularity of DNA barcoding and the sequencing of universally amplified short highly variable DNA markers, Makarova *et al.* (2012) assessed the suitability of the *tuf* gene for this application. Alignment of all the publicly available *tuf* gene sequences from the NCBI, representing groups 16SrI, III, IV, V, VII, VIII, X, and XII, allowed the identification of conserved regions within the *tuf* gene and allowed amplification of as many groups as possible. The approach of designing primer cocktails was used, whereby different variants of the same primer are mixed in equimolar amounts. Primers Tuf340 (two variants) and Tuf890 (three variants) were designed for the first-round PCR, and primers Tuf400 (five variants) and Tuf835 (three variants) for the nested PCR. A comparison of phylogenetic trees constructed from the resulting 420–444 bp tuf barcode and the standard 1.2 kbp fragment of the 16S rRNA gene was completed. This revealed that the *tuf* gene tree was highly congruent with the 16Sr RNA tree, had higher inter- and intragroup sequence divergence, and allowed separation of the main ribosomal groups and most of their subgroups. Ninety-one strains were tested; however, these represented only 13 of the 16Sr groups, so the true universality of these primers remains unclear, and the authors stressed the sequences were intended for identification of phytoplasmas and not for phylogenetic analysis.

A second elongation factor gene, *fus*, encoding EF-G, is adjacent to *tuf* in the phytoplasma genome and has also been used for the design of PCR primers (Berg and Seemüller, 1999). However, these primers, which were designed based on an apple proliferation sequence, were only able to amplify from 16SrI, III, X, and XII group phytoplasmas and not from 16SrII, V, or VII group phytoplasmas. There are currently no records of the use of these or other *fus* gene primers on any other phytoplasmas.

secY gene

The *secY* gene, which encodes a protein involved in the protein secretion mechanism from bacteria (in conjunction with SecA and other gene products), was initially identified in phytoplasmas in a randomly cloned 16SrV flavescence dorée DNA fragment that had originally been used as a hybridization probe (Daire *et al.*, 1997). Based on the sequencing of this fragment, a set of primers FD9f3/FD9r2 was designed to amplify a 1150 bp fragment that was found to be partly homologous to the *secY* translocase gene from mycoplasmas (Angelini *et al.*, 2003). This randomly cloned DNA fragment was subsequently used in a suppressive subtractive hybridization and genome walking approach to identify a larger clone that encoded the 3′ end of the *rplO* gene (encoding ribosomal protein L15), the entire *secY* gene along with the adjacent *map* (methionine amino peptidase) gene, and the 5′

end of the *infA* gene, encoding initiation factor If-1 (Arnaud *et al.*, 2007). This genome walking strategy was also used in the same study on a second randomly cloned PCR product to identify a clone containing the second half of the *uvrB* gene, encoding the subunit B of exonuclease ABC, and the nearly complete *degV* gene, encoding a protein of the DegV family.

Based on these sequences, primers were designed that could be used in an analysis of the 16SrV phytoplasmas infecting grapevine and alders in Europe (Arnaud *et al.*, 2007). Nested primer sets were used to amplify a 1174 bp fragment of the *secY* gene, an 803 bp fragment of the *map* gene, and a 1126 bp fragment of the *uvrB–degV* region from over 40 16SrV isolates. Phylogenetic trees were constructed from the sequences obtained. The resultant analysis consistently supported the existence of three distinct flavescence dorée strain clusters. Strain cluster FD1 had little variability and was present mainly in south-western France. Cluster FD2 showed no variability and was found in France and Italy, whilst cluster FD3 was highly variable and detected only in Italy. These results suggested that the FD2 strains may have been spread through the distribution of infected propagation material between France and Italy. In addition, the analyses showed that the alder yellows phytoplasmas and German Palatinate grapevine yellows phytoplasmas belong to the same phylogenetic subclade.

The *secY* gene has also been used for the differentiation of the aster yellows group phytoplasmas (Lee *et al.*, 2006). In this study, primers were designed based on the published aster yellows and onion yellows sequences to amplify a 1.4 kb near-full-length *secY* gene. As far as these authors are aware, these primers have only been tested on aster yellows group phytoplasmas, so they are probably specific for this group. Twenty representative 16SrI isolates were used in this study, representing the ten 16SrI subgroups (-A, -B, -C, -D, -E, -F, -K, -L, -M, and -N). Sequencing and RFLP analysis of the resultant PCR products delineated the isolates into ten SecY groups. Two of these SecY subgroups corresponded to the 16SrI-A subgroup. Members of the 16SrI-B subgroup were also split into two SecY subgroups, and the isolates from 16SrI-L and 16SrI-M also fell into one of these two SecY groupings, consistent with the findings from the earlier *tuf* gene analysis of Marcone *et al.* (2000). The remaining 16SrI subgroups and SecY subgroups coincided with each other.

Unlike previous studies using the *secY* gene, which only analyzed a single 16Sr group, a study by Lee *et al.* (2010) constructed a *secY* gene phylogeny including 83 strains from 12 different 16Sr groups, finding the results to be nearly congruent with that inferred by 16S rRNA gene-based phylogeny, and delineating distinct lineages that were irresolvable by 16S rRNA gene phylogeny. Nucleic acid sequence similarity between strains of different phytoplasma groups ranged from 53.5% to 77.9% for the *secY* gene compared with 85.1–96.9% for 16S rRNA gene sequences, and the analysis demonstrated that strains within groups 16SrII, III, and V each form two distinct genetically divergent lineages. Whilst this permitted finer differentiation of phytoplasma strains, the approach used was convoluted and entailed sequencing large segments of the *spc* ribosomal protein operon, using iterative rounds of primer design and primers located in the flanking ribosomal protein *rpl15* (or *rplO*) and *map* genes. This was combined with cloning and primer walking to obtain the 2.2 kb (in 16SrII, III, IV, V, VI, VII, and X groups) or 2.8 kb (in 16SrI, XII, XIII, and XVI-II groups) multigene amplicon. The analysis demonstrated different gene orders between the differently sized amplicons and the absence of the *adk* gene in the 16Sr groups with the shorter sequence.

secA gene

The *secA* gene, which encodes SecA, the ATP-dependent force generator in the bacterial precursor protein translocation cascade system, was originally sequenced from the onion yellows phytoplasma as part of the genome sequencing project (Kakizawa *et al.*, 2001). In these studies, the 2505 bp gene, which encodes an 835 amino acid peptide, was cloned into an expression vector to produce tagged proteins that were used to raise antibodies for immunohistochemical studies. In subsequent work, Hodgetts *et al.* (2008) aligned the onion yellows *secA* gene sequence with that of aster yellows witches'-broom (16SrI group) and the equivalent gene from coconut lethal yellowing (16SrIV group). Based on these alignments, three degenerate primers were designed, SecAfor1 (5′-GARATGAAAACTGGRGAAGG-3′), SecAfor2 (5′-GAYGARGSWAGAAC KCCT-3′), and SecArev3 (5′-GTTTTRGCAGTTCCTGTCATNCC-3′) for use in a semi-nested PCR assay in an attempt to amplify a fragment of the *secA* gene from a broad range of phytoplasma phylogenetic groups.

When the primer pair SecAfor1/SecArev3 was used in PCR, products of the expected size (about 840 bp) were generated from approximately 90% of the 34 phytoplasma DNA samples tested, which included representative isolates from 13 16Sr groups (16SrI through to 16SrXIII with the exception of group 16SrVIII, and 16SrXXII). Some of the PCR products were of a low concentration, probably reflecting a low titre of phytoplasma DNA in these samples or possibly poor primer binding, but when a semi-nested approach was used, in which phytoplasma DNA samples were first amplified with SecAfor1/SecArev3 and then reamplified with SecAfor2/SecArev3, the expected PCR product (approximately 480 bp) was readily obtained from all 34 phytoplasmas. Subsequently, it has also been shown that these primers work on 16SrXIV group phytoplasmas and more targeted primers have been developed for improved amplification with strains from 16SrXI and XIV (Bekele *et al.*, 2011a). Sequencing of these PCR products confirmed that they all encode the SecA protein and that the region for the 16SrI group phytoplasmas is two amino acids longer than that from all the other phylogenetic groups.

Phylogenetic analysis based on the *secA* gene sequences has supported results obtained for other genes and provided improved resolution of groups and subgroups, particularly when compared with the 16S rRNA gene. The subgroups of 16SrI are clearly defined, and 16SrII clearly splits into a cluster that contains 16SrII-B and 16SrII-C isolates and a cluster that contains the 16SrII-D isolates. These results support the *in silico* results of Wei *et al.* (2007), who classified the 16SrII-B strains as '*Ca.* Phytoplasma aurantifolia' whilst classifying the 16SrII-D strains as '*Ca.* P. australasiae.' In addition, the results using the *secA* gene have clearly distinguished strains within the coconut lethal yellowing-type disease group. These phytoplasmas cause economically important lethal diseases of coconuts, all characterized by similar syndromes, which include premature fruit drop, floral necrosis, leaf discoloration, and decline. Subgroup 16SrIV-A phytoplasmas are associated with lethal yellowing of coconut and other palm species in the Americas, whereas phytoplasmas that induce similar symptoms on coconut in Africa are referred to by other names, such as Cape St Paul wilt in Ghana, Awka in Nigeria, and lethal decline in Tanzania. Strain differences were previously identified through 16S rRNA gene analysis. Mpunami *et al.* (1999) also showed that it was possible to differentiate East African from West African coconut-associated phytoplasmas by selective amplification of 16S rRNA gene sequences during PCR or by RFLP analysis of 16S rRNA gene PCR products, such that they were classified as groups 16SrIV-B and 16SrIV-C, respectively. In the *in silico* work of Wei *et al.* (2007), the 16SrIV-C Nigerian coconut lethal decline phytoplasma or Awka, which has an almost identical 16S rRNA gene sequence to the Cape St Paul wilt phytoplasma from Ghana, was allocated a distinct 16Sr group, 16SrXXII-A. The analysis by Hodgetts *et al.* (2008) of the *secA* gene confirmed the high degree of divergence between the different coconut phytoplasmas and supported the separation of the coconut lethal yellows diseases of the Americas from the coconut lethal decline diseases of Africa. It also indicated that the Tanzanian lethal decline phytoplasma should be put in a separate group to the Nigerian and Ghanaian lethal declines, which was subsequently reflected in the description of three species: (i) '*Ca.* P. palmae' with strains from the Americas, previously 16SrIV-A (Bertaccini *et al.*, 2022); (ii) '*Ca.* P. palmicola' with strains from Mozambique as 16SrXXII-A and Ghana and Côte d'Ivoire as 16SrXXII-B (Harrison *et al.*, 2014); and (iii) '*Ca.* P. cocostanzaniae' with strains from Tanzania, previously 16SrIV-B (Bertaccini *et al.*, 2022).

Typically, work has been moving away from *in vitro* RFLP. However, efforts were made to design new primers for RFLP analysis of a selection of phytoplasmas by combining the more variable *secA* gene with only two restriction enzymes (DdeI, MaeIII) but which have more complex recognition sites (Sneideris *et al.*, 2021). Sixteen primers were designed using publicly available *secA* gene sequences, and over 50 primer combinations were tested in efforts to amplify *secA* gene fragments from phytoplasmas belonging to

16SrI-C, I-L, I-M, III-A, V-B, and XXI-A subgroups. Despite the efforts, the difficulties of universal primer design were evident, as only two primer sets were successful at amplifying two strains (one primer set amplified only Canada peach X-disease, and a second primer set only amplified '*Ca.* P. ziziphi'), and a full-length sequence of the *secA* gene was only obtained for '*Ca.* P. asteris' (16SrI-C), and partial fragments for Canada peach X-disease (1997 bp) and for '*Ca.* P. ziziphi' JWB phytoplasma strains (1327 bp). These sequences were then used to design primer pair secAdF1/secAdR1, which amplify a *secA* gene fragment (approximately 1200 bp) of 14 phytoplasma strains. The resultant RFLP method allowed the identification and discrimination of phytoplasma strains from five groups (16SrI, III, V-B, XII-B, and XXI-A) but only a few of the subgroups within some groups (16SrI-A, -B, -C, -D, and 16SrIII-A, -B, -E, -F, -H), and the utility of these primers beyond this study remain unclear.

cpn60 gene

The *chaperonin 60* (*cpn60*) gene, also known as *groEL* or *hsp60*, is a single copy gene that has been used in a wide range of bacterial genera for strain characterization, with a 550 bp fragment of the gene proposed as a DNA barcode for the domain bacteria (Links *et al.*, 2012), demonstrating its potential for strain discrimination. The first study of *cpn60* in phytoplasmas was to evaluate its effectiveness at detecting variability in '*Ca.* P. asteris' related strains (Mitrović *et al.*, 2011). Alignments used publicly available sequences from '*Ca.* P. asteris'-related strains alongside '*Ca.* P. mali' and '*Ca.* P. australiense' to design primer AYgroelF (5′-GGCAAAGAAGCAAGAAAAG-3′) to be universal for all the phytoplasmas, and primer AYgroelR (5′-TTTAAGGGTTGTAAAAGTTG-3′) to be specific to aster yellows strains, producing a partial gene fragment of 1397 bp specifically in aster yellows phytoplasmas. Specificity testing against 16SrII-D, V-C, X-B, XII-A, and XII confirmed the primers only amplified 16SrI strains. Phylogenetic analyses using 27 strains demonstrated consistency in the clustering and relationships inferred from the *groEL* gene compared with the 16S rRNA, *rp*, and *secY* genes. However, RFLP analysis showed mixed results: with differentiation achieved among 16SrI-A, -B, -C, -F, and -P subgroups, along with further differentiation within strains 16SrI-A, -B, and -C, yielding nine cpn60 subgroups compared with five 16SrI subgroups, although 16SrI-B, -L, and -M subgroups could not be discriminated by the region.

Following this study, Dumonceaux *et al.* (2014) attempted to access the *cpn60* gene from a more genetically diverse range of phytoplasmas to expand the utility of studying genes in phytoplasmas. This was possible due to the advances in the study of other bacterial genera and the creation of universal primers to allow the use of the region as a DNA barcode. Two sets of highly degenerate universal primers were developed – H279 (5′-GAIIIIGCIGGIGAYGGIACIACIAC-3′), H280 (5′-YKIYKITCICCRAAICCIGGIGCYTT-3′), H1612 (5′-GAIIIIGCIGGYGACGGYACSACSAC-3′), and H1613 (5′-CGRCGRTCRCCGAAGCCSGGIGCCTT-3′) – amplifying a 549–567 bp region of the gene, named the cpn60 universal target (cpn60 UT), in a wide variety of bacterial groups (Hill *et al.*, 2006). Due to phytoplasma strains always being present in a plant/insect background, with the likely co-presence of other non-target bacterial species, these universal primers would not be suitable (specific) for phytoplasmas. Therefore, the corresponding primer annealing sites were analyzed in published phytoplasma sequences, and two new sets of primers designed, H279p (5′-GATIIIGCAGGIGATGGAACMACIAC-3′) and H280p (5′-TGRTTITCICCAAAACCAGGIGCATT-3′) primarily to 16SrI strains, and D0317 (5′-GATIIIKCIGGIGAYGGIACIACIAC-3′) and D0318 (5′-TGRTKITCICCAAAACYWGGIGCWTC-3′) to the peanut witches'-broom (PnWB) phytoplasma. Primer set H279p/H280p successfully generated amplicons from strains corresponding to '*Ca.* P. asteris,' '*Ca.* P. solani,' '*Ca.* P. mali,' '*Ca.* P. prunorum,' and '*Ca.* P. pyri' (16SrI, X and XII), but failed to amplify from other '*Ca.* Phytoplasma' species. The primer set D0317/D0318 successfully amplified from '*Ca.* Phytoplasma' species, including '*Ca.* P. phoenicium,' '*Ca.* P. fraxini,' and '*Ca.* P. ulmi' (16SrV, VII, and IX), which could not be amplified with the H279p/H280p primer pair. Combined, the primers successfully amplified the cpn60 UT from 43 strains representing seven 16Sr groups, which were highly divergent in sequence (61–98% identity), and phylogenetic trees generated using the cpn60 sequences were congruent with the 16S rRNA gene, with 12 distinct groupings. A later update demonstrated the primers also successfully amplified 16SrIV, XI, XIV, and XV, although it was not stated with which primer pair each group was amplified (Muirhead *et al.*, 2019). However, it has been reported that 16SrIII ('*Ca.* P. pruni') phytoplasmas lack the *cpn60* gene (in multiple strains with genome sequences), and with the large degree of sequence variation seen within the seven 16Sr groups analyzed, how universal these primers are across all 16Sr groups beyond the 12 tested (16SrI, II, IV, V, VII, and IX through to XV) remains to be seen.

The utility of *cpn60* gene sequences was further developed by Pérez-López *et al.* (2016), who used the increasing number of publicly available sequences for the region to develop and validate an *in silico* RFLP method for the gene, later updated into a web-based software named the CpnClassiPhyR (Muirhead *et al.*, 2019). Bioinformatic analysis with seven restriction enzymes of 96 sequences from strains representing 19 16Sr subgroups resulted in 25 distinctive RFLP profiles, and the authors established threshold similarity coefficients to delineate new cpn60 UT groups and subgroups. Phylogenetic analysis also showed a clear delineation of the three phytoplasma clades, with the gene providing a complementary classification. However, the lack of universal primers remains, despite the promise of the gene for classification.

RNA polymerase subunit B (*rpoB*) gene

The DNA-dependent RNA polymerases (DpRp) are complex enzymes that catalyze gene transcription in cells, with a structural core of four subunits (named α, β, β′, and ω) in bacteria. The β-subunit gene (*rpoB*) is a multidomain protein which exists as a single copy in the phytoplasma genome (Valiunas *et al.*, 2013). From the end of the 1990s onwards, publications successfully using the *rpoB* gene for classification and phylogenetic analysis studies in a wide range of prokaryote species became commonplace, where primers were typically designed in each study for the genus of interest.

The first attempt to study the *rpoB* gene in phytoplasmas (Valiunas *et al.*, 2013) took a similar approach. Using the strain clover phyllody (CPh) ('*Ca.* P. asteris,' 16SrI), DNA was extracted from phytoplasma-enriched preparations of plant sieve cells, partially digested with a restriction enzyme, and then cloned, with recombinant plasmids screened and inserts sequenced. A locus containing ribosomal protein genes (*rplA* (*rpl1*), *rplJ* (*rpl10*), and *rplL* (*rpl7*/*rpl12*)), a complete *rpoB* gene, and a partial *rpoC* gene were identified. Alignments of the *rpoB* gene sequence from CPh and a limited number of other genomes ('*Ca.* P. asteris' strains OY-M and AYWB, '*Ca.* P. mali' (16SrX), '*Ca.* P. australiense' (16SrXII) and PnWB (16SrII)) were used to identify conserved regions and six primers were designed for amplification of a segment of the *rpoB* gene. Primers rpoBF1 (5′-TGCCCAATTTAATTGAAATTC-3′) and rpoBR1 (5′-AAGACCAATTCGAAATTGG-3′) amplify a 1432 bp fragment and could be used as first-round primers, with three other primer combinations then used to amplify various strains, either directly or as nested/semi-nested PCRs. Primers rpoBF2 (5′-GATT GGTTTTTAAAACACGG-3′) and rpoBR1 amplified a 1396 bp fragment in 16SrI strains, rpoBF1 and rpoBR3 (5′-TTACCTAAATGATCGATATCATC-3′) amplified a 1377 bp fragment in 16SrIII, X, and XII (apart from XII-E) strains, and rpoBF4 (5′-TTTCTCAAATTGTACGTTCC-3′) and rpoBR2 (5′-ATTGGTTTTTTAACAATTCTCC-3′) amplified a 1090 bp fragment in 16SrXII-E strains. Partial *rpoB* gene sequences were obtained from 17 phytoplasma strains from groups 16SrI (subgroups -A, -B, -C, -L, and -M), III (subgroups -A, -E, -F, and -H), X-A, and XII (subgroups -A and -E). Phylogenetic analysis of both the full gene and the partial gene PCR amplicon of the *rpoB* sequences, along with deduced amino acid sequences, yielded trees congruent with those inferred from the 16S rRNA gene. Topologies were similar, with branch lengths indicating clear distinctions between closely related phytoplasma strains and the more variable *rpoB* gene sequences more clearly distinguishing among phytoplasmas. The partial 1.3 kb fragment of the *rpoB* gene (between primer pair rpoBF2/rpoBR3) showed 61.6–79.4% sequence identity between 16Sr groups and 95.1–98.8% sequence identity between strains of a given 16Sr subgroup (where present). Furthermore, virtual (*in silico*) RFLP analysis allowed improved distinction among closely related strains. Based on the demonstrated specificities of the primers and the number of primers required to amplify only five phytoplasma 16Sr groups, it would seem likely that application to other 16Sr groups will be challenging, and further primers would be required. While this study indicates the potential of the *rpoB* gene region for classification, the lack of further reports utilizing this gene perhaps reflects the limitations of the primers to broader phytoplasma groups.

Other genes

Over the years, numerous attempts have been made to identify other genes that could be used to design universal primers for phytoplasma amplification utilizing a range of different methods. Goodwin *et al.* (1994) used a plasmid that had been isolated from a 16SrI severe strain of western aster yellows by cesium chloride (CsCl) ethidium bromide gradients to design primers that could amplify a 235 bp fragment of the plasmid DNA. These primers were shown to work on several other strains of aster yellows, 16SrX pear decline, 16SrV elm yellows, and 16SrVI beet leafhopper-transmitted virescence from both plant and insect samples, and the PCR approach was at least 500 times more sensitive than previous tests using the plasmid DNA as a probe in dot blots. It was also possible to use restriction enzyme digests to differentiate between strains, and sequencing showed around 93% homology between the elm yellows and aster yellows plasmid sequence. However, plasmids have not been found in all phytoplasma strains, the gene from which the sequence came has not been identified, and whether this technique and these primers would be universally applicable is unknown.

At about the same time, Jarausch *et al.* (1994) sequenced a clone derived from the apple proliferation phytoplasma and found that one of the open reading frames encoded a protein with significant homology to bacterial nitroreductases. The primers derived from this sequence successfully amplified the same gene from other 16SrX phytoplasmas, allowing the researchers to use RFLP analysis to assign the isolates to different subgroups. However, when the same primers were used on DNA from isolates belonging to groups 16SrI, III, V, VII, and XII, no PCR products were obtained. Similarly, Streten and Gibb (2003) obtained random clones from 16SrII tomato big bud phytoplasma DNA and sequenced them to identify the putative genes present. Primers were then designed by aligning these sequences with those of equivalent genes from other phytoplasmas, mycoplasmas, and bacteria. Four sets of primers were designed targeting specific regions of the DNA polymerase beta II chain gene, the peptide chain release factor (RF-1) gene, the ATP-dependent RNA helicase gene, and the *FtsH* gene, which encodes a protein involved in cell division. These primers were tested on three isolates of tomato big bud phytoplasma and two isolates of the closely related 16SrII sweet potato little leaf, but none of the four primer sets were able to amplify from all five isolates. The authors of this chapter have subsequently tested these primers on a range of other phytoplasmas and confirmed that they only amplify from selected 16SrII isolates and are not universal phytoplasma primers. Wagner *et al.* (2001) used a different approach and designed primers based on known mycoplasma sequences for the RNase P gene, which encodes a protein essential for tRNA maturation. Following the sequencing of the PCR product that they obtained from apple proliferation, they refined the primers and were able to get an improved PCR product for complete sequencing. However, there are no reports of these primers being used on any other phytoplasma groups.

A more circuitous route has been used to obtain DNA sequences and primers for immunodominant membrane proteins of phytoplasmas. These proteins are on the outer surface of the phytoplasmas and are believed to be involved in mediating the specific attachment of these bacteria to their insect vectors. They were first identified when monoclonal antibodies were prepared against 16SrII sweet potato witches'-broom-infected plant material that had been enriched for phytoplasma by using leaf midribs (Shen and Lin, 1993). These antibodies were shown to react to an 18.4 kDa antigenic protein in western blots only from infected plants. They were then used to screen a genomic library made from CsCl–bisbenzimide-enriched phytoplasma DNA (Yu *et al.*, 1998). Protein expression was induced, and clones that expressed the protein were identified and sequenced. Using this approach, a 519 bp sequence that encoded the antigenic protein was identified, and primers were designed from this sequence and shown to amplify a PCR product of the correct size from the closely related PnWB phytoplasma. However, these primers failed to amplify anything from the other phytoplasma-infected plants tested, which included 16SrI, V, VIII, and XI. The same approach was used to obtain the apple proliferation immunodominant protein (IDP) (Berg *et al.*, 1999) and also that of Western X-disease (Blomquist *et al.*, 2001), but sequencing of these genes showed that there was no significant homology between these nor between them and other phytoplasma IDP genes. No primers were developed from this work for isolating IDP genes from other phytoplasmas.

For the 16SrI-B chlorante aster yellows (AY-C) phytoplasma, the gene was cloned by isolating the affinity-purified IDP on SDS-PAGE, fragmenting the protein chemically and enzymatically, and sequencing the cleavage products. Degenerate primers were then designed based on these amino acid sequences and used to amplify the AY-C *amp* gene (Barbara *et al.*, 2002). Primers designed for this gene were then used to clone the equivalent gene from the 16SrI-C clover phyllody phytoplasma. Although the intergenic regions flanking the gene and parts of the open reading frame showed high levels of sequence similarity between the two isolates, the predicted large hydrophilic domain exhibited low nucleic acid and amino acid conservation.

It is worth noting that the terminology for these major antigenic proteins can be confusing and that the terms antigenic membrane protein (Amp), immunodominant membrane protein (Imp), and immunodominant protein (Idp) are sometimes being used for the same protein and sometimes for different ones. In onion yellows, the *amp* gene (Amp) is located between the *GroEL* gene and the NAD+ synthase gene, whilst in aster yellows witches'-broom and '*Ca.* P. australiense' the gene in the same location is referred to as the immunodominant membrane protein. In '*Ca.* P. mali' there is no gene in an equivalent genomic position, and the identified immunodominant protein is located between the *dnaD* gene and the *pyrG* gene. In contrast, a gene occupies the same position in the onion yellows, aster yellows witches'-broom, and '*Ca.* P. australiense' genomes, but it is annotated as encoding a hypothetical protein. Therefore, the design of universal primers for phytoplasma immunogenic proteins is unlikely since they occur in at least two locations in phytoplasma genomes and are very dissimilar in sequence. The authors of this chapter have recently attempted to design primers from the sequences of genes flanking both of these major immunogenic proteins in the published phytoplasma genome sequences to determine whether this could be a strategy for isolating these genes from other phytoplasmas. However, none of the primers that have been designed have been successful.

These examples, spanning the mid-1990s to mid-2000s, have used molecular biology techniques that are now becoming less commonplace in laboratories, and attempts to identify new gene targets for phylogeny and/or identification are now primarily focused on exploitation of phytoplasma genome sequences. Some of these attempts have been to amplify the full-length gene sequence for specific genes from all phytoplasmas other than those already completely sequenced. To do this requires knowledge of the flanking genes so that primers can be designed from these to amplify into and through the gene of interest. However, studies have demonstrated that gene orders and synteny vary between phytoplasmas (Jomantiene *et al.*, 2007), including closely related strains such as AYWB and OY-M (Bai *et al.*, 2006), so it is therefore not possible, as yet, to predict which flanking genes to use for such a strategy to ensure success in amplifying complete genes from diverse groups of phytoplasmas. Instead, an alternate approach is to amplify parts of targeted genes using primers designed from well-conserved regions identified by aligning gene sequences from available genome data, including phylogenetically distant phytoplasmas. However, even with this strategy, it is often challenging to identify well-conserved sequences with balanced AT nucleotide content suitable for primer design and minimizing primer degeneracy, which can otherwise lead to the amplification of non-specific products.

With more partial/complete genome sequences becoming available, it is increasingly common for researchers to use these to attempt to design new primers for known or unknown targets. As a result, many other gene regions have now been studied in detail. Typically, the objective is to analyze a single 16Sr group (or, at most, a few), or a '*Ca.* Phytoplasma' species, often within an epidemiological study. This approach has shown success for strains within a group/species. For example, a detailed analysis of '*Ca.* P. solani' strains from 142 grapevines in Italy (northern, central, and southern regions) and the Republic of North Macedonia used sequencing of eight previously uncharacterized genomic fragments (*cbiQ-glyA*, *rplS-tyrS-csdB*, *trxA-truB-rsuA*, *hlyC*, *potC-potD*, *pnp*, *gyrA-gyrB*, and *aspS-mesJ*), alongside the *tuf* gene. Sequence variants in four of the eight genomic fragments (*hlyC*, *cbiQ-glyA*, *trxA-truB-rsuA*, and *rplS-tyrS-csdB*) mirrored those seen in the *tuf* gene, with two types identified. The variants were differentially distributed in the geographic regions studied and were found to be correlated with the prevalence of nettle and bindweed, providing new insights into the genetic diversity and ecology of the populations within vineyards (Passera *et al.*, 2020). Understandably, the primers were not tested against any other 16Sr groups, and interestingly, the other four genomic fragments showed no sequence variation, demonstrating the importance of target gene selection. Presumably due to the availability of successful primers, it is not uncommon to see particular genes being used repeatedly in various studies of a particular '*Ca.* Phytoplasma' species once primers have been published. For example, numerous reports utilize sequence analysis of the genes *vmp1* and *stamp*, both encoding membrane proteins, to study '*Ca.* P. solani' bois noir strains in vineyards, such as in the Republic of North Macedonia (Kostadinovska *et al.*, 2014), Bosnia and Herzegovinia (Delić *et al.*, 2016), in different European geographic areas over a period of two decades (Contaldo *et al.*, 2021), and in Slovenia and Austria over multiple years (Mehle *et al.*, 2022).

It is uncommon for attempts to develop universal or semi-universal primers to be successful despite recurrent attempts over the years and with the increased availability of sequence data. An effect of this is that numerous publications will use a given alternate gene(s) for their specific study purposes. Consequently, the number of sequences for different genes is increasing. However, it is not possible to compare these genes across more diverse phytoplasmas or between studies. While useful in the specific application, the limited specificity of primers does not allow pan-genus analysis, and the applicability is thus restricted.

Multigene analysis

Multilocus sequence typing (MLST) is a well-established technique that identifies and groups bacterial strains based on DNA sequence data, typically from seven to eight housekeeping genes, producing unique allelic profiles (called sequence types or haplotypes) that represent every single nucleotide polymorphism to allow easy comparison of strains (Maiden, 2006). A similar but distinct approach that evolved from MLST is multilocus sequence analysis (MLSA). Instead of assigning allelic profiles, MLSA concatenates DNA sequences from partial gene fragments to determine phylogenetic relationships, typically via the construction of phylogenetic trees (Glaeser and Kämpfer, 2015). MLST is well established in bacteria, with over 130 different microbial species and genera schemes available and curated on the PubMLST.org website (Jolley *et al.*, 2018). The standardization of genes and databases of entries with metadata enables worldwide comparison between different strains and among strains from different studies. The exact set of genes used may vary and is determined for each genus/species as appropriate. MLST has been used for plant pathogens, such as the bacterial pathogen *Xylella fastidiosa*, where MLST of the *leuA*, *petC*, *malF*, *cysG*, *holC*, *nuoL*, and *gltT* genes is used for strain discrimination between and within the subspecies (Scally *et al.*, 2005). Due to performing analysis of sequences from the same set of seven or more housekeeping genes, the terms MLST and MLSA are often used interchangeably, despite the analysis method (allelic profiles and concatenated phylogeny, respectively) being different. As described in this chapter, for phytoplasmas, there are still very few genes with primers capable of amplifying most species, which has hindered the creation of a uniform MLST/MLSA for the phytoplasma genus. Nonetheless, the principles of MLST/MLSA have been applied to numerous studies, as discussed here.

What is described as the first MLST scheme for phytoplasmas (actually MLSA) analyzed eight strains of four phytoplasmas of the PnWB group (16SrII-A) from *Arachis hypogaea*, *Crotalaria pallida*, *Tephrosia purpurea*, and *Cleome viscosa* from Hainan Province, China (Li *et al.*, 2014). Comparison of the 16S rRNA gene was performed with a concatenated tree from partial sequences of the *rplV-rpsC*, *rpoB*, *gyrB*, *dnaK*, *dnaJ*, *recA*, and *secY* genes. Where the 16S rRNA gene showed 99.9–100% sequence identity, different degrees of genetic variability were observed across four of the genes (*rplV-rpsC*, *secY*, *dnaJ*, and *dnaK*), and the analysis revealed minor differences between the strains.

To investigate the genetic diversity among 132 samples originating from Ghana, Nigeria, and Mozambique, another MLST scheme was developed for '*Ca.* P. palmicola' (16SrXXII), where based on the 16S rRNA, subgroup 16SrXXII-A includes strains from Mozambique and Nigeria, and subgroup 16SrXXII-B strains are from Ghana and Côte d'Ivoire (Pilet *et al.*, 2019). Based on eight housekeeping genes, *dnaC*, *gyrB*, *leuS*, *lpd*, *secA*, *recA*, *rsmI*, and *rplV*, with three genes (*gyrB*, *recA*, and *rplV*) in common with the scheme of Li *et al.* (2014), new primers were designed for seven of the genes included in the study and the primers were all found to be specific for '*Ca.* P. palmicola.' The concatenated sequence of 4512 bp for each isolate allowed the differentiation of eight sequence types (STs) based on 205 segregating nucleotides, with four STs present in Ghana (ST1–ST4), three different STs in Mozambique (ST5–ST7), and a unique ST (ST8) in Nigeria. Analysis of the genes individually showed that the least discriminatory gene (*secA*) revealed three STs, each corresponding to a different country of origin, and the most discriminatory gene (*lpd*) differentiated into six STs, with one in Nigeria, two in Mozambique, and three in Ghana.

These examples illustrate the benefits of MLST schemes, although the applications are for a single '*Ca.* Phytoplasma' species, demonstrating the technical challenges posed by the requirement for primers capable of amplifying at least seven genes from all the strains and species of interest, which is further supported by the lack of multispecies/multi-16Sr group MLST/MLSA schemes. Consequently, in an analogous approach that is considered more feasible, applying the same principles of MLST/MLSA, it is very common for studies to include analysis of several genes, often to address taxonomic or epidemiological questions. However, different genes are used in different studies depending on the phytoplasma(s) of interest (that is, lacking the gene standardization associated with MLST/MLSA for a genus/species), often less than seven genes are included (typically three to five), and genes may not always be housekeeping genes. In these instances, rather than designating sequence types, subgroups are often assigned numbers and letters after a prefix of the gene (e.g. secY(IX)-C1 and secY(IX)-C2, or tuf-b1 and tuf-b2), in a manner analogous to the 16Sr group and subgroup system.

Multigene analysis has often been applied to the delineation of strains, or description of new '*Ca.* Phytoplasma' species. Comparative phylogenetic analyses of the 16SrRNA gene and the *rplV (rpl22)–rpsC(rps3)* and *secY* genes indicated that the pigeon pea witches'-broom group (16SrIX) consists of at least six distinct genetic lineages, some of which could not be readily differentiated based on the 16S rRNA gene alone (Lee *et al.*, 2012). The species description of '*Ca.* P. stylosanthis' from *Stylosanthes* little leaf phytoplasma found in Australia utilized phylogenetic analysis of *tuf*, *secA*, and partial *rp* operon (*rps19-rpl22-rps3*) genes, in combination with the 16S rRNA, with all genes producing similar branching patterns (Rodrigues Jardim *et al.*, 2021). Indeed, examples of analyzing multiple genes, typically in combination with the 16S rRNA gene, are numerous. Other applications, amongst many, include: (i) epidemiological studies such as those looking at strain prevalence (using *tuf*, *vmp1*, and *stamp* genes) on bois noir strains in the Republic of North Macedonia (Kostadinovska *et al.*, 2014); (ii) host distribution and timing of infection (using *tuf*, *vmp1*, and *stamp* genes) on bois noir strains from nine European countries over two decades (Contaldo *et al.*, 2021); (iii) population diversity (using the PR-1, PR-2, and PR-3 non-ribosomal region, *rplV-rpsC* and *secY* genes) on '*Ca.* P. mali' populations in Italy (Casati *et al.*, 2011); (iv) geographic and temporal diversity (using *tuf*, *secY*, *stamp*, and *vmp1* genes) on bois noir strains from Slovenia and Austria (Mehle *et al.*, 2022); and (v) analyzing strain native ranges and evolutionary origins (using *rpl22-rps3*, *secY*, and *map* genes) on '*Ca.* P. ulmi' strains from Serbia (Jović *et al.*, 2011).

In an extension of MLSA into phylogenomics, where whole-genome sequences are analyzed, a study of 42 phytoplasma strains from 16SrII, all with at least 97.59% 16S rRNA sequence similarity, was undertaken to identify the species' boundaries within the group (Rodrigues Jardim *et al.*, 2023). Based on phylogenomic analyses and using ANI and AAI, eight distinct taxa equivalent to species were identified, six of which were novel, and each having distinct biological and/or ecological features. Strains within the same species had ANI and AAI values of >97% and shared ≥80% of their genomic segments based on the ANI analysis. Interestingly, the clades observed in the phylogenomic and whole-genome comparative analyses were consistent with previous MLSA studies of 16SrII strains, providing strong evidence for the utility of the MLST/MLSA for strain differentiation.

Despite the merit of MLST/MLSA and multigene studies, the primers used are generally species-specific, preventing the transfer of the schemes to other phytoplasma species or the broader phytoplasma genus (Pilet *et al.*, 2019). Whilst phylogenomics offers the same, or increased, resolving power, this is reliant upon more technically demanding laboratory and bioinformatic analysis, and the challenges of creating high-quality phytoplasma genomes mean widescale, pan-genus analysis in this manner is unlikely in the near future. Therefore, the benefits of multigene analysis are currently restricted to a limited range of 16Sr groups or species.

Diagnostics Developments Based on Alternative Genes

Phytoplasma diagnostics has been routinely based on the 16S rRNA gene, using conventional PCR (Smart *et al.*, 1996; Firrao *et al.*, 2005), arrays (Nicolaisen and Bertaccini, 2007), and real-time PCR (Christensen *et al.*, 2004). However, when using the 16S rRNA gene, there are sometimes problems with false positives due to the amplification of non-target species. In addition, it is often difficult to diagnose the presence of mixed infections since RFLP profiles can become complicated to interpret when overlapping patterns occur in a single sample or when assays are group-specific. As a result, primers based on alternative genes have been devised for conventional PCR diagnostics, as already discussed in this chapter. In addition, a range of alternative diagnostic methods based on these other gene sequences have been established, such as heteroduplex mobility assays (HMA) (Wang and Hiruki, 2005), single-strand conformation polymorphisms (SSCP) (Music *et al.*, 2008), terminal RFLP (T-RFLP) (Hodgetts *et al.*, 2007), real-time PCR (Wei *et al.*, 2004; Hren *et al.*, 2007; Hodgetts *et al.*, 2009), and LAMP (summarized in Hodgetts *et al.*, 2019). As a result, there are a small number of truly universal tests (which detect all known phytoplasmas) and numerous tests for different subsets of phytoplasmas, such as a particular species, a 16Sr group, or a subset of groups. The landscape of diagnostic test methods and individual assays is very large and complex, can be confusing to the non-specialist, and requires careful determination of the most appropriate method/test for each application.

HMA is a technique in which PCR products from different samples are denatured and re-annealed prior to electrophoresis through acrylamide gels. Homoduplexes (annealing of complementary strands that match) and heteroduplexes (annealing of strands with mismatches) are then observed, and variations between sequences can be detected. This technique has been developed for phytoplasmas, based on the 16S–23S spacer region, and was used to analyze variability between 62 phytoplasmas collected from North America, Europe, and Asia (Wang and Hiruki, 2005). The results were useful for detecting subgroups and provided a rapid and sensitive test alternative to RFLP analysis.

SSCP is a similar method to HMA, in which PCR products are denatured to produce single-stranded DNA, which is then separated on polyacrylamide gels such that sequences with different primary structures fold differently and migrate differently. Polymorphisms between sequences can then be observed, and because the technique analyzes the whole sequence rather than just the restriction endonuclease sites that are analyzed in RFLP analyses, band shifts and polymorphisms between isolates are more readily observed. The technique has been used with primers for the 16S rRNA, *tuf* gene, and *dnaB* gene (Music *et al.*, 2008) and has been shown to distinguish between 16SrXII isolates collected from Croatia and the Republic of North

Macedonia, which could not be separated into separate subgroups using conventional RFLP analysis. However, although the assay is easy to perform, conditions for separation of bands have to be established empirically for each set of fragments, making the technique quite time-consuming for routine assays.

T-RFLP has also been developed to identify phytoplasmas *in planta* and assign them to phylogenetic groups (Hodgetts *et al.*, 2007). In this method, one of the PCR primers is labeled with a fluorescent tag, such that following PCR and restriction enzyme digestion with an appropriate enzyme, it is the size of the terminal restriction fragment (TRF) containing the labeled primer, which can be analyzed on an automated DNA-sequencing machine (by capillary electrophoresis). Hodgetts *et al.* (2007) developed primers based on the 23S rRNA gene that could amplify simultaneously from the plant host chloroplast DNA and any phytoplasma present in a sample. This meant that following restriction enzyme digestion, the chloroplast TRF acted as an internal control to show that PCR had worked, whilst any phytoplasma present gave a TRF of a specific size, depending on which group it belonged to. For example, if the chloroplast TRF and a phytoplasma TRF appeared in a sample, the group to which the phytoplasma belonged could be ascertained. If the chloroplast TRF alone appeared, the sample could be regarded as negative for phytoplasma. While if no chloroplast TRF appeared following PCR, it indicated that the DNA sample was unable to support PCR and presumably required a clean-up step to remove PCR inhibitors prior to retesting. In addition, the method can also be used to monitor mixed infections if phytoplasmas from more than one taxonomic group are present in the same plant (indicated by multiple phytoplasma TRFs), and it can also be used as a semi-quantitative method for measuring fluxes in phytoplasma populations between samples.

A better automated, more accurate, and sensitive method for detecting phytoplasma is real-time PCR, which can also be used to quantify levels. This method was initially developed primarily based on the 16S rRNA gene, but subsequently assays based on other genes have become established. For example, Wei *et al.* (2004) developed a specific real-time PCR assay to detect onion yellows phytoplasma based on the *tuf* gene. Similarly, Linck *et al.* (2017) and Kogej *et al.* (2020) developed specific real-time PCR assays based on the *secY* gene to detect elm yellows phytoplasmas and '*Ca.* P. pruni,' respectively. In addition, Hren *et al.* (2007) developed specific assays for bois noir based on the *secY* gene and for flavescence doreé based on a random stolbur genomic fragment. Whilst all the previously mentioned examples use TaqMan® chemistry (with fluorogenic probes in addition to primers providing additional specificity), Galetto *et al.* (2005) developed specific real-time PCR assays using the primer-based SYBR® Green chemistry coupled with melting curve analysis. These assays were designed to detect apple proliferation based on the phytoplasma nitroreductase gene, bois noir based on non-ribosomal DNA, and flavescence doreé based on ribosomal DNA. Bloch *et al.* (2022) also developed a qPCR-HRMA assay (high-resolution melt-curve analysis, similar to SYBR® Green using a proprietary double-stranded nucleic acid stain) based on the *secA* gene. This assay enables the identification and discrimination of '*Ca.* P. palmae', '*Ca.* P. aculeata', and '*Ca.* P. hispanola' affecting palms in the Caribbean basin. Hodgetts *et al.* (2009) also developed a universal real-time PCR assay to detect all phytoplasmas based on the 23S rRNA gene, along with specific assays for different groups and subgroups (16SrI, II, IV-A and -B, XII, and XXII). Interestingly, several of these assays utilize TaqMan® minor groove binder (MGB) probes to allow shorter DNA sequences as probes, demonstrating the challenges of designing specific assays for phytoplasma DNA.

Another method that has gained popularity is LAMP, which uses strand-displacing polymerases (typically *Bst*) combined with four or six primers (designed to six or eight nucleotide locations) to allow very rapid amplification of target DNA, often in a matter of minutes. LAMP can be optimized to the testing scenario, with detection possible via numerous methods, including end-point (e.g. nucleic-acid detection lateral flow devices (LFDs), intercalating dyes, colorimetric change, or agarose gels) or in real-time (e.g. the portable Genie® platform or any real-time PCR platform). The assay is also very robust in the presence of PCR inhibitors and can often be combined with crude sample lysis protocols instead of DNA extraction. Combined, these features mean LAMP is highly amenable to performing testing in the field or where resources are limited (such as in developing countries). However, one drawback of LAMP is the requirement for each assay to have more primers (compared to PCR or real-time PCR), which, combined with the low GC nucleotide content of phytoplasma DNA, can make the design of assays with the desired specificity challenging. Therefore, the use of genes other than the 16S rRNA that provide the required sequence specificity has been commonplace. The difficulties of using the 16S rRNA as a target can be illustrated with two examples. Obura *et al.* (2011) intended to design an assay to detect Napier stunt phytoplasma (16SrXI and 16SrIII) in eastern Africa; however, testing demonstrated that the assay detected 16SrVI, X, XII, and XIV in addition to the target 16SrXI, whilst it did not amplify 16SrI, II, III, and IV. De Jonghe *et al.* (2017) designed an assay using the 16S rRNA gene to specifically detect 16SrX phytoplasmas (apple proliferation, pear decline, and European stone fruit yellows), which was shown to be specific to 16SrX when tested against DNA from 16SrI, II, III, V, VII, XI, and XII. However, it is not only the choice of target gene that affects assay specificity, as seen by Wang *et al.* (2017), who designed a LAMP assay for 16SrI phytoplasmas based on the *tuf* gene, but this detected only five of the eight 16SrI strains tested. Numerous examples of assays designed using other gene targets achieved the desired specificity level, either species, group, or subgroup. Alič *et al.* (2022) used the *ftsH/hflB* gene region to design an assay specific to a '*Ca.* P. asteris' 16SrI-B subgroup strain from grapevines in South Africa, whilst Sugawara *et al.* (2012) used the *groEL* gene to design an assay for '*Ca.* P. asteris.' Assays designed for the specific detection of seven individual 16Sr groups (I, II, III, V, VI, X, and XII) based on the leucyl tRNA synthetase gene (*leuS*) demonstrated the targeted specificity when tested against an additional seven 16Sr groups (16SrIV, VII, IX, XI, XIII, XIV, and XXII) (Aljafer and Dickinson 2021). In separate studies, the 16S–23S intergenic spacer (IGS) gene sequence was used to design specific assays for 16SrI and 16SrXXII phytoplasmas (Tomlinson *et al.*, 2010), as well as 16SrII and 16SrXII phytoplasmas (Bekele *et al.*, 2011b). Kogovšek *et al.* (2015) used the 23S rRNA gene to design a specific assay for flavescence dorée detection, designing and optimizing sample preparation and testing for on-site detection, which could be completed in 1 hour. Additionally, the *secA* gene was used to develop and validate a rapid detection assay for '*Ca.* P. solani' (bois noir), suitable for both laboratory and on-site detection (Kogovšek *et al.*, 2017).

What is perhaps more important in the future is to integrate these new and improved methods into rapid and easy-to-perform detection and classification systems, suitable for both low- and high-resource laboratories, and for in-field use. Creating methods applicable to laboratories working throughout the worldwide distribution of phytoplasmas and making harmonized identification and classification schemes suited for the non-specialist would allow the phytoplasmology field to progress in a unified manner. The ambition of a rapid, specific, sensitive test that not only identifies whether a phytoplasma is present but also what taxonomic group it belongs to, which is effective across all the species within the phytoplasma genus, remains a challenge for researchers to solve.

Conclusion

The use of DNA sequences to build phylogenetic trees is widespread and recognized as a valid approach for identifying taxonomic relationships between organisms. Early studies in most organisms have been based on well-conserved genes that are easy to analyze, particularly ribosomal RNA gene sequences. However, it has long been recognized that there are limitations to basing any classification system on a single, well-conserved gene, and in other organisms, such as the fungi, taxonomic classifications have now moved on and are based on combined datasets for multiple genes including some that encode proteins and are therefore less well conserved. Finding similar genes to improve phylogenetics in the phytoplasmas has proven difficult because of the inability to culture the organisms, the limited number of published sequences, and problems in finding universal primers that can amplify a specified gene from a broad range of different phytoplasmas. However, semi-universal primers are now available for part of the *rp* operon, the *cpn60* and *rpoB* genes, primers that may be universal for part of the *tuf* gene, and universal primers for part of the *secA* gene. It is therefore now possible to start combining datasets for different genes into a more complete and robust phylogenetic analysis of these organisms, resulting in better definition of species, groups, and subgroups. The use of multiple genes for in-depth studies within a particular 16Sr group has also become widespread, allowing fine strain differentiation. However, despite much research, a phytoplasma genus-wide multigene classification system has remained elusive.

Whilst increasing amounts of sequence information have become available over the years, developing further universal primers has remained a major challenge in the study of phytoplasmas. This is compounded by the lack of availability of DNA from all 16Sr groups, hindering the assessment of new primers, and the need to evaluate existing primer specificity with each newly described '*Ca.* Phytoplasma' species. Despite rapid advances in whole-genome sequencing, leveraging the full benefits of these technologies to phytoplasmas has been difficult. Indeed, in many laboratories, Sanger sequencing of gene(s) is the standard approach used for phytoplasma identification/classification, and, particularly in low-resource settings, *in vitro* RFLP remains in use. There is also a dichotomy between research focused on developing methods for phytoplasma genus-wide identification, which is essential for screening in applications such as biosecurity, and research with a very specific focus, for example on a given 16Sr group, crop/plant, or within a region/country. This has resulted in a wealth of different methods and primers and requires the end user to ensure the suitability of a given test to the desired application. The benefits of studying a broader range of genes in phytoplasmas are clear, and already, the results have led to improved diagnostic methods via the development of new assays for conventional PCR, real-time PCR, and LAMP. The aim in the future must be to build on these advances to develop rapid, simple diagnostic systems that cannot only show that a phytoplasma is present in a sample but also identify the strain.

References

Akahori, M., Miyazaki, A., Koinuma, H., Tokuda, R., Iwabuchi, N. *et al.* (2024) Use of the 23S rRNA gene as a target template in the universal loop-mediated isothermal amplification (LAMP) of genomic DNA from phytoplasmas. *Microbiology Spectrum* 12, e00106-24.

Alič, Š., Dermastia, M., Burger, J., Dickinson, M., Pietersen, G., Pietersen, G., and Dreo, T. (2022) Genome-informed design of a LAMP assay for the specific detection of the strain of '*Candidatus* Phytoplasma asteris' phytoplasma occurring in grapevines in South Africa. *Plant Disease* 106, 2927–2939.

Aljafer, N. and Dickinson, M. (2021) Evaluating LAMP assays for detection of phytoplasmas classified in different ribosomal groups. *Journal of Plant Pathology* 103, 1315–1321.

Andersen, M.T., Liefting, L.W., Havukkala, I., and Beever, R.E. (2013) Comparison of the complete genome sequence of two closely related isolates of '*Candidatus* Phytoplasma australiense' reveals genome plasticity. *BMC Genomics* 14, 1–15.

Angelini, E., Negrisolo, E., Clair, D., Borgo, M., and Boudon-Padieu, E. (2003) Phylogenetic relationships among Flavescence dorée strains and related phytoplasmas determined by heteroduplex mobility assay and sequence of ribosomal and non-ribosomal DNA. *Plant Pathology* 52, 663–672.

Anthony, R.M., Brown, T.J., and French, G.L. (2000) Rapid diagnosis of bacteremia by universal amplification of 23S ribosomal DNA followed by hybridisation to an oligonucleotide array. *Journal of Clinical Microbiology* 38, 781–788.

Arnaud, G., Malembic-Maher, S., Salar, P., Bonnet, P., Maixner, M. *et al.* (2007) Multilocus sequence typing confirms the close genetic interrelatedness of three distinct flavescence dorée phytoplasma strain clusters and group 16SrV phytoplasmas infecting grapevine and alder in Europe. *Applied and Environmental Microbiology* 73, 4001–4010.

Bai, X., Zhang, J., Ewing, A., Miller, S.A., Radek, A.J. *et al.* (2006) Living with genome instability: The adaptation of phytoplasmas to diverse environments of their insect and plant hosts. *Journal of Bacteriology* 188, 3682–3696.

Barbara, D.J., Morton, A., Clark, M.F., and Davies, D.L. (2002) Immunodominant membrane proteins from two phytoplasmas in the aster yellows clade (chlorante aster yellows and clover phyllody) are highly divergent in the major hydrophilic region. *Microbiology* 148, 157–167.

Bekele, B., Abeysinghe, S., Hoat, T.X., Hodgetts, J., and Dickinson, M. (2011a) Development of specific *secA*-based diagnostics for the 16SrXI and 16SrXIV phytoplasmas of the Gramineae. *Bulletin of Insectology* 64 (Suppl. 1), 15–16.

Bekele, B., Hodgetts, J., Tomlinson, J., Boonham, N., Nikolić, P., Swarbrick, P., and Dickinson, M. (2011b) Use of a real-time LAMP isothermal assay for detecting 16SrII and XII phytoplasmas in fruit and weeds of the Ethiopian Rift Valley. *Plant Pathology* 60, 345–355.

Berg, M. and Seemüller, E. (1999) Chromosomal organization and nucleotide sequence of the genes coding for the elongation factors G and Tu of the apple proliferation phytoplasma. *Gene* 226, 103–109.

Berg, M., Davies, D.L., Clark, M.F., Vetten, H.J., Maier, G., Marcone, C., and Seemüller, E. (1999) Isolation of the gene encoding an immunodominant membrane protein of the apple proliferation phytoplasma, and expression and characterization of the gene product. *Microbiology* 145, 1937–1943.

Bertaccini, A., Arocha-Roseto, Y., Contaldo, N., Duduk, B., Fiore, N. *et al.* (2022) Revision of the '*Candidatus* Phytoplasma' species description guidelines. *International Journal of Systematic and Evolutionary Microbiology* 72, 005353.

Bloch, M., Helmick, E.E., and Bahder, B.W. (2022) Differentiation of palm-infecting phytoplasmas in the Caribbean basin using high resolution melt curve analysis of the *secA* gene. *Plant Disease* 106, 2480–2489.

Blomquist, C.L., Barbara, D.J., Davies, D.L., Clark, M.F., and Kirkpatrick, B.C. (2001) An immunodominant membrane protein gene from the Western X-disease phytoplasma is distinct from those of other phytoplasmas. *Microbiology* 147, 571–580.

Casati, P., Quaglino, F., Stern, A.R., Tedeschi, R., Alma, A., and Bianco, P.A. (2011) Multiple gene analyses reveal extensive genetic diversity among '*Candidatus* Phytoplasma mali' populations. *Annals of Applied Biology* 158, 257–266.

Christensen, N.M., Nicolaisen, M., Hansen, M., and Schulz, A. (2004) Distribution of phytoplasmas in infected plants as revealed by real-time PCR and bioimaging. *Molecular Plant-Microbe Interactions* 17, 1175–1184.

Contaldo, N., Stepanović, J., Pacini, F., Bertaccini, A., and Duduk, B. (2021) Molecular variability and host distribution of '*Candidatus* Phytoplasma solani' strains from different geographic origins. *Microorganisms* 9, 2530.

Daire, X., Clair, D., Reinert, W., and Boudon-Padieu, E. (1997) Detection and differentiation of grapevine yellows phytoplasma belonging to the elm yellows group and to the stolbur subgroup by PCR amplification of non-ribosomal DNA. *European Journal of Plant Pathology* 103, 507–514.

De Jonghe, K., De Roo, I., and Maes, M. (2017) Fast and sensitive on-site isothermal assay (LAMP) for diagnosis and detection of three fruit tree phytoplasmas. *European Journal of Plant Pathology* 147, 749–759.

De La Rue, S., Padovan, A., and Gibb, K. (2001) *Stylosanthes* is a host for several phytoplasmas, one of which shows unique 16S–23S intergenic spacer region heterogeneity. *Journal of Phytopathology* 149, 613–619.

Delić, D., Balech, B., Radulović, M., Lolić, B., Karačić, A. *et al.* (2016) *Vmp1* and *stamp* genes variability of '*Candidatus* phytoplasma solani' in Bosnian and Herzegovinian grapevine. *European Journal of Plant Pathology* 145, 221–225.

Dumonceaux, T.J., Green, M., Hammond, C., Perez, E., and Olivier, C. (2014) Molecular diagnostic tools for detection and differentiation of phytoplasmas based on chaperonin-60 reveal differences in host plant infection patterns. *PLoS One* 9, e116039.

Firrao, G., Gibb, K., and Streten, C. (2005) Short taxonomic guide to the genus '*Candidatus* phytoplasma'. *Journal of Plant Pathology* 87, 249–263.

Galetto, L., Bosco, D., and Marzachì, C. (2005) Universal and group-specific real-time PCR diagnosis of flavescence dorée (16Sr-V), bois noir (16Sr-XII) and apple proliferation (16Sr-X) phytoplasmas from field-collected plant hosts and insect vectors. *Annals of Applied Biology* 147, 191–201.

Glaeser, S.P. and Kämpfer, P. (2015) Multilocus sequence analysis (MLSA) in prokaryotic taxonomy. *Systematic and Applied Microbiology* 38, 237–245.

Goodwin, P.H., Xue, B.G., Kuske, C.R., and Sears, M.K. (1994) Amplification of plasmid DNA to detect plant pathogenic mycoplasmalike organisms. *Annals of Applied Biology* 124, 27–36.

Gundersen, D.E., Lee, I.-M., Rehner, S.A., Davis, R.E., and Kingsbury, D.T. (1994) Phylogeny of mycoplasmalike organisms (Phytoplasmas): a basis for their classification. *Journal of Bacteriology* 176, 5244–5254.

Harrison, J., Hussain, R.M., Aspin, A., Grant, M.R., Vicente, J.G., and Studholme, D.J. (2023) Phylogenomic analysis supports the transfer of 20 pathovars from *Xanthomonas campestris* into *Xanthomonas euvesicatoria*. *Taxonomy* 3, 29–45.

Harrison, N.A., Womack, M., and Carpio, M.L. (2002) Detection and characterization of a lethal yellowing (16SrIV) group phytoplasma in Canary Island date palms affected by lethal decline in Texas. *Plant Disease* 86, 676–681.

Harrison, N.A., Davis, R.E., Oropeza, C., Helmick, E.E., Narvaez, M. *et al.* (2014) '*Candidatus* Phytoplasma palmicola', associated with a lethal yellowing-type disease of coconut (*Cocos nucifera* L.) in Mozambique. *International Journal of Systematic and Evolutionary Microbiology* 64, 1890–1899.

Hill, J.E., Town, J.R., and Hemmingsen, S.M. (2006) Improved template representation in cpn60 polymerase chain reaction (PCR) product libraries generated from complex templates by application of a specific mixture of PCR primers. *Environmental Microbiology* 8, 741–746.

Hodgetts, J. (2019) Rapid sample preparation and LAMP for phytoplasma detection. *Phytoplasmas: Methods and Protocols. Methods in Molecular Biology* 1875, 187–201.

Hodgetts, J., Ball, T., Boonham, N., Mumford, R., and Dickinson, M. (2007) Use of terminal restriction fragment length polymorphism (T-RFLP) for identification of phytoplasmas in plants. *Plant Pathology* 56, 357–365.

Hodgetts, J., Boonham, N., Mumford, R., Harrison, N., and Dickinson, M. (2008) Phytoplasma phylogenetics based on analysis of *secA* and 23S rRNA gene sequences for improved resolution of candidate species of '*Candidatus* Phytoplasma'. *International Journal of Systematics and Evolutionary Microbiology* 58, 1826–1837.

Hodgetts, J., Boonham, N., Mumford, R., and Dickinson, M. (2009) Panel of 23S rRNA gene-based real-time PCR assays for improved universal and group-specific detection of phytoplasmas. *Applied and Environmental Microbiology* 75, 2945–2950.

Hren, M., Boben, J., Rotter, A., Kralj, P., Gruden, K., and Ravnikar, M. (2007) Real-time PCR detection systems for Flavescence doreé and Bois noir phytoplasmas in grapevine: comparison with conventional PCR detection and application in diagnostics. *Plant Pathology* 56, 785–796.

IRPCM (International Research Programme on Comparative Mycoplasmology) Phytoplasma/Spiroplasma Working Team – Phytoplasma Taxonomy Group (2004) '*Candidatus* Phytoplasma', a taxon for the wall-less, non-helical prokaryotes that colonize plant phloem and insects. *International Journal of Systematics and Evolutionary Microbiology* 54, 1243–1255.

James, T.Y., Kauff, F., Schoch, C.L., Brandon Matheny, P., Hofstetter, V. *et al.* (2006) Restructuring the early evolution of *Fungi* using a six-gene phylogeny. *Nature* 443, 818–822.

Jarausch, W., Saillard, C., Dosba, F., and Bové, J.-M. (1994) Differentiation of mycoplasmalike organisms (MLOs) in European fruit trees by PCR using specific primers derived from the sequence of a chromosomal fragment of the apple proliferation MLO. *Applied and Environmental Microbiology* 60, 2916–2923.

Jolley, K.A., Bray, J.E., and Maiden, M.C. (2018) Open-access bacterial population genomics: BIGSdb software, the PubMLST. org website and their applications. *Wellcome Open Research* 3, 124.

Jomantiene, R., Zhao, Y., and Davis, R.E. (2007) Sequence-variable mosaics: Composites of recurrent transposition characterizing the genomes of phylogenetically diverse phytoplasmas. *DNA and Cell Biology* 26, 557–564.

Jović, J., Cvrković, T., Mitrović, M., Petrović, A., Krstić, O., Krnjajić, S., and Toševski, I. (2011) Multigene sequence data and genetic diversity among '*Candidatus* Phytoplasma ulmi' strains infecting *Ulmus* spp. in Serbia. *Plant Pathology* 60, 356-368.

Kakizawa, S., Oshima, K., Kuboyama, T., Nishigawa, H., Jung, H. *et al.* (2001) Cloning and expression analysis of phytoplasma protein translocation genes. *Molecular Plant–Microbe Interactions* 14, 1043–1050.

Keeling, P.J., Burger, G., Durnford, D.G., Lang, B.F., Lee, R.W. *et al.* (2005) The tree of eukaryotes. *Trends in Ecology and Evolution* 20, 670–676.

Kogej, Z., Dermastia, M., and Mehle, N. (2020) Development and validation of a new TaqMan real-time PCR for detection of '*Candidatus* Phytoplasma pruni'. *Pathogens* 9, 642.

Kogovšek, P., Hodgetts, J., Hall, J., Prezelj, N., Nikolić, P. *et al.* (2015) LAMP assay and rapid sample preparation method for on-site detection of flavescence dorée phytoplasma in grapevine. *Plant Pathology* 64, 286–296.

Kogovšek, P., Mehle, N., Pugelj, A., Jakomin, T., Schroers, H.J., Ravnikar, M., and Dermastia, M. (2017) Rapid loop-mediated isothermal amplification assays for grapevine yellows phytoplasmas on crude leaf-vein homogenate has the same performance as qPCR. *European Journal of Plant Pathology* 148, 75-84.

Kostadinovska, E., Quaglino, F., Mitrev, S., Casati, P., Bulgari, D., and Bianco, P. (2014) Multiple gene analyses identify distinct "bois noir" phytoplasma genotypes in the Republic of Macedonia. *Phytopathologia Mediterranea* 53, 491–501.

Kube M., Schneider, B., Kuhl, H., Dandekar, T., Heitmann, K. *et al.* (2008) The linear chromosome of the plant pathogenic mycoplasma '*Candidatus* Phytoplasma mali'. *BMC Genomics* 9, 306.

Lee, I.-M., Gundersen-Rindal, D.E., Davis, R.E., and Bartoszyk I.M. (1998) Revised classification scheme of phytoplasmas based on RFLP analyses of 16S rRNA and ribosomal protein gene sequences. *International Journal of Systematic Bacteriology* 48, 1153–1169.

Lee, I.-M., Zhao, Y., and Bottner, K.D. (2006) *SecY* gene sequence analysis for finer differentiation of diverse strains in the aster yellows phytoplasma group. *Molecular and Cellular Probes* 20, 87–91.

Lee, I.M., Bottner-Parker, K.D., Zhao, Y., Davis, R.E., and Harrison, N.A. (2010) Phylogenetic analysis and delineation of phytoplasmas based on *secY* gene sequences. *International Journal of Systematic and Evolutionary Microbiology* 60, 2887–2897.

Lee, I.M., Bottner-Parker, K.D., Zhao, Y., Bertaccini, A., and Davis, R.E. (2012) Differentiation and classification of phytoplasmas in the pigeon pea witches'-broom group (16SrIX): an update based on multiple gene sequence analysis. *International Journal of Systematic and Evolutionary Microbiology* 62, 2279–2285.

Li, Y., Piao, C.G., Tian, G.Z., Liu, Z.X., Guo, M.W., Lin, C.L., and Wang, X.Z. (2014) Multilocus sequences confirm the close genetic relationship of four phytoplasmas of peanut witches'-broom group 16SrII-A. *Journal of Basic Microbiology* 54, 818–827.

Liefting, L.W., Andersen, M.T., Beever, R.E., Gardner, R.C., and Forster, R.L. (1996) Sequence heterogeneity in the two 16S rRNA genes of Phormium yellow leaf phytoplasma. *Applied and Environmental Microbiology* 62, 3133–3139.

Lim, P.-O. and Sears, B.B. (1991) DNA sequence of the ribosomal protein genes *rpl2* and *rps19* from a plant pathogenic mycoplasma-like organism. *FEMS Microbiology Letters* 84, 71–74.

Lim, P.-O. and Sears, B.B. (1992) Evolutionary relationships of a plant-pathogenic mycoplasmalike organism and *Acholeplasma laidlawii* deduced from two ribosomal protein gene sequences. *Journal of Bacteriology* 174, 2606–2611.

Linck, H., Krüger, E., and Reineke, A. (2017) A multiplex TaqMan qPCR assay for sensitive and rapid detection of phytoplasmas infecting *Rubus* species. *PLoS One* 12, e0177808.

Links, M.G., Dumonceaux, T.J., Hemmingsen, S.M., and Hill, J.E. (2012) The chaperonin-60 universal target is a barcode for bacteria that enables *de novo* assembly of metagenomic sequence data. *PLoS One* 7, e49755.

Maiden, M.C. (2006) Multilocus sequence typing of bacteria. *Annual Reviews of Microbiology* 60, 561–588.

Makarova, O., Contaldo, N., Paltrinieri, S., Kawube, G., Bertaccini, A., and Nicolaisen, M. (2012) DNA barcoding for identification of '*Candidatus* Phytoplasmas' using a fragment of the elongation factor Tu gene. *PLoS One* 7, e52092.

Marcone, C., Lee, I.-M., Davis, R.E., Ragozzino, A., and Seemüller, E. (2000) Classification of aster yellows-group phytoplasmas based on combined analyses of rRNA and *tuf* gene sequences. *International Journal of Systematics and Evolutionary Microbiology* 50, 1703–1713.

Martini, M., Lee, I.-M., Bottner, K.D., Zhao, Y., Botti, S. *et al.* (2007) Ribosomal protein gene based phylogeny for finer differentiation and classification of phytoplasmas. *International Journal of Systematics and Evolutionary Microbiology* 57, 2037–2051.

Mehle, N., Kavčič, S., Mermal, S., Vidmar, S., Pompe Novak, M. *et al.* (2022) Geographical and temporal diversity of '*Candidatus* Phytoplasma solani' in wine-growing regions in Slovenia and Austria. *Frontiers in Plant Science* 13, 889675.

Mitrović, J., Kakizawa, S., Duduk, B., Oshima, K., Namba, S., and Bertaccini, A. (2011) The *groEL* gene as an additional marker for finer differentiation of '*Candidatus* Phytoplasma asteris'-related strains. *Annals of Applied Biology* 159, 41–48.

Mpunami, A.A., Tymon, A., Jones, P., and Dickinson, M.J. (1999) Genetic diversity in the coconut lethal yellowing disease phytoplasmas of East Africa. *Plant Pathology* 48, 109–114.

Muirhead, K., Pérez-López, E., Bahder, B.W., Hill, J.E., and Dumonceaux, T.J. (2019) The CpnClassiPhyR facilitates phytoplasma classification and taxonomy using cpn60 universal target sequences. In: Olivier, C.Y., Dumonceaux, T.J., and Pérez-López, E. (eds) *Sustainable Management of Phytoplasma Diseases in Crops Grown in the Tropical Belt: Biology and Detection*. Springer, pp. 1-27.

Music, M.Š., Krajačić, M., and Škorić, D. (2008) The use of SSCP analysis in the assessment of phytoplasma gene variability. *Journal of Microbiological Methods* 73, 69–72.

Musić, M.Š., Samarzija, I., Hogenhout, S.A., Haryono, M., Cho, S.T., and Kuo, C.H. (2019) The genome of '*Candidatus* Phytoplasma solani' strain SA-1 is highly dynamic and prone to adopting foreign sequences. *Systematic and Applied Microbiology* 42, 117–127.

Nei, M. and Li, W.-H. (1979) Mathematical model for studying genetic variation in terms of restriction endonucleases. *Proceedings of the National Academy of Sciences of the USA* 76, 5269–5273.

Nicolaisen, M. and Bertaccini, A. (2007) An oligonucleotide microarray-based assay for identification of phytoplasma 16S ribosomal groups. *Plant Pathology* 56, 332–336.

Obura, E., Masiga, D., Wachira, F., Gurja, B., and Khan, Z.R. (2011) Detection of phytoplasma by loop-mediated isothermal amplification of DNA (LAMP). *Journal of Microbiological Methods* 84, 312–316.

Ogier, J.C., Pagès, S., Galan, M., Barret, M., and Gaudriault, S. (2019) *rpoB*, a promising marker for analyzing the diversity of bacterial communities by amplicon sequencing. *BMC Microbiology* 19, 1–16.

Oshima, K., Kakizawa, S., Nishigawa, H., Jung, H.-Y., Wei, W. *et al.* (2004) Reductive evolution suggested from the complete genome sequence of a plant pathogenic phytoplasma. *Nature Genetics* 36, 27–29.

Passera, A., Zhao, Y., Murolo, S., Pierro, R., Arsov, E. *et al.* (2020) Multilocus genotyping reveals new molecular markers for differentiating distinct genetic lineages among "*Candidatus* Phytoplasma Solani" strains associated with grapevine bois noir. *Pathogens* 9, 970.

Pérez-López, E., Olivier, C.Y., Luna-Rodríguez, M., and Dumonceaux, T.J. (2016) Phytoplasma classification and phylogeny based on *in silico* and *in vitro* RFLP analysis of cpn60 universal target sequences. *International Journal of Systematic and Evolutionary Microbiology* 66, 5600–5613.

Pilet, F., Quaicoe, R.N., Osagie, I.J., Freire, M., and Foissac, X. (2019) Multilocus sequence analysis reveals three distinct populations of "*Candidatus* Phytoplasma palmicola" with a specific geographical distribution on the African continent. *Applied and Environmental Microbiology* 85, e02716-18.

Ricker, N., Qian, H., and Fulthorpe, R.R. (2012) The limitations of draft assemblies for understanding prokaryotic adaptation and evolution. *Genomics* 100, 167–175.

Rodrigues Jardim, B., Kinoti, W.M., Tran-Nguyen, L.T., Gambley, C., Rodoni, B., and Constable, F.E. (2021) '*Candidatus* Phytoplasma stylosanthis', a novel taxon with a diverse host range in Australia, characterised using multilocus sequence analysis of 16S rRNA, *secA*, *tuf*, and *rp* genes. *International Journal of Systematic and Evolutionary Microbiology* 71, 004589.

Rodrigues Jardim, B., Tran-Nguyen, L.T., Gambley, C., Al-Sadi, A.M., Al-Subhi, A.M. *et al.* (2023) The observation of taxonomic boundaries for the 16SrII and 16SrXXV phytoplasmas using genome-based delimitation. *International Journal of Systematic and Evolutionary Microbiology* 73, 005977.

Rubio, S., Barnes, A., Webb, K., and Hodgetts, J. (2017) A real-time PCR assay for improved rapid, specific detection of *Cryphonectria parasitica*. *Annals of Applied Biology* 171, 52–61.

Scally, M., Schuenzel, E.L., Stouthamer, R., and Nunney, L. (2005) Multilocus sequence type system for the plant pathogen *Xylella fastidiosa* and relative contributions of recombination and point mutation to clonal diversity. *Applied and Environmental Microbiology* 71, 8491–8499.

Schneider, B., Gibb, K.S., and Seemüller, E. (1997) Sequence and RFLP analysis of the elongation factor Tu gene used in differentiation and classification of phytoplasmas. *Microbiology* 143, 3381–3389.

Shen, W.C. and Lin, C.P. (1993) Production of monoclonal antibodies against a mycoplasmalike organism associated with sweet potato witches' broom. *Phytopathology* 83, 671–675.

Smart, C.D., Schneider, B., Blomquist, C.L., Guerra, L.J., Harrison, N.A. *et al.* (1996) Phytoplasma-specific PCR primers based on sequence of the 16S–23S rRNA spacer region. *Applied and Environmental Microbiology* 62, 2988–2993.

Sneideris, D., Ivanauskas, A., Zizyte, M., and Valiunas, D. (2021) *secA* gene suitability for fast and easy identification of Phytoplasmas by RFLP analysis. *European Journal of Plant Pathology* 160, 737–743.

Streten, C. and Gibb, K.S. (2003) Identification of genes in the tomato big bud phytoplasma and comparison to those in sweet potato little leaf-V4 phytoplasma. *Microbiology* 149, 1797–1805.

Streten, C. and Gibb, K.S. (2005) Genetic variation in '*Candidatus* Phytoplasma australiense'. *Plant Pathology* 54, 8–14.

Sugawara, K., Himeno, M., Keima, T., Kitazawa, Y., Maejima, K., Oshima, K., and Namba, S. (2012) Rapid and reliable detection of phytoplasma by loop-mediated isothermal amplification targeting a housekeeping gene. *Journal of General Plant Pathology* 78, 389–397.

Tan, C.M., Lin, Y.C., Li, J.R., Chien, Y.Y., Wang, C.J. *et al.* (2021) Accelerating complete phytoplasma genome assembly by immunoprecipitation-based enrichment and MinION-based DNA sequencing for comparative analyses. *Frontiers in Microbiology* 12, 766221.

Tomlinson, J.A., Boonham, N., and Dickinson, M. (2010) Development and evaluation of a one-hour DNA extraction and loop-mediated isothermal amplification assay for rapid detection of phytoplasmas. *Plant Pathology* 59, 465–471.

Toth, K.F., Harrison, N., and Sears, B.B. (1994) Phylogenetic relationships among members of the class *Mollicutes* deduced from *rps3* gene sequences. *International Journal of Systematics and Evolutionary Microbiology* 44, 119–124.

Tran-Nguyen, L.T.T., Kube, M., Schneider, B., Reinhardt, R., and Gibb, K.S. (2008) Comparative genome analysis of '*Candidatus* Phytoplasma australiense' (sub group tuf-Australia 1: rp-A) and '*Ca*. P. asteris' strains OY-M and AY-WB. *Journal of Bacteriology* 190, 3979–3991.

Valiunas, D., Jomantiene, R., and Davis, R.E. (2013) Evaluation of the DNA-dependent RNA polymerase β-subunit gene (*rpoB*) for phytoplasma classification and phylogeny. *International Journal of Systematic and Evolutionary Microbiology* 63, 3904–3914.

Wagner, M., Fingerhut, C., Gross, H.J., and Schön, A. (2001) The first phytoplasma RNaseP RNA provides new insights into the sequence requirements of this ribozyme. *Nucleic Acids Research* 29, 2661–2665.

Wang, K. and Hiruki, C. (2005) Distinctions between phytoplasmas at the subgroup level detected by heteroduplex mobility assay. *Plant Pathology* 54, 625–633.

Wang, K., Hiruki, C., and Yeh, F. (2003) Molecular evolution of phytoplasmas based on polymorphisms in the 16S rRNA genes and the 16/23S spacer regions. *Proceedings of the Japanese Academy Series B* 79, 155–162.

Wang, S., Wang, S., Lin, C., Yu, S., Wang, L. *et al*. (2017) Loop-mediated isothermal amplification assay for detection of five phytoplasmas belonging to 16Sr I group based on target *tuf* gene. *Scientia Silvae Sinicae* 53, 54–63.

Wei, W. and Zhao, Y. (2022) Phytoplasma taxonomy: Nomenclature, classification, and identification. *Biology* 11, 1119.

Wei, W., Kakizawa, S., Suzuki, S., Jung, H.Y., Nishigawa, H. *et al*. (2004) *In planta* dynamic analysis of onion yellows phytoplasma using localised inoculation by insect transmission. *Phytopathology* 94, 244–250.

Wei, W., Davis, R.E., Lee, I.-M., and Zhao, Y. (2007) Computer-simulated RFLP analysis of 16S rRNA genes: Identification of ten new phytoplasma groups. *International Journal of Systematics and Evolutionary Microbiology* 57, 1855–1867.

Wei, W., Lee, I.-M., Davis, R.E., Suo, X., and Zhao, Y. (2008) Automated RFLP pattern comparison and similarity coefficient calculation for rapid delineation of new and distinct phytoplasma 16Sr subgroup lineages. *International Journal of Systematics and Evolutionary Microbiology* 58, 2368–2377.

Yu, Y.-L., Yeh K.-W., and Lin C.-P. (1998) An antigenic protein gene of a phytoplasma associated with sweet potato witches' broom. *Microbiology* 144, 1257–1262.

Zhao, Y., Wei, W., Lee, I.M., Shao, J., Suo, X., and Davis, R.E. (2013) The *i*PhyClassifier, an interactive online tool for phytoplasma classification and taxonomic assignment. *Phytoplasma: Methods and Protocols*. *Methods in Molecular Biology* 938, 329–338.

6 Phytoplasma Transmission via Seeds: A Hypothesis Awaiting Definitive Evidence

Wei Wei* **and Yan Zhao**
Molecular Plant Pathology Laboratory, ARS-USDA, Beltsville, Maryland, USA

Abstract

The potential for seed transmission of phytoplasmas remains an area of active research, with recent studies suggesting this route as a possible mechanism for pathogen spread. However, there is currently no definitive evidence to prove the viability and infectivity of phytoplasmas transmitted via seeds. This review explores the biology of phytoplasmas, current knowledge on seed transmission, the mechanisms by which pathogens are transmitted through seeds, and the unique challenges of establishing seed transmission in phytoplasmas. Additionally, it addresses the policy and regulatory implications of these findings and provides recommendations for future research, including cross-validation by multiple laboratories to ensure robust and reproducible evidence.

Introduction

Seed transmission is a crucial pathway for the spread of plant pathogens, with significant implications for agriculture, forestry, and ecosystems worldwide. Pathogens can survive in seeds during dormancy and be transported across regions through the global seed trade (Dell'Olmo *et al.*, 2023). Infected seeds often escape detection during inspections and quarantine measures, serving as hidden sources of infection that can result in disease outbreaks in new regions. Once seeds germinate, pathogens can infect plants during their early developmental stages, which are particularly vulnerable to disease. This often leads to systemic infections and the establishment of persistent disease cycles (Maude, 1996; Gitaitis and Walcott, 2007). These cycles may spread further through secondary routes, including wind, water, and insect vectors (Correa *et al.*, 2023; Hernandez *et al.*, 2023). Documented examples of seed-transmitted pathogens include fungi such as *Fusarium oxysporum* and *Aspergillus flavus* (Mehan *et al.*, 1991; Gullino *et al.*, 2024), bacteria like *Xanthomonas campestris* pv. *campestris* and *Pseudomonas syringae* pv. *tomato* (Gullino *et al.*, 2009; Chen *et al.*, 2021), and viruses such as tobacco mosaic virus (TMV) and cucumber mosaic virus (CMV) (Ali and Kobayashi, 2010; Ellis *et al.*, 2020). These pathogens enter seeds externally through contamination or via vascular tissues during seed development. They persist through dormancy and reactivate germination, infecting the emerging plant (Elmer, 2001; Dombrovsky and Smith, 2017; Gebeyaw, 2020).

In contrast, the potential for seed transmission in phytoplasmas remains a controversial

*Corresponding author: wei.wei@usda.gov

 Phytoplasmas: Genomes, Plant Hosts and Vectors, 2nd Edition. (Ed. W. Wei)
DOI: 10.1079/9781800627031.0006

and unresolved hypothesis. Phytoplasmas are unculturable, wall-less bacteria belonging to the class Mollicutes, and they are restricted to the phloem tissues of plants (Lee *et al.*, 2000; Wei and Zhao, 2022). With highly reduced genomes and limited metabolic functions, they rely heavily on plant hosts and insect vectors for nutrition and replication (Oshima *et al.*, 2004; Tan *et al.*, 2021). These pathogens induce a variety of economically important plant diseases, with symptoms including yellowing, stunting, phyllody, witches'-broom, and decline. Phytoplasmas infect a broad range of crops, such as sugarcane, coconut, and grapevines. Their global distribution significantly impacts agriculture, often leading to reduced yields and diminished crop quality (MacLean *et al.*, 2011; Sugio *et al.*, 2011; Wei *et al.*, 2013, 2019).

Phytoplasmas are primarily transmitted through insect vectors, particularly sap-feeding hemipterans such as leafhoppers, planthoppers, and psyllids (Weintraub and Beanland, 2006). These vectors acquire phytoplasma by feeding on the phloem of infected plants, and the bacteria multiply within the insect's body, particularly in the midgut and salivary glands. The pathogens are then transmitted to healthy plants during subsequent feeding. This circulative and propagative transmission mechanism ensures efficient spread, with specific insect species often associated with phytoplasma strains (Weintraub and Beanland, 2006). Another significant mode of transmission is vegetative propagation. Infected plant materials, such as cuttings, grafts, or bulbs, can carry and spread the pathogen. This mode is important for crops like grapevines and potatoes that are propagated vegetatively (Wei *et al.*, 2013).

Although phloem tissues are present in reproductive structures and developing seeds, they are absent in mature seeds (Werner *et al.*, 2011; Feurtado, 2017), making the mechanism of phytoplasma transmission through seeds biologically complex. Recent studies have detected phytoplasma DNA in seeds and seedlings (Çağlar *et al.*, 2019; Satta *et al.*, 2019; Kirdat *et al.*, 2022; Lu *et al.*, 2024) and Kirdat *et al.* (2022) reported visualizing phytoplasmas in seeds using scanning electron microscopy (SEM). These findings demonstrate the potential presence of phytoplasmas in seed tissues and raise the possibility of vertical transmission. However, the presence of DNA or even the detection of phytoplasma cells alone does not confirm their viability or infectivity, leaving the question of seed transmission unresolved.

This chapter explores the biology of phytoplasmas, current knowledge on potential seed transmission, and the broader mechanisms by which pathogens are transmitted through seeds. It addresses the unique challenges of confirming seed transmission in phytoplasmas, particularly their confinement to phloem, which is absent in mature seeds, and the methodological difficulties in proving viability and infectivity. As a phytoplasma research group, these authors were requested by a regulatory agency to provide scientific insights on this issue, emphasizing the urgent need to evaluate existing data, establish standardized validation protocols, and fill key knowledge gaps to support evidence-based policy decisions and to strengthen agricultural biosecurity.

Seed Formation and Structure and the Absence of Phloem in Mature Seeds

A seed is the mature product of sexual reproduction in flowering plants, originating from the ovules within a flower's ovary after fertilization (Primack, 1987; Glover, 2014; Steeves and Sawhney, 2017). Its development begins with the interaction of key floral organs, followed by the formation of the fruit, which encloses and protects the developing seeds. A typical flower has four main organs: sepals, petals, stamens, and carpels. Among these, carpels (or pistils) are responsible for seed formation. The carpel comprises the stigma, style, and ovary (Glover, 2014; Steeves and Sawhney, 2017). After pollination, pollen grains land on the stigma and travel through the style to reach the ovules inside the ovary. Once fertilization occurs, the ovule develops into a seed, while the ovary matures into a fruit (Glover, 2014; Steeves and Sawhney, 2017). The stamens, which consist of the anther and filament, produce pollen grains containing male gametes. Fertilization occurs when a pollen grain fertilizes an ovule inside the ovary, initiating seed development. Following fertilization, the flower's ovary develops into a fruit, which protects the seeds and aids in their dispersal (Primack, 1987; Glover, 2014; Steeves and Sawhney, 2017). Depending on the plant species, fruits can be fleshy (such as berries and drupes) or dry (such as legumes and capsules). In berries, the fruit consists of a fully fleshy pericarp, including the exocarp (outer skin), mesocarp (middle fleshy layer), and endocarp (inner fleshy layer), with seeds embedded within the pulp (Bobrov and Romanov, 2019). Examples of berries include tomatoes, grapes, and blueberries.

A fully developed seed consists of three main parts: the seed coat, endosperm, and embryo (Boesewinkel and Bouman, 1984; Ohto *et al.*, 2009; Doll and Ingram, 2022). The seed coat forms the outer protective layer, shielding the seed from physical damage, desiccation, and microbial invasion. It develops from the integuments of the ovule and also regulates water and oxygen intake during germination. The endosperm is a nutrient-rich tissue that provides energy for the developing embryo. In some seeds, such as cereals, the endosperm remains prominent, while in others, such as legumes, it is absorbed during development and stored in the cotyledons. The embryo, which is the core of the seed and will eventually grow into a new plant, consists of the plumule (future shoot), radicle (future root), and cotyledons (seed leaves), which store food and support early growth during germination (Boesewinkel and Bouman, 1984; Ohto *et al.*, 2009; Doll and Ingram, 2022).

During seed development, the seed remains attached to the parent plant via the funiculus, a stalk-like structure that contains vascular tissues, including phloem, which transports nutrients from the parent plant to the developing seed (Larsson *et al.*, 2017; Barro-Trastoy *et al.*, 2020). However, as the seed matures, the vascular connection is severed, and the seed becomes independent, lacking functional phloem (Werner *et al.*, 2011; Feurtado, 2017). This absence of phloem in mature seeds is a key reason why phloem-restricted pathogens, such as phytoplasmas and liberibacter, are unlikely to be transmitted through seeds.

Commonly Used Procedure to Prove Seed-borne Pathogen Transmission

Proving seed-borne transmission of pathogens is a critical step in understanding the role of seeds in the spread of plant diseases. To establish this transmission mode, it must be demonstrated that a pathogen is associated with seeds, can survive seed dormancy, infects seedlings during germination, and causes disease (Hiddink *et al.*, 2023). The process begins with carefully collecting seeds from symptomatic and asymptomatic plants, properly labeling and handling them to prevent cross-contamination. Seeds collected from symptomatic plants are presumed to harbor the pathogen, while seeds from asymptomatic plants serve as controls (Hiddink *et al.*, 2023).

To determine whether the pathogen is located on the seed surface or internally within the seed tissues, surface sterilization is performed. This involves treating seeds with sterilizing agents, such as sodium hypochlorite, to remove external contaminants (McGrath, 2005; Parsa *et al.*, 2016). Following sterilization, testing methods are applied to detect the pathogen. Molecular techniques like PCR and quantitative PCR (qPCR) are highly sensitive for detecting pathogen DNA or RNA (Ellis *et al.*, 2020; Thierry *et al.*, 2020; Schumpp *et al.*, 2021). ELISA also offers a rapid and specific method for detecting pathogen proteins, making it ideal for large-scale screening (Salomone and Roggero, 2002; Sevik and Kose-Tohumcu, 2011). Microscopy methods, such as transmission electron microscopy (TEM), can provide visual confirmation of the pathogen's location (Yang *et al.*, 1997; Chen *et al.*, 2021; Martinez *et al.*, 2021). If the pathogen is culturable, isolation on selective growth media can further validate its presence and viability (Arias *et al.*, 2020; Kim *et al.*, 2022).

Controlled germination of the seeds is an essential step in proving seed-borne transmission. Seeds are germinated under sterile conditions in a controlled environment, such as an insect-proof greenhouse or in sterile growth media. This ensures that any pathogen detected in the resulting seedlings originates from the seed itself, ruling out external contamination or vector-mediated transmission (Antignus *et al.*, 2007; Andreason *et al.*, 2021). Seedlings are closely monitored for disease symptoms, and further testing is conducted to detect the pathogen in seedling tissues, confirming its transmission from the seed to the new plant (Antignus *et al.*, 2007; Andreason *et al.*, 2021). Healthy seeds that have undergone surface sterilization are included as controls to confirm the absence of pathogens in the experimental environment. These controls help validate that the detected pathogen is transmitted through the seed, not external sources.

Reproduction of the disease in healthy plants is a critical step in fulfilling Koch's postulates, which are used to establish a causal relationship between a pathogen and a disease. The pathogen isolated from the seed or seedling must be shown to cause the same disease symptoms when introduced into a healthy plant. It must then be re-isolated from the newly infected plant and identified as the same pathogen. This process confirms the pathogen's ability to infect plants and spread disease through seed-borne transmission (Kabeere *et al.*, 1997).

For pathogens that cannot be cultured, fulfilling the traditional Koch's postulates may not be feasible. In such cases, modified versions of Koch's postulates can be applied (Falkow, 1988; Fredericks and Relman, 1996; Byrd and Segre, 2016; Prescott *et al.*, 2017). Molecular methods play a key role in these modifications, where the pathogen's genetic material can be detected and shown to be consistently associated with the disease. Techniques such as PCR or next-generation sequencing enable precise pathogen identification at the genetic level (Sibley *et al.*, 2012; Hilt and Ferrieri, 2022). In addition to molecular methods, high-resolution imaging techniques, such as TEM, SEM, or confocal microscopy, can provide visual evidence of the pathogen within the seed or seedling tissues. These lens-based methods allow direct visualization of the pathogen and its interaction with the host, supporting the validation of modified Koch's postulates when molecular evidence alone is insufficient (Franke *et al.*, 2014; Sanchez Pina *et al.*, 2021).

Seed Transmission Modes and Mechanisms of Different Pathogens

Seed transmission of pathogens occurs through diverse modes and mechanisms, reflecting each pathogen group's unique biology and survival strategies. Fungal pathogens such as *Fusarium* spp. and *Aspergillus* spp. can cause significant seed-borne diseases by infecting seeds externally or internally (Pietro *et al.*, 2003; Singh and Mathur, 2004; Blanco and Aveling, 2016; Goko *et al.*, 2021; Martín *et al.*, 2022). External infections occur when fungal spores adhere to the seed coat during harvest or storage, especially under humid conditions (Martín *et al.*, 2022). On the other hand, internal infections result from fungal infiltration of vascular tissues during seed development, where the fungi settle in the embryo or endosperm. These internally infected seeds often serve as a reservoir, enabling fungi to survive dormancy and reactivate during germination, leading to diseases such as seed rot, seedling blight, and damping-off (Pietro *et al.*, 2003; Singh and Mathur, 2004; Blanco and Aveling, 2016). For example, *Fusarium verticillioides* commonly infects corn (maize) and is responsible for seedling blight and root rot in many cereals. This pathogen often infiltrates seeds through the vascular system, remaining dormant until the seed germinates. Upon reactivation, it compromises seedling growth, resulting in stunted or weakened plants. Additionally, *F. verticillioides* produces fumonisins, toxic secondary metabolites that can contaminate the food chain (Wilke *et al.*, 2007; Duncan and Howard, 2010).

Bacterial pathogens like *Xanthomonas campestris* and *Pseudomonas syringae* use diverse strategies to colonize seeds (Roberts *et al.*, 1999; Dutta *et al.*, 2014). They can be externally transmitted when seeds come into contact with contaminated plant surfaces, soil, or water. Occasionally, internal contamination occurs when bacteria infiltrate developing seeds through natural openings or cracks in the seed coat. Once present on or within seeds, these bacteria can persist during dormancy. Upon germination, they spread apoplastically through intercellular spaces, leading to systemic infections that cause bacterial blights, spots, or wilts (Roberts *et al.*, 1999; Dutta *et al.*, 2014).

Plant viruses are transmitted through seeds by two primary mechanisms: external contamination of the seed coat and internal invasion of the embryo (Johansen *et al.*, 1994; Maule and Wang, 1996; Sastry, 2013; Dombrovsky and Smith, 2017; Cobos *et al.*, 2019). External contamination occurs when the virus adheres to the seed surface during plant growth or harvest, with infection initiated during germination through abrasions caused by soil particles. This mechanism is well documented for tobamoviruses like TMV (Sastry, 2013; Dombrovsky and Smith, 2017). Viruses can infect seeds by entering reproductive organs (ovules or pollen) before fertilization or moving from maternal tissues to the embryo after fertilization during early seed development. For successful transmission, the virus must infiltrate reproductive structures without affecting gamete or embryo viability. Once inside, the virus persists in the seed coat, embryo, or endosperm during dormancy. Upon germination, it becomes active and spreads systemically in the seedling, starting a new infection cycle (Johansen *et al.*, 1994; Maule and Wang, 1996). This efficient transmission method, observed in viruses like TMV and CMV, presents a significant challenge in managing viral crop diseases.

Nematodes, though less commonly associated with seeds, also exploit seed transmission. Species like *Anguina tritici* infest seeds by embedding their larvae or eggs within developing pods or grains. These nematodes survive storage in a dormant state and become active during germination. They then infect the emerging plant, causing diseases such as seed gall, which can severely affect crop quality and yield (Holajjer *et al.*, 2020).

Challenges in Proving Seed Transmission of Phytoplasmas

Phytoplasma transmission through seeds has long been a topic of debate, primarily due to the unique biological characteristics of these pathogens. Phytoplasmas are phloem-restricted pathogens, meaning they reside exclusively in the phloem tissue of plants, where they move and multiply (Lee *et al.*, 2000; Wei and Zhao, 2022). Since mature seeds lack functional phloem (Werner *et al.*, 2011; Feurtado, 2017), it is unclear how phytoplasmas could persist within seeds or establish an infection in germinating seedlings. This anatomical limitation represents a significant obstacle to confirming seed transmission of phytoplasmas.

Phytoplasmas cannot be cultured in conventional artificial media (Wei and Zhao, 2022). Despite isolated reports of limited success with specialized formulations (Contaldo *et al.*, 2012), these findings have yet to be independently verified or consistently reproduced. Inability to culture phytoplasmas makes it difficult to design experiments that could conclusively demonstrate seed transmission. Therefore, studies must rely on indirect methods, such as molecular detection or microscopy studies. Another significant barrier to demonstrating seed transmission is that phytoplasmas cannot be mechanically inoculated or transferred (Lee *et al.*, 2000; Weintraub and Beanland, 2006). Many plant pathogens can be transmitted by simple mechanical methods, such as rubbing or sap inoculation (Laidlaw, 1986; Hsu *et al.*, 1993). In contrast, phytoplasmas require insect vectors, typically phloem-feeding insects such as leafhoppers or psyllids, to transmit them from one plant to another in nature (Wei and Zhao, 2022). This reliance on specific vectors complicates experimental efforts to artificially induce infection through seeds and confirm whether phytoplasmas can indeed be transmitted via this route.

Given these significant hurdles, proving phytoplasma transmission through seeds requires a systematic approach. This begins with surface sterilization and precise dissection of seed tissues, followed by molecular detection methods, microscopy-based visualization, and grow-out tests (details provided in the following section). These steps are essential to determine whether phytoplasmas can truly persist in seeds (embryos) and establish viable infections in subsequent generations.

Studies and Current Evidence on the Potential Seed Transmission of Phytoplasmas

Although the primary mode of phytoplasma transmission involves insect vectors, the potential for seed transmission has sparked considerable debate. Some studies revealed that phytoplasma cannot be transmitted through seeds (Faghihi *et al.*, 2011; Akhtar *et al.*, 2013; Ranebennur *et al.*, 2022). However, some studies have detected phytoplasma DNA in seeds and seedlings, suggesting a potential vertical transmission route (Cordova *et al.*, 2003; Nečas *et al.*, 2008; Çağlar *et al.*, 2019). This uncertainty arises from the challenges in detecting phytoplasmas in seeds, distinguishing true seed-borne infection from surface contamination, and demonstrating the ability of phytoplasmas to survive and remain infectious across plant generations. However, the absence of phloem connections in mature seeds raises questions about how phytoplasmas might survive in seeds. While some reports indicate their presence in non-vascular tissues like the endosperm or seed coat, the viability and infectivity of phytoplasmas in seedlings remain unproven. This uncertainty continues to spark debate among scientists and points to the need for more research in this area.

In this chapter, four studies on the potential phytoplasma transmission through seeds are reviewed, summarized, and compared (Table 6.1), covering: (i) aster yellows and stolbur phytoplasmas in corn; (ii) Bogia coconut syndrome phytoplasma in coconut and betel nut; (iii) lethal yellowing phytoplasma in coconut palms; and (iv) sandalwood spike phytoplasma in sandalwood (Cordova *et al.* 2003; Satta *et al.*, 2019; Kirdat *et al.*, 2022; Lu *et al.*, 2024).

Satta *et al.* (2019) investigated the potential for seed transmission of phytoplasmas in corn by analyzing seedlings grown from seeds of symptomatic and phytoplasma-infected cobs. Seeds were sterilized and germinated *in vitro*, and the resulting seedlings were grown in insect-proof cages. Molecular detection using nested PCR on 16S ribosomal DNA identified phytoplasma DNA (aster yellows and stolbur phytoplasmas) in 17 out of 79 seedlings. Viability tests involved isolating phytoplasmas in chemically defined media (PivL®), with color changes and colony formation observed in multiple trials. Re-isolation of phytoplasmas confirmed their presence in one seedling 90 days post germination, and colonies were propagated in liquid and solid media over subsequent passages (Table 6.1). Despite these findings, the study did not address whether the detected phytoplasmas were located within the seed embryo, confined to maternal tissues, or present as surface contaminants. Moreover, the low transmission rate (17 out of 79 seedlings) and the preliminary nature of the viability data suggest that further research is required to determine whether phytoplasmas can establish systemic infections via seeds. While the results hint at the possibility of seed-borne phytoplasmas, conclusive evidence is lacking, particularly as the reported *in vitro* isolation and propagation of phytoplasmas have yet to be replicated or confirmed by other research groups.

The study conducted by Lu *et al.* (2024) focused on the seed transmission of '*Ca.* Phytoplasma noviguineense,' associated with Bogia coconut syndrome (BCS), a lethal disease affecting coconut and betel nut palms in Papua New Guinea. Researchers collected drupes from symptomatic and asymptomatic palms, dissected various seed tissues, and analyzed them for phytoplasma DNA using nested PCR, loop-mediated isothermal amplification (LAMP) and phylogenetic analysis. Phytoplasma DNA was detected in multiple seed-associated tissues, including the husk, shell, flesh, and embryos, with embryos showing a detection rate of 11.3%. However, DNA detection alone is insufficient to prove phytoplasma colonization in the seed embryo.

Seedlings germinated in insect-proof cages were tested at 6 months of age. Among these, 25.6% of coconut and betel nut seedlings tested positive for phytoplasma DNA. Phylogenetic analysis confirmed that the DNA sequences matched BCS-associated phytoplasma strains. However, the study did not include experiments to assess the viability of phytoplasmas or their ability to cause systemic infections in mature plants. Additionally, the detection of phytoplasma DNA in embryos and seedlings suggests the possibility of seed-borne transmission, but the lack of functional validation limits its epidemiological relevance (Table 6.1).

The study by Cordova *et al.* (2003) presented preliminary evidence suggesting the potential for seed transmission of lethal yellowing (LY) phytoplasma in coconut palms, based on the detection of phytoplasma DNA in 13 out of 72 embryos from infected palms. Nevertheless, key limitations remain. The low frequency of detection implies that transmission, if possible, may be uncommon or dependent on specific factors, such as the developmental stage of the fruit during infection. While *in situ* PCR localized phytoplasma DNA to regions like the plumule, indicating possible internal colonization of embryos, the study was unable to confirm whether the detected phytoplasmas were viable or capable of causing disease. Additionally, seed transmission remains speculative without direct proof that phytoplasma-positive embryos can develop into infected plants. The lack of clear mechanisms for phytoplasma movement into the embryo and the absence of

Table 6.1. The findings and methodologies of the studies on seed transmission of phytoplasmas in corn, coconut, and sandalwood.

Study details	Study 1 Satta *et al.*, 2019	Study 2 Lu *et al.*, 2024	Study 3 Cordova *et al.*, 2003	Study 4 Kirdat *et al.*, 2022
Phytoplasma	Aster yellows and stolbur phytoplasmas	'*Ca.* Phytoplasma noviguineense' (Bogia coconut syndrome (BCS) phytoplasma)	Lethal yellowing (LY) phytoplasma	Sandalwood spike disease (SSD) phytoplasma
Plant host	Corn (*Zea mays*)	Coconut (*Cocos nucifera*), betel nut (*Areca catechu*)	Coconut (*Cocos nucifera*)	Sandalwood (*Santalum album*)
Detection methods	Nested PCR and cultivation in media*	Nested PCR, LAMP, phylogenetic analysis	Nested PCR, *in situ* PCR, DAPI staining	Nested PCR, qPCR, and SEM†
Tissue types analyzed	Seeds and seedlings	Husk, shell, flesh, embryo, seedlings	Embryos, seedlings	Seeds and seedlings
Controlled environment	Insect-proof cages	Insect-proof cages	Not explicitly mentioned	Insect-proof cages
Evidence of phytoplasma DNA	17 out of 79 seedlings tested positive for phytoplasma DNA	10 out of 39 coconut seedlings and 15 out of 59 betel nut seedlings tested positive	13 out of 72 embryos tested positive for phytoplasma DNA	Phytoplasma DNA detected in 35.14% of seeds and 38.66% of seedlings
Viability testing performed	Yes, isolation and growth in media; colony formation observed	No direct viability testing	No direct viability testing	No direct viability testing
Embryo colonization	Not confirmed	Phytoplasma DNA detected in 11.3% of embryos	Phytoplasma DNA is localized specifically in the plumule and surrounding cells	Not confirmed
Seedling infection	Positive for phytoplasma DNA, but systemic infection unproven	25.6% of coconut and betel nut seedlings were positive for phytoplasma DNA	Positive seedlings observed, but systemic infection not demonstrated	38.66% of seedlings tested positive for phytoplasma DNA
Functional transmission	Preliminary evidence; no mature plant infection shown	Preliminary evidence; systemic infection not demonstrated	Preliminary evidence; systemic infection not demonstrated	Preliminary evidence; viability beyond seedlings not proven
Transmission efficiency	Low (17 out of 79 seedlings tested positive)	Low (25.6% of seedlings tested positive)	Low (18% of embryos tested positive)	Moderate (38.66% of seedlings tested positive)
Conclusion	Evidence of seed-borne phytoplasma DNA; viability remains unproven	Evidence of seed-borne phytoplasma DNA; viability remains unproven	Evidence of seed-borne phytoplasma DNA; viability remains unproven	Evidence of seed-borne phytoplasma DNA; viability remains unproven

**In vitro* culture of phytoplasmas has not been verified by other independent research groups.
†Mature seeds lack phloem, where phytoplasmas reside. SEM visualization of phytoplasma cells in seeds need to be confirmed by other independent research groups.
DAPI, 4′,6-diamidino-2-phenylindole; LAMP, loop-mediated isothermal amplification; PCR, polymerase chain reaction; SEM, scanning electron microscopy.

viability-specific assays further complicate the interpretation of results. The authors recommended bioassays, controlled germination experiments, and field trials to establish the viability of seeds and infectivity of phytoplasmas. This would provide the definitive evidence needed to confirm seed transmission (Table 6.1).

The study conducted by Kirdat *et al.* (2022) investigates the potential seed transmission of sandalwood spike disease (SSD) caused by phytoplasma in sandalwood (*Santalum album*). Seeds from symptomatic trees were tested using nested PCR, nested-qPCR, and SEM. The results indicated phytoplasma DNA in 35.14% of seeds and 38.66% of early-stage seedlings. While the authors reported visualizing phytoplasma-like cells in seed tissues using SEM, the provided figure only showed such cells in the fruit mesocarp (not part of the seed) and the hypocotyl tissue of the seedling (Kirdat *et al.*, 2022).

The study was conducted under insect-proof conditions, ensuring that any detected phytoplasma originated from seeds, not external vectors. While these findings suggest seed-borne transmission, key limitations prevent definitive proof. No viability testing was conducted, leaving it uncertain whether the detected phytoplasmas could cause systemic infection. Seedlings were monitored for up to 120 days, and no systemic infection in mature plants was demonstrated. The low germination rate of symptomatic seeds (21.9%) and early death of seedlings further question the ecological relevance of seed-borne phytoplasmas. Moreover, the results remain inconclusive without evidence of phytoplasma viability beyond the early stages (Table 6.1). The study acknowledged significant gaps, particularly in proving whether phytoplasmas infect the embryo or remain restricted to maternal tissues. The absence of viability assays also limits the distinction between live and dead phytoplasmas. Although nested PCR confirmed the presence of phytoplasma DNA, further research involving viability tests, inoculation experiments, and long-term field studies is necessary to validate seed-borne transmission and its role in SSD spread.

Based on these four studies, seed-borne phytoplasma transmission, if it occurs, appears inefficient and likely constitutes a minor pathway compared with vector-mediated or vegetative transmission. While phytoplasma DNA has been detected in seeds, critical gaps remain in understanding whether these pathogens can infect and persist within the embryo and are viable and capable of causing systemic infections. These unresolved questions highlight the complexity of studying seed-borne phytoplasma transmission and its potential role in disease epidemiology.

Circumstantial evidence, such as the use of insect-proof environments to exclude secondary infections, supports the possibility of seed transmission. However, conclusive proof requires: (i) demonstrating viable and infective phytoplasmas within seeds; (ii) directly observing phytoplasma colonization pathways into seed tissues, particularly the embryo; and (iii) experimentally reproducing infection cycles from seeds to mature plants under controlled conditions. Until these criteria are met, seed-borne phytoplasma transmission remains an unproven hypothesis, underscoring the need for further research to determine its broader implications for seed pathology and disease management.

Standardized Protocols for Seed Transmission of Phytoplasmas

A standardized protocol, which includes sterilization, diagnostics, microscopy, and grow-out tests, is crucial for examining seed transmission of phytoplasmas.

1. Sterilization. This step is essential for ensuring that detected phytoplasmas originate from internal tissues, as it eliminates external contaminants. Seeds should undergo treatment with 70% ethanol for 1 min, followed by 2% sodium hypochlorite for 10 min, after which they are rinsed and dissected under a microscope to analyze the tissues.

2. DNA analysis. DNA is extracted and analyzed using PCR or qPCR with specific primers. A higher phytoplasma load in the embryo indicates internal infection; however, DNA detection alone does not confirm cell viability. Viable cells can be identified through microscopy based on their shape and the absence of a rigid cell wall.

3. Grow-out test. This test confirms vertical transmission by monitoring progeny from infected seeds in sterile conditions. Researchers observe symptoms and conduct tissue sampling to test for phytoplasma presence using PCR/qPCR. Controls with healthy seeds are included to rule out contamination.

4. Validation. Consistent detection of phytoplasmas across generations provides evidence of seed-borne transmission. It is crucial to control environmental factors for accuracy and to involve cross-validation by different laboratories. Clear and robust evidence is necessary for policymakers to validate seed certification and impose trade restrictions, ensuring that regulations are scientifically sound and not unnecessarily restrictive.

Conclusion

Phytoplasmas face significant barriers to seed transmission due to the absence of a phloem environment in seeds, which is essential for their survival and multiplication. While phytoplasma DNA has been detected in seeds, this does not confirm the presence of viable or infective forms capable of causing systemic infection in seedlings. If seed transmission occurs, it is likely rare and dependent on specific host–pathogen interactions, representing exceptions rather than a common pathway. The lack of direct evidence for embryonic infection suggests seed transmission is unlikely to play a significant role in the phytoplasma life cycle, although further research is warranted to clarify its feasibility and implications.

References

Akhtar, K.P., Dickinson, M., Shah, T.M., and Sarwar, N. (2013) Natural occurrence, identification and transmission of the phytoplasma associated with flax phyllody and stem fasciation in Pakistan. *Phytoparasitica* 41, 383–389. DOI: 10.1007/s12600-013-0299-8.

Ali, A. and Kobayashi, M. (2010) Seed transmission of Cucumber mosaic virus in pepper. *Journal of Virological Methods* 163, 234–237. DOI: 10.1016/j.jviromet.2009.09.026.

Andreason, S.A., Olaniyi, O.G., Gilliard, A.C., Wadl, P.A., Williams, L.H. *et al.* (2021) Large-scale seedling grow-out experiments do not support seed transmission of sweet potato leaf cur virus in sweet potato. *Plants* 10, 139. DOI: 10.3390/plants10010139.

Antignus, Y., Lachman, O., and Pearlsman, M. (2007) Spread of Tomato apical stunt viroid (TASVd) in greenhouse tomato crops is associated with seed transmission and bumble bee activity. *Plant Disease* 91, 47–50. DOI: 10.1094/PD-91-0047.

Arias, S.L., Block, C.C., Mayfield, D.A., Santillana, G., Stulberg, M.J. *et al.* (2020) Occurrence in seeds and potential seed transmission of Xanthomonas vasicola pv. vasculorum in maize in the United States. *Phytopathology* 110, 1139–1146. DOI: 10.1094/PHYTO-08-19-0306-R.

Barro-Trastoy, D., Gómez, M.D., Tornero, P., and Pérez-Amador, M.A. (2020) On the way to ovules: The hormonal regulation of ovule development. *Critical Reviews in Plant Sciences* 39, 431–456. DOI: 10.1080/07352689.2020.1820203.

Blanco, R. and Aveling, T.A.S. (2016) Seed-borne *Fusarium* pathogens in agricultural crops. *Acta Horticulturae* 1204, 161–170. DOI:10.17660/ActaHortic.2018.1204.21

Bobrov, A.V. and Romanov, M.S. (2019) Morphogenesis of fruits and types of fruit of angiosperms. *Botany Letters* 166, 366–399. DOI: 10.1080/23818107.2019.1663448.

Boesewinkel, F.D. and Bouman, F. (1984) The seed: structure. In: Johri, B.M. (ed.) *Embryology of Angiosperms*. Springer, Berlin, pp. 567–610. DOI: 10.1007/978-3-642-69302-1_12.

Byrd, A.L. and Segre, J.A. (2016) Adapting Koch's postulates. *Science* 351, 224–226. DOI: 10.1126/science.aad6753.

Çağlar, B.K., Satar, S., Bertaccini, A., and Elbeaino, T. (2019) Detection and seed transmission of Bermudagrass phytoplasma in maize in Turkey. *Journal of Phytopathology* 167, 248–255. DOI: 10.1111/jph.12792.

Chen, N.W.G., Ruh, M., Darrasse, A., Foucher, J., Briand, M. *et al.* (2021) Common bacterial blight of bean: A model of seed transmission and pathological convergence. *Molecular Plant Pathology* 22, 1464–1480. DOI: 10.1111/mpp.13067.

Cobos, A., Montes, N., López-Herranz, M., Gil-Valle, M., and Pagán, I. (2019) Within-host multiplication and speed of colonization as infection traits associated with plant virus vertical transmission. *Journal of Virology* 93, e01078-19. DOI: 10.1128/JVI.01078-19.

Contaldo, N., Bertaccini, A., Paltrinieri, S., Windsor, H.M., and Windsor, G.D. (2012) Axenic culture of plant pathogenic phytoplasmas. *Phytopathologia Mediterranea* 51, 607–617. DOI: 10.14601/Phytopathol_Mediterr-11773.

Cordova, I., Jones, P., Harrison, N.A., and Oropeza, C. (2003) In situ PCR detection of phytoplasma DNA in embryos from coconut palms with lethal yellowing disease. *Molecular Plant Pathology* 4, 99–108. DOI: 10.1046/j.1364-3703.2003.00152.x.

Correa, D.F., Stevenson, P.R., Umaña, M.N., Coelho, L.D.S., Lima Filho, D.D.A. *et al.* (2023) Geographic patterns of tree dispersal modes in Amazonia and their ecological correlates. *Global Ecology and Biogeography* 32, 49–69. DOI: 10.1111/geb.13648.

Dell'Olmo, E., Tiberini, A., and Sigillo, L. (2023) Leguminous seedborne pathogens: Seed health and sustainable crop management. *Plants* 12, 2040. DOI: 10.3390/plants12102040.

Doll, N.M. and Ingram, G.C. (2022) Embryo–endosperm interactions. *Annual Review of Plant Biology* 73, 293–321. DOI: 10.1146/annurev-arplant-102820-091838.

Dombrovsky, A. and Smith, E. (2017) Seed transmission of Tobamoviruses: Aspects of global disease distribution. In: Jimenez-Lopez, J.C. (ed.) *Advances in Seed Biology*. IntechOpen, pp. 233–260. DOI: 10.5772/intechopen.70244.

Duncan, K.E. and Howard, R.J. (2010) Biology of maize kernel infection by Fusarium verticillioides. *Molecular Plant-Microbe Interactions* 23, 6–16. DOI: 10.1094/MPMI-23-1-0006.

Dutta, B., Gitaitis, R., Smith, S., and Langston Jr, D. (2014) Interactions of seedborne bacterial pathogens with host and non-host plants in relation to seed infestation and seedling transmission. *PLoS One* 9, e99215. DOI: 10.1371/journal.pone.0099215.

Ellis, M.D., Hoak, J.M., Ellis, B.W., Brown, J.A., Sit, T.L. *et al.* (2020) Quantitative real-time PCR analysis of individual flue-cured tobacco seeds and seedlings reveals seed transmission of tobacco mosaic virus. *Phytopathology* 110, 194–205. DOI: 10.1094/PHYTO-07-19-0245-R.

Elmer, W.H. (2001) Seeds as vehicles for pathogen importation. *Biological Invasions* 3, 263–271. DOI: 10.1023/A:1015244004722.

Faghihi, M.M., Bagheri, A.N., Bahrami, H.R., Hasanzadeh, H., Rezazadeh, R. *et al.* (2011) Witches'-broom disease of lime affects seed germination and seedling growth but is not seed transmissible. *Plant Disease* 95, 419–422. DOI: 10.1094/PDIS-07-10-0503.

Falkow, S. (1988) Molecular Koch's postulates applied to microbial pathogenicity. *Reviews of Infectious Disease* 10 (Suppl. 2), S274–S276. DOI: 10.1093/clinids/10.Supplement_2.S274.

Feurtado, J.A. (2017) Physiology of maturation. In: Thomas, B., Murray, B.G., and Murphy, D.J. (eds) *Encyclopedia of Applied Plant Sciences*, 2nd edn. Academic Press, London, pp. 521–531.

Franke, J.L., Geary, B., and Meyer, S.E. (2014) Identification of the infection route of a *Fusarium* seed pathogen into nondormant *Bromus tectorum* seeds. *Phytopathology* 104, 1306–1313. DOI: 10.1094/PHYTO-03-14-0077-R.

Fredericks, D.N. and Relman, D.A. (1996) Sequence-based identification of microbial pathogens: A reconsideration of Koch's postulates. *Clinical Microbiology Reviews* 9, 18–33. DOI: 10.1128/CMR.9.1.18.

Gebeyaw, M. (2020) Review on: impact of seed-borne pathogens on seed quality. *American Journal of Plant Biology* 5, 77–81. DOI: 10.11648/j.ajpb.20200504.11.

Gitaitis, R. and Walcott, R. (2007) The epidemiology and management of seedborne bacterial diseases. *Annual Review of Phytopathology* 45, 371–397. DOI: 10.1146/annurev.phyto.45.062806.094321.

Glover, B. (2014) *Understanding Flowers and Flowering*, 2nd edn. Oxford University Press, Oxford, UK.

Goko, M.L., Murimwa, J.C., Gasura, E., Rugare, J.T. and Ngadze, E. (2021) Identification and characterisation of seed-borne fungal pathogens associated with maize (*Zea mays* L.). *International Journal of Microbiology* 2021, 6702856. DOI: 10.1155/2021/6702856.

Gullino, M.L., Gilardi, G., Sanna, M., and Garibaldi, A. (2009) Epidemiology of *Pseudomonas syringae* pv. *syringae* on tomato. *Phytoparasitica* 37, 461–466. DOI: 10.1007/s12600-009-0055-2.

Gullino, M.L., Gilardi, G., and Pugliese, M. (2024) Management of *Fusarium* wilt of lettuce. *Plant Health Cases* 2024, phcs20240025. DOI: 10.1094/PHCS-2024-0025.

Hernandez, J.O., Naeem, M., and Zaman, W. (2023) How does changing environment influence plant seed movements as populations of dispersal vectors decline? *Plants* 12, 1462. DOI: 10.3390/plants12071462.

Hiddink, G.A., Willmann, R., Woudenberg, J.H.C., and Souza-Richards, R. (2023) Seed health testing: Doing things right. *PhytoFrontiers* 3, 71–74. DOI: 10.1094/PHYTOFR-02-23-0014-FI.

Hilt, E.E. and Ferrieri, P. (2022). Next generation and other sequencing technologies in diagnostic microbiology and infectious diseases. *Genes* 13, 1566. DOI: 10.3390/genes13091566.

Holajjer, P., Jadon, K.S., Chandrawat, B.S., and Gawade, B. (2020) Seed-borne and seed-associated nematodes: An overview. In: Kumar, R. and Gupta, A. (eds) *Seed-borne Diseases of Agricultural Crops: Detection, Diagnosis and Management*. Springer, Singapore, pp. 355–368.

Hsu, S.T., Hong, W.F., Tzeng, K.C., and Chen, C.C. (1993) Bacterial wilt of perilla caused by *Pseudomonas solanacearum* and its transmission. *Plant Disease* 77, 674–677.

Johansen, E., Edwards, M.C., and Hampton, R.O. (1994) Seed transmission of viruses: Current perspectives. *Annual Review of Phytopathology* 32, 363–386. DOI: 10.1146/annurev.py.32.090194.002051.

Kabeere, F., Hill, M.J., and Hampton, J.G. (1997) The transmission of *Fusarium subglutinans* from maize seeds to seedlings. *Australasian Plant Pathology* 26, 126–130. DOI: 10.1071/AP97025.

Kim, J., Roy, M., Ahn, S.H., Shanmugam, G., Yang, J.S., Jung, H.W., and Jeon, J. (2022) Culturable endophytes associated with soybean seeds and their potential for suppressing seed-borne pathogens. *Plant Pathology Journal* 38, 313–321. DOI: 10.5423/PPJ.OA.02.2022.0025.

Kirdat, K., Tiwarekar, B., Swetha, P., Padma, S., Thorat, V. *et al.* (2022) Nested real-time PCR assessment of vertical transmission of sandalwood spike phytoplasma ('Ca. phytoplasma asteris'). *Biology* 11, 1494. DOI: 10.3390/biology11101494.

Laidlaw, W.M.R. (1986) Mechanical aids to improve the speed and sensitivity of plant virus diagnosis by the biological test method. *Annals of Applied Biology* 108, 309–318. DOI: 10.1111/j.1744-7348.1986.tb07666.x.

Larsson, E., Vivian-Smith, A., Offringa, R., and Sundberg, E. (2017) Auxin homeostasis in *Arabidopsis* ovules is anther-dependent at maturation and changes dynamically upon fertilization. *Frontiers in Plant Science* 8, 1735. DOI: 10.3389/fpls.2017.01735.

Lee, I.M., Davis, R.E., and Gundersen-Rindal, D.E. (2000) Phytoplasma: phytopathogenic mollicutes. *Annual Review of Microbiology* 54, 221–255. DOI: 10.1146/annurev.micro.54.1.221.

Lu, H., Wilson, B., Zhang, H., Woruba, S.B., Feng, B., Johnson, A.C. *et al.* (2024) Detection and identification of Bogia coconut syndrome phytoplasma from seed-associated tissues and seedlings of coconut (*Cocos nucifera*) and betel nut (*Areca catechu*). *Scientific Reports* 14, 11542. DOI: 10.1038/s41598-024-11542-0.

MacLean, A.M., Sugio, A., Makarova, O.V., Findlay, K.C., Grieve, V.M. *et al.* (2011) Phytoplasma effector SAP54 induces indeterminate leaf-like flower development in *Arabidopsis* plants. *Plant Physiology* 157, 831–841. DOI: 10.1104/pp.111.181586.

Martín, I., Gálvez, L., Guasch, L., and Palmero, D. (2022) Fungal pathogens and seed storage in the dry state. *Plants* 11, 3167. DOI: 10.3390/plants11223167.

Martinez, S.I., Wegner, A., Bohnert, S., Schaffrath, U., and Perelló, A. (2021) Tracing seed to seedling transmission of the wheat blast pathogen *Magnaporthe oryzae* pathotype *Triticum*. *Plant Pathology* 70, 1562–1571. DOI: 10.1111/ppa.13415.

Maude, R.I. (1996) *Seedborne Diseases and their Control: Principles and Practice*. CAB International, Wallingford, UK.

Maule, A.J. and Wang, D. (1996) Seed transmission of plant viruses: a lesson in biological complexity. *Trends in Microbiology* 4, 153–158. DOI: 10.1016/0966-842X(96)30028-0.

McGrath, M.T. (2005) *Treatments for managing bacterial pathogens in vegetable seed. Vegetable MD Online*. Cornell University, Ithaca, NY.

Mehan, V.K., Mayee, C.D., Jayanthi, S., and McDonald, D. (1991) Preharvest seed infection by *Aspergillus flavus* group fungi and subsequent aflatoxin contamination in groundnuts in relation to soil types. *Plant and Soil*, 136, 239–248. DOI: 10.1007/BF00010818.

Nečas, T., Mašková, V., and Krška, B. (2008) The possibility of ESFY phytoplasma transmission through flowers and seeds. *Acta Horticulturae* 781, 443–448. DOI: 10.17660/ActaHortic.2008.781.56.

Ohto, M.A., Floyd, S.K., Fischer, R.L., Goldberg, R.B., and Harada, J.J. (2009) Effects of APETALA2 on embryo, endosperm, and seed coat development determine seed size in *Arabidopsis*. *Sexual Plant Reproduction* 22, 277–289. DOI: 10.1007/s00497-009-0116-1.

Oshima, K., Kakizawa, S., Nishigawa, H., Jung, H.Y., Wei, W. *et al.* (2004) Reductive evolution suggested from the complete genome sequence of a plant-pathogenic phytoplasma. *Nature Genetics* 36, 27–29. DOI: 10.1038/ng1277.

Parsa, S., García-Lemos, A.M., Castillo, K., Ortiz, V., López-Lavalle, L.A.B., Braun, J., and Vega, F.E. (2016) Fungal endophytes in germinated seeds of the common bean, *Phaseolus vulgaris*. *Fungal Biology* 120, 783–790. DOI: 10.1016/j.funbio.2016.03.003.

Pietro, A.D., Madrid, M.P., Caracuel, Z., Delgado-Jarana, J., and Roncero, M.I.G. (2003) *Fusarium oxysporum*: Exploring the molecular arsenal of a vascular wilt fungus. *Molecular Plant Pathology* 4, 315–325. DOI: 10.1046/j.1364-3703.2003.00180.x.

Prescott, J., Feldmann, H., and Safronetz, D. (2017) Amending Koch's postulates for viral disease: When "growth in pure culture" leads to a loss of virulence. *Antiviral Research* 137, 1–5. DOI: 10.1016/j.antiviral.2016.11.001.

Primack, R.B. (1987) Relationships among flowers, fruits, and seeds. *Annual Review of Ecology, Evolution, and Systematics* 18, 409–430. DOI:10.1146/annurev.es.18.110187.002205.

Ranebennur, H., Rawat, K., Rao, A., Kumari, P., Chalam, V.C., Meshram, N., and Rao, G.P. (2022) Transmission efficiency of a 'Candidatus Phytoplasma australasia' (16SrII-D) related strain associated with sesame phyllody by dodder, grafting and leafhoppers. *European Journal of Plant Pathology* 164, 193–208. DOI: 10.1007/s10658-022-02548-2.

Roberts, S.J., Hiltunen, L.H., Hunter, P.J., and Brough, J. (1999) Transmission from seed to seedling and secondary spread of *Xanthomonas campestris* pv. *campestris* in Brassica transplants: Effects of dose and watering regime. *European Journal of Plant Pathology* 105, 879–889. DOI: 10.1023/A:1008733404382.

Salomone, A. and Roggero, P. (2002) Host range, seed transmission and detection by ELISA and lateral flow of an Italian isolate of *Pepino mosaic virus*. *Journal of Plant Pathology* 84, 65–68.

Sanchez Pina, M.A., Gómez-Aix, C., Méndez-López, E., Gosalvez Bernal, B., and Aranda, M.A. (2021) Imaging techniques to study plant virus replication and vertical transmission. *Viruses* 13, 358. DOI: 10.3390/v13030358.

Sastry, K.S. (2013) *Seed-borne Plant Virus Diseases*. Springer, Dordrecht, the Netherlands. DOI: 10.1007/978-94-007-7822-5.

Satta, E., Paltrinieri, S., and Bertaccini, A. (2019) Phytoplasma transmission by seed. In: Bertaccini, A., Weintraub, P.G., Rao, G.P., and Mori, N. (eds) *Phytoplasmas: Plant Pathogenic Bacteria – II*. Springer, Singapore, pp. 131–147. DOI: 10.1007/978-981-13-9632-8_6.

Schumpp, O., Bréchon, A., Brodard, J., Dupuis, B., Farinelli, L. *et al.* (2021) Large-scale RT-qPCR diagnostics for seed potato certification. *Potato Research* p2021, 1–17. DOI: 10.1007/s11540-021-09509-5.

Sevik, M.A. and Kose-Tohumcu, E. (2011) The ELISA analysis results in tomato (*Lycopersicon esculentum* Mill.) seed health testing for *Tobacco mosaic virus*. *Zemdirbyste* 98, 301–306.

Sibley, C.D., Peirano, G., and Church, D.L. (2012) Molecular methods for pathogen and microbial community detection and characterization: Current and potential application in diagnostic microbiology. *Infection Genetics and Evolution* 12, 505–521. DOI: 10.1016/j.meegid.2012.02.009.

Singh, D. and Mathur, S.B. (2004) Location of fungal hyphae in seeds. In: Singh, D. and Mathur, S.B. (eds) *Histopathology of Seed-borne Infections*. CRC Press, Boca Raton, FL, pp. 101–168.

Steeves, T.A. and Sawhney, V.K. (2017) *Essentials of Developmental Plant Anatomy*. Oxford University Press, Oxford, UK.

Sugio, A., MacLean, A.M., Grieve, V.M., and Hogenhout, S.A. (2011) Phytoplasma protein effector SAP11 enhances insect vector reproduction by manipulating plant development and defense hormone biosynthesis. *Proceedings of the National Academy of Sciiences of the USA* 108, 1254–1263. DOI: 10.1073/pnas.1014456108.

Tan, Y., Li, Q., Zhao, Y., Wei, H., Wang, J. *et al.* (2021) Integration of metabolomics and existing omics data reveals new insights into phytoplasma-induced metabolic reprogramming in host plants. *PLoS One* 16, e0246203. DOI: 10.1371/journal.pone.0246203.

Thierry, M., Chatet, A., Fournier, E., Tharreau, D., and Ioos, R. (2020) A PCR, qPCR, and LAMP toolkit for the detection of the wheat blast pathogen in seeds. *Plants* 9, 277. DOI: 10.3390/plants9020277.

Wei, W. and Zhao, Y. (2022) Phytoplasma taxonomy: nomenclature, classification, and identification. *Biology* 11, 1119. DOI: 10.3390/biology11081119.

Wei, W., Davis, R.E., Nuss, D.L., and Zhao, Y. (2013) Phytoplasmal infection derails genetically preprogrammed meristem fate and alters plant architecture. *Proceedings of the National Academy of Sciiences of the USA* 110, 19149–19154. DOI: 10.1073/pnas.1310807110.

Wei, W., Davis, R.E., Bauchan, G.R., and Zhao, Y. (2019) New symptoms identified in phytoplasma-infected plants reveal extra stages of pathogen-induced meristem fate-derailment. *Molecular Plant-Microbe Interactions* 32, 1314–1323. DOI: 10.1094/MPMI-03-19-0069-R.

Weintraub, P.G. and Beanland, L. (2006) Insect vectors of phytoplasmas. *Annual Review of Entomology* 51, 91–111. DOI: 10.1146/annurev.ento.51.110104.151039.

Werner, D., Gerlitz, N., and Stadler, R. (2011) A dual switch in phloem unloading during ovule development in *Arabidopsis*. *Protoplasma* 248, 225–235. DOI: 10.1007/s00709-010-0164-x.

Wilke, A.L., Bronson, C.R., Tomas, A., and Munkvold, G.P. (2007) Seed transmission of *Fusarium verticillioides* in maize plants grown under three different temperature regimes. *Plant Disease* 91, 1109–1115. DOI: 10.1094/PDIS-91-9-1109.

Yang, Y., Kim, K.S., and Anderson, E.J. (1997) Seed transmission of cucumber mosaic virus in spinach. *Phytopathology* 87, 924–931. DOI: 10.1094/PHYTO.1997.87.9.924.

7 Overview of RNA Interference Methods to Hamper Phytoplasma Transmission by Phloem-Feeding Vectors

Cecilia Parise,[1,2] Luciana Galetto,[2] Sabrina Palmano,[2] and Cristina Marzachì[2*]
[1]*DISAFA, Università degli Studi di Torino, Torino, Italy;* [2]*Istituto per la Protezione Sostenibile delle Piante, CNR, Torino, Italy*

Abstract

RNA interference (RNAi) is a biological process that regulates gene expression by degrading specific mRNA. This process occurs in most eukaryotes. Initially exploited as a functional genomic tool, RNAi has emerged as a powerful and specific approach for protecting crops against plant pests. Regulatory and technical challenges continue to limit its use at a large scale in the field. RNAi technology has been applied to four hemipteran vectors of phytoplasmas belonging to 16S rRNA groups I and V. The chapter summarizes different strategies for the delivery of silencing molecules (double-stranded RNAs) to these insect species, with special focus on the rationale behind the selection of target genes. For phytoplasma vectors, abolishing or even reducing transmission efficiency can lower the need for insecticide burden in infected crop areas. RNAi functions in some phytoplasma leafhopper vectors and offers an environmentally and ecologically safe approach to control these economically important diseases.

RNA Interference and Pest Control

RNA interference (RNAi) is a sequence-specific gene regulation process found in eukaryotes (Baulcombe, 2004; Fire *et al.*, 1998), and it serves as a primary defense mechanism against foreign nucleic acids, such as those from transposons or viruses (Santos *et al.*, 2019; Bonning and Saleh, 2021). Initially, RNAi strategies were exploited as molecular tools for studying gene function. Today, they are emerging as a powerful and precise approach to control plant pathogens, including fungi, viruses, and insect pests (Christiaens *et al.*, 2020; Taning *et al.*, 2020). RNAi technology uses double-stranded RNAs (dsRNAs) to silence essential genes of targeted pests. Three RNAi pathways have been described in insects: small interfering RNA (siRNA), micro RNA (miRNA), and piwi-interacting RNA (piRNA), reviewed in Ortolá and Daròs (2024). These differ by the length and origin of the guiding small RNAs, enzymatic components, and functional roles, but all involve Argonaute (Ago) family proteins to cleave sequence-specific target RNAs, which are then degraded. In the case of RNAi control of pests, exogenous dsRNAs are processed through the siRNA pathway. This includes Dicer-2 cleavage of long dsRNAs into siRNAs, which are then loaded into RNA-induced silencing complexes (RISC) together with Ago2, resulting in RNA degradation and consequent post-transcriptional gene silencing. This basic mechanism is known in arthropods, but knowledge gaps remain. Systemic RNAi is

*Corresponding author: cristina.marzachi@cnr.it

 Phytoplasmas: Genomes, Plant Hosts and Vectors, 2nd Edition. (Ed. W. Wei)

DOI: 10.1079/9781800627031.0007

well documented for plants, fungi, and nematodes, in which RNA-dependent RNA polymerases (RdRPs) amplify the silencing signal by generating secondary siRNAs that spread throughout the organism (Pinzón *et al.*, 2019; Liu *et al.*, 2020). Insects lack RdRPs (Tomoyasu *et al.*, 2008; Zong *et al.*, 2009; Li *et al.*, 2018; Christiaens *et al.*, 2020), leaving the systemic RNAi signal in arthropods largely unexplained. Another unsolved issue is the variable efficiency of the RNAi mechanism across insect species (Silver *et al.*, 2021). Coleoptera species display robust RNAi responses (Baum *et al.*, 2007; Zhu *et al.*, 2011; Palli, 2014), whereas lepidopterans and dipterans mainly respond only to naked dsRNAs through injection, and, even then, in limited cell types (Miller *et al.*, 2008; Whyard *et al.*, 2009; Terenius *et al.*, 2011; Zhu and Palli, 2020). Moreover, several constraints still hamper the field application of RNAi in agriculture: (i) efficient delivery of dsRNAs to sap-sucking insects (Christiaens and Smagghe, 2014; Coleman *et al.*, 2015; Yu *et al.*, 2016); (ii) weathering stability of dsRNA molecules (Yu *et al.*, 2013); (iii) the absence of harmonized legislation for using exogenous dsRNAs, currently approved only in some non-EU countries (Mezzetti *et al.*, 2020; OECD, 2020; Dietz-Pfeilstetter *et al.*, 2021); and (iv) a lack of risk assessment protocols for unintended off-target effects (Papadopoulou *et al.*, 2020). Current RNAi strategies for crop protection rely on either "spray-induced" gene silencing (SIGS) or "host-induced" gene silencing (HIGS) (Zotti *et al.*, 2018; Christiaens *et al.*, 2020; Das and Sherif, 2020; Lim *et al.*, 2020, Liu *et al.*, 2020; Taning *et al.*, 2020; Zhu and Palli, 2020; Pugsley *et al.*, 2021; Rank and Koch, 2021). The former exploits exogenous dsRNAs, either naked or protected inside nanoparticles (Long *et al.*, 2024; Lu *et al.*, 2024; Ma *et al.*, 2024; Zheng *et al.*, 2024), to silence an essential gene of a given pest, while the latter relies on genetically modified (GM) plants that produce dsRNAs for the same purpose. The uptake of dsRNA of at least 60 bp or longer is crucial for efficient RNAi of the target pest (Bolognesi *et al.*, 2012). In the case of phloem-feeding insects, targeting exogenous silencing RNAs (sRNAs) and their dsRNA precursors to the phloem may be challenging, depending on the plant species and the delivery strategy (Biedenkopf *et al.*, 2020; Sidorova *et al.*, 2021). However, when exogenous dsRNAs are applied directly to the plant vascular system by petiole absorption or trunk injection, they distribute systemically through the apoplast and xylem and potentially trigger RNAi in piercing-sucking insects (Dalakouras *et al.*, 2018; Bennett *et al.*, 2020). The development of nanocarriers may further enhance spray delivery of dsRNAs, especially in the case of sap-sucking insects (Ma *et al.*, 2024; Tang *et al.*, 2024). Finally, selection of silencing target genes is crucial for the development of species-specific RNAi-based control strategies, but a general approach may be proposed through large-scale RNAi screenings in model organisms (Buer *et al.*, 2025) and subsequent transfer to appropriate pest species, since orthologous genes differ in their efficacy across species (Ulrich *et al.*, 2015; Mehlhorn *et al.*, 2021; Silver *et al.*, 2021).

RNAi technology has been explored for approximately 38 insect species belonging to 18 families in the Hemiptera order, including aphids, whiteflies, hoppers, psyllids, plant bugs, stink bugs, and shield bugs (Niebres and Alviar, 2023). Silencing has been applied to several xylem and phloem feeders, which are important plant pathogen vectors and pose serious threats to agriculture (Chougule and Bonning, 2012; Jain *et al.*, 2020). Aphids and whiteflies are mainly plant virus vectors, whereas leafhoppers, planthoppers, and psyllids are also able to transmit bacteria, including phytoplasmas (Kanakala and Ghanim, 2016). RNAi has been investigated for controlling several major agricultural pests, including aphids (Yu *et al.*, 2016), whiteflies (Saurabh *et al.*, 2021; Mahmood *et al.*, 2023), planthoppers (*Nilaparvata lugens* (Yang *et al.*, 2022)), psyllids (*Diaphorina citri* (Yu and Killiny, 2020; Marques *et al.*, 2021)), and leafhoppers (*Nephotettix cincticeps* (Matsumoto and Hattori, 2016)).

Main Constraints of RNAi Application to Hemipteran Vectors

Some challenges need to be solved before dsRNA-based insecticides can be widely used in crop management. Accurate target gene selection is crucial for a successful RNAi strategy. Target gene silencing should influence insect fitness either by inducing lethal phenotypes (Mehlhorn *et al.*, 2021) or by targeting detoxification genes (Eakteiman *et al.*, 2018). In the case of plant pathogen vectors, whose direct damage is often secondary to that caused by the disease, another possible alternative could be reducing pathogen transmission capability (Kanakala and Ghanim, 2016; Niebres and Alviar, 2023).

Another critical factor is the design of the dsRNA molecule, including its sequence, base-pair length, and required dosage. To prevent unintended effects on beneficial insects and other animals, each dsRNA sequence should be designed using *in silico* comparison with off-target transcriptomes (Jain *et al.*, 2021). dsRNAs between 200 and 500 bp are usually efficient in inducing silencing (De Andrade and Hunter, 2016; Mehlhorn *et al.*, 2021). In RNAi technology, the strength of phenotype alteration generally correlates with the amount of administered dsRNAs (Mehlhorn *et al.*, 2021). Fusion of selected target sequences from different genes may also increase RNAi-based control of aphids (Wang *et al.*, 2024). Moreover, transgenic tobacco plants resistant to *Bemisia tabaci* and *Myzus persicae* have been obtained by artificial design of dsRNAs targeting orthologs of the different species (Dong *et al.*, 2024). However, several factors can affect the dsRNA effective dosage, such as insect species, developmental stage, and the delivery method (Jain *et al.*, 2020).

Degradation of dsRNAs in the environment and/or within the insect body is an additional constraint, as RNA molecules can be destabilized by UV radiation, pH variation, and nucleases of the insect digestive tract (Christiaens and Smagghe, 2014).

Silencing dsRNA-degrading genes (dsRNase) can be a further strategy to improve RNAi efficiency (Sharma *et al.*, 2021; Shu *et al.*, 2024). Uncertainties about the mechanisms of dsRNA uptake in insects need to be resolved as well. The transmembrane family member 1 gene (*SID1*), encoding a protein of the dsRNA-gated channels required for systemic RNAi, is a core component of dsRNA uptake in nematodes (Feinberg and Hunter, 2003). SID1-like orthologs are present in arthropods (Maruekawong *et al.*, 2018) and most insects (Cagliari *et al.*, 2019; Jain *et al.*, 2021), except dipterans (Feinberg and Hunter, 2003). Their implication in dsRNA uptake has been elucidated for some species (Xu *et al.*, 2013; Miyata *et al.*, 2014; Cappelle *et al.*, 2016), but not for others (Luo *et al.*, 2012; Wynant *et al.*, 2014), probably due to unresolved orthology problems (Tomoyasu *et al.*, 2008). Recent findings conclusively established that this gene is involved in dsRNA uptake in the model aphid *Acyrthosiphon pisum* (Ye *et al.*, 2023). Uptake and systemization of dsRNAs may also occur through scavenger receptor-mediated clathrin-dependent endocytic mechanisms, as observed in hemipteran and other insect orders (Cooper *et al.*, 2019).

One of the main drawbacks in applying RNAi to insects belonging to the Hemiptera order is the delivery method. In many assays, dsRNA molecules are directly microinjected into the target insect or administered by feeding. Unfortunately, for insect vectors, such techniques are limited to laboratory settings and are poorly effective due to the feeding behavior of sap-sucking insects. To overcome environmental barriers and directly reach plant vascular vessels, dsRNA molecules can be administered by trunk injection (Dalakouras *et al.*, 2018), root soaking (Andrade and Hunter, 2017), or foliar spray following nanoparticle absorption (e.g. Bio-Clay, star polycation, amphiphilic double hydroxide; Mitter *et al.*, 2017). Another solution could be *in planta* dsRNA production (Christiaens and Smagghe, 2014; Jain *et al.*, 2021; Niebres and Alviar, 2023). Additionally, genetically expressed dsRNAs can be processed by the plant RNAi machinery (Christiaens and Smagghe, 2014) and may intensify selection pressure, fostering the development of resistance to RNAi-based biopesticides (Jain *et al.*, 2021). Indeed, after the commercialization of the first effective dsRNA-based insecticide to limit Colorado purple beet populations (Rodrigues *et al.*, 2021), resistance to RNAi developed under laboratory selection experiments (Mishra *et al.*, 2021). This autosomal, polygenic, and recessive resistance was also effective against dsRNA targets different from those used for selection, but did not generate evident fitness cost for the resistant population (Pinto *et al.*, 2023).

Additional challenges are presented by insect vectors of fastidious phloem-limited bacteria, such as '*Candidatus* Liberibacter' and '*Candidatus* Phytoplasma' species. Functional studies on these pathogens are hampered by difficulties in establishing axenic cultures and a shortage of fully annotated complete genomes.

RNAi and Phytoplasma Insect Vectors

Systematic literature review

A search conducted on November 20, 2023, in the Web of Science (WoS) core collection database (https://clarivate.com/products/scientific-and-academic-research/research-discovery-and-workflow-solutions/webofscience-platform/web-of-science-core-collection/; last accessed June 2025) with the search string "(RNAi OR RNA interference) AND Hemipter* AND vector*" yielded 108 papers. About 36% of the retrieved documents – which focused on human diseases, direct plant damages, and applications in the frame of functional genomics – were excluded from further analyses. Of the remaining ones, 33% were devoted to RNAi applications targeting plant or insect viruses, and 31% centered on insect vectors of bacteria. In particular, 26 papers discussed RNAi in psyllids, 19 in leafhoppers and planthoppers, 12 in whiteflies, eight in aphids, and four in mealybugs. Regarding the vector-transmitted bacteria, about 79% of the manuscripts addressed RNAi applications related to different '*Candidatus* Liberibacter' species, 12% involved '*Candidatus* Phytoplasma' species, and 6% examined *Xylella fastidiosa* and spiroplasmas. A subsequent search using the same string across all the WoS databases on November 27, 2023, retrieved 473 papers. After refining the results with the keyword "Phytoplasma," the pool was narrowed down to nine relevant papers. Due to the limited availability of articles on phytoplasmas, the search was extended to Google Scholar (https://scholar.google.com/; last accessed June 2025) with "RNAi AND Phytoplasma." The resulting ten relevant papers are summarized in Table 7.1.

Insect species and transmitted phytoplasmas

Most RNAi-based studies on phytoplasma vectors (Table 7.1) have focused on the leafhoppers *Scaphoideus titanus* Ball and *Euscelidius variegatus* Kirschbaum (Hemiptera: Cicadellidae), which are the natural and laboratory vectors of flavescence dorée phytoplasma (FDp), respectively. This pathogen belongs to the 16SrV phytoplasma group, subgroups -C and -D, and is associated with flavescence dorée, an economically important disease of grapevine that affects many wine-producing regions of Europe (Debonneville *et al.*, 2022). Flavescence dorée is considered an A2 quarantine pest and a candidate priority pest in the EU (EFSA *et al.*, 2019). Infected plants show severe leaf and vascular symptoms (Rizzoli *et al.*, 2022), often leading to bunch drying and plant death (EFSA *et al.*, 2020, p. 202). The persistent and propagative transmission of FDp is driven by molecular interactions between insect vector proteins and FDp membrane proteins (Arricau-Bouvery *et al.*, 2018, 2021, 2023; Trivellone *et al.*, 2019; Canuto *et al.*, 2023). Additionally, a specific phytoplasma protein (variable membrane protein A (VmpA)) is associated with epidemic FDp strains transmitted by *S. titanus* (Malembic-Maher *et al.*, 2020). Other epidemiological cycles involve different vectors, such as *Dictyophara europaea*, *Orientus ishidae*, *Oncopsis alni*, and *Allygus* spp. (Malembic-Maher *et al.*, 2020), as well as wild vegetation like *Ailanthus altissima*, *Clematis vitalba*, *Corylus avellana*, and *Alnus* spp. (EFSA Panel on Plant Health *et al.*, 2016; Mehle *et al.*, 2019). These alternative hosts contribute to maintaining the natural inoculum of this Palearctic phytoplasma. Conventional management of flavescence dorée relies on control of vector populations through compulsory insecticide applications, removal of infected plants to reduce inoculum source, and planting of grafted cuttings that are phytoplasma-free (Oliveira *et al.*, 2020). Functional studies of FDp genes also utilize a laboratory pathosystem involving Vicia *faba*/*E. variegatus* (Caudwell *et al.*, 1972), which is easier to handle than the natural woody host plant (grapevine) and the natural monovoltine and monophagous vector (*S. titanus*).

Under field and laboratory conditions, *E. variegatus* can also transmit the chrysanthemum yellows phytoplasma (CYp), which is a '*Candidatus* Phytoplasma asteris'-related strain, and belongs to the 16SrI-B phytoplasma subgroup. In addition to *E. variegatus*, *Macrosteles quadripunctulatus* Kirschbaum, and *Euscelis incisus* Kirschbaum have also been identified as efficient natural vectors of CYp, with all three leafhoppers supporting pathogen multiplication

Table 7.1. Extraction table of the retrieved relevant literature on RNAi applied to phytoplasma vectors, with special focus on the main features of each pathosystem, including dsRNA delivery methods, experimental purposes, selected target genes, validation methods, and observed phenotypes.

dsRNA delivery methods	Insect species	Transmitted phytoplasmas and their 16Sr subgroups	Experimental aims	Target genes of insect vectors	Molecular methods of validation	Observed phenotypes	Reference
Cell transfection	*Euscelidius variegatus*	Flavescence dorée (FDp, 16SrV-C/D)	Validation of interaction with FD VmpA membrane protein	Unknown1_ Leucin Rich Repeat (*UK1_LRR*)	qPCR for transcript reduction	Reduced adhesion of VmpA-coated beads to insect cells	Canuto *et al.*, 2023
Abdominal microinjection	*Euscelidius variegatus*	Flavescence dorée (FDp, 16SrV-C/D), Chrysanthemum yellows (CYp, 16SrI-B)	Demonstration of RNAi occurrence, inducing lethality for vector control	Actin	qPCR for transcript reduction	Increased mortality	Abbà *et al.*, 2019
			Validation of interaction with phytoplasma Amp membrane protein	ATP synthase beta	qPCR for transcript reduction, smallRNAseq, cognate protein depletion	Delayed increase of mortality, reduced phytoplasma multiplication, female sterility	Abbà *et al.*, 2019; Galetto *et al.*, 2021a, 2021b
	Scaphoideus titanus	Flavescence dorée (FDp, 16SrV-C/D)	Demonstration of RNAi occurrence	ATP synthase beta	qPCR for transcript reduction, cognate protein depletion	Ovary morphology alteration	Ripamonti *et al.*, 2022a
	Dalbulus maidis	Maize bushy stunt (MBS, 16SrI-B)	Demonstration of RNAi occurrence	Bicaudal C	qPCR for transcript reduction	Female reduced fecundity, ovary morphology alteration	Dalaisón-Fuentes *et al.*, 2022

Feeding on artificial diet	*Dalbulus maidis*	Maize bushy stunt (MBS, 16SrI-B)	Demonstration of RNAi occurrence	Bicaudal C	qPCR for transcript reduction	Female reduced fecundity, ovary morphology alteration	Dalaisón-Fuentes *et al.*, 2022
	Dalbulus maidis	Maize bushy stunt (MBS 16SrI-B)	Demonstration of RNAi occurrence, inducing lethality for vector control	ATP synthase subunit B, ATP synthase subunit D	qPCR for transcript reduction	Increased mortality and wing deformation in emerging adults from treated nymphs	Jones *et al.*, 2021
	Euscelidius variegatus	Flavescence dorée (FDp, 16SrV-C/D)	Validation of interaction with FD VmpA membrane protein	Clathrin heavy chain	qPCR for transcript reduction	Decreased phytoplasma colonization of insect tissues	Arricau-Bouvery *et al.*, 2023
	Psammotettix striatus	Wheat blue dwarf (WBD, 16SrI-C)	Validation of interaction with WBD Imp membrane protein	Alpha-tubulin	qPCR for transcript reduction	Reduced phytoplasma transmission	Ding *et al.*, 2022
Feeding on dsRNA-treated plants by petiole absorption	*Euscelidius variegatus*	Flavescence dorée (FDp, 16SrV-C/D), '*Ca.* P. asteris' (CYp, 16SrI-B)	Optimization of plant-mediated dsRNA delivery strategy	Natterin, legumain, ATP synthase beta	qPCR for transcript reduction	Not evaluated	Rossi *et al.*, 2023
	Scaphoideus titanus	Flavescence dorée (FDp 16SrV-C/D)		ATP synthase beta			

Amp, antigenic membrane protein; ATP synthase beta, adenosine triphosphate synthase beta; FDp, flavescence dorée phytoplasma; qPCR, quantitative polymerase chain reaction; VmpA, variable membrane protein A.

(Bosco *et al.*, 2007). Interestingly, unlike the detrimental effects observed in the insect vectors of FDp (Bressan *et al.*, 2005a, 2005b; Galetto *et al.*, 2018), CYp has a beneficial effect on the fitness of these insects (D'Amelio *et al.*, 2008; Galetto *et al.*, 2018). It is noteworthy that the transmission specificity of '*Ca.* P. asteris'-related strains are driven by molecular interactions occurring between phytoplasma membrane proteins and actin, myosin, ATP synthase beta, and other vector components with unknown roles (Suzuki *et al.*, 2006; Galetto *et al.*, 2011; Neriya *et al.*, 2014; Rashidi *et al.*, 2015).

The corn leafhopper *Dalbulus maidis* (DeLong & Wolcott) (Hemiptera: Cicadellidae) is one of the most important pests to affect corn (maize) crops throughout the Americas (Oliveira and Frizzas, 2022; Pozebon *et al.*, 2022). It is responsible for persistent and propagative transmission of maize rayado fino virus, maize bushy stunt phytoplasma (MBSP), and *Spiroplasma kunkelii* (Gámez, 1973; Nault, 1980; Jones and Medina, 2020). These bacterial pathogens, either alone or in combination, cause the disease complex known as corn stunt, which severely affects corn crops. The severity of the disease is closely tied to infestation levels, which can reach 100% in some areas. In such cases, crop losses may reach up to 90%, highlighting the significant threat this disease poses to agriculture (Dalaisón-Fuentes *et al.*, 2022). MBSP ('*Ca.* P. asteris' 16SrI-B; Lee, 2004) is associated with a corn disease characterized by leaf reddening, shortening of internodes, plant height reduction, lower grain yield, and lateral shoot production (Nault, 1980). Several secreted effectors identified in the MBSP draft genome (Sugio and Hogenhout, 2012) have been associated with differential symptom severity in plants (Orlovskis *et al.*, 2017). These effectors have also been shown to alter vector behavior, facilitating phytoplasma spreading to healthy plants in the crop (García Gonzalez *et al.*, 2018; Ramos *et al.*, 2020).

The wheat blue dwarf (WBD) phytoplasma is also a member of the 16SrI-C phytoplasma subgroup, but it clearly represents an ecologically distant lineage from '*Ca.* P. asteris' (Chen *et al.*, 2014) and it has been recently recognized as a novel taxon, namely '*Ca.* P. tritici' (Zhao *et al.*, 2021). This pathogen is associated with dwarfism, witches'-broom, and sterility, leading to severe losses in wheat production in northwest China (Wu *et al.*, 2010). Since the 1960s, at least ten disease outbreaks have been caused by the WBD phytoplasma in the Shaanxi Province of China, resulting in a loss of approximately 50,000 tons of wheat grains per epidemic year (Huang *et al.*, 2023). WBD phytoplasma is transmitted by the leafhopper *Psammotettix striatus* L. (Hemiptera: Cicadellidae) and infects a variety of plant species, including members of the Poaceae, Brassicaceae, and Solanaceae families (Ding *et al.*, 2022). Several secreted proteins from the WBD phytoplasma have been identified as effectors involved in plant infection (Wang *et al.*, 2018). Interestingly, one of these proteins functions as an RNA silencing suppressor in the model plants *Nicotiana benthamiana* and *Arabidopsis thaliana* (Wang *et al.*, 2021).

Essential genes for RNAi

The first step to demonstrate RNAi occurrence was to confirm the presence of core components of RNAi machinery in the assembled transcriptomes of *E. variegatus* (Abbà *et al.*, 2019), *S. titanus* (Ripamonti *et al.*, 2022a), and *D. maidis* (Dalaisón-Fuentes *et al.*, 2022). In particular, in the first two species, ten genes involved in the three main insect RNAi pathways were retrieved: Argonaute1, Dicer1, Loquacious, Drosha, and Pasha (miRNA pathway); Argonaute2, Dicer2, and *R2D2* (siRNA pathway); and Argonaute3 and Piwi (piRNA pathway) (Abbà *et al.*, 2019; Ripamonti *et al.*, 2022a). Similarly, a search for the RNAi-related genes on the *D. maidis* adult transcriptome revealed the presence of Dicer2, Argonaute2, R2D2, and Sid1 (Dalaisón-Fuentes *et al.*, 2022).

Methods of dsRNA delivery

The main strategies for delivering dsRNA to silence genes in phytoplasma vectors include cell transfection (Canuto *et al.*, 2023), abdominal microinjection (Abbà *et al.*, 2019; Galetto *et al.*, 2021a, 2021b; Dalaisón-Fuentes *et al.*, 2022; Ripamonti *et al.*, 2022a), and feeding on an artificial diet (Jones *et al.*, 2021; Dalaisón-Fuentes

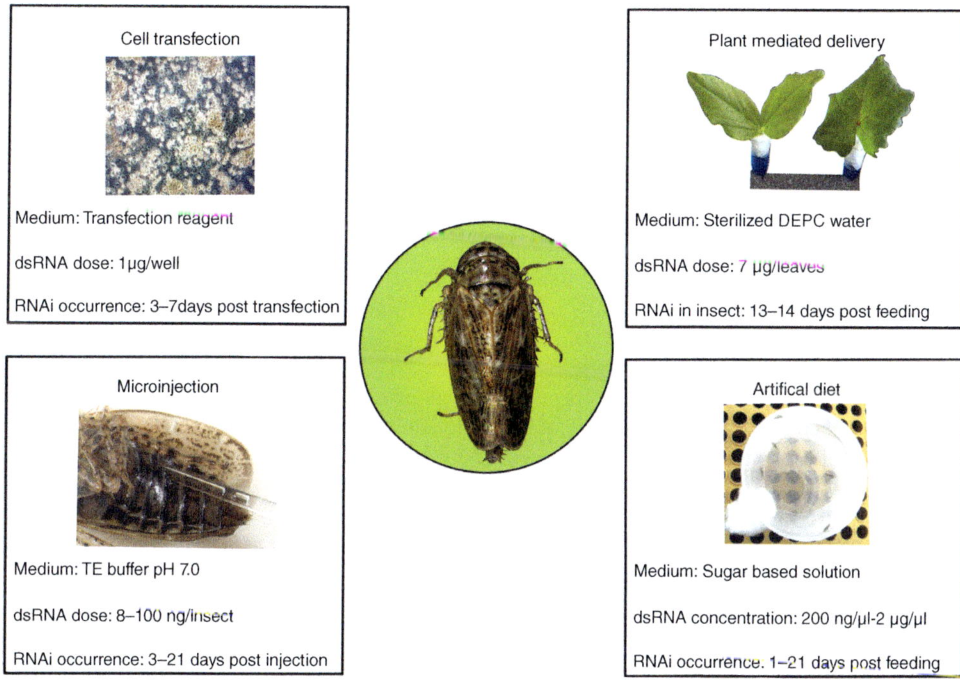

Fig. 7.1. Visual summary of dsRNA delivery methods.

et al., 2022; Ding *et al.*, 2022; Arricau-Bouvery *et al.*, 2023) or on dsRNA-treated plants by petiole absorption (Rossi *et al.*, 2023) (Fig. 7.1; Table 7.1).

According to one conventional procedure, *E. variegatus* cell cultures (nearly 80% confluence) were transfected with 1 µg of dsRNA using FuGENE® HD Transfection Reagent (Promega). A combinatorial approach using specific dsRNAs mixed together successfully silenced up to 13 target genes simultaneously. This method was equally effective at both 3 and 7 days post-transfection (Canuto *et al.*, 2023).

Microinjection has been proposed to administer dsRNAs into the abdomen of newly emerged adults of *E. variegatus* (Abbà *et al.*, 2019), *S. titanus* (Ripamonti *et al.*, 2022a), and *D. maidis* (Dalaisón-Fuentes *et al.*, 2022). Briefly, insects were anesthetized with a CO_2 flush and then injected with glass needles, premade with puller devices, mounted on a microinjector or syringe. The amount of injected dsRNAs ranged from 50 to 100 ng per insect in *D. maidis* (Dalaisón-Fuentes *et al.*, 2022) and from 8 to 80 ng per insect in *E. variegatus*, whereas a standard dose of 80 ng per insect was used in *S. titanus* (Abbà *et al.*, 2019; Ripamonti *et al.*, 2022a). The average injection volume was 0.5 µl. All the tested doses significantly reduced specific transcript levels and efficiently produced the expected phenotypes.

Insects fed on an artificial diet supplemented with dsRNAs showed an efficient silencing of specific transcripts. Nymphs of *D. maidis* (Jones *et al.*, 2021) and *P. striatus* (Ding *et al.*, 2022) as well as adults of *D. maidis* (Dalaisón-Fuentes *et al.*, 2022) and *E. variegatus* (Arricau-Bouvery *et al.*, 2023) were fed sugar-based solutions containing dsRNAs. The feeding duration ranged from a minimum of 48 h (Arricau-Bouvery *et al.*, 2023; Ding *et al.*, 2022) to 3 days (Dalaisón-Fuentes *et al.*, 2022) or even 4 days (Jones *et al.*, 2021). In all cases, the liquid diet was placed between two layers of parafilm stretched over the top of either truncated cone-shaped plastic cages (Jones *et al.*, 2021; Dalaisón-Fuentes *et al.*, 2022; Ding *et al.*, 2022) or the cap of 1.5 ml microtubes (Arricau-Bouvery *et al.*, 2023). The concentration of dsRNAs in the feeding solution ranged from 200 ng/μl (Dalaisón-Fuentes *et al.*, 2022) to 500 ng/μl (Jones *et al.*, 2021; Ding *et al.*, 2022), with daily renewal of the solution. Alternatively, concentrations up to 2 μg/μl were used without replacing the feeding diet (Arricau-Bouvery *et al.*, 2023).

Recently, a plant-mediated delivery method (petiole absorption) was proposed to administer dsRNAs to nymphs of *E. variegatus* and *S. titanus* through feeding on treated leaves (Rossi *et al.*, 2023). In this procedure, the petioles of detached leaves were immersed in the dsRNA solution (7 μg/leaf), until there was complete absorption in the xylem. Each dsRNA-treated leaf was then used to feed about 20 fifth-instar nymphs for 48 h. Three different gene transcripts, two gut-specific (natterin and legumain) and one ubiquitously expressed (ATP synthase beta), were significantly reduced in insects fed on leaves treated with cognate-specific dsRNAs (Rossi *et al.*, 2023). The petiole absorption method has been successful in delivering dsRNAs systemically into plant xylem and apoplast, without triggering the plant's own RNAi machinery (Ghosh *et al.*, 2017; Dalakouras *et al.*, 2018). Interestingly, phloem-feeding insects, including *S. titanus*, do not feed strictly on phloem only. Electropenetrography data have shown that they also probe the xylem (Chuche *et al.*, 2017; Ripamonti *et al.*, 2022b). This behavior supports the petiole absorption as a natural method to deliver dsRNAs in leafhopper vectors that transmit phytoplasmas.

Soaking nymphs in a solution containing dsRNAs together with a co-adjuvant has proven to be a successful delivery method to trigger RNAi in *Diaphorina citri* and *Aedes aegypti* (Killiny *et al.*, 2014; Yu *et al.*, 2017; Arshad *et al.*, 2021). However, this method was ineffective in *E. variegatus* (Rossi *et al.*, 2023) and *Drosophila melanogaster* (Powell *et al.*, 2017).

Overall, microinjection is more efficient than delivery through feeding (Dalaisón-Fuentes *et al.*, 2022; Rossi *et al.*, 2023), although both methods have their drawbacks. Microinjection is commonly regarded as the gold standard technique, but it is labor-intensive, induces mortality, and is not well suited for treating nymphal stages. On the other hand, feeding delivery methods are limited to a laboratory-scale dimension.

Experimental aims and target genes

Most RNAi-based studies on phytoplasma vectors (Table 7.1) have exploited RNAi as a functional genomic tool, mainly to elucidate the role of insect genes that interact with phytoplasma proteins to disrupt pathogen transmission (Galetto *et al.*, 2021a; Ding *et al.*, 2022; Arricau-Bouvery *et al.*, 2023; Canuto *et al.*, 2023) (Fig. 7.2). Other RNAi studies have focused on understanding fundamental metabolic insect pathways (Galetto *et al.*, 2021b; Dalaisón-Fuentes *et al.*, 2022), some of which can be potentially targeted to control the vector population by silencing essential genes (Abbà *et al.*, 2019; Jones *et al.*, 2021).

For RNAi-based approaches aiming to reduce vector populations, promising phenotypes were observed in *D. maidis* by silencing bicaudal C. This led to alterations in oocyte development and reduced oviposition (Dalaisón-Fuentes *et al.*, 2022). Silencing of vacuolar ATP synthase subunits B and D increased mortality and induced wing deformation (Jones *et al.*, 2021). Similarly, silencing actin and ATP synthase beta in *E. variegatus* adults by dsRNA injection reduced the survival rate (Abbà *et al.*, 2019). Furthermore, in *E. variegatus*, when ATP synthase beta was silenced, complete female sterility was also observed due to a lack of egg development. This was accompanied by altered expression of genes involved in oocyte maturation, such as hexamerin and cathepsin L, at both transcript and protein levels (Galetto *et al.*, 2021b). Consistently in *S. titanus*, silencing ATP synthase beta impaired egg development, and led to overexpression of hexamerin and cathepsin L. However, insect survival was not affected, probably due to the slower turnover of cognate protein in this monovoltine species (Ripamonti *et al.*, 2022a).

ATP synthase beta of *E. variegatus*, known to interact with the antigenic membrane protein (Amp) of CY phytoplasma (Galetto *et al.*, 2011), was also investigated as a putative target gene to disrupt phytoplasma transmission by the RNAi approach. A significant reduction of both CYp and FDp pathogens was measured in silenced *E. variegatus*, even if no difference in acquisition frequency was observed between insects treated with dsATP (dsRNAs targeting ATP synthase

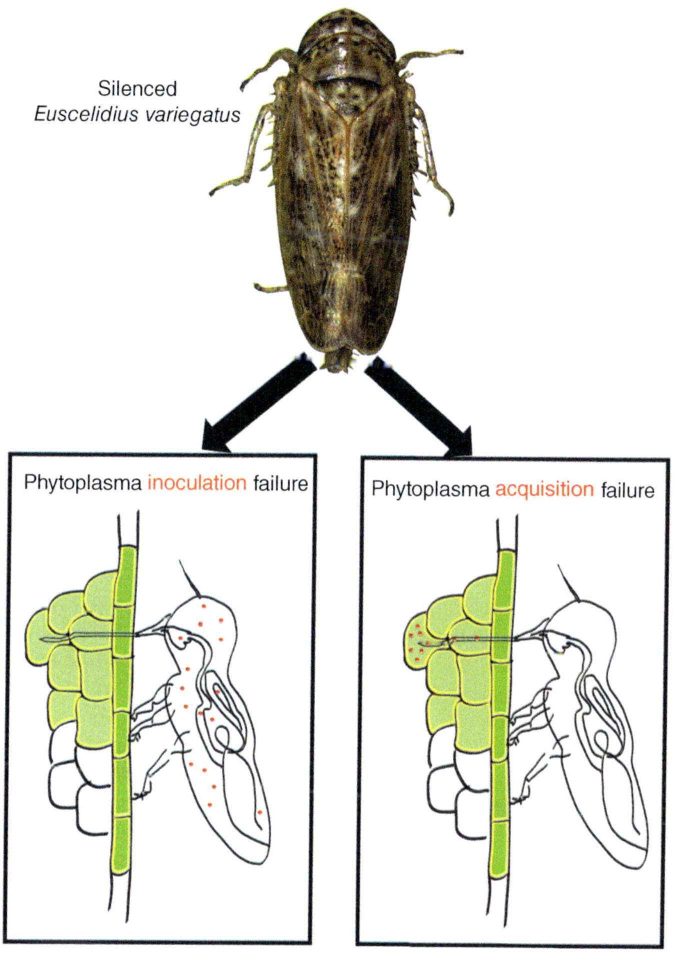

Fig. 7.2. dsRNA-mediated phytoplasma transmission disruption. Following exogenous dsRNA application, hampered recognition between vector and phytoplasma proteins leads to transmission failure, which can result in a lack of acquisition or of inoculation.

beta) and dsGFP (dsRNAs targeting green fluorescent protein), following feeding on infected plants (Galetto *et al.*, 2021a). A combinatorial RNAi-based approach was used as a functional validation platform to screen 13 *E. variegatus* genes encoding interaction partners of the FDp VmpA. Among these, an unknown protein with leucin rich repeat domains (UK1_LRR) was identified as playing a role in the internalization of FDp in vector cell cultures. Silencing the gut-specific *UK1_LRR* gene reduced the adhesion of VmpA-coated beads to the *E. variegatus* cell monolayer (Canuto *et al.*, 2023). Similarly, silencing the insect gene of clathrin heavy chain, which interacts with FDp VmpA, reduced endocytosis-mediated internalization of phytoplasma cells within *Drosophila* S2 phagocytic lines, as well as the FDp titre in silenced *E. variegatus*, following feeding on infected plants (Arricau-Bouvery *et al.*, 2023). In the WBD pathosystem, *in vitro* approaches identified α-tubulin of *P. striatus* as an interaction partner of the immunodominant membrane protein (Imp) of WBD phytoplasma. Silencing this gene reduced the vector's capability to transmit the pathogen (Ding *et al.*, 2022).

Methods to assess target gene silencing

Quantifying the reduction of target gene transcripts by qPCR is essential after dsRNA administration to confirm successful RNAi activation in insects (Mehlhorn *et al.*, 2021). This method was routinely used for validation in the studies summarized in Table 7.1. Gene expression analysis is a routine laboratory procedure, but in this case, some protocol features deserve particular attention. First, primers used for gene expression (qPCR validation) should be designed outside of the dsRNA targeted sequence to avoid false negatives or underestimation of gene knockdown due to amplification of the dsRNAs themselves (Onchuru and Kaltenpoth, 2019). Second, the timing of sampling after dsRNA treatment is also crucial to detect the effective gene silencing. Optimal sampling times should be evaluated on a case-by-case basis through time-course analysis, as transcript abundance and/or turnover vary among genes. For phytoplasma leafhopper vectors, efficient gene silencing generally becomes evident between 1 and 4 days after dsRNA treatment (Abbà *et al.*, 2019; Jones *et al.*, 2021; Dalaisón-Fuentes *et al.*, 2022; Ripamonti *et al.*, 2022; Canuto *et al.*, 2023). Third, tissue-specific gene expression analysis may be required to evaluate the distribution of dsRNAs within the insect and identify which organs are most affected by RNAi treatment. Finally, the guidelines of minimum information for publication of quantitative real-time PCR experiments (MIQE) need to be followed (Bustin *et al.*, 2009). This includes treating samples with DNase and ensuring the stability of reference genes (preferably at least two) for accurate normalization (Vandesompele *et al.*, 2002). However, in some cases, gene expression analysis alone could be misleading as RNAi-triggered compensatory upregulation of target gene paralogs may be observed (Kitzmann *et al.*, 2013).

Evaluation of cognate protein expression is a method for assessing specific induced gene silencing and further exploring the resulting phenotype changes. Commercial as well as custom-made synthesized antibodies raised against specific gene products targeted by the dsRNAs are helpful tools for quantitative western blots, following normalization against housekeeping proteins or total protein levels. Time-course analysis is also crucial in this case, as protein turnover rate strongly impacts the results. For example, the slow protein turnover of mitochondrial respiratory complex components (Bogenhagen and Haley, 2020) explains the delayed reduction in ATP synthase beta protein levels upon silencing in leafhoppers. Significant protein depletion in *E. variegatus* was observed 15 days after dsRNA injection (Galetto *et al.*, 2021a) and it took 20 days in *S. titanus* (Ripamonti *et al.*, 2022a). However, the specificity of the antibody may affect the correct estimation of gene silencing due to the possible presence of several isoforms (Guillem-Amat *et al.*, 2023; Wu *et al.*, 2022).

Finally, sequencing of sRNA libraries obtained from dsRNA-treated insects is a straightforward procedure to confirm RNAi occurrence. The presence of total sRNA reads mapping to the administered dsRNA sequence proves efficient RNAi activation and proper synthesis of sRNAs derived from the dsRNA molecules. In *E. variegatus*, the length distribution of sRNAs peaking at 21 nt is consistent with the processing of the dsRNA into siRNA by the RNAi pathway. Moreover, these sRNA reads map exclusively to the fragment of transcript targeted by the dsRNA sequence (Galetto *et al.*, 2021a). The absence of read coverage outside this region suggests that secondary siRNA synthesis, observed in plants and nematodes, does not occur in this case (Christiaens *et al.*, 2020).

Further Perspectives

Novel strategies can be explored for phytoplasma vector control, such as the use of artificial microRNA (AmiRNA; Bally *et al.*, 2020), plastid-mediated RNAi (Li *et al.*, 2023), and production of dsRNA expressing transgenic rootstocks (Choudry *et al.*, 2024; Saakre *et al.*, 2024). This latter approach may also overcome public concerns about GM organisms. Field-scale application of dsRNA through HIGS needs further optimization as probably the most suitable delivery method to sap-sucking insects (Dong *et al.*, 2024; Yousaf *et al.*, 2024). This approach can be coupled to further development of nanoparticle-based strategies (Verma and Modgil, 2024; Zarrabian and Sherif, 2024), which will likely reduce dsRNAs degradation, endosomal escape, and inadequate cellular absorption.

RNAi effects on off-target arthropod species should be carefully evaluated by *in silico* and *in vivo* methods. In particular, analyses of phylogenetic conservation of target homolog sequences among pests and beneficial species and bioassay-based evaluation of RNAi impact on fitness of off-target species must be optimized. Particular attention should be given to the number and the length of identical consecutive nucleotides longer than 16 bases among gene fragments of target and non-target organisms. Indeed, a perfect match of at least 16 nucleotides in the central part of the siRNA is required to efficiently drive mRNA recognition and cleavage (Du, 2005). It is noteworthy that identical stretches of 19 nt could lead to non-target effects in *Drosophila* (Kulkarni *et al.*, 2006), but also that the presence of at least three contiguous 21 nt identical sequences in the dsRNA region is required to observe a significant lethal or sublethal effect following diet bioassays in the western corn rootworm (Bachman *et al.*, 2013). Therefore, a case-by-case study is required to consider the exposure of the appropriate beneficial fauna to dsRNAs in each agronomic context.

In any case, besides its possible use as an innovative insecticide, RNAi is a powerful molecular tool to characterize the transmission mechanism of fastidious plant pathogens by non-model vectors.

Conclusion

The demonstration that RNAi functions in some phytoplasma leafhopper vectors paves the way toward its application in population control. However, dsRNA delivery to sap-sucking insects, especially under field conditions, is still challenging. Currently, the exogenous application of dsRNAs is not approved worldwide. Nevertheless, these tools offer a promising approach for environmental and ecological safety, particularly as an alternative to pesticides in the context of global trade and climate change, which contribute to the spread of exotic pests.

Acknowledgments

Funded by the European Union NextGenerationEU mission 4, component 2, investment 1.1 under the PRIN call 2022 (Project HEMINT, CUP code B53D23017470006, project code 2022BPB5A8), and partly by the MIDI Project granted by Regione Piemonte.

References

Abbà, S., Galetto, L., Ripamonti, M., Rossi, M., and Marzachì, C. (2019) RNA interference of muscle actin and ATP synthase beta increases mortality of the phytoplasma vector *Euscelidius variegatus*. *Pest Management Science* 75, 1425–1434. DOI: 10.1002/ps.5263.

Andrade, E.C. and Hunter, W.B. (2017) RNAi feeding bioassay: Development of a non-transgenic approach to control Asian citrus psyllid and other hemipterans. *Entomologia Experimentalis et Applicata* 162, 389–396. DOI: 10.1111/eea.12544.

Arricau-Bouvery, N., Dubrana, M.-P., Canuto, F., Duret, S., Brocard, L. *et al.* (2023) Flavescence dorée phytoplasma enters insect cells by a clathrin-mediated endocytosis allowing infection of its insect vector. *Scientific Reports* 13, 2211. DOI: 10.1038/s41598-023-29341-1.

Arricau-Bouvery, N., Duret, S., Dubrana, M.-P., Batailler, B., Desqué, D. *et al.* (2018) Variable membrane protein A of flavescence dorée phytoplasma binds the midgut perimicrovillar membrane of *Euscelidius variegatus* and promotes adhesion to its epithelial cells. *Applied and Environmental Microbiology* 84, 8–17. DOI: 10.1128/AEM.02487-17.

Arricau-Bouvery, N., Duret, S., Dubrana, M.-P., Desqué, D., Eveillard, S. *et al.* (2021) Interactions between the Flavescence dorée phytoplasma and its insect vector indicate lectin-type adhesion mediated by the adhesin VmpA. *Scientific Reports* 11, 11222. DOI: 10.1038/s41598-021-90809-z.

Arshad, F., Sharma, A., Lu, C., and Gulia-Nuss, M. (2021) RNAi by soaking *Aedes aegypti* pupae in dsRNA. *Insects* 12, 634. DOI: 10.3390/insects12070634.

Bachman, P.M., Bolognesi, R., Moar, W.J., Mueller, G.M., Paradise, M.S. *et al.* (2013) Characterization of the spectrum of insecticidal activity of a double-stranded RNA with targeted activity against Western Corn Rootworm (*Diabrotica virgifera virgifera* LeConte). *Transgenic Research* 22, 1207–1222. DOI: 10.1007/s11248-013-9716-5.

Bally, J., Fishilevich, E., Doran, R.L., Lee, K., Campos, S.B. *et al.* (2020) Plin-amiR, a pre-microRNA-based technology for controlling herbivorous insect pests. *Plant Biotechnology Journal* 18, 1925–1932. DOI: 10.1111/pbi.13352.

Baulcombe, D. (2004) RNA silencing in plants. *Nature* 431, 356–363. DOI: 10.1038/nature02874.

Baum, J.A., Bogaert, T., Clinton, W., Heck, G.R., Feldmann, P. *et al.* (2007) Control of coleopteran insect pests through RNA interference. *Nature Biotechnology* 25, 1322–1326. DOI: 10.1038/nbt1359.

Bennett, M., Deikman, J., Hendrix, B., and Iandolino, A. (2020) Barriers to efficient foliar uptake of dsRNA and molecular barriers to dsRNA activity in plant cells. *Frontiers in Plant Science* 11, 816. DOI: 10.3389/fpls.2020.00816.

Biedenkopf, D., Will, T., Knauer, T., Jelonek, L., Furch, A.C.U., Busche, T., and Koch, A. (2020) Systemic spreading of exogenous applied RNA biopesticides in the crop plant *Hordeum vulgare*. *ExRNA* 2, 12. DOI: 10.1186/s41544-020-00052-3.

Bogenhagen, D.F. and Haley, J.D. (2020) Pulse-chase SILAC-based analyses reveal selective oversynthesis and rapid turnover of mitochondrial protein components of respiratory complexes. *Journal of Biological Chemistry* 295, 2544–2554. DOI: 10.1074/jbc.RA119.011791.

Bolognesi, R., Ramaseshadri, P., Anderson, J., Bachman, P., Clinton, W. *et al.* (2012) Characterizing the mechanism of action of double-stranded RNA activity against Western corn rootworm (*Diabrotica virgifera virgifera* LeConte). *PLoS One* 7, e47534. DOI: 10.1371/journal.pone.0047534.

Bonning, B.C. and Saleh, M.-C. (2021) The interplay between viruses and RNAi pathways in insects. *Annual Review of Entomology* 66, 61–79. DOI: 10.1146/annurev-ento-033020-090410.

Bosco, D., Galetto, L., Leoncini, P., Saracco, P., Raccah, B., and Marzachì, C. (2007) Interrelationships between "*Candidatus* Phytoplasma asteris" and its leafhopper vectors (Homoptera: Cicadellidae). *Journal of Economic Entomology* 100, 1504–1511. DOI: 10.1603/0022-0493-100.5.1504.

Bressan, A., Clair, D., Sémétey, O., and Boudon-Padieu, É. (2005a) Effect of two strains of Flavescence dorée phytoplasma on the survival and fecundity of the experimental leafhopper vector *Euscelidius variegatus* Kirschbaum. *Journal of Invertebrate Pathology* 89, 144–149. DOI: 10.1016/j.jip.2005.03.001.

Bressan, A., Girolami, V., and Boudon-Padieu, E. (2005b) Reduced fitness of the leafhopper vector *Scaphoideus titanus* exposed to Flavescence doree phytoplasma. *Entomologia Experimentalis et Applicata* 115, 283–290. DOI: 10.1111/j.1570-7458.2005.00240.x.

Buer, B., Dönitz, J., Milner, M., Mehlhorn, S., Hinners, C. *et al.* (2025) Superior target genes and pathways for RNAi-mediated pest control revealed by genome-wide analysis in the beetle Tribolium castaneum. *Pest Management Science* 81, 1026–1036.

Bustin, S.A., Benes, V., Garson, J.A., Hellemans, J., Huggett, J. *et al.* (2009) The MIQE guidelines: Minimum information for publication of quantitative real-time PCR experiments. *Clinical Chemistry* 55, 611–622. DOI: 10.1373/clinchem.2008.112797.

Cagliari, D., Dias, N.P., Galdeano, D.M., dos Santos, E.Á., Smagghe, G., and Zotti, M.J. (2019) Management of pest insects and plant diseases by non-transformative RNAi. *Frontiers in Plant Science* 10, 1319. DOI: 10.3389/fpls.2019.01319.

Canuto, F., Duret, S., Dubrana, M.-P., Claverol, S., Malembic-Maher, S., Foissac, X., and Arricau-Bouvery, N. (2023) A knockdown gene approach identifies an insect vector membrane protein with Leucin Rich Repeats as one of the receptors for the VmpA adhesin of flavescence dorée phytoplasma. *Frontiers in Cellular and Infection Microbiology* 13, 1289100. DOI: 10.3389/fcimb.2023.1289100.

Cappelle, K., De Oliveira, C.F.R., Van Eynde, B., Christiaens, O., and Smagghe, G. (2016) The involvement of clathrin-mediated endocytosis and two Sid-1-like transmembrane proteins in double-stranded RNA uptake in the Colorado potato beetle midgut. *Insect Molecular Biology* 25, 315–323. DOI: 10.1111/imb.12222.

Caudwell, A., Kuszala, C., Larrue, J., and Bachelier, J. (1972) Transmission de la Flavescence dorée de la fève à la fève par des cicadelles des genres *Euscelis* et *Euscelidius*. *Annales de Phytopathology* no. hors série, 181–189.

Chen, W., Li, Y., Wang, Q., Wang, N., and Wu, Y. (2014) Comparative genome analysis of wheat blue dwarf phytoplasma, an obligate pathogen that causes wheat blue dwarf disease in China. *PLoS One* 9, e96436. DOI: 10.1371/journal.pone.0096436.

Choudry, M.W., Nawaz, P., Jahan, N., Riaz, R., Ahmed, B. *et al.* (2024) RNA based gene silencing modalities to control insect and fungal plant pests – Challenges and future prospects. *Physiological and Molecular Plant Pathology* 130, 102241. DOI: 10.1016/j.pmpp.2024.102241.

Chougule, N.P. and Bonning, B.C. (2012) Toxins for transgenic resistance to hemipteran pests. *Toxins* 4, 405–429. DOI: 10.3390/toxins4060405.

Christiaens, O. and Smagghe, G. (2014) The challenge of RNAi-mediated control of hemipterans. *Current Opinion in Insect Science* 6, 15–21. DOI: 10.1016/j.cois.2014.09.012.

Christiaens, O., Whyard, S., Vélez, A.M., and Smagghe, G. (2020) Double-stranded RNA technology to control insect pests: current status and challenges. *Frontiers in Plant Science* 11, 451. DOI: 10.3389/fpls.2020.00451.

Chuche, J., Sauvion, N., and Thiéry, D. (2017) Mixed xylem and phloem sap ingestion in sheath-feeders as normal dietary behavior: Evidence from the leafhopper *Scaphoideus titanus*. *Journal of Insect Physiology* 102, 62–72. DOI: 10.1016/j.jinsphys.2017.01.014.

Coleman, A.D., Wouters, R.H.M., Mugford, S.T., and Hogenhout, S.A. (2015) Persistence and transgenerational effect of plant-mediated RNAi in aphids. *Journal of Experimental Botany* 66, 541–548. DOI: 10.1093/jxb/eru450.

Cooper, A.M., Silver, K., Zhang, J., Park, Y., and Zhu, K.Y. (2019) Molecular mechanisms influencing efficiency of RNA interference in insects. *Pest Management Science* 75, 18–28. DOI: 10.1002/ps.5126.

Dalaisón-Fuentes, L.I., Pascual, A., Gazza, E., Welchen, E., Rivera-Pomar, R., and Catalano, M.I. (2022) Development of efficient RNAi methods in the corn leafhopper *Dalbulus maidis*, a promising application for pest control. *Pest Management Science* 78, 3108–3116. DOI: 10.1002/ps.6937.

Dalakouras, A., Jarausch, W., Buchholz, G., Bassler, A., Braun, M. *et al.* (2018) Delivery of hairpin RNAs and small RNAs into woody and herbaceous plants by trunk injection and petiole absorption. *Frontiers in Plant Science* 9, 1253. DOI: 10.3389/fpls.2018.01253.

D'Amelio, R., Palermo, S., Marzachì, C., and Bosco, D. (2008) Influence of Chrysanthemum yellows phytoplasma on the fitness of two of its leafhopper vectors, *Macrosteles quadripunctulatus* and *Euscelidius variegatus*. *Bulletin of Insectology* 61, 349–354.

Das, P.R. and Sherif, S.M. (2020) Application of exogenous dsRNAs-induced RNAi in agriculture: Challenges and triumphs. *Frontiers in Plant Science* 11, 946. DOI: 10.3389/fpls.2020.00946.

De Andrade, E.C. and Hunter, W.B. (2016) RNA Interference – Natural gene-based technology for Highly Specific Pest Control (HiSPeC). In: Abdurakhmonov, I.Y. (ed.) *RNA Interference*. InTech, pp.391–409. DOI: 10.5772/61612.

Debonneville, C., Mandelli, L., Brodard, J., Groux, R., Roquis, D., and Schumpp, O. (2022) The complete genome of the "Flavescence dorée" phytoplasma reveals characteristics of low genome plasticity. *Biology* 11, 953. DOI: 10.3390/biology11070953.

Dietz-Pfeilstetter, A., Mendelsohn, M., Gathmann, A., and Klinkenbuß, D. (2021) Considerations and regulatory approaches in the USA and in the EU for dsRNA-based externally applied pesticides for plant protection. *Frontiers in Plant Science* 12, 682387. DOI: 10.3389/fpls.2021.682387.

Ding, L., Lu, W., Yang, Y., Zhong, Q., Zhou, T. *et al.* (2022) Immunodominant membrane protein (Imp) promotes the transmission of wheat blue dwarf (WBD) phytoplasma by directly interacting with α-tubulin in leafhoppers. *European Journal of Plant Pathology* 162, 357–367. DOI: 10.1007/s10658-021-02408-3.

Dong, Y., Zhang, Q., Mao, Y., Wu, M., Wang, Z., Chang, L., and Zhang, J. (2024) Control of two insect pests by expression of a mismatch corrected double-stranded RNA in plants. *Plant Biotechnology Journal* 22, 2010–2019. DOI: 10.1111/pbi.14021.

Du, Q. (2005) A systematic analysis of the silencing effects of an active siRNA at all single-nucleotide mismatched target sites. *Nucleic Acids Research* 33, 1671–1677. DOI: 10.1093/nar/gki312.

Eakteiman, G., Moses-Koch, R., Moshitzky, P., Mestre-Rincon, N., Vassão, D.G. *et al.* (2018) Targeting detoxification genes by phloem-mediated RNAi: A new approach for controlling phloem-feeding insect pests. *Insect Biochemistry and Molecular Biology* 100, 10–21. DOI: 10.1016/j.ibmb.2018.05.008.

EFSA (European Food Safety Authority), Baker, R., Gilioli, G., Behring, C., Candiani, D. *et al.* (2019) *Grapevine Flavescence dorée Pest Report and Datasheet to Support Ranking of EU Candidate Priority Pests*. Available at: https://doi.org/10.5281/ZENODO.2789595 (accessed June 2025).

EFSA (European Food Safety Authority), Tramontini, S., Delbianco, A., and Vos, S. (2020) Pest survey card on Flavescence dorée phytoplasma and its vector *Scaphoideus titanus*. *EFSA Supporting Publication* 17, 1909E. DOI: 10.2903/sp.efsa.2020.EN-1909.

EFSA Panel on Plant Health (PLH), Jeger, M., Bragard, C., Caffier, D., Candresse, T. *et al.* (2016) Risk to plant health of Flavescence dorée for the EU territory. *EFSA Journal* 14, e4603. DOI: 10.2903/j.efsa.2016.4603.

Feinberg, E.H. and Hunter, C.P. (2003) Transport of dsRNA into cells by the transmembrane protein SID-1. *Science* 301, 1545–1547. DOI: 10.1126/science.1087117.

Fire, A., Xu, S., Montgomery, M.K., Kostas, S.A., Driver, S.E., and Mello, C.C. (1998) Potent and specific genetic interference by double-stranded RNA in *Caenorhabditis elegans*. *Nature* 391, 806–811. DOI: 10.1038/35888.

Galetto, L., Bosco, D., Balestrini, R., Genre, A., Fletcher, J., and Marzachì, C. (2011) The major antigenic membrane protein of "*Candidatus* Phytoplasma asteris" selectively interacts with ATP synthase and actin of leafhopper vectors. *PLoS One* 6, e22571. DOI: 10.1371/journal.pone.0022571.

Galetto, L., Abbà, S., Rossi, M., Vallino, M., Pesando, M. *et al.* (2018) Two phytoplasmas elicit different responses in the insect vector *Euscelidius variegatus* Kirschbaum. *Infection and Immunity* 86, iai.00042-18. DOI: 10.1128/IAI.00042-18.

Galetto, L., Abbà, S., Rossi, M., Ripamonti, M., Palmano, S., Bosco, D., and Marzachì, C. (2021) Silencing of ATP synthase β reduces phytoplasma multiplication in a leafhopper vector. *Journal of Insect Physiology* 128, 104176. DOI: 10.1016/j.jinsphys.2020.104176.

Galetto, L., Ripamonti, M., Abbà, S., Rossi, M., Manfredi, M., Bosco, D., and Marzachì, C. (2021b) Silencing of ATP synthase β induces female sterility in a leafhopper phytoplasma vector. *Entomologia Generalis* 41, 497–510. DOI: 10.1127/entomologia/2021/1267.

Gámez, R. (1973) Transmission of rayado fino virus of maize (Zea mays) by *Dalbulus maidis*. *Annals of Applied Biology* 73, 285–292. DOI: 10.1111/j.1744-7348.1973.tb00935.x.

García Gonzalez, J., Giraldo Jaramillo, M., and Roberto Spotti Lopes, J. (2018) Undetected infection by Maize Bushy Stunt phytoplasma enhances host-plant preference to *Dalbulus maidis* (Hemiptera: Cicadellidae). *Environmental Entomology* 47, 396–402. DOI: 10.1093/ee/nvy001.

Ghosh, S.K.B., Hunter, W.B., Park, A.L., and Gundersen-Rindal, D.E. (2017) Double strand RNA delivery system for plant-sap-feeding insects. *PLoS One* 12, e0171861. DOI: 10.1371/journal.pone.0171861.

Guillem-Amat, A., López-Errasquín, E., García-Ricote, I., Barbero, J.L., Sánchez, L., Casas-Tintó, S., and Ortego, F. (2023) Immunodetection of truncated forms of the α6 subunit of the nAChR in the brain of Spinosad resistant *Ceratitis capitata* phenotypes. *Insects* 14, 857. DOI: 10.3390/insects14110857.

Huang, Y.-K., Wang, X.-Y., Zhang, R.-Y., Li, J., Li, Y.-H. *et al.* (2023) The diversity, distribution, and status of phytoplasma diseases in China. In: Tiwari, A.K., Caglayan, K., Al-Sadi, A.M., Azadvar, M., and Abeysinghe, S. (eds) *Phytoplasma Diseases in Asian Countries. Diversity, Distribution, and Current Status*. Academic Press, pp. 121–147. DOI: 10.1016/B978-0-323-91896-1.00018-0.

Jain, R.G., Robinson, K.E., Fletcher, S.J., and Mitter, N. (2020) RNAi-based functional genomics in Hemiptera. *Insects* 11, 557. DOI: 10.3390/insects11090557.

Jain, R.G., Robinson, K.E., Asgari, S., and Mitter, N. (2021) Current scenario of RNAi-based hemipteran control. *Pest Management Science* 77, 2188–2196. DOI: 10.1002/ps.6153.

Jones, T.L. and Medina, R.F. (2020) Corn stunt disease: An ideal insect–microbial–plant pathosystem for comprehensive studies of vector-borne plant diseases of corn. *Plants* 9, 747. DOI: 10.3390/plants9060747.

Jones, T.-K.L., Bernal, J.S., and Medina, R.F. (2021) Assessing the functionality of RNA interference (RNAi) in the phloem-feeding maize pest *Dalbulus maidis* (preprint). *bioRxiv*. DOI: 10.1101/2021.09.29.462424.

Kanakala, S. and Ghanim, M. (2016) RNA interference in insect vectors for plant viruses. *Viruses* 8, 329. DOI: 10.3390/v8120329.

Killiny, N., Hajeri, S., Tiwari, S., Gowda, S., and Stelinski, L.L. (2014) Double-stranded RNA uptake through topical application mediates silencing of five CYP4 genes and suppresses insecticide resistance in *Diaphorina citri*. *PLoS One* 9, e110536. DOI: 10.1371/journal.pone.0110536.

Kitzmann, P., Schwirz, J., Schmitt-Engel, C., and Bucher, G. (2013) RNAi phenotypes are influenced by the genetic background of the injected strain. *BMC Genomics* 14, 5. DOI: 10.1186/1471-2164-14-5.

Kulkarni, M.M., Booker, M., Silver, S.J., Friedman, A., Hong, P., Perrimon, N., and Mathey-Prevot, B. (2006) Evidence of off-target effects associated with long dsRNAs in *Drosophila melanogaster* cell-based assays. *Nature Methods* 3, 833–838. DOI: 10.1038/nmeth935.

Lee, I.-M. (2004) "*Candidatus* Phytoplasma asteris", a novel phytoplasma taxon associated with aster yellows and related diseases. *International Journal of Systematic and Evolutionary Microbiology* 54, 1037–1048. DOI: 10.1099/ijs.0.02843-0.

Li, H., Bowling, A.J., Gandra, P., Rangasamy, M., Pence, H.E. *et al.* (2018) Systemic RNAi in Western corn rootworm, *Diabrotica virgifera virgifera*, does not involve transitive pathways: No transitive RNAi in Western corn rootworm. *Insect Science* 25, 45–56. DOI: 10.1111/1744-7917.12382.

Li, S., Kim, D.S., and Zhang, J. (2023) Plastid-mediated RNA interference: A potential strategy for efficient pest control. *Plant Cell and Environment* 46, 2595–2605. DOI: 10.1111/pce.14652.

Lim, F.-H., Rasid, O.A., As'wad, A.W.M., Vadamalai, G., Parveez, G.K.A., and Wong, M.-Y. (2020) The future is now: Revolution of RNA-mediated gene silencing in plant protection against insect pests and diseases. *Plant Biotechnology Reports* 14, 643–662. DOI: 10.1007/s11816-020-00640-7.

Liu, S., Jaouannet, M., Dempsey, D.A., Imani, J., Coustau, C., and Kogel, K.-H. (2020) RNA-based technologies for insect control in plant production. *Biotechnology Advances* 39, 107463. DOI: 10.1016/j.biotechadv.2019.107463.

Long, G.-J., Liu, X.-Z., Guo, H., Zhang, M.-Q., Gong, L.-L. *et al.* (2024) Oral-based nanoparticle-wrapped dsRNA delivery system: A promising approach for controlling an urban pest, *Blattella germanica*. *Journal of Pest Science* 97, 739–755. DOI: 10.1007/s10340-023-01677-7.

Lu, Z., Xia, T., Zhang, C., He, Q., Zhong, H. *et al.* (2024) Characterization of an RR-2 cuticle protein *DcCP8* and its potential application based on SPc nanoparticle-wrapped dsRNA in *Diaphorina citri*. *Pest Management Science* 80, 6262–6275. DOI: 10.1002/ps.8355.

Luo, Y., Wang, X., Yu, D., and Kang, L. (2012) The SID-1 double-stranded RNA transporter is not required for systemic RNAi in the migratory locust. *RNA Biology* 9, 663–671. DOI: 10.4161/rna.19986.

Ma, Y.-F., Zhao, Y.-Q., Zhou, Y., Feng, H.-Y., Gong, L.-L. *et al.* (2024) Nanoparticle delivered RNAi-based pesticide target screening for the rice pest white-backed planthopper and risk assessment for a natural predator. *Science of the Total Environment* 926, 171286. DOI: 10.1016/j.scitotenv.2024.171286.

Mahmood, M.A., Naqvi, R.Z., Siddiqui, H.A., Amin, I., and Mansoor, S. (2023) Current knowledge and implementations of *Bemisia tabaci* genomic technologies for sustainable control. *Journal of Pest Science* 96, 427–440. DOI: 10.1007/s10340-022-01520-5.

Malembic-Maher, S., Desqué, D., Khalil, D., Salar, P., Bergey, B. *et al.* (2020) When a Palearctic bacterium meets a Nearctic insect vector: Genetic and ecological insights into the emergence of the grapevine Flavescence dorée epidemics in Europe. *PLoS Pathogenes* 16, e1007967. DOI: 10.1371/journal.ppat.1007967.

Marques, V.V., Angelotti-Mendonça, J., and Roberto, S.R. (2021) Advances and challenges in RNA interference technology for citrus Huanglongbing vector control. *Horticulturae* 7, 277. DOI: 10.3390/horticulturae7090277.

Maruekawong, K., Tirasophon, W., Panyim, S., and Attasart, P. (2018) Involvement of Lv SID-1 in dsRNA uptake in *Litopenaeus vannamei*. *Aquaculture* 482, 65–72. DOI: 10.1016/j.aquaculture.2017.09.027.

Matsumoto, Y. and Hattori, M. (2016) Gene silencing by parental RNA interference in the green rice leafhopper, *Nephotettix cincticeps* (Hemiptera: Cicadellidae): Parental RNAi in *Nephotettix cincticeps*. *Archives of Insect Biochemistry and Physiology* 91, 152–164. DOI: 10.1002/arch.21315.

Mehle, N., Jakoš, N., Mešl, M., Miklavc, J., Matko, B. *et al.* (2019) Phytoplasmas associated with declining of hazelnut (*Corylus avellana*) in Slovenia. *European Journal of Plant Pathology* 155, 1117–1132. DOI: 10.1007/s10658-019-01839-3.

Mehlhorn, S., Hunnekuhl, V.S., Geibel, S., Nauen, R., and Bucher, G. (2021) Establishing RNAi for basic research and pest control and identification of the most efficient target genes for pest control: A brief guide. *Frontiers in Zoology* 18, 60. DOI: 10.1186/s12983-021-00444-7.

Mezzetti, B., Smagghe, G., Arpaia, S., Christiaens, O., Dietz-Pfeilstetter, A. *et al.* (2020) RNAi: What is its position in agriculture? *Journal of Pest Science* 93, 1125–1130. DOI: 10.1007/s10340-020-01238-2.

Miller, S.C., Brown, S.J., and Tomoyasu, Y. (2008) Larval RNAi in Drosophila? *Development Genes and Evolution* 218, 505–510. DOI: 10.1007/s00427-008-0238-8.

Mishra, S., Dee, J., Moar, W., Dufner-Beattie, J., Baum, J. *et al.* (2021) Selection for high levels of resistance to double-stranded RNA (dsRNA) in Colorado potato beetle (*Leptinotarsa decemlineata* Say) using non-transgenic foliar delivery. *Science Reports* 11, 6523. DOI: 10.1038/s41598-021-85876-1.

Mitter, N., Worrall, E.A., Robinson, K.E., Li, P., Jain, R.G. *et al.* (2017) Clay nanosheets for topical delivery of RNAi for sustained protection against plant viruses. *Nature Plants* 3, 16207. DOI: 10.1038/nplants.2016.207.

Miyata, K., Ramaseshadri, P., Zhang, Y., Segers, G., Bolognesi, R., and Tomoyasu, Y. (2014) Establishing an *in vivo* assay system to identify components involved in environmental RNA interference in the Western Corn Rootworm. *PLoS One* 9, e101661. DOI: 10.1371/journal.pone.0101661.

Nault, L.R. (1980) Maize Bushy Stunt and Corn Stunt: A comparison of disease symptoms, pathogen host ranges, and vectors. *Phytopathology* 70, 659. DOI: 10.1094/Phyto-70-659.

Neriya, Y., Maejima, K., Nijo, T., Tomomitsu, T., Yusa, A. *et al.* (2014) Onion yellow phytoplasma P38 protein plays a role in adhesion to the hosts. *FEMS Microbiology Letters* 361, 115–122. DOI: 10.1111/1574-6968.12620.

Niebres, C. and Alviar, K.B. (2023) Disruption of transmission of plant pathogens in the insect order Hemiptera using recent advances in RNA interference biotechnology. *Archives of Insect Biochemistry and Physiology* 113, e22023. DOI: 10.1002/arch.22023.

OECD (Organisation for Economic Co-operation and Development) (2020) *Considerations for the Environmental Risk Assessment of the Application of Sprayed or Externally Applied ds-RNA-Based Pesticides*. Series on Pesticides No. 104. OECD, Paris..

Oliveira, C.M.D. and Frizzas, M.R. (2022) Eight decades of *Dalbulus maidis* (DeLong & Wolcott) (Hemiptera, Cicadellidae) in Brazil: What we know and what we need to know. *Neotropical Entomology* 51, 1–17. DOI: 10.1007/s13744-021-00932-9.

Oliveira, M.J.R.A., Castro, S., Paltrinieri, S., Bertaccini, A., Sottomayor, M. *et al.* (2020) "Flavescence dorée" impacts growth, productivity and ultrastructure of *Vitis vinifera* plants in Portuguese "Vinhos Verdes" region. *Scientia Horticulturae* 261, 108742. DOI: 10.1016/j.scienta.2019.108742.

Onchuru, T.O. and Kaltenpoth, M. (2019) Quantitative PCR primer design affects quantification of dsRNA-mediated gene knockdown. *Ecology and Evolution* 9, 8187–8192. DOI: 10.1002/ece3.5387.

Orlovskis, Z., Canale, M.C., Haryono, M., Lopes, J.R.S., Kuo, C.-H., and Hogenhout, S.A. (2017) A few sequence polymorphisms among isolates of maize bushy stunt phytoplasma associate with organ proliferation symptoms of infected maize plants. *Annals of Botony* 119, 869–884. DOI: 10.1093/aob/mcw213

Ortolá, B. and Daròs, J.-A. (2024) RNA interference in insects: From a natural mechanism of gene expression regulation to a biotechnological crop protection promise. *Biology* 13, 137. DOI: 10.3390/biology13030137.

Palli, S.R. (2014) RNA interference in Colorado potato beetle: Steps toward development of dsRNA as a commercial insecticide. *Current Opinion in Insect Science* 6, 1–8. DOI: 10.1016/j.cois.2014.09.011.

Papadopoulou, N., Devos, Y., Álvarez-Alfageme, F., Lanzoni, A., and Waigmann, E. (2020) Risk assessment considerations for genetically modified RNAi plants: EFSA's activities and perspective. *Frontiers in Plant Science* 11, 445. DOI: 10.3389/fpls.2020.00445.

Pinto, M.M.D., Ferreira Dos Santos, R., De Bortoli, S.A., Moar, W., and Jurat-Fuentes, J.L. (2023) Lack of fitness costs in dsRNA-resistant *Leptinotarsa decemlineata* (Coleoptera: Chrysomelidae). *Journal of Economic Entomology* 116, 1352–1359. DOI: 10.1093/jee/toad095.

Pinzón, N., Bertrand, S., Subirana, L., Busseau, I., Escrivá, H., and Seitz, H. (2019) Functional lability of RNA-dependent RNA polymerases in animals. *PLoS Genetics* 15, e1007915. DOI: 10.1371/journal.pgen.1007915.

Powell, M., Pyati, P., Cao, M., Bell, H., Gatehouse, J.A., and Fitches, E. (2017) Insecticidal effects of dsRNA targeting the Diap1 gene in dipteran pests. *Science Reports* 7, 15147. DOI: 10.1038/s41598-017-15534-y.

Pozebon, H., Stürmer, G.R., and Arnemann, J.A. (2022) Corn stunt pathosystem and its leafhopper vector in Brazil. *Journal of Economic Entomology* 115, 1817–1833. DOI: 10.1093/jee/toac147.

Pugsley, C.E., Isaac, R.E., Warren, N.J., and Cayre, O.J. (2021) Recent advances in engineered nanoparticles for RNAi-mediated crop protection against insect pests. *Frontiers in Agronomy* 3, 652981. DOI: 10.3389/fagro.2021.652981.

Ramos, A., Esteves, M.B., Cortés, M.T.B., and Lopes, J.R.S. (2020) Maize Bushy Stunt phytoplasma favors its spread by changing host preference of the insect vector. *Insects* 11, 600. DOI: 10.3390/insects11090600.

Rank, A.P. and Koch, A. (2021) Lab-to-field transition of RNA spray applications – How far are we? *Frontiers in Plant Science* 12, 755203. DOI: 10.3389/fpls.2021.755203.

Rashidi, M., Galetto, L., Bosco, D., Bulgarelli, A., Vallino, M., Veratti, F., and Marzachì, C. (2015) Role of the major antigenic membrane protein in phytoplasma transmission by two insect vector species. *BMC Microbiology* 15, 193. DOI: 10.1186/s12866-015-0522-5.

Ripamonti, M., Cerone, L., Abbà, S., Rossi, M., Ottati, S., Palmano, S., Marzachì, C., and Galetto, L. (2022a) Silencing of ATP synthase β impairs egg development in the leafhopper *Scaphoideus titanus*, vector of the phytoplasma associated with grapevine Flavescence dorée. *International Journal of Molecular Sciences* 23, 765. DOI: 10.3390/ijms23020765.

Ripamonti, M., Maron, F., Cornara, D., Marzachì, C., Fereres, A., and Bosco, D. (2022b) Leafhopper feeding behaviour on three grapevine cultivars with different susceptibilities to Flavescence dorée. *Journal of Insect Physiology* 137, 104366. DOI: 10.1016/j.jinsphys.2022.104366.

Rizzoli, A., Jelmini, L., Pezzatti, G.B., Jermini, M., Schumpp, O. *et al.* (2022) Impact of the "Flavescence dorée" phytoplasma on xylem growth and anatomical characteristics in trunks of 'Chardonnay' grapevines (*Vitis vinifera*). *Biology* 11, 978. DOI: 10.3390/biology11070978.

Rodrigues, T.B., Mishra, S.K., Sridharan, K., Barnes, E.R., Alyokhin, A. *et al.* (2021) First sprayable double-stranded RNA-based biopesticide product targets proteasome subunit beta type-5 in Colorado potato beetle (*Leptinotarsa decemlineata*). *Frontiers in Plant Science* 12, 728652. DOI: 10.3389/fpls.2021.728652.

Rossi, M., Ottati, S., Bucci, L., Fusco, A., Abbà, S. *et al.* (2023) Lab-scale method for plant-mediated delivery of dsRNAs to phloem-feeding leafhoppers. *Journal of Pest Science* DOI: 10.1007/s10340-023-01670-0.

Saakre, M., Jaiswal, S., Rathinam, M., Raman, K.V., Tilgam, J. *et al.* (2024) Host-delivered RNA interference for durable pest resistance in plants: Advanced methods, challenges, and applications. *Molecular Biotechnology* 66, 1786–1805. DOI: 10.1007/s12033-023-00833-9.

Santos, D., Mingels, L., Vogel, E., Wang, L., Christiaens, O. *et al.* (2019) Generation of virus- and dsRNA-derived siRNAs with species-dependent length in insects. *Viruses* 11, 738. DOI: 10.3390/v11080738.

Saurabh, S., Mishra, M., Rai, P., Pandey, R., Singh, J. *et al.* (2021) Tiny flies: A mighty pest that threatens agricultural productivity – A case for next-generation control strategies of whiteflies. *Insects* 12, 585. DOI: 10.3390/insects12070585.

Sharma, R., Taning, C.N.T., Smagghe, G., and Christiaens, O. (2021) Silencing of double-stranded ribonuclease improves oral RNAi efficacy in Southern green stinkbug *Nezara viridula*. *Insects* 12, 115. DOI: 10.3390/insects12020115.

Shu, Q., Liu, G., He, J., Hu, P., Dong, Z. *et al.* (2024) RNAi efficiency is enhanced through knockdown of double-stranded RNA-degrading enzymes in butterfly *Papilio xuthus*. *Archives of Insect Biochemistry and Physiology* 115, e22113. DOI: 10.1002/arch.22113.

Sidorova, T., Miroshnichenko, D., Kirov, I., Pushin, A., and Dolgov, S. (2021) Effect of grafting on viral resistance of non-transgenic plum scion combined with transgenic PPV-resistant rootstock. *Frontiers in Plant Science* 12, 621954. DOI: 10.3389/fpls.2021.621954.

Silver, K., Cooper, A.M., and Zhu, K.Y. (2021) Strategies for enhancing the efficiency of RNA interference in insects. *Pest Management Science* 77, 2645–2658. DOI: 10.1002/ps.6277.

Sugio, A. and Hogenhout, S.A. (2012) The genome biology of phytoplasma: Modulators of plants and insects. *Current Opinion in Microbiology* 15, 247–254. DOI: 10.1016/j.mib.2012.04.002.

Suzuki, S., Oshima, K., Kakizawa, S., Arashida, R., Jung, H.-Y. *et al.* (2006) Interaction between the membrane protein of a pathogen and insect microfilament complex determines insect-vector specificity. *Proceedings of the National Academy of Sciences* 103, 4252–4257. DOI: 10.1073/pnas.0508668103.

Tang, X., Jiang, X., Chen, Q., and Lin, X. (2024) Amphiphilic layered double hydroxides enhance plant-mediated delivery of dsRNAs to phloem-feeding planthoppers. *Chemical Engineering Journal* 491, 151953. DOI: 10.1016/j.cej.2024.151953.

Taning, C.N., Arpaia, S., Christiaens, O., Dietz-Pfeilstetter, A., Jones, H. *et al.* (2020) RNA-based biocontrol compounds: Current status and perspectives to reach the market. *Pest Management Science* 76, 841–845. DOI: 10.1002/ps.5686.

Terenius, O., Papanicolaou, A., Garbutt, J.S., Eleftherianos, I., Huvenne, H. *et al.* (2011) RNA interference in Lepidoptera: An overview of successful and unsuccessful studies and implications for experimental design. *Journal of Insect Physiology* 57, 231–245. DOI: 10.1016/j.jinsphys.2010.11.006

Tomoyasu, Y., Miller, S.C., Tomita, S., Schoppmeier, M., Grossmann, D., and Bucher, G. (2008) Exploring systemic RNA interference in insects: A genome-wide survey for RNAi genes in *Tribolium*. *Genome Biology* 9, R10. DOI: 10.1186/gb-2008-9-1-r10.

Trivellone, V., Ripamonti, M., Angelini, E., Filippin, L., Rossi, M., Marzachí, C., and Galetto, L. (2019) Evidence suggesting interactions between immunodominant membrane protein Imp of Flavescence dorée phytoplasma and protein extracts from distantly related insect species. *Journal of Applied Microbiology* 127, 1801–1813. DOI: 10.1111/jam.14445.

Ulrich, J., Dao, V.A., Majumdar, U., Schmitt-Engel, C., Schwirz, J. *et al.* (2015) Large scale RNAi screen in *Tribolium* reveals novel target genes for pest control and the proteasome as prime target. *BMC Genomics* 16, 674. DOI: 10.1186/s12864-015-1880-y.

Vandesompele, J., De Preter, K., Pattyn, F., Poppe, B., Van Roy, N., De Paepe, A., and Speleman, F. (2002) Accurate normalization of real-time quantitative RT-PCR data by geometric averaging of multiple internal control genes. *Genome Biology* 3, 34.

Verma, K. and Modgil, M. (2024) RNA interference (RNAi) mediated technique for combating plant diseases: Harnessing nanoparticles for effective delivery and enhanced efficacy. *Plant Cell, Tissue and Organ Culture* 157, 53. DOI: 10.1007/s11240-024-02773-x.

Wang, L., Chen, W., Ma, H., Li, J., Hao, X., and Wu, Y. (2021) Identification of RNA silencing suppressor encoded by wheat blue dwarf (WBD) phytoplasma. *Plant Biology Journal* 23, 843–849. DOI: 10.1111/plb.13257.

Wang, N., Li, Y., Chen, W., Yang, H.Z., Zhang, P.H., and Wu, Y.F. (2018) Identification of wheat blue dwarf phytoplasma effectors targeting plant proliferation and defence responses. *Plant Pathology* 67, 603–609. DOI: 10.1111/ppa.12786.

Wang, Z., Qin, C., Chen, Y., Yu, X., Chen, R., Niu, J., and Wang, J. (2024) Fusion dsRNA designs incorporating multiple target sequences can enhance the aphid control capacity of an RNAi-based strategy. *Pest Management Science* 80, 2689–2697. DOI: 10.1002/ps.7975.

Whyard, S., Singh, A.D., and Wong, S. (2009) Ingested double-stranded RNAs can act as species-specific insecticides. *Insect Biochemistry and Molecular Biology* 39, 824–832. DOI: 10.1016/j.ibmb.2009.09.007.

Wu, J.-J., Cheng, M.-D., Ze, L.-J., Shen, C.-H., Jin, L., and Li, G.-Q. (2022) Dissecting the isoform-specific roles of FTZ-F1 in the larval–larval and larval–pupal ecdyses in *Henosepilachna vigintioctopunctata*. *Insects* 13, 228. DOI: 10.3390/insects13030228.

Wu, Y., Hao, X., Li, Z., Gu, P., An, F. *et al.* (2010) Identification of the phytoplasma associated with Wheat Blue Dwarf disease in China. *Plant Disease* 94, 977–985. DOI: 10.1094/PDIS-94-8-0977.

Wynant, N., Santos, D., Van Wielendaele, P., and Vanden Broeck, J. (2014) Scavenger receptor-mediated endocytosis facilitates RNAinterference in the desert locust, *Schistocerca gregaria*. *Insect Molecular Biology* 23, 320–329. DOI: 10.1111/imb.12083.

Xu, H., Chen, T., Ma, X., Xue, J., Pan, P. *et al.* (2013) Genome-wide screening for components of small interfering RNA (siRNA) and micro-RNA (miRNA) pathways in the brown planthopper, *Nilaparvata lugens* (Hemiptera: Delphacidae). *Insect Molecular Biology* 22, 635–647. DOI: 10.1111/imb.12051.

Yang, X., Liu, S., Lu, W., Du, M., Qiao, Z. *et al.* (2022) Delta and jagged are candidate target genes of RNAi biopesticides for the control of *Nilaparvata lugens*. *Frontiers in Bioengineering and Biotechnology* 10, 1023729. DOI: 10.3389/fbioe.2022.1023729.

Ye, C., An, X., Xie, B., Ding, B., Niu, J., and Wang, J. (2023) The involvement of systemic RNA interference deficient-1-like (SIL1) in cellular dsRNA uptake in *Acyrthosiphon pisum*. *Insect Science* 30, 1393–1404. DOI: 10.1111/1744-7917.13167.

Yousaf, I., Tabassum, B., Jabbar, B., Amjad, M.A., Qaisar, U. *et al.* (2024) Efficacy of arginine kinase as a promising RNAi target in *Aphis gossypii* genome as revealed through aphid bioassay on field-grown transgenic cotton plants. *Journal of Plant Protection Research* 64, 242–252. DOI: 10.24425/jppr.2024.151257.

Yu, N., Christiaens, O., Liu, J., Niu, J., Cappelle, K. *et al.* (2013) Delivery of dsRNA for RNAi in insects: An overview and future directions. *Insect Science* 20, 4–14. DOI: 10.1111/j.1744-7917.2012.01534.x.

Yu, X., Gowda, S., and Killiny, N. (2017) Double-stranded RNA delivery through soaking mediates silencing of the muscle protein and increases mortality to the Asian citrus psyllid, *Diaphorina citri*. *Pest Management Science* 73, 1846–1853. DOI: 10.1002/ps.4549.

Yu, X. and Killiny, N. (2020) RNA interference-mediated control of Asian citrus psyllid, the vector of the huanglongbing bacterial pathogen. *Tropical Plant Pathology* 45, 298–305. DOI: 10.1007/s40858-020-00356-7.

Yu, X.-D., Liu, Z.-C., Huang, S.-L., Chen, Z.-Q., Sun, Y.-W. *et al.* (2016) RNAi-mediated plant protection against aphids. *Pest Management Science* 72, 1090–1098. DOI: 10.1002/ps.4258.

Zarrabian, M. and Sherif, S.M. (2024) Silence is not always golden: A closer look at potential environmental and ecotoxicological impacts of large-scale dsRNA application. *Science of the Total Environment* 950, 175311. DOI: 10.1016/j.scitotenv.2024.175311.

Zhao, Y., Wei, W., Davis, R.E., Lee, I.-M., and Bottner-Parker, K.D. (2021) The agent associated with blue dwarf disease in wheat represents a new phytoplasma taxon, "*Candidatus* Phytoplasma tritici." *International Journal of Systematic and Evolutionary Microbiology* 71. DOI: 10.1099/ijsem.0.004604

Zheng, W., Xu, X., Huang, X., Peng, J., Ma, W. *et al.* (2024) Spray-induced and nanocarrier-delivered gene silencing system targeting juvenile hormone receptor components: Potential application as fertility inhibitors for *Adelphocoris suturalis* management. *Pest Management Science* 80, 3743–3751. DOI: 10.1002/ps.8077.

Zhu, F., Xu, J., Palli, R., Ferguson, J., and Palli, S.R. (2011) Ingested RNA interference for managing the populations of the Colorado potato beetle, *Leptinotarsa decemlineata*. *Pest Management Science* 67, 175–182. DOI: 10.1002/ps.2048.

Zhu, K.Y. and Palli, S.R. (2020) Mechanisms, applications, and challenges of insect RNA interference. *Annual Review of Entomology* 65, 293–311. DOI: 10.1146/annurev-ento-011019-025224.

Zong, J., Yao, X., Yin, J., Zhang, D., and Ma, H. (2009) Evolution of the RNA-dependent RNA polymerase (RdRP) genes: Duplications and possible losses before and after the divergence of major eukaryotic groups. *Gene* 447, 29–39. DOI: 10.1016/j.gene.2009.07.004.

Zotti, M., dos Santos, E.A., Cagliari, D., Christiaens, O., Taning, C.N.T., and Smagghe, G. (2018) RNA interference technology in crop protection against arthropod pests, pathogens and nematodes. *Pest Management Science* 74, 1239–1250. DOI: 10.1002/ps.4813.

8 Functional Analysis of Membrane and Secretory Proteins Mediating Host–Phytoplasma Interactions

Kensaku Maejima,[1] Yugo Kitazawa,[2] Shigeyuki Kakizawa,[3] Kenro Oshima,[4*] and Shigetou Namba[1]

[1]The University of Tokyo, Tokyo, Japan; [2]Yamaguchi University, Yamaguchi, Japan; [3]National Institute of Advanced Industrial Science and Technology (AIST), Ibaraki, Japan; [4]Hosei University, Tokyo, Japan

Abstract

Phytoplasmas are plant pathogenic bacteria that reside intracellularly within the plant phloem. Immunodominant membrane proteins account for a major proportion of the total cell membrane proteins in phytoplasmas, and some of them have been shown to be involved in phytoplasma interactions with both insect and plant hosts. With the accumulation of genomic information on phytoplasmas, many candidates of secretory protein genes have been reported. Among them, TENGU, phyllogen, SAP11, and SAP05 have been well studied in terms of molecular mechanisms and how they manipulate hosts. Further, our understanding of membrane and secretory proteins will enable the development of strategies to control phytoplasma diseases by targeting these proteins.

Introduction

Phytoplasmas (genus '*Candidatus* Phytoplasma') are plant pathogenic bacteria that reside intracellularly within the plant phloem (Hogenhout *et al.*, 2008; Maejima *et al.*, 2014b; Namba, 2019). Phytoplasma-infected plants exhibit a wide range of unique symptoms, including flower malformation, yellowing, dwarfing, witches'-broom, purple tops, and phloem necrosis (Arashida *et al.*, 2008; Himeno *et al.*, 2014; Namba, 2019). Since the infected plants eventually decline and die in most cases, phytoplasmas cause great damage to agricultural production and forest trees (Iwabuchi *et al.*, 2017). Phytoplasmas can also reside endocellularly within insects such as leafhoppers, planthoppers, and psyllids, and are transmitted by these insect vectors (Oshima *et al.*, 2011; Maejima *et al.*, 2014b). It is of interest how phytoplasmas can adapt to two diverse intracellular environments (i.e. plant and insect cells).

Phytoplasmas induce unique symptoms in host plants such as phyllody and witches'-broom, and how they manipulate plants has been of much interest. Many Gram-negative pathogens of plants and animals possess type III secretion systems (T3SSs) that can inject bacterial virulence effector proteins into host cells (Grant *et al.*, 2006). Although phytoplasmas lack genes for the T3SS, they possess genes for the protein translocation Sec system in their genomes

*Corresponding author: kenro@hosei.ac.jp

DOI: 10.1079/9781800627031.0008

(Kakizawa *et al.*, 2001, 2004). Since phytoplasmas are cell wall-less and reside inside host cells, their membrane proteins and secretory proteins via the Sec system may function in the cytoplasm of the host cell and are predicted to have some crucial roles in host–parasite interactions and/or virulence as effector proteins (Oshima *et al.*, 2013, 2023). This chapter summarizes the roles of membrane and secretory proteins of phytoplasmas in host–phytoplasma interactions.

Protein Secretion System in Phytoplasmas

Bacterial secretion systems

There are several independent protein secretion systems in bacteria (Economou, 1999). In many bacteria, including *Escherichia coli*, these systems secrete various proteins such as toxins, adhesins, and hydrolytic enzymes. Of these systems, only the Sec system is essential for cell viability and cannot be knocked out (Baba *et al.*, 2006, 2020). In *Bacillus subtilis*, the Sec pathway is considered the most important of four distinct transport pathways (Tjalsma *et al.*, 2000). The Sec protein translocation system is best characterized in *E. coli*, comprising at least 11 proteins and one RNA species (Economou, 1999). Among these proteins, SecY, SecE, SecG, and SecA form a translocase complex that acts as export machinery at the cytoplasmic membrane. SecA, SecY, and SecE are required for protein translocation and cell viability in *E. coli* (Economou, 1999) and protein translocation activity can be reconstituted *in vitro* with only these three proteins (Akimaru *et al.*, 1991), whereas SecG is not essential. Secretory proteins have an N-terminal signal peptide. SecA binds the secretory protein and transports it through the SecYEG pore.

Presence of the Sec system in phytoplasmas

The first description of the Sec system in phytoplasma was in the '*Candidatus* Phytoplasma asteris' OY-M strain. Genes encoding the SecA, SecY, and SecE proteins, the essential components of the Sec system, have been identified (Kakizawa *et al.*, 2001, 2004), and the expression of SecA protein in phytoplasma-infected plants has been confirmed (Kakizawa *et al.*, 2001; Wei *et al.*, 2004). Whole-genome sequencing of the OY-M strain confirmed that there are single-copy genes for *SecA*, *SecY*, and *SecE* (Oshima *et al.*, 2004). These three genes were also identified in the '*Ca.* P. asteris' aster yellows witches'-broom (AY-WB) strain genome (Bai *et al.*, 2006), and the *SecY* genes were cloned from several phytoplasmas (Lee *et al.*, 2006).

With the accumulation of whole-genome sequences of phytoplasmas, the presence of *SecA* and *SecY* genes has been confirmed in all 11 completely sequenced genomes (Hogenhout *et al.*, 2008; Kube *et al.*, 2012; Garcion *et al.*, 2021). SecE is a short protein that is difficult to detect through sequence homology searches and may be overlooked in initial genome analyses; however, it is likely to be widely conserved across phytoplasmas. These results strongly suggest that the Sec system is a common feature in phytoplasmas. On the other hand, some symbiotic bacteria of insects, including *Carsonella rudii*, a symbiotic bacterium of psyllids, have been reported to lack the Sec system (Nakabachi *et al.*, 2006; Moran *et al.*, 2008). The Sec system may not be essential for bacteria that live exclusively within host cells.

Functional evidence of the Sec system in phytoplasmas

Phytoplasmas are difficult to manipulate genetically, making it challenging to study the function of the Sec system directly. However, several observations suggest that the Sec system is functional in phytoplasmas. Firstly, it has been confirmed that an antigenic membrane protein (Amp), a type of immunodominant membrane protein found in phytoplasmas, has a signal sequence of the Sec system at its N-terminus, and this signal sequence is cleaved in the OY phytoplasma. By producing anti-Amp antibodies and performing western blotting on Amp-expressing *E. coli* cells and phytoplasma-infected plants, the signal sequence was detected to be cleaved (Kakizawa *et al.*, 2004). Additionally, as mentioned later, numerous effector proteins with signal sequences at their N-terminus have been identified, and their functional analysis strongly suggests that these effectors are secreted via the

Sec system. These findings indicate that the Sec system in phytoplasmas is functional.

Signal sequence prediction and effector identification

The membrane or secretory proteins of cell wall-less phytoplasmas likely function directly in the cytoplasm of host plants and insect cells. It is speculated that proteins secreted via the Sec system may influence pathogenicity. Therefore, identifying these proteins in the phytoplasma genome is crucial for understanding host–phytoplasma interactions. Generally, secretory proteins passing through the Sec system have an N-terminal signal peptide. This consensus signal sequence consists of three parts: a positively charged N-terminal domain with at least one arginine or lysine residue (*n*-region), a hydrophobic core domain forming an α-helix that penetrates the membrane (*h*-region), and an A-X-A sequence serving as the signal peptidase I (SPaseI) cleavage site (*c*-region) (Tjalsma *et al.*, 2000; Garcion *et al.*, 2021). In this sequence, the alanine residues can be replaced by other small, preferably uncharged residues, such as valine, leucine, or isoleucine (Tjalsma *et al.*, 2000).

The initial survey of signal sequences was conducted on the Amp of the OY-M strain, as mentioned earlier. The predicted export signal sequence of OY Amp closely matched the general Sec system signal peptides, suggesting that OY Amp is exported by the Sec system of both *E. coli* and phytoplasma, with its signal sequence being processed. This indicates a commonality in signal sequence recognition mechanisms between phytoplasma and *E. coli* (Kakizawa *et al.*, 2004). Consequently, prediction programs such as SignalP (Nielsen *et al.*, 1997), PSORT (Nakai and Kanehisa, 1991), or Phobius (Käll *et al.*, 2004, 2007) could be used to recognize signal sequences in phytoplasma proteins and identify their secretory proteins.

As more phytoplasma genome sequences become available, comprehensive surveys of effector candidate genes with secretion signals can be conducted, allowing their expression in plants to study their functions. Various methods for predicting secretion signals have been widely discussed, particularly regarding the software and parameter combinations. For example, the prediction of secretory proteins in AY-WB phytoplasma was mainly performed using the SignalP v3.0 hidden Markov model (HMM) (Bai *et al.*, 2009). Comprehensive and detailed analyses were conducted using various existing software packages for signal peptide prediction to identify the optimal software and parameters for detecting signal peptides from phytoplasma genes, and several effective methods were proposed (Garcion *et al.*, 2021). Additionally, the PhyEffector algorithm pipeline, which has higher accuracy and sensitivity, was introduced to detect non-classic effectors (Carreón-Anguiano *et al.*, 2023), as discussed later.

With the establishment of these signal prediction methods and the accumulation of phytoplasma genome sequences, it is expected that effector candidate genes can be identified more accurately, leading to increased insights into host–phytoplasma interactions through functional analysis.

Non-classic secretion system in phytoplasmas

Non-classic secretion systems have been reported in both bacteria and eukaryotes (Bendtsen *et al.*, 2005). These secretory proteins lack typical secretion signals; most of their secretion pathways are still unknown. Numerous non-classic secretory proteins that affect the host have been detected outside the bacterial cell. For example, in the human pathogen *Mycobacterium tuberculosis*, the secretion of glutamine synthetase (GlnA) (Harth *et al.*, 1994; Harth and Horwitz, 1997), superoxide dismutase (SodA) (Harth and Horwitz, 1999), and several antigenic proteins (Brodin *et al.*, 2004) were suggested to be involved in pathogenicity. A protein necessary for the secretion of SodA in *M. tuberculosis* has been identified and named SecA2 (Braunstein *et al.*, 2003). Other examples include staphylococcal nuclease in *M. smegmatis* (Recchi *et al.*, 2002). In *Listeria monocytogenes*, many proteins have been detected extracellularly, including central metabolic enzymes pyruvate dehydrogenase and enolase, and elongation factors G and Tu (Antelmann *et al.*, 2001; Lenz *et al.*, 2003; Schaumburg *et al.*, 2004). A pore-forming cytotoxic protein, ClyA, from *E. coli* has also been reported; it lacks an N-terminal signal peptide but is released via membrane vesicles, suggesting that vesicle-mediated secretion is important in *E. coli* (Miller *et al.*, 2003; Wai *et al.*, 2003). These secretory proteins were summarized comprehensively, and a program called SecretomeP was developed to identify them using neural network algorithms (Bendtsen *et al.*, 2005). Many of these proteins are considered "moonlighting" proteins, possessing multiple functions as they also play roles within the cytoplasm.

Non-classic secretory proteins were investigated in '*Ca.* P. ziziphi' (Gao *et al.*, 2023). Using both Gram-positive and Gram-negative settings of SecretomeP, multiple candidates were identified. These were then experimentally validated using the alkaline phosphatase system (phoA system) in *E. coli* to confirm extracellular transport, followed by expression in plants to identify several factors that suppress the hypersensitive response (HR). Additionally, the PhyEffector pipeline includes the SecretomeP program, allowing the detection of non-classic secretion proteins. Such methods are expected to continue identifying novel effectors in phytoplasmas and further elucidate host–phytoplasma interactions.

Other protein secretion systems in phytoplasmas

Many Gram-negative pathogens possess T3SSs and type IV secretion systems (T4SSs) with pilus-like structures for injecting virulence proteins into host cells (Cornelis and Van Gijsegem, 2000; Grohmann *et al.*, 2003). However, phytoplasmas lack both T3SS and T4SS, and no genes or pili-like structures for these systems have been observed in phytoplasmas. Phytoplasmas retain a YidC integration system for membrane protein insertion (Oshima *et al.*, 2004; Bai *et al.*, 2006), similar to that in *E. coli* (Samuelson *et al.*, 2000), suggesting it may play an important role in host–phytoplasma interactions.

Host-Interacting Membrane Proteins in Phytoplasmas

The major membrane proteins of phytoplasmas

Previous studies have identified a group of membrane proteins, known as immunodominant membrane proteins (IDPs), as a significant component of the total cellular membrane proteins in most phytoplasmas (Shen and Lin, 1993). These IDPs are located on the exterior surface of the cell membrane and play a significant role in the attachment of bacteria to host cell surfaces, implicating them in host–phytoplasma interactions (Milne *et al.*, 1995; Kakizawa *et al.*, 2006b; Konnerth *et al.*, 2016).

IDP genes were isolated from several phytoplasmas (Berg *et al.*, 1999; Blomquist *et al.*, 2001; Barbara *et al.*, 2002; Morton *et al.*, 2003; Kakizawa *et al.*, 2004, 2006b), and they are classified into three types: (i) Imp (immunodominant membrane protein), (ii) IdpA (immunodominant membrane protein A), and (iii) Amp (antigenic membrane protein) (Kakizawa *et al.*, 2006b; Konnerth *et al.*, 2016). Each type is genetically distinct, with no amino acid similarity among them. These three types of IDPs and VmpA (variable membrane protein A) are summarized in Table 8.1. Each membrane protein was reported to interact with the host, and detailed information is provided in this chapter. The gene encoding Imp has been found in multiple phytoplasmas, suggesting it may be ancestral, while orthologs of IdpA and Amp are more specific to certain phytoplasma groups (Kakizawa *et al.*, 2006b; Konnerth *et al.*, 2016).

Genetic variability and evolutionary significance of IDPs in phytoplasmas

Immunodominant membrane protein sequences from various phytoplasma strains are highly variable, with their sequence similarity between phytoplasmas being lower than that of surrounding genes or non-coding regions, indicating strong divergent selective pressures (Kakizawa *et al.*, 2006a). The extracellular domains of IDPs are more variable than the transmembrane domains, suggesting host–phytoplasma interactions promote this variability (Barbara *et al.*, 2002; Kakizawa *et al.*, 2004). The sequence identities of *imp* genes do not correlate with those of 16S rDNA, indicating that the variability of IDPs reflects some factors other than evolutionary time (Morton *et al.*, 2003).

Positive selection has been observed on Amps, where substitutions provide fitness advantages, implying the crucial roles of Amp in host–pathogen interactions (Kakizawa *et al.*, 2006a). Most positively selected amino acids are

Table 8.1. Key phytoplasma membrane proteins that mediate host interactions.

Name	Phytoplasma species (strain)	Genetic context	Target and function	References
Amp (antigenic membrane protein)	'*Ca.* P. asteris' (OY, ROL, etc.) '*Ca.* P. japonicum' '*Ca.* P. solani' '*Ca.* P. meliae' '*Ca.* P. australiense'	*groEL-amp-nadE*	Forms a complex with insect microfilament proteins (actin, myosin heavy chain, myosin light chain) and interacts with ATP synthase in a vector-specific manner Suppresses plant defenses	Suzuki *et al.*, 2006; Galetto *et al.*, 2011; Wang *et al.*, 2023
IdpA (immunodominant membrane protein A)	'*Ca.* P. pruni' (WX, PoiBI)	*dnaX-idpA-tRNA*^{Ser}	Remains poorly understood	Blomquist *et al.*, 2001; Neriya *et al.*, 2011; Villamor & Eastwell, 2019
Imp (immunodominant membrane protein)	Many '*Ca.* Phytoplasma' spp.	*rnc-dnaD-imp-pryG*	Interacts with several insect proteins, including actin, myosin heavy chain, ATP synthase, α-tubulin, as well as plant F-actin	Kakizawa *et al.*, 2009; Siampour *et al.*, 2011; Ding *et al.*, 2022; Boonrod *et al.*, 2023
VmpA (variable membrane protein A)	FD-related phytoplasmas '*Ca.* P. rubi'	*VmpA-tRNA*^{Tyr}*-16S rRNA*	Acts as an adhesin to glycoconjugates on insect sell surface, and Uk1_LRR is identified as a receptor Interacts with adaptor protein complexes AP-1 and AP-2 in clathrin-mediated endocytosis	Arricau-Bouvery *et al.*, 2018, 2021, 2023; Malembic-Maher *et al.*, 2020; Canuto *et al.*, 2023

ATP, adenosine triphosphate; *Ca.* P., *Candidatus* Phytoplasma; OY, onion yellows; PoiBI, poinsettia branch-inducing; ROL, rice orange leaf; WX, Western X-disease.

in the central hydrophilic domain of Amp, likely due to interactions with the extracellular environment. Similar positive selection was observed in several *imp* genes (Kakizawa *et al.*, 2009; Neriya *et al.*, 2011; Konnerth *et al.*, 2016), highlighting their importance in host–phytoplasma interactions. These variations could be an adaptation to evade the insect vector's immune system or to better attach to host proteins. Similar mechanisms have been observed in other bacterial membrane proteins, such as OspC of *Borrelia burgdorferi* (Wilske *et al.*, 1996).

Interactions between membrane proteins and insect vectors

Phytoplasmas are typically transmitted by specific insect vectors, even though they can infect a wide range of plant hosts. For example, OY phytoplasma is transmitted by *Macrosteles striifrons* but not by *Nephotettix cincticeps*, which transmits rice yellow dwarf (RYD) phytoplasma instead. This specificity influences the host range and extent of crop damage.

Phytoplasma infection in insect hosts involves multiple steps after acquisition feeding from plant phloem: (i) adhesion, uptake, and multiplication in gut epithelial cells of the alimentary canal; (ii) entry into the hemocoel by crossing the intestinal wall and circulation in the hemolymph; and (iii) adhesion, uptake, and multiplication in salivary glands, resulting in successful transmission to new plants through infection feeding (Lefol *et al.*, 1994; Koinuma *et al.*, 2020). The alimentary canal and salivary glands are the major barriers to the transmission of phytoplasmas, and the ability to pass through them is crucial for vector determination (Purcell *et al.*, 1981). The adhesion, uptake, and intracellular dynamics of phytoplasmas in insect cells are considered to involve proteins on the surface of phytoplasma cells, with research primarily focusing on IDPs.

Amp is the first membrane protein for which interactions with host factors have been elucidated. Amp of OY phytoplasma forms a complex with three insect microfilament proteins: actin, myosin heavy chain, and myosin light chain, both *in vitro* and *in vivo* (Suzuki *et al.*, 2006). This complex formation was detected in vector leafhoppers but not in non-vector leafhoppers, suggesting that the Amp–microfilament complex (AM complex) formation plays a significant role in insect vector specificity. Also, the Amp of chrysanthemum yellows (CY) phytoplasma was found to interact with the ATP synthase and actin of its vectors but not with the homologous proteins of non-vectors (Galetto et al., 2011), and to be involved in infection of the insect vector (Rashidi *et al.*, 2015). Amp of rice orange leaf (ROL) phytoplasma was reported to interact with the actin of its vectors, and to enhance phytoplasma accumulation and inhibit defense responses in rice plants (Wang *et al.*, 2023). Therefore, Amp could be involved in phytoplasma interactions with both insect and plant hosts.

As in the case of Amp, Imp is involved in interactions between phytoplasmas and their insect vectors (Siampour *et al.*, 2011; Trivellone *et al.*, 2019). Imp of lime witches'-broom (LWB) phytoplasma interacts with multiple insect proteins, including actin, myosin heavy chain, and ATP synthase (Siampour *et al.*, 2011). Ding *et al.* (2022) showed that interaction between Imp of wheat blue dwarf (WBD) phytoplasma and α-tubulin of the vector (*Psammotettix striatus*), and RNA interference (RNAi) of α-tubulin in *P. striatus*, results in a decrease of transmission efficiency of WBD phytoplasma. Furthermore, it is reported that Imp of '*Ca.* P. mali' co-localized with the actin cytoskeleton in plant cells (Boonrod *et al.*, 2012) and interacted with plant F-actin (Boonrod *et al.*, 2023). Thus, Imp could also be involved in phytoplasma interactions with both plants and insects. Imp shares structural homology with an F-actin-binding protein, talin rod domain-containing protein 1 (TLNRD1) (Liu *et al.*, 2024). The C-terminus of the exposed region of Imp was suggested to be an actin-binding region (Boonrod *et al.*, 2023; Liu *et al.*, 2024).

IdpA was identified as an immunodominant membrane protein of Western X-disease (WX) phytoplasma (Blomquist *et al.*, 2001). IdpA has been found in phytoplasmas closely related to WX phytoplasma, which also have Imp (Neriya *et al.*, 2011; Villamor and Eastwell, 2019). Although the sequence diversity and expression levels of IdpA have been investigated, its function is poorly understood.

In flavescence dorée (FD) phytoplasma, the sequence diversity of the variable membrane proteins VmpA and VmpB has been shown to correlate with their vector specificity (Malembic-Maher *et al.*, 2020). VmpA is an adhesin that specifically interacts with glycoconjugates on the surface of insect vector cells (Arricau-Bouvery *et al.*, 2018, 2021). Proteomics approaches have identified candidate insect proteins that interact with VmpA. The adaptor protein complexes AP-1 and AP-2 are involved in clathrin-mediated endocytosis, which is vital for the FD phytoplasma to enter insect cells (Arricau-Bouvery *et al.*, 2023). Uk1_LRR is a transmembrane protein with leucine-rich repeat domains and is involved in the attachment of VmpA to insect cells (Canuto *et al.*, 2023).

A plasmid-encoded membrane protein, ORF3, of OY phytoplasma has been suggested to be necessary for insect transmission (Oshima *et al.*, 2001; Nishigawa *et al.*, 2002) because the *orf3* gene or its promoter was lost from the plasmids of the non-insect-transmissible line of OY phytoplasma (Ishii *et al.*, 2009a, 2009b). The preferential expression of ORF3 protein in OY-infected insects compared with OY-infected plants supports the notion that ORF3 plays a role in insect hosts (Ishii *et al.*, 2009a). P38 is another membrane protein of OY phytoplasma. It contains a mollicutes adhesin motif (MAM) conserved in adhesin P40 of *Mycoplasma agalactiae* and P89 of *Spiroplasma citri* and interacts with insect extracts in a MAM-dependent manner (Neriya *et al.*, 2014). The insect molecules targeted by these membrane proteins are unknown, and their identification is expected to elucidate the molecular mechanisms of insect transmission of phytoplasmas.

Secretory Proteins (Effectors) in Phytoplasmas

Importance of secretory proteins

Phytoplasmas are expected to secrete proteins directly into the host cytoplasm via the Sec system or a non-classic secretion system, as described earlier. These proteins, called effectors, are thought to play critical roles in host–phytoplasma interactions and disease symptom induction. Therefore, their functional analysis has become one of the most interesting topics for phytoplasma researchers.

With the accumulation of genomic information on phytoplasmas, many effector candidate genes have been reported (Carreón-Anguiano *et al.*, 2023). While the functions of a few effectors have been suggested (Table 8.2), the roles of most of the other effectors remain unknown. However, even those known functions, such as influencing plant morphogenesis and attracting insect vectors, seem to tell us what is essential for the survival of phytoplasmas. Further, our understanding of the molecular mechanisms by which effectors manipulate hosts will enable the development of strategies to control phytoplasma diseases by targeting these proteins.

Conservation and/or horizontal transfer of secretory proteins

Some effector genes are retained in diverse phytoplasmas. Does this indicate that phytoplasmas acquired these effectors before their speciation? Many of them are encoded within clusters of repeated gene sequences called potential mobile units (PMUs). The horizontal transfer of PMUs may have occurred between divergent phytoplasma lineages (Chung *et al.*, 2013; Ku *et al.*, 2013). Analysis of the synteny and flanking regions of the PMUs indicates that at least one effector family, phyllogen/SAP54 homologs, has been horizontally transferred between distantly related phytoplasmas, accompanied by PMUs (Tokuda *et al.*, 2023). This suggests that for effector genes of phytoplasmas, horizontal transfer as well as conservation along phylogenetic evolution must be taken into account.

Research on phytoplasma effectors is still in its early stages and is dominated by studies on individual phytoplasma strains, so even homologs can have very different names (Table 8.2). This may hinder future integrated understanding of effectors in phytoplasmas. To address this issue, it would be beneficial to adopt a generic naming system for effectors (by family or category as shown in Table 8.2) or to create a database. In addition, it is necessary to address in which *Candidatus* species or strains of phytoplasmas each effector gene has been found. This

Table 8.2. Phytoplasma effector proteins, host target, and functional roles.

Family/ category	Name	Phytoplasma species (strain)*	Target	Other host factors	Function	References
Phyllogen/ SAP54 homolog	PHYL/SAP54/ SJP3/Zaofeng3	'*Ca*. P. asteris' (several strains) '*Ca*. P. aurantifolia' '*Ca*. P. fragariae' '*Ca*. P. japonicum' '*Ca*. P. luffae' '*Ca*. P. oryzae' '*Ca*. P. phoenicium' '*Ca*. P. pruni' '*Ca*. P. solani' '*Ca*. P. trifolii' '*Ca*. P. ziziphi'	Floral MADS domain transcription factors (MTFs)	Rad23	Degrades A- and E-class MTFs via ubiquitin-independent pathway and induces phyllody symptom	MacLean *et al.*, 2011, 2014; Maejima *et al.*, 2014a, 2015; Yang *et al.*, 2015; Kitazawa *et al.*, 2017, 2022; Iwabuchi *et al.*, 2020; Tokuda *et al.*, 2023; Suzuki *et al.*, 2024
SAP11 homolog	SAP11/SWP1/ SJP1/SJP2/ Zaofeng6	'*Ca*. P. asteris' (AY-WB and other several strains) '*Ca*. P. aurantifolia' '*Ca*. P. mali' '*Ca*. P. pruni' '*Ca*. P. solani' '*Ca*. P. tritici' '*Ca*. P. ziziphi'	CIN-TCPs TB/ CYC-TCPs		Destabilizes TCPs, downregulates jasmonic acid biosynthesis, and increases the production of stems	Sugio *et al.*, 2011; Wang *et al.*, 2018; Zhou *et al.*, 2021; Chen *et al.*, 2022; Correa Marrero *et al.*, 2024; Mittelberger *et al.*, 2024
SWP12 homolog	SWP12	'*Ca*. P. asteris' '*Ca*. P. australasiae' '*Ca*. P. phoenicium' '*Ca*. P. tritici'	WRKY74 of *Triticum aestivum*		Degrades TaWRKY74 via ubiquitin-dependent pathway and weakens plant resistance	Bai *et al.*, 2022, 2023
TENGU	TENGU	'*Ca*. P. asteris' (several strains)	Unknown		Suppresses auxin signaling, increases production of stems, and induces sterility	Hoshi *et al.*, 2009; Sugawara *et al.*, 2013; Minato *et al.*, 2014
-	PM19_00185	'*Ca*. P. mali'	E2 UBC10 of *Malus domestica*		Possesses E3 ligase activity and suppresses plant basal defense	Strohmayer *et al.*, 2019
-	SAP05	'*Ca*. P. asteris' (AY-WB, OY)	SPL and GATA transcription factors	RPN10	Degrades SPL/GATA via ubiquitin-independent pathway and increases production of stems	Huang *et al.*, 2021; Liu *et al.*, 2023; Yan *et al.*, 2024; Zhang *et al.*, 2024a
-	SPR1	'*Ca*. P. asteris' (ROL)	GS2		Interacts with GS2 and induces chlorosis that attracts insect vectors	Zhang *et al.*, 2024b
-	SWP16	'*Ca*. P. tritici'	Unknown		inhibits RNA silencing in plant	Wang *et al.*, 2021

*Many other phytoplasma species possess effector homologs. However, only those that have been functionally studied are listed here.
Ca. P., *Candidatus* Phytoplasma; GS, glutamine synthetase; ROL, rice orange leaf; SPL, squamosa promoter-binding protein-like protein.

will help to understand functional conservation and differences among effector homologs in terms of phytoplasma pathogenicity (Iwabuchi *et al.*, 2020; Suzuki *et al.*, 2024) and is also valuable for understanding the mechanisms of effector evolution and acquisition (Iwabuchi *et al.*, 2020; Tokuda *et al.*, 2023).

Specific effectors

Tengu

A small secretory peptide, TENGU, encoded by OY phytoplasma, was the first identified phytoplasma pathogenic factor (Hoshi *et al.*, 2009). TENGU is predicted to be produced as a 70 amino acid (aa) preprotein, with a 32 aa signal peptide at its N-terminus. The C-terminal 38 aa peptide is then secreted into the plant cytoplasm via the Sec system. Transient or transgenic expression of this C-terminal 38 aa peptide in *Nicotiana benthamiana* and *Arabidopsis thaliana* induces short and bushy phenotypes that resemble the symptoms of phytoplasma-infected plants (Hoshi *et al.*, 2009; Omenge *et al.*, 2021). Although phytoplasmas can infect both plant and insect hosts (Strauss, 2009; Oshima *et al.*, 2011), the expression level of TENGU in the plant host is approximately five times higher than that in the insect host, suggesting that TENGU may play an essential role in plant hosts. Interestingly, the N-terminal 11 aa of the TENGU secreted region are sufficient for its function (Sugawara *et al.*, 2013). Microarray analyses showed that many auxin-related genes were significantly downregulated in TENGU-transgenic plants, suggesting that TENGU inhibits auxin signaling or biosynthesis pathways (Hoshi *et al.*, 2009). It has also been reported that phytoplasma-infected periwinkles can show remission of disease symptoms when cultured in a medium containing a high concentration of auxin (Curković Perica *et al.*, 2007; Curković Perica, 2008). Thus, auxin may have a significant effect on the plant–phytoplasma interaction.

TENGU has also been shown to act as a sterility inducer (Minato *et al.*, 2014). Transgenic expression of TENGU-induced developmental defects in *A. thaliana* is similar to those observed in *auxin response factor 6* (*ARF6*) and *ARF8* double mutants. The expression levels of *ARF6* and *ARF8* genes were significantly decreased in TENGU-expressing transgenic plants, and jasmonic acid (JA) and auxin syntheses were reduced in TENGU-transgenic buds. These data suggest that TENGU suppresses the JA and auxin biosynthesis pathways by repressing *ARF6* and *ARF8*, resulting in impaired flower maturation (Minato *et al.*, 2014).

Phyllogen

Flower malformation, such as phyllody (transformation of floral organs into leaf-like structures), virescence (greening of floral organs), and proliferation (development of shoots from flowers), is also a characteristic symptom of phytoplasma infection (Himeno *et al.*, 2011). It has been reported that a novel gene family of phytoplasma effectors, termed phyllody-inducing genes or the phyllogen family, induces floral malformation in *A. thaliana* (MacLean *et al.*, 2011; Maejima *et al.*, 2014a; Yang *et al.*, 2015). Phyllogen has been shown to cause phyllody phenotypes in several other eudicot species belonging to three different families (Kitazawa *et al.*, 2017). Phyllogen genes have been identified in a number of '*Ca.* Phytoplasma' species, and are proposed to be horizontally transferred interspecifically (Iwabuchi *et al.*, 2020) via horizontal transfer of PMUs (Tokuda *et al.*, 2023). These reports suggest the phyllogen family is a universally effective phyllody-inducing effector common to phytoplasmas.

Phyllogen family proteins, such as SAP54 and PHYL, target the products of floral homeotic genes constituting the floral quartet model (MacLean *et al.*, 2014; Maejima *et al.*, 2014a; Singh and Lakhanpaul, 2020), which in turn encode MADS domain transcription factors (MTFs). Phyllogens degrade angiosperm A- and E-class MTFs in a proteasome-dependent manner (MacLean *et al.*, 2014; Maejima *et al.*, 2014a, 2015; Kitazawa *et al.*, 2017), suggesting that phyllogens recognize a structure that is highly conserved among these MTFs. Crystal structure analyses have revealed that the K domain of MTFs, a phyllogen-binding region (MacLean *et al.*, 2014), has two α-helices with conserved hydrophobic residues that are important for MTF multimerization (Puranik *et al.*, 2014). These conserved residues in the K domain are critical for recognition by phyllogens (Kitazawa *et al.*, 2022). This explains why phyllogens can target MTFs from a wide range of plants. Two phyllogens – $PHYL_{OY}$ and $PHYL_{PnWB}$, phyllogens of '*Ca.* P. asteris' OY and '*Ca.* P. aurantifolia' peanut witches'-broom (PnWB) strains, respectively – also consist of two α-helices, which are important for phyllody-inducing activity (Iwabuchi *et al.*, 2019; Liao *et al.*, 2019). Sequence comparison of phyllogens showed that their phyllody-inducing activity is determined by a single amino acid polymorphism on the second α-helix (Iwabuchi *et al.*, 2020). The residues responsible for interaction with MTFs, identified from polymorphisms and artificial mutations, are concentrated on one surface of the phyllogen structure, which is a putative MTF-binding interface (Kitazawa *et al.*, 2023; Tokuda *et al.*, 2023).

Phyllogen-mediated phyllody induction requires Rad23 (RADIATION SENSITIVE23). Initially identified for its role in DNA lesion recognition, Rad23 is one of the most studied shuttle proteins that translocate ubiquitinated proteins to the proteasome system (MacLean *et al.*, 2014). Phyllogens translocate MTFs to the proteasome system by mediating the interaction between MTF and Rad23 to form a ternary complex (Kitazawa *et al.*, 2022). This ternary complex is assembled in the order MTF/phyllogen to MTF/phyllogen/Rad23 in a ubiquitin-independent manner. This indicates that phyllogens functionally mimic ubiquitin to translocate their targets into the proteasome system. The target specificity of phyllogen is regulated by both the MTF-binding ability (MacLean *et al.*, 2014; Iwabuchi *et al.*, 2019, 2020; Liao *et al.*, 2019) and the Rad23 recruitment ability of the MTF/phyllogen complex (Suzuki *et al.*, 2024). Further structural insights into the interaction between phyllogen family proteins, MTFs, and Rad23s will further elucidate the molecular mechanisms of how phytoplasmas manipulate the morphology of plant hosts.

In addition to the induction of phyllody, there are reports that witches'-broom-like phenotypes were induced by phyllogen from paulownia witches'-broom (PaWB) phytoplasma and jujube witches'-broom (JWB) phytoplasma (Cao *et al.*, 2021; Chen *et al.*, 2024), which are called PaWB-SAP54, SJP3, and Zaofeng3 but are the same protein as with $PHYL_{IIP}$ (Tokuda *et al.*, 2023). It has been discussed that these phyllogens interact with and inhibit the accumulation of SPL (squamosa promoter-binding protein-like protein), which negatively regulates axillary bud formation and branching development (Cao *et al.*, 2021), or TCP (TEOSINTE BRANCHED 1/CYCLOIDEA/PCF) as in the case of SAP11 described next (Chen *et al.*, 2024). In addition, SAP54 has been reported as a factor that attracts insects (Orlovskis and Hogenhout, 2016). This process is thought to involve the degradation of SVP, a type of MTF (Orlovskis *et al.*, 2024).

SAP11 and homologs

SAP11 is a secretory protein that affects plant morphogenesis and was first identified from AY-WB phytoplasma (Bai *et al.*, 2009). SAP11 of AY-WB ($SAP11_{AYWB}$)-expressing *A. thaliana* exhibits a bushy phenotype akin to the witches'-broom symptom induced by phytoplasmas (Sugio *et al.*, 2011). Transgenic *A. thaliana* expressing either $SAP11_{AYWB}$ or $SAP11_{CaPM}$ (SAP11 of '*Ca.* P. mali') shows crinkled leaves, but $SAP11_{PnWB}$ (SAP11 of PnWB phytoplasma) or $SAP11_{OYM}$ (SAP11 of OY-M phytoplasma) expressing transgenic lines display very weakly altered leaf phenotypes (Chang *et al.*, 2018).

SAP11 and its homologs interact with and destabilize TEOSINTE BRANCHED1, CYCLOIDEA, PROLIFERATING CELL FACTORs (TCPs) transcription factors (Sugio *et al.*, 2011, 2014; Janik *et al.*, 2017; Chang *et al.*, 2018; Wang *et al.*, 2018). However, the target range differed among SAP11 homologs (Correa Marrero *et al.*, 2024). For example, $SAP11_{AYWB}$ shows a strong ability to destabilize diverse class II CINCINNATA (CIN) related TCPs (CIN-TCPs) and TB/CYC-TCPs but not class I PCF-TCPs (Sugio *et al.*, 2011; Chang *et al.*, 2018). Conversely, $SAP11_{CaPM}$ cannot destabilize some of the class II CIN-TCPs, and $SAP11_{PnWB}$ and $SAP11_{OYM}$ show only a weak ability to destabilize one class II CIN-TCPs (TCP2). These different abilities may be related to the observation that $SAP11_{AYWB}$ transgenic plants have severely crinkled leaves (Chang *et al.*, 2018). CIN-TCPs are involved in the upregulation of *LOX2*, which encodes a lipoxygenase that mediates the first step of the JA synthesis pathway (Schommer *et al.*, 2008). Therefore, the destabilization of CIN-TCPs by SAP11 leads to a downregulation of JA biosynthesis (Sugio *et al.*, 2011). The fecundity of insect vectors increased on the SAP11-expressing plants compared with normal plants, which may be advantageous for the fitness of phytoplasmas and their insect vectors (Sugio *et al.*, 2011). It has also been reported that SAP11 suppresses salicylic acid-mediated defense responses and enhances the growth of a bacterial pathogen (Lu *et al.*, 2014). The SAP11-mediated TCP destabilization process is not yet known. Proteasome inhibitors reportedly inhibit the destabilization of SAP11-mediated TCP degradation in a previous study (Wang *et al.*, 2018) but not in another study (Sugio *et al.*, 2014).

The effects of SAP11 in diverse plants have also been elucidated. Transgenic *Nicotiana benthamiana* lines expressing SAP11 exhibit an altered aroma phenotype that correlates with defects in glandular trichome development and 3-isobutyl-2-methoxypyrazine biosynthesis (Tan *et al.*, 2016). $SAP11_{CaPM}$ expression in *N. occidentalis* leaves leads to the downregulation of genes involved in defense response and photosynthetic processes-related genes. In contrast, the expression of genes involved in energy production was enhanced (Mittelberger *et al.*, 2024). SAP11 from '*Ca.* P. ziziphi' (called SJP1/2 in Zhou *et al.* (2021) or ZaoFeng6 in Chen *et al.* (2022)) induces witches'-broom-like shoot proliferation in jujube and affects the expression and accumulation of jujube TCPs (Chen *et al.*, 2024; Ma *et al.*, 2024a, 2024b).

SAP05

SAP05 is a secretory protein of AY-WB phytoplasma, which induces the development of witches'-broom-like proliferations of leaves and sterile shoots in *A. thaliana* (Huang *et al.*, 2021). In addition, SAP05 has been shown to delay *A. thaliana* senescence. Targets of SAP05 are GATA and SPL transcription factors (GATA/SPL): GATAs regulate plant organ development, timing of flowering, and branching patterns (Ding *et al.*, 2015), whereas SPLs regulate developmental phase transitions of vascular plants (Huijser and Schmid, 2011). SAP05 directly interacts with *A. thaliana* GATA/SPL, resulting in their degradation. By degrading the targets, SAP05 seems to induce inhibition of plant developmental phase transitions, delayed plant aging, and witches'-broom-like excessive vegetative tissue. Some phytoplasmas possess one or two SAP05 homologs, some of which have the property of degrading either SPL or GATA.

SAP05-mediated degradation of GATA/SPL is regulated by the plant proteasome (Huang *et al.*, 2021). SAP05 interacts with *Arabidopsis* RPN10, a 26S proteasome subunit, simultaneously with GATA/SPL. RPN10 naturally acts as the ubiquitin receptor, recruiting ubiquitinated proteins for proteasomal degradation. However, SAP05-induced GATA/SPL degradation is ubiquitination-independent because SAP05 intermediates the interaction between GATA/SPL and RPN10. Structural analyses revealed that SAP05 displays a bimodular architecture with two distinct interfaces for the interaction with either GATA/SPL or RPN10 (Liu *et al.*, 2023; Yan *et al.*, 2024). Although GATA and RPN10 are found in diverse eukaryotes, SAP05 specifically interacts with those from plants but not from insects or animals (Huang *et al.*, 2021; Zhang *et al.*, 2024a), suggesting that SAP05 plays a crucial role in phytoplasma infection in plants. Several key residues in SAP05 involved in its interaction with host proteins have been identified (Liu *et al.*, 2023; Yan *et al.*, 2024; Zhang *et al.*, 2024a). Huang *et al.* (2021) generated an RPN10 mutant with a 2 aa substitution that prevented interaction with SAP05. *A. thaliana* carrying the RPN10 mutant instead of the wild type did not produce severely deformed leaves or an increased number of lateral shoots upon phytoplasma infection, suggesting that the blocking SAP05 activity enhances host tolerance toward phytoplasmas.

Other effectors of phytoplasmas

PM19_00185 is a secretory protein identified in '*Ca.* P. mali,' the causal agent of apple proliferation (Strohmayer *et al.*, 2019). Yeast two-hybrid screenings reveal that PM19_00185 interacts with E2 ubiquitin-conjugating enzymes of *A. thaliana*. An *in vitro* ubiquitination assay demonstrates that PM19_00185 functions as an E3 ligase with E2 UBC09 of *A. thaliana* and E2 UBC10 of *Malus domestica*. The growth of a non-host bacterium (*Pseudomonas syringae* pv. *tabaci*) in PM19_00185-expressing transgenic *A. thaliana* suggests that PM19_00185 may suppress basal plant defenses.

SWP16 identified in '*Ca.* P. tritici' WBD strain acts as an RNA silencing suppressor (RSS) that specifically suppresses systemic RNA silencing and promotes the accumulation of potato virus X (PVX) in *N. benthamiana* (Wang *et al.*, 2021). Generally, RSSs are encoded in the genomes of plant viruses but are also found in some bacteria, oomycetes, and fungi (Mosher and Baulcombe, 2008). RSSs from *Phytophthora sojae* have been shown to disrupt the accumulation and biogenesis of small RNAs (Qiao *et al.*, 2013). Similarly, SWP16 inhibits the production of miRNAs in *A. thaliana* (Wang *et al.*, 2021).

SWP12 is another secretory protein from the WBD phytoplasma. The expression of SWP12 in wheat (*Triticum aestivum*) facilitates phytoplasma colonization (Bai *et al.*, 2022). SWP12 reduces the basal immunity, such as Bax-induced cell death, of *N. benthamiana*, promoting leaf colonization by fungi (*Phytophthora parasitica* and *Sclerotinia sclerotiorum*) and a virus (tobacco mild green mosaic virus). TaWRKY74 (*T. aestivum*) and NbWRKY17 (*N. benthamiana*) were identified as host targets of SWP12 through yeast two-hybrid screening and GST (glutathione *S*-transferase) pulldown analysis (Bai *et al.*, 2022). SWP12 degrades TaWRKY74 via the ubiquitin-dependent pathway, suggesting that SWP12 weakens plant resistance to facilitate phytoplasma colonization. Some nonclassic secretory proteins (ncSecP3, ncSecP9, ncSecP12, ncSecP14, ncSecP16, and ncSecP22) in '*Ca.* P. ziziphi' also suppress Bax- and INF1-induced cell death responses in *N. benthamiana* (Gao *et al.*, 2023). They are novel types of secretory proteins without signal peptides or translocation signals.

SPR1 in the ROL phytoplasma translocates to chloroplasts and interacts with the chloroplastic glutamine synthetase (GS2) (Zhang *et al.*, 2024a). The direct interaction between SRP1 and GS2 disrupts the decamer formation of the GS2 holoenzyme, inducing chlorosis that attracts insect vectors.

Acknowledgements

KM acknowledges support of JSPS KAKENHI (Grant Numbers 20KK0132, 21H04722, 23K05237, 25H02311) and JST FOREST Program (Grant Number JPMJFR243Q). YK acknowledges support of JSPS KAKENHI (Grant Number 24K08908). KO acknowledges support of JSPS KAKENHI (Grant Number 24K01761).

References

Akimaru, J., Matsuyama, S., Tokuda, H., and Mizushima, S. (1991) Reconstitution of a protein translocation system containing purified SecY, SecE, and SecA from *Escherichia coli*. *Proceedings of the National Academy of Sciences of the USA* 88, 6545–6549. DOI: 10.1073/pnas.88.15.6545.

Antelmann, H., Tjalsma, H., Voigt, B., Ohlmeier, S., Bron, S. *et al.* (2001) A proteomic view on genome-based signal peptide predictions. *Genome Research* 11, 1484–1502. DOI: 10.1101/gr.182801.

Arashida, R., Kakizawa, S., Ishii, Y., Hoshi, A., Jung, H.Y. *et al.* (2008) Cloning and characterization of the antigenic membrane protein (Amp) gene and *in situ* detection of Amp from malformed flowers infected with Japanese hydrangea phyllody phytoplasma. *Phytopathology* 98, 769–775. DOI: 10.1094/PHYTO-98-7-0769.

Arricau-Bouvery, N., Duret, S., Dubrana, M.P., Batailler, B., Desqué, D. *et al.* (2018) Variable membrane protein a of flavescence dorée phytoplasma binds the midgut perimicrovillar membrane of *Euscelidius variegatus* and promotes adhesion to its epithelial cells. *Applied and Environmental Microbiology* 84, 02487-17. DOI: 10.1128/aem.02487-17.

Arricau-Bouvery, N., Duret, S., Dubrana, M.P., Desqué, D., Eveillard, S. *et al.* (2021) Interactions between the flavescence dorée phytoplasma and its insect vector indicate lectin-type adhesion mediated by the adhesin VmpA. *Science Reports* 11, 11222. DOI: 10.1038/s41598-021-90809-z.

Arricau-Bouvery, N., Dubrana, M.P., Canuto, F., Duret, S., Brocard, L. *et al.* (2023) Flavescence dorée phytoplasma enters insect cells by a clathrin-mediated endocytosis allowing infection of its insect vector. *Science Reports* 13, 2211. DOI: 10.1038/s41598-023-29341-1.

Baba, T., Ara, T., Hasegawa, M., Takai, Y., Okumura, Y. *et al.* (2006) Construction of *Escherichia coli* K-12 in-frame, single-gene knockout mutants: The Keio collection. *Molecular and Systematic Biology* 2, 2006.0008. DOI: 10.1038/msb4100050.

Baba, T., Kakizawa, S., Mori, H., Kuruma, Y., Kurokawa, K. *et al.* (2020) Minimal genomes: How many genes are required for living cells. *Journal of Geography (Chigaku Zasshi)* 129, 805–824. DOI: 10.5026/jgeography.129.805.

Bai, B., Zhang, G., Li, Y., Wang, Y., Sujata, S. *et al.* (2022) The '*Candidatus* Phytoplasma tritici' effector SWP12 degrades the transcription factor TaWRKY74 to suppress wheat resistance. *Plant Journal* 112, 1473–1488. DOI: 10.1111/tpj.16029.

Bai, B., Zhang, G., Pei, B., Song, Q., Hao, X.A. *et al.* (2023) The function of the phytoplasma effector SWP12 depends on the properties of two key amino acids. *Journal of Biological Chemistry* 299, 103052. DOI: 10.1016/j.jbc.2023.103052.

Bai, X., Zhang, J., Ewing, A., Miller, S.A., Jancso Radek, A. *et al.* (2006) Living with genome instability: The adaptation of phytoplasmas to diverse environments of their insect and plant hosts. *Journal of Bacteriology* 188, 3682–3696. DOI: 10.1128/jb.188.10.3682-3696.2006.

Bai, X., Correa, V.R., Toruño, T.Y., Ammar, E.-D., Kamoun, S. *et al.* (2009) AY-WB phytoplasma secretes a protein that targets plant cell nuclei. *Molecular Plant-Microbe Interactions* 22, 18–30. DOI: 10.1094/mpmi-22-1-0018.

Barbara, D.J., Morton, A., Clark, M.F., and Davies, D.L. (2002) Immunodominant membrane proteins from two phytoplasmas in the aster yellows clade (chlorante aster yellows and clover phyllody) are highly divergent in the major hydrophilic region. *Microbiology* 148, 157–167. DOI: 10.1099/00221287-148-1-157.

Bendtsen, J.D., Kiemer, L., Fausbøll, A., and Brunak, S. (2005) Non-classical protein secretion in bacteria. *BMC Microbiology* 5, 58. DOI: 10.1186/1471-2180-5-58.

Berg, M., Davies, D.L., Clark, M.F., Vetten, H.J., Maie, G. *et al.* (1999) Isolation of the gene encoding an immunodominant membrane protein of the apple proliferation phytoplasma, and expression and characterization of the gene product. *Microbiology* 145, 1937–1943. DOI: 10.1099/13500872-145-8-1937.

Blomquist, C.L., Barbara, D.J., Davies, D.L., Clark, M.F., and Kirkpatrick, B.C. (2001) An immunodominant membrane protein gene from the Western X-disease phytoplasma is distinct from those of other phytoplasmas. *Microbiology* 147, 571–580. DOI: 10.1099/00221287-147-3-571.

Boonrod, K., Munteanu, B., Jarausch, B., Jarausch, W., and Krczal, G. (2012) An immunodominant membrane protein (Imp) of '*Candidatus* Phytoplasma mali' binds to plant actin. *Molecular Plant-Microbe Interactions* 25, 889–895. DOI: 10.1094/mpmi-11-11-0303.

Boonrod, K., Kuaguim, L., Braun, M., Müller-Renno, C., Ziegler, C. *et al.* (2023) Identification of the actin-binding region and binding to host plant apple actin of immunodominant transmembrane protein of '*Candidatus* phytoplasma mali'. *International Journal of Molecular Science* 24, 968. DOI: 10.3390/ijms24020968.

Braunstein, M., Espinosa, B.J., Chan, J., Belisle, J.T., and Jacobs, W.R., Jr. (2003) SecA2 functions in the secretion of superoxide dismutase A and in the virulence of *Mycobacterium tuberculosis*. *Molecular Microbiology* 48, 453–464. DOI: 10.1046/j.1365-2958.2003.03438.x.

Brodin, P., Rosenkrands, I., Andersen, P., Cole, S.T., and Brosch, R. (2004) ESAT-6 proteins: protective antigens and virulence factors? *Trends in Microbiology* 12, 500–508. DOI: 10.1016/j.tim.2004.09.007.

Canuto, F., Duret, S., Dubrana, M.P., Claverol, S., Malembic-Maher, S. *et al.* (2023) A knockdown gene approach identifies an insect vector membrane protein with leucin-rich repeats as one of the receptors for the VmpA adhesin of flavescence dorée phytoplasma. *Frontiers in Cellular Infection Microbiology* 13, 1289100. DOI: 10.3389/fcimb.2023.1289100.

Cao, Y., Sun, G., Zhai, X., Xu, P., Ma, L. *et al.* (2021) Genomic insights into the fast growth of paulownias and the formation of Paulownia witches' broom. *Molecular Plant* 14, 1668–1682. DOI: 10.1016/j.molp.2021.06.021.

Carreón-Anguiano, K.G., Vila-Luna, S.E., Sáenz-Carbonell, L., and Canto-Canche, B. (2023) PhyEffector, the first algorithm that identifies classical and non-classical effectors in phytoplasmas. *Biomimetics* 8, 550. DOI: 10.3390/biomimetics8070550.

Chang, S.H., Tan, C.M., Wu, C.T., Lin, T.H., Jiang, S.Y. *et al.* (2018) Alterations of plant architecture and phase transition by the phytoplasma virulence factor SAP11. *Journal of Experimental Botany* 69, 5389-5401. DOI: 10.1093/jxb/ery318.

Chen, P., Chen, L., Ye, X., Tan, B., Zheng, X. *et al.* (2022) Phytoplasma effector Zaofeng6 induces shoot proliferation by decreasing the expression of ZjTCP7 in Ziziphus jujuba. *Horticulture Research* 9. DOI: 10.1093/hr/uhab032.

Chen, P., Zhang, Y., Li, Y., Yang, Q., Li, Q. *et al.* (2024) Jujube witches' broom phytoplasma effector Zaofeng3, a homologous effector of SAP54, induces abnormal floral organ development and shoot proliferation. *Phytopathology* 114, 200–210. DOI: 10.1094/phyto-10-21-0448-r.

Chung, W.C., Chen, L.L., Lo, W.S., Lin, C.P., and Kuo, C.H. (2013) Comparative analysis of the peanut witches'-broom phytoplasma genome reveals horizontal transfer of potential mobile units and effectors. *PLoS One* 8, e62770. DOI: 10.1371/journal.pone.0062770.

Cornelis, G.R. and Van Gijsegem, F. (2000) Assembly and function of type III secretory systems. *Annual Review of Microbiology* 54, 735–774. DOI: 10.1146/annurev.micro.54.1.735.

Correa Marrero, M., Capdevielle, S., Huang, W., Al-Subhi, A.M., Busscher, M. *et al.* (2024) Protein interaction mapping reveals widespread targeting of development-related host transcription factors by phytoplasma effectors. *Plant Journal* 117, 1281–1297. DOI: 10.1111/tpj.16546.

Curković Perica, M. (2008) Auxin-treatment induces recovery of phytoplasma-infected periwinkle. *Journal of Applied Microbiology* 105, 1826–1834. DOI: 10.1111/j.1365-2672.2008.03946.x.

Curković Perica, M., Lepedus, H., and Seruga Musić, M. (2007) Effect of indole-3-butyric acid on phytoplasmas in infected *Catharanthus roseus* shoots grown in vitro. *FEMS Microbiology Letters* 268, 171–177. DOI: 10.1111/j.1574-6968.2006.00577.x.

Ding, L., Yan, S., Jiang, L., Liu, M., Zhang, J. *et al.* (2015) HANABA TARANU regulates the shoot apical meristem and leaf development in cucumber (*Cucumis sativus* L.). *Journal of Experimental Botany* 66, 7075–7087. DOI: 10.1093/jxb/erv409.

Ding, L., Lu, W., Yang, Y., Zhong, Q., Zhou, T. *et al.* (2022) Immunodominant membrane protein (Imp) promotes the transmission of wheat blue dwarf (WBD) phytoplasma by directly interacting with α-tubulin in leafhoppers. *European Journal of Plant Pathology* 162, 357–367. DOI: 10.1007/s10658-021-02408-3.

Economou, A. (1999) Following the leader: bacterial protein export through the Sec pathway. *Trends in Microbiology* 7, 315–320. DOI: 10.1016/s0966-842x(99)01555-3.

Galetto, L., Bosco, D., Balestrini, R., Genre, A., Fletcher, J. *et al.* (2011) The major antigenic membrane protein of "Candidatus Phytoplasma asteris" selectively interacts with ATP synthase and actin of leafhopper vectors. *PLoS One* 6, e22571. DOI: 10.1371/journal.pone.0022571.

Gao, X., Ren, Z., Zhao, W., and Li, W. (2023) *Candidatus* Phytoplasma ziziphi encodes non-classically secreted proteins that suppress hypersensitive cell death response in *Nicotiana benthamiana*. *Phytopathology Research* 5, 11. DOI: 10.1186/s42483-023-00166-0.

Garcion, C., Béven, L., and Foissac, X. (2021) Comparison of current methods for signal peptide prediction in Phytoplasmas. *Frontiers in Microbiology* 12, 661524. DOI: 10.3389/fmicb.2021.661524.

Grant, S.R., Fisher, E.J., Chang, J.H., Mole, B.M., and Dangl, J.L. (2006) Subterfuge and manipulation: Type III effector proteins of phytopathogenic bacteria. *Annual Review of Microbiology* 60, 425–449. DOI: 10.1146/annurev.micro.60.080805.142251.

Grohmann, E., Muth, G., and Espinosa, M. (2003) Conjugative plasmid transfer in gram-positive bacteria. *Microbiology and Molecular Biology Reviews* 67, 277–301, table of contents. DOI: 10.1128/mmbr.67.2.277-301.2003.

Harth, G. and Horwitz, M.A. (1997) Expression and efficient export of enzymatically active *Mycobacterium tuberculosis* glutamine synthetase in *Mycobacterium smegmatis* and evidence that the information for export is contained within the protein. *Journal of Biological Chemistry* 272, 22728-22735. DOI: 10.1074/jbc.272.36.22728

Harth, G. and Horwitz, M.A. (1999) Export of recombinant *Mycobacterium tuberculosis* superoxide dismutase is dependent upon both information in the protein and mycobacterial export machinery. A model for studying export of leaderless proteins by pathogenic mycobacteria. *Journal of Biological Chemistry* 274, 4281–4292. DOI: 10.1074/jbc.274.7.4281.

Harth, G., Clemens, D.L., and Horwitz, M.A. (1994) Glutamine synthetase of *Mycobacterium tuberculosis*: Extracellular release and characterization of its enzymatic activity. *Proceedings of the National Academy of Science of the USA* 91, 9342–9346. DOI: 10.1073/pnas.91.20.9342.

Himeno, M., Neriya, Y., Minato, N., Miura, C., Sugawara, K. *et al.* (2011) Unique morphological changes in plant pathogenic phytoplasma-infected petunia flowers are related to transcriptional regulation of floral homeotic genes in an organ-specific manner. *Plant Journal* 67, 971–979. DOI: 10.1111/j.1365-313X.2011.04650.x.

Himeno, M., Kitazawa, Y., Yoshida, T., Maejima, K., Yamaji, Y. *et al.* (2014) Purple top symptoms are associated with reduction of leaf cell death in phytoplasma-infected plants. *Scientific Reports* 4, 4111. DOI: 10.1038/srep04111.

Hogenhout, S.A., Oshima, K., Ammar, E.-D., Kakizawa, S., Kingdom, H.N. *et al.* (2008) Phytoplasmas: Bacteria that manipulate plants and insects. *Molecular Plant Pathology* 9, 403–423. DOI: 10.1111/j.1364-3703.2008.00472.x.

Hoshi, A., Oshima, K., Kakizawa, S., Ishii, Y., Ozeki, J. *et al.* (2009) A unique virulence factor for proliferation and dwarfism in plants identified from a phytopathogenic bacterium. *Proceedings of the National Academy of Science of the USA* 106, 6416–6421. DOI: 10.1073/pnas.0813038106.

Huang, W., MacLean, A.M., Sugio, A., Maqbool, A., Busscher, M. *et al.* (2021) Parasitic modulation of host development by ubiquitin-independent protein degradation. *Cell* 184, 5201–5214.e5212. DOI: 10.1016/j.cell.2021.08.029.

Huijser, P. and Schmid, M. (2011) The control of developmental phase transitions in plants. *Development* 138, 4117–4129. DOI: 10.1242/dev.063511.

Ishii, Y., Kakizawa, S., Hoshi, A., Maejima, K., Kagiwada, S. *et al.* (2009a) In the non-insect-transmissible line of onion yellows phytoplasma (OY-NIM), the plasmid-encoded transmembrane protein ORF3 lacks the major promoter region. *Microbiology* 155, 2058–2067. DOI: 10.1099/mic.0.027409-0.

Ishii, Y., Oshima, K., Kakizawa, S., Hoshi, A., Maejima, K. *et al.* (2009b) Process of reductive evolution during 10 years in plasmids of a non-insect-transmissible phytoplasma. *Gene* 446, 51–57. DOI: 10.1016/j.gene.2009.07.010.

Iwabuchi, N., Endo, A., Kameyama, N., Satoh, M., Miyazaki, A. *et al.* (2017) First report of 'Candidatus Phytoplasma malaysianum' associated with Elaeocarpus yellows of Elaeocarpus zollingeri. *Journal of General Plant Pathology* 84, 160–164. DOI: 10.1007/s10327-017-0761-4.

Iwabuchi, N., Maejima, K., Kitazawa, Y., Miyatake, H., Nishikawa, M. *et al.* (2019) Crystal structure of phyllogen, a phyllody-inducing effector protein of phytoplasma. *Biochemical and Biophysical Research Communications* 513, 952–957. DOI: 10.1016/j.bbrc.2019.04.060.

Iwabuchi, N., Kitazawa, Y., Maejima, K., Koinuma, H., Miyazaki, A. *et al.* (2020) Functional variation in phyllogen, a phyllody-inducing phytoplasma effector family, attributable to a single amino acid polymorphism. *Molecular Plant Pathology* 21, 1322–1336. DOI: 10.1111/mpp.12981.

Janik, K., Mithöfer, A., Raffeiner, M., Stellmach, H., Hause, B. *et al.* (2017) An effector of apple proliferation phytoplasma targets TCP transcription factors-a generalized virulence strategy of phytoplasma? *Molecular Plant Pathology* 18, 435–442. DOI: 10.1111/mpp.12409.

Kakizawa, S., Oshima, K., Kuboyama, T., Nishigawa, H., Jung, H. *et al.* (2001) Cloning and expression analysis of *Phytoplasma* protein translocation genes. *Molecular Plant-Microbe Interactions* 14, 1043-1050. DOI: 10.1094/mpmi.2001.14.9.1043.

Kakizawa, S., Oshima, K., Nishigawa, H., Jung, H.Y., Wei, W. *et al.* (2004) Secretion of immunodominant membrane protein from onion yellows phytoplasma through the Sec protein-translocation system in *Escherichia coli*. *Microbiology* 150, 135–142. DOI: 10.1099/mic.0.26521-0.

Kakizawa, S., Oshima, K., Jung, H.Y., Suzuki, S., Nishigawa, H. *et al.* (2006a) Positive selection acting on a surface membrane protein of the plant-pathogenic phytoplasmas. *Journal of Bacteriology* 188, 3424–3428. DOI: 10.1128/jb.188.9.3424-3428.2006.

Kakizawa, S., Oshima, K., and Namba, S. (2006b) Diversity and functional importance of phytoplasma membrane proteins. *Trends in Microbiology* 14, 254–256. DOI: 10.1016/j.tim.2006.04.008.

Kakizawa, S., Oshima, K., Ishii, Y., Hoshi, A., Maejima, K. *et al.* (2009) Cloning of immunodominant membrane protein genes of phytoplasmas and their in planta expression. *FEMS Microbiology Letters* 293, 92–101. DOI: 10.1111/j.1574-6968.2009.01509.x.

Käll, L., Krogh, A., and Sonnhammer, E.L.L. (2004) A combined transmembrane topology and signal peptide prediction method. *Journal of Molecular Biology* 338, 1027–1036. DOI: 10.1016/j.jmb.2004.03.016.

Käll, L., Krogh, A., and Sonnhammer, E.L.L. (2007) Advantages of combined transmembrane topology and signal peptide prediction – The Phobius web server. *Nucleic Acids Research* 35, W429–W432. DOI: 10.1093/nar/gkm256.

Kitazawa, Y., Iwabuchi, N., Himeno, M., Sasano, M., Koinuma, H. *et al.* (2017) Phytoplasma-conserved phyllogen proteins induce phyllody across the Plantae by degrading floral MADS domain proteins. *Journal of Experimental Botany* 68, 2799–2811. DOI: 10.1093/jxb/erx158.

Kitazawa, Y., Iwabuchi, N., Maejima, K., Sasano, M., Matsumoto, O. *et al.* (2022) A phytoplasma effector acts as a ubiquitin-like mediator between floral MADS-box proteins and proteasome shuttle proteins. *Plant Cell* 34, 1709–1723. DOI: 10.1093/plcell/koac062.

Kitazawa, Y., Iwabuchi, N., Maejima, K., Matsumoto, O., Suzuki, M. *et al.* (2023) Random mutagenesis-based screening of the interface of phyllogen, a bacterial phyllody-inducing effector, for interaction with plant MADS-box proteins. *Frontiers in Plant Science* 14, 1058059. DOI: 10.3389/fpls.2023.1058059.

Koinuma, H., Maejima, K., Tokuda, R., Kitazawa, Y., Nijo, T. *et al.* (2020) Spatiotemporal dynamics and quantitative analysis of phytoplasmas in insect vectors. *Scientific Reports* 10, 4291. DOI: 10.1038/s41598-020-61042-x.

Konnerth, A., Krczal, G., and Boonrod, K. (2016) Immunodominant membrane proteins of phytoplasmas. *Microbiology* 162, 1267–1273. DOI: 10.1099/mic.0.000331.

Ku, C., Lo, W.-S., and Kuo, C.-H. (2013) Horizontal transfer of potential mobile units in phytoplasmas. *Mobile Genetic Elements* 3, e26145. DOI: 10.4161/mge.26145.

Kube, M., Mitrovic, J., Duduk, B., Rabus, R., and Seemuller, E. (2012) Current view on phytoplasma genomes and encoded metabolism. *Scientific World Journal* 2012, 185942. DOI: 10.1100/2012/185942.

Lee, I.M., Zhao, Y., and Bottner, K.D. (2006) SecY gene sequence analysis for finer differentiation of diverse strains in the aster yellows phytoplasma group. *Molecular and Cellular Probes* 20, 87–91. DOI: 10.1016/j.mcp.2005.10.001.

Lefol, C., Lherminier, J., Boudon-Padieu, E., Larrue, J., Louis, C. *et al.* (1994) Propagation of flavescence dorée MLO (mycoplasma-like organism) in the leafhopper vector *Euscelidius variegatus* Kbm. *Journal of Invertebrate Pathology* 63, 285–293. DOI: 10.1006/jipa.1994.1053.

Lenz, L.L., Mohammadi, S., Geissler, A., and Portnoy, D.A. (2003) SecA2-dependent secretion of autolytic enzymes promotes *Listeria monocytogenes* pathogenesis. *Proceedings of the National Academy of Science of the USA* 100, 12432–12437. DOI: 10.1073/pnas.2133653100.

Liao, Y.T., Lin, S.S., Lin, S.J., Sun, W.T., Shen, B.N. *et al.* (2019) Structural insights into the interaction between phytoplasmal effector causing phyllody 1 and MADS transcription factors. *Plant Journal* 100, 706–719. DOI: 10.1111/tpj.14463.

Liu, C.Y., Cheng, H.P., Lin, C.P., Liao, Y.T., Ko, T.P. *et al.* (2024) Structural insights into the molecular mechanism of phytoplasma immunodominant membrane protein. *IUCrJ* 11, 384–394. DOI: 10.1107/s2052252524003075.

Liu, Q., Maqbool, A., Mirkin, F.G., Singh, Y., Stevenson, C.E.M. *et al.* (2023) Bimodular architecture of bacterial effector SAP05 that drives ubiquitin-independent targeted protein degradation. *Proceedings of the National Academy of Science of the USA* 120, e2310664120. DOI: 10.1073/pnas.2310664120.

Lu, Y.T., Li, M.Y., Cheng, K.T., Tan, C.M., Su, L.W. *et al.* (2014) Transgenic plants that express the phytoplasma effector SAP11 show altered phosphate starvation and defense responses. *Plant Physiology* 164, 1456–1469. DOI: 10.1104/pp.113.229740.

Ma, F., Huang, X., Zhou, J., Zhang, N., Deng, M. *et al.* (2024a) The '*Candidatus* phytoplasma ziziphi' effectors SJP1 and SJP2 destabilise the bifunctional regulator ZjTCP7 to modulate floral transition and shoot branching. *Plant, Cell and Environment* DOI: 10.1111/pce.14913.

Ma, F., Zheng, Y., Zhang, N., Deng, M., Zhao, M. *et al.* (2024b) The '*Candidatus* Phytoplasma ziziphi' effectors SJP1/2 negatively control leaf size by stabilizing the transcription factor ZjTCP2 in jujube. *Journal of Experimental Botany* 75, 3054–3069. DOI: 10.1093/jxb/erae042.

MacLean, A.M., Sugio, A., Makarova, O.V., Findlay, K.C., Grieve, V.M. *et al.* (2011) Phytoplasma effector SAP54 induces indeterminate leaf-like flower development in Arabidopsis plants. *Plant Physiology* 157, 831–841. DOI: 10.1104/pp.111.181586.

MacLean, A.M., Orlovskis, Z., Kowitwanich, K., Zdziarska, A.M., Angenent, G.C. *et al.* (2014) Phytoplasma effector SAP54 hijacks plant reproduction by degrading MADS-box proteins and promotes insect colonization in a RAD23-dependent manner. *PLoS Biology* 12, e1001835. DOI: 10.1371/journal.pbio.1001835.

Maejima, K., Iwai, R., Himeno, M., Komatsu, K., Kitazawa, Y. *et al.* (2014a) Recognition of floral homeotic MADS domain transcription factors by a phytoplasmal effector, phyllogen, induces phyllody. *Plant Journal* 78, 541–554. DOI: 10.1111/tpj.12495.

Maejima, K., Oshima, K., and Namba, S. (2014b) Exploring the phytoplasmas, plant pathogenic bacteria. *Journal of General Plant Pathology* 80, 210–221. DOI: 10.1007/s10327-014-0512-8.

Maejima, K., Kitazawa, Y., Tomomitsu, T., Yusa, A., Neriya, Y. *et al.* (2015) Degradation of class E MADS-domain transcription factors in Arabidopsis by a phytoplasmal effector, phyllogen. *Plant Signaling and Behavior* 10, e1042635. DOI: 10.1080/15592324.2015.1042635.

Malembic-Maher, S., Desqué, D., Khalil, D., Salar, P., Bergey, B. *et al.* (2020) When a Palearctic bacterium meets a Nearctic insect vector: Genetic and ecological insights into the emergence of the grapevine Flavescence dorée epidemics in Europe. *PLoS Pathogens* 16, e1007967. DOI: 10.1371/journal.ppat.1007967.

Miller, S.I., Bader, M., and Guina, T. (2003) Bacterial vesicle formation as a mechanism of protein transfer to animals. *Cell* 115, 2–3. DOI: 10.1016/s0092-8674(03)00767-0.

Milne, R.G., Ramasso, E., Lenzi, R., Masenga, V., Sarindu, N. *et al.* (1995) Preembedding and postembedding immunogold labeling and electron-microscopy in plant host tissues of 3 antigenically unrelated MLOs – Primula yellows, tomato big bud and bermudagrass white leaf. *European Journal of Plant Pathology* 101, 57–67. DOI: 10.1007/BF01876094.

Minato, N., Himeno, M., Hoshi, A., Maejima, K., Komatsu, K. *et al.* (2014) The phytoplasmal virulence factor TENGU causes plant sterility by downregulating of the jasmonic acid and auxin pathways. *Science Reports* 4, 7399. DOI: 10.1038/srep07399.

Mittelberger, C., Moser, M., Hause, B., and Janik, K. (2024) '*Candidatus* Phytoplasma mali' SAP11-Like protein modulates expression of genes involved in energy production, photosynthesis, and defense in *Nicotiana occidentalis* leaves. *BMC Plant Biology* 24, 393. DOI: 10.1186/s12870-024-05087-4.

Moran, N.A., McCutcheon, J.P., and Nakabachi, A. (2008) Genomics and evolution of heritable bacterial symbionts. *Annual Review of Genetics* 42, 165–190. DOI: 10.1146/annurev.genet.41.110306.130119.

Morton, A., Davies, D.L., Blomquist, C.L., and Barbara, D.J. (2003) Characterization of homologues of the apple proliferation immunodominant membrane protein gene from three related phytoplasmas. *Molecular Plant Pathology* 4, 109-114. DOI: 10.1046/j.1364-3703.2003.00155.x.

Mosher, R.A. and Baulcombe, D.C. (2008) Bacterial pathogens encode suppressors of RNA-mediated silencing. *Genome Biology* 9, 237. DOI: 10.1186/gb-2008-9-10-237.

Nakabachi, A., Yamashita, A., Toh, H., Ishikawa, H., Dunbar, H.E. *et al.* (2006) The 160-kilobase genome of the bacterial endosymbiont *Carsonella*. *Science* 314, 267. DOI: 10.1126/science.1134196.

Nakai, K. and Kanehisa, M. (1991) Expert system for predicting protein localization sites in gram-negative bacteria. *Proteins* 11, 95–110. DOI: 10.1002/prot.340110203.

Namba, S. (2019) Molecular and biological properties of phytoplasmas. *Proceedings of the Japan Academy, Series B* 95, 401–418. DOI: 10.2183/pjab.95.028.

Neriya, Y., Sugawara, K., Maejima, K., Hashimoto, M., Komatsu, K. *et al.* (2011) Cloning, expression analysis, and sequence diversity of genes encoding two different immunodominant membrane proteins in poinsettia branch-inducing phytoplasma (PoiBI). *FEMS Microbiology Letters* 324, 38–47. DOI: 10.1111/j.1574-6968.2011.02384.x.

Neriya, Y., Maejima, K., Nijo, T., Tomomitsu, T., Yusa, A. *et al.* (2014) Onion yellow phytoplasma P38 protein plays a role in adhesion to the hosts. *FEMS Microbiology Letters* 361, 115-122. DOI: 10.1111/1574-6968.12620

Nielsen, H., Engelbrecht, J., Brunak, S., and von Heijne, G. (1997) Identification of prokaryotic and eukaryotic signal peptides and prediction of their cleavage sites. *Protein Engineering* 10, 1-6. DOI: 10.1093/protein/10.1.1.

Nishigawa, H., Oshima, K., Kakizawa, S., Jung, H.Y., Kuboyama, T. *et al.* (2002) A plasmid from a non-insect-transmissible line of a phytoplasma lacks two open reading frames that exist in the plasmid from the wild-type line. *Gene* 298, 195–201. DOI: 10.1016/s0378-1119(02)00975-7.

Omenge, K.M., Rümpler, F., Kathalingam, S.S., Furch, A.C.U., and Theißen, G. (2021) Studying the function of phytoplasma effector proteins using a chemical-inducible expression system in transgenic plants. *International Journal of Molecular Science* 22, 13582. DOI: 10.3390/ijms222413582.

Orlovskis, Z. and Hogenhout, S.A. (2016) A bacterial parasite effector mediates insect vector attraction in host plants independently of developmental changes. *Frontiers in Plant Science* 7, 885. DOI: 10.3389/fpls.2016.00885.

Orlovskis, Z., Singh, A., Kliot, A., Huang, W., and Hogenhout, S.A. (2024): Phytoplasma targeting of MADS-Box factor SVP suppresses leaf responses to insect vector males, promoting female attraction and colonization. *eLife* 13, RP98992. DOI: 10.7554/elife.98992.1.

Oshima, K., Kakizawa, S., Nishigawa, H., Kuboyama, T., Miyata, S. *et al.* (2001) A plasmid of phytoplasma encodes a unique replication protein having both plasmid- and virus-like domains: Clue to viral ancestry or result of virus/plasmid recombination? *Virology* 285, 270–277. DOI: 10.1006/viro.2001.0938.

Oshima, K., Kakizawa, S., Nishigawa, H., Jung, H.Y., Wei, W. *et al.* (2004) Reductive evolution suggested from the complete genome sequence of a plant-pathogenic phytoplasma. *Nature Genetics* 36, 27–29. DOI: 10.1038/ng1277.

Oshima, K., Ishii, Y., Kakizawa, S., Sugawara, K., Neriya, Y. *et al.* (2011) Dramatic transcriptional changes in an intracellular parasite enable host switching between plant and insect. *PLoS One* 6, e23242. DOI: 10.1371/journal.pone.0023242.

Oshima, K., Maejima, K., and Namba, S. (2013) Genomic and evolutionary aspects of phytoplasmas. *Frontiers in Microbiology* 4, 230. DOI: 10.3389/fmicb.2013.00230.

Oshima, K., Maejima, K., Isobe, Y., Endo, A., Namba, S. *et al.* (2023) Molecular mechanisms of plant manipulation by secreting effectors of phytoplasmas. *Physiological and Molecular Plant Pathology* 125, 102009. DOI: 10.1016/j.pmpp.2023.102009.

Puranik, S., Acajjaoui, S., Conn, S., Costa, L., Conn, V. *et al.* (2014) Structural basis for the oligomerization of the MADS domain transcription factor SEPALLATA3 in Arabidopsis. *Plant Cell* 26, 3603–3615. DOI: 10.1105/tpc.114.127910.

Purcell, A., Richardson, J., and Finlay, A. (1981) Multiplication of the agent of X-disease in a non-vector leafhopper *Macrosteles fascifrons*. *Annals of Applied Biology* 99, 283–289. DOI: 10.1111/j.1744-7348.1981.tb04797.x.

Qiao, Y., Liu, L., Xiong, Q., Flores, C., Wong, J. *et al.* (2013) Oomycete pathogens encode RNA silencing suppressors. *Nature Genetics* 45, 330–333. DOI: 10.1038/ng.2525.

Rashidi, M., Galetto, L., Bosco, D., Bulgarelli, A., Vallino, M. *et al.* (2015) Role of the major antigenic membrane protein in phytoplasma transmission by two insect vector species. *BMC Microbiology* 15, 193. DOI: 10.1186/s12866-015-0522-5.

Recchi, C., Rauzier, J., Gicquel, B., and Reyrat, J.M. (2002) Signal-sequence-independent secretion of the staphylococcal nuclease in *Mycobacterium smegmatis*. *Microbiology* 148, 529–536. DOI: 10.1099/00221287-148-2-529.

Samuelson, J.C., Chen, M., Jiang, F., Möller, I., Wiedmann, M. *et al.* (2000) YidC mediates membrane protein insertion in bacteria. *Nature* 406, 637–641. DOI: 10.1038/35020586.

Schaumburg, J., Diekmann, O., Hagendorff, P., Bergmann, S., Rohde, M. *et al.* (2004) The cell wall subproteome of *Listeria monocytogenes*. *Proteomics* 4, 2991–3006. DOI: 10.1002/pmic.200400928.

Schommer, C., Palatnik, J.F., Aggarwal, P., Chételat, A., Cubas, P. *et al.* (2008) Control of jasmonate biosynthesis and senescence by miR319 targets. *PLoS Biology* 6, e230. DOI: 10.1371/journal.pbio.0060230.

Shen, W.C. and Lin, C.P. (1993) Production of monoclonal-antibodies against a mycoplasmalike organism associated with sweet-potato witches-broom. *Phytopathology* 83, 671–675. DOI: 10.1094/Phyto-83-671.

Siampour, M., Galetto, L., Bosco, D., Izadpanah, K., and Marzachi, C. (2011) *In vitro* interactions between immunodominant membrane protein of lime witches' broom phytoplasma and leafhopper vector proteins. *Bulletin of Insectology* 64, S149–S150.

Singh, A. and Lakhanpaul, S. (2020) Detection, characterization and evolutionary aspects of S54LP of SP (SAP54 Like Protein of Sesame Phyllody): A phytoplasma effector molecule associated with phyllody development in sesame (*Sesamum indicum* L.). *Physiology and Molecular Biology of Plants* 26, 445–458. DOI: 10.1007/s12298-020-00764-8.

Strauss, E. (2009) Phytoplasma research begins to bloom. *Science* 325, 388–390. DOI: 10.1126/science.325_388.

Strohmayer, A., Moser, M., Si-Ammour, A., Krczal, G., and Boonrod, K. (2019) '*Candidatus* Phytoplasma mali' genome encodes a protein that functions as an E3 ubiquitin ligase and could inhibit plant basal defense. *Molecular Plant-Microbe Interaction* 32, 1487–1495. DOI: 10.1094/mpmi-04-19-0107-r

Sugawara, K., Honma, Y., Komatsu, K., Himeno, M., Oshima, K. *et al.* (2013) The alteration of plant morphology by small peptides released from the proteolytic processing of the bacterial peptide TENGU. *Plant Physiology* 162, 2005–2014. DOI: 10.1104/pp.113.218586.

Sugio, A., Kingdom, H.N., MacLean, A.M., Grieve, V.M., and Hogenhout, S.A. (2011) Phytoplasma protein effector SAP11 enhances insect vector reproduction by manipulating plant development and defense hormone biosynthesis. *Proceedings of the National Academy of the USA* 108, E1254–1263. DOI: 10.1073/pnas.1105664108

Sugio, A., MacLean, A.M., and Hogenhout, S.A. (2014) The small phytoplasma virulence effector SAP11 contains distinct domains required for nuclear targeting and CIN-TCP binding and destabilization. *New Phytologist* 202, 838–848. DOI: 10.1111/nph.12721.

Suzuki, M., Kitazawa, Y., Iwabuchi, N., Maejima, K., Matsuyama, J. *et al.* (2024) Target degradation specificity of phytoplasma effector phyllogen is regulated by the recruitment of host proteasome shuttle protein. *Molecular Plant Pathology* 25, e13410. DOI: 10.1111/mpp.13410.

Suzuki, S., Oshima, K., Kakizawa, S., Arashida, R., Jung, H.Y. *et al.* (2006) Interaction between the membrane protein of a pathogen and insect microfilament complex determines insect-vector specificity. *Proceedings of the National Academy of the USA* 103, 4252–4257. DOI: 10.1073/pnas.0508668103.

Tan, C.M., Li, C.H., Tsao, N.W., Su, L.W., Lu, Y.T. *et al.* (2016) Phytoplasma SAP11 alters 3-isobutyl-2-methoxypyrazine biosynthesis in *Nicotiana benthamiana* by suppressing NbOMT1. *Journal of Experimental Botany* 67, 4415–4425. DOI: 10.1093/jxb/erw225.

Tjalsma, H., Bolhuis, A., Jongbloed, J.D., Bron, S., and van Dijl, J.M. (2000) Signal peptide-dependent protein transport in Bacillus subtilis: A genome-based survey of the secretome. *Microbiology and Molecular Biology Reviews* 64, 515–547. DOI: 10.1128/mmbr.64.3.515-547.2000.

Tokuda, R., Iwabuchi, N., Kitazawa, Y., Nijo, T., Suzuki, M. *et al.* (2023) Potential mobile units drive the horizontal transfer of phytoplasma effector. *Frontiers in Genetics* 14, 1132432. DOI: 10.3389/fgene.2023.1132432.

Trivellone, V., Ripamonti, M., Angelini, E., Filippin, L., Rossi, M. *et al.* (2019) Evidence suggesting interactions between immunodominant membrane protein Imp of Flavescence dorée phytoplasma and protein extracts from distantly related insect species. *Journal of Applied Microbiology* 127, 1801–1813. DOI: 10.1111/jam.14445.

Villamor, D.E.V. and Eastwell, K.C. (2019) Multilocus characterization, gene expression analysis of putative immunodominant protein coding regions, and development of recombinase polymerase amplification assay for detection of '*Candidatus* Phytoplasma pruni' in *Prunus avium*. *Phytopathology* 109, 983–992. DOI: 10.1094/PHYTO-09-18-0326-R.

Wai, S.N., Lindmark, B., Söderblom, T., Takade, A., Westermark, M. *et al.* (2003) Vesicle-mediated export and assembly of pore-forming oligomers of the enterobacterial ClyA cytotoxin. *Cell* 115, 25–35. DOI: 10.1016/s0092-8674(03)00754-2.

Wang, L., Chen, W., Ma, H., Li, J., Hao, X. *et al.* (2021) Identification of RNA silencing suppressor encoded by wheat blue dwarf (WBD) phytoplasma. *Plant Biology (Stuttgart)* 23, 843–849. DOI: 10.1111/plb.13257.

Wang, N., Yang, H., Yin, Z., Liu, W., Sun, L. *et al.* (2018) Phytoplasma effector SWP1 induces witches' broom symptom by destabilizing the TCP transcription factor BRANCHED1. *Molecular Plant Pathology* 19, 2623–2634. DOI: 10.1111/mpp.12733.

Wang, Z., Yang, X., Zhou, S., Zhang, X., Zhu, Y. *et al.* (2023) The antigenic membrane protein (amp) of rice orange leaf phytoplasma suppresses host defenses and is involved in pathogenicity. *International Journal of Molecular Science* 24, 4494. DOI: 10.3390/ijms24054494.

Wei, W., Kakizawa, S., Jung, H.Y., Suzuki, S., Tanaka, M. *et al.* (2004) An antibody against the SecA membrane protein of one phytoplasma reacts with those of phylogenetically different phytoplasmas. *Phytopathology* 94, 683–686. DOI: 10.1094/phyto.2004.94.7.683.

Wilske, B., Busch, U., Fingerle, V., Jauris-Heipke, S., Preac Mursic, V. *et al.* (1996) Immunological and molecular variability of OspA and OspC. Implications for *Borrelia* vaccine development. *Infection* 24, 208–212. DOI: 10.1007/bf01713341.

Yan, X., Yuan, X., Lv, J., Zhang, B., Huang, Y. *et al.* (2024) Molecular basis of SAP05-mediated ubiquitin-independent proteasomal degradation of transcription factors. *Nature Communications* 15, 1170. DOI: 10.1038/s41467-024-45521-7.

Yang, C.Y., Huang, Y.H., Lin, C.P., Lin, Y.Y., Hsu, H.C. *et al.* (2015) MicroRNA396-targeted SHORT VEGETATIVE PHASE is required to repress flowering and is related to the development of abnormal flower symptoms by the phyllody symptoms1 effector. *Plant Physiology* 168, 1702–1716. DOI: 10.1104/pp.15.00307.

Zhang, L., Du, Y., Qu, Q., and Zheng, Q. (2024a) Structure basis for recognition of plant Rpn10 by phytoplasma SAP05 in ubiquitin-independent protein degradation. *iScience* 27, 108892. DOI: 10.1016/j.isci.2024.108892.

Zhang, X.-F., Li, Z., Lin, H., Cheng, Y., Wang, H. *et al.* (2024b) A phytoplasma effector destabilizes chloroplastic glutamine synthetase inducing chlorotic leaves that attract leafhopper vectors. *Proceedings of the National Academy of the USA* 121, e2402911121. DOI: doi:10.1073/pnas.2402911121.

Zhou, J., Ma, F., Yao, Y., Deng, M., Chen, M. *et al.* (2021) Jujube witches' broom phytoplasma effectors SJP1 and SJP2 induce lateral bud outgrowth by repressing the ZjBRC1-controlled auxin efflux channel. *Plant, Cell and Environment* 44, 3257–3272. DOI: 10.1111/pce.14141.

9 Molecular Aspects of Phytoplasma Infection: Ubiquitin-Proteasome System and Transcription Factors Under Attack

Christa Malfertheiner,[1†] Simone Egger,[2†] Cecilia Mittelberger,[1] Katrin Janik,[1] Wei Wei,[3] and Mattia Tabarelli[1*]

[1]*Molecular Biology and Microbiology, Group of Functional Genomics, Laimburg Research Centre, Pfatten (Vadena), Italy;* [2]*Molecular Biology and Microbiology, Group of Breeding Genomics, Laimburg Research Centre, Pfatten (Vadena), Italy;* [3]*Molecular Plant Pathology Laboratory, ARS-USDA, Beltsville, Maryland, USA*

Abstract

Phytoplasma infections involve molecular mechanisms, including interference with the host ubiquitin-proteasome system and transcription factors. Bacterial effectors manipulate key host proteins and pathways, disrupting immune and developmental processes to promote plant colonization. Nearly two decades of research have illuminated how these strategies used by phytoplasmas determine phenotypic changes and suppress host defenses. Upon examining these complex interactions, this chapter provides an up-to-date overview of the state of the art of research in this field, with the aim of highlighting knowledge gaps and possible research directions to achieve a well-rounded understanding of this fascinating and intricate pathosystem.

Introduction

One of the most fascinating aspects of phytoplasmas is their amazing ability to induce a wide array of molecular changes within the host plant to facilitate their colonization. Phytoplasmas possess genomes that rank among the smallest of all living organisms, ranging from approximately 1300 kb to as little as 550 kb. This extreme genome reduction results from their parasitic behavior, which has led to the loss of metabolic pathways not essential for survival (Oshima *et al.*, 2013). For example, the phytoplasma chromosome lacks genes critical for fatty acid metabolism, oxidative phosphorylation, and even ATP synthase (Oshima *et al.*, 2004). Despite this significant gene loss, the genome has preserved a set of genes encoding putative effectors or virulence factors, emphasizing the importance of these proteins for phytoplasma survival and pathogenicity. Since the first effectors were identified in the late 2000s (Bai *et al.*, 2009), several effectors have been described, such as phyllogens, SAP11-like, and SAP05-like proteins (Bai *et al.*, 2009; Huang *et al.*, 2021; Carreón-Anguiano *et al.*, 2023). Although many aspects of the function of these effectors remain obscure, some pathways have been comprehensively

*Corresponding author: mattia.tabarelli@laimburg.it

†Christa Malfertheiner and Simone Egger have contributed equally to the work.

DOI: 10.1079/9781800627031.0009

described. These proteins exert their influence by mimicking, binding, deactivating, stabilizing, and relocating different plant targets, thereby inducing substantial physiological and morphological changes in plants.

This chapter describes recent findings regarding the molecular mechanism underlying phytoplasma infection and colonization. It explains how phytoplasmas, like other plant pathogens, manipulate the ubiquitin-proteasome system (UPS) to degrade specific host proteins. This chapter also highlights how phytoplasmas interfere with various plant transcription factor families, altering plant responses and impacting disease progression.

The Ubiquitin-Proteasome System during Infection

The plant immune system is constantly exposed to various stimuli and defends against a wide range of pathogens, including bacteria, viruses, fungi, and other microorganisms. To manage these pathogen attacks, plants have developed a fine-tuned, two-layered immune system (Jones and Dangl, 2006; Boller and He, 2009; Yuan *et al.*, 2021). The first layer of defense involves the recognition of microbe- or pathogen-associated molecular patterns (MAMPs/PAMPs), that is, conserved, microbe-derived molecules. These are detected by pattern recognition receptors (PRRs) located on the plasma membrane, triggering a response known as PAMP-triggered immunity (PTI) (Jones and Dangl, 2006; Couto and Zipfel, 2016; Yuan *et al.*, 2021). Some phytopathogens can evade the PTI and release effector molecules into the plant cells. These effectors are recognized by the second layer of the plant immune system, which initiates effector-triggered immunity (ETI) (Jones and Dangl, 2006; Yuan *et al.*, 2021). Both the PTI and ETI require the activation of downstream reactions involving hormone signaling pathways and regulating numerous proteins. The UPS plays a crucial role in these processes by facilitating the adaptability and turnover of proteins required for a robust immune response.

The UPS: overview and function

The UPS is essential for protein degradation in eukaryotic cells, coordinating key processes like immune responses, growth, abiotic stress, hormonal signaling, embryogenesis, and senescence in plants (Jones and Dangl, 2006; Vierstra, 2009; Trujillo and Shirasu, 2010; Marino *et al.*, 2012; Sharma *et al.*, 2016; Langin *et al.*, 2023). A key component of the UPS system is the highly conserved protein ubiquitin, which consists of approximately 76 amino acids. Ubiquitin contains seven different lysine (K) residues, K6, K11, K27, K29, K33, K48, and K63 where ubiquitin molecules can attach, forming polyubiquitin chains that regulate various cellular functions (Zhou and Zeng, 2017; Xu and Xue, 2019; Su *et al.*, 2020).

The process of ubiquitination requires the activation of a cascade of enzymes, encompassing the three enzymes E1 (ubiquitin-activating), E2 (ubiquitin-conjugating), and E3 (ubiquitin ligase) (Zhou and Zeng, 2017; Su *et al.*, 2020; Langin *et al.*, 2023). E1 catalyzes the ATP-dependent activation of ubiquitin through the adenylation of the carboxyl group of ubiquitin. The activated ubiquitin is then transferred to an E2, which interacts with an E3 ligase. The E3 ligase mediates the transfer of the ubiquitin residue from E2 to a specific target/substrate. Activating this enzymatic ubiquitination cascade can result in different patterns, such as mono- or polyubiquitination (Langin *et al.*, 2023). Polyubiquitination refers to the attachment of multiple ubiquitin proteins to K48, targeting proteins for degradation by the 26S proteasome, a 2.5 MDa complex composed of a 20S catalytic core and one or two 19S regulatory particles (Zhou and Zeng, 2017; Su *et al.*, 2020; Langin *et al.*, 2023).

The 26S proteasome recognizes ubiquitinated proteins through direct binding to ubiquitin receptors (Rpn1, Rpn10, or Rpn13), or via shuttle proteins (e.g. RAD23) (Vierstra, 2009; Poot *et al.*, 2017; Marshall and Vierstra, 2019). Deubiquitinating enzymes (DUB) cleave ubiquitin chains, recycling ubiquitin for reuse in future ubiquitination (Sharma *et al.*, 2016; Marshall and Poot *et al.*, 2017; Vierstra, 2019). *Arabidopsis thaliana* has more than 1600 genes related to the 26S proteasome machinery, with most encoded E3 ubiquitin ligases or subunits of a multiprotein E3 ligase complex (Sharma *et al.*, 2016; Su *et al.*, 2020).

E3 ubiquitin ligases are key factors of the ubiquitination process. Due to their diversity, the UPS can react and adapt to environmental changes and respond to different signals in multiple cellular locations. E3 enzymes are categorized into four main classes: homologous to the E6-AP carboxyl terminus (HECT), Really Interesting New Gene (RING), Plant U-Box (PUB), and Cullin-RING ligases (CRLs) (Trujillo and Shirasu, 2010; Sharma *et al.*, 2016; Langin *et al.*, 2023). This diversity in E3 ligases highlights their essential role in regulating protein degradation and maintaining cellular balance.

Virulence Strategies of Plant-Pathogenic Bacteria: Exploiting the Plant UPS

The UPS has emerged as a prominent target for pathogens. Plant-pathogenic bacteria employ different strategies to interfere with the UPS by either mimicking components of the UPS or hijacking the existing machinery to manipulate host cellular functions (Ramachandran *et al.*, 2021). One major strategy phytopathogens employ involves using a type III secretion system to deliver type III effector molecules (T3Es) into plant host cells. These effectors are translocated to special compartments within the host cell, where they interact with various components of the immune response. By targeting elements of the UPS and other immune pathways, T3Es disrupt host defense responses, thereby enhancing pathogen survival and virulence (Singer *et al.*, 2013; Büttner, 2016; Couto and Zipfel, 2016).

A well-documented example of UPS manipulation is *Pseudomonas syringae* pv. *tomato*, the causative agent of bacterial speck disease in tomatoes. This pathogen delivers approximately 30 type III effector proteins into the host cells, including AvrPtoB. AvrPtoB is a conserved protein with a molecular mass of *c.*59 kDa, triggering immunity and the hypersensitive response (HR) in tomato plants expressing the tomato resistance (R) protein Pto (Abramovitch and Martin, 2005). The effector has a modular structure, with a C-terminal domain that functions as an E3 ligase and an N-terminal region comprising multiple target-binding domains (Rosebrock *et al.*, 2007; Lei *et al.*, 2020). In tomatoes, AvrPtoB targets the plant kinase Fen, a key regulator of basal immune defense. In *Arabidopsis*, AvrPtoB targets additional immune components, including pattern recognition receptors (PRRs) such as FLS2 (FLAGELLING SENSING 2) and CERK1 (CHITIN ELICITOR RECEPTOR KINASE 1). It also degrades NPR1 (Non-expressor of PR genes 2), the master regulator of salicylic acid signaling. These activities collectively disable key aspects of the plant's innate immune system, enhancing bacterial virulence (Rosebrock *et al.*, 2007; Vierstra, 2009; Chen *et al.*, 2017; Wang *et al.*, 2019; Ramachandran *et al.*, 2021).

Effectors from the *Xanthomonas* species also exploit UPS machinery by mimicking E3 ligase activity. The effector XopL from *Xanthomonas campestris* pv. *vesicatoria* has been found to interfere with components of the host autophagy machinery (Leong *et al.*, 2022). Through its ligase activity, XopL ubiquitinates SH3P2 (a protein essential for autophagosome biogenesis), marking it for proteasomal degradation. The release of the bacterial effector dampens plant immunity by impeding autophagosome biogenesis and consequently enhancing *Xanthomonas*'s virulence (Singer *et al.*, 2013; Leong *et al.*, 2022). Even though mimicking E3 ligase activity represents an efficient strategy to degrade host proteins, bacteria have also developed other strategies to interfere with the plant immune system. For example, *P. syringae* pv. *syringae* produces syringolin A (SylA), a bacterial toxin that directly inhibits the eukaryotic proteasome. By blocking proteasome activity, SylA disrupts the salicylic acid signaling, impairs immune signaling in adjacent tissue, and thus facilitates bacterial motility and colonization (Groll *et al.*, 2008; Misas-Villamil *et al.*, 2013).

Meddling in Defense: Phytoplasma Effector Proteins' Influence on UPS and Plant Immunity

Similar to *Pseudomonas* and *Xanthomonas* that manipulate the UPS to initiate proteasomal degradation of host proteins, phytoplasmas also employ effector proteins to hijack or interfere with UPS components. Several sec-secreted effector proteins, including SAP54, $PHYL_{OY}$, SAP05, PM19_00185, and SWP12, exhibit unique strategies for targeting the UPS to modulate plant immunity and development (MacLean *et al.*, 2014; Strohmayer *et al.*, 2019; Huang *et al.*, 2021; Bai *et al.*, 2022; Kitazawa *et al.*, 2022).

SAP54 is an effector protein of '*Candidatus* Phytoplasma asteris' strain, known as aster yellows witches'-broom (AY-WB), which has been found to induce the transformation of flowers into leaf-like structures (MacLean *et al.*, 2014). Through its interaction with RAD23, a family of proteasomal shuttle and ubiquitin-binding proteins, SAP54 mediates the degradation of MADS domain transcription factors (MTFs) through the 26S proteasome in *Arabidopsis*. The degradation of the MTFs disrupts floral development, leading to the formation of leafy flowers. These morphological alterations enhance the plant's attractiveness to insect vectors, facilitating the spread of phytoplasmas (MacLean *et al.*, 2014).

$PHYL_{OY}$, a phyllogen from the '*Ca.* P. asteris' onion yellow (OY) strain also targets *Arabidopsis* MTF SEP3. Localization and interaction studies showed that the bacterial effector protein $PHYL_{OY}$ mediates the interaction between SEP3 and RAD23, leading to the recruitment of SEP3 to the proteasome. Furthermore, data suggest that the degradation of the MTF SEP3 is independent of the *in planta* ubiquitination. $PHYL_{OY}$ itself takes over the role of ubiquitin by recognizing both the K domain of MTFs and the UBA2 domain of RAD23 and thus inducing the ubiquitin-independent degradation of MTFs (Kitazawa *et al.*, 2022). Both SAP54 and $PHYL_{OY}$ belong to the phyllogen family of proteins and target MTFs, critical regulators of floral development, as they coordinate the floral transition and the formation of floral organs. Both effectors are independent of ubiquitination and interact with the proteasomal shuttle protein RAD23 to induce the degradation of their targets.

SAP05 is another effector from the '*Ca.* P. asteris' AY-WB strain. Huang *et al.* (2021) found that SAP05 targets transcription factors of the GATA and SPL families, which are essential for plant developmental transitions. Unlike traditional UPS-mediated degradation, SAP05 coordinates the degradation of transcription factors of the SPL/GATA family by directly hijacking the proteasomal receptor RPN10, bypassing ubiquitination. SAP05-induced degradation of SPL and GATA transcription factors influences the transitions of the developmental phases of the plant, resulting in the delay of plant aging, promotes the formation of the characteristic vegetative tissue (witches'-brooms), and causes the sterility of adult shoots. These changes benefit phytoplasmas by generating more habitats and resources. Interestingly, SAP05 does not interact with the RPN10 of the insect vectors, despite only two amino acid differences between the plant and insect RPN10 proteins, highlighting its specificity (Huang *et al.*, 2021).

PM19_00185, an effector secreted by '*Ca.* P. mali,' the causal agent of apple proliferation, has been shown to interact with different E2 ligases in *A. thaliana. In vitro* ubiquitination assays revealed that PM19_00185 exhibits E3 ligase activity in conjunction with E2 ligases UBC10 from *Malus domestica* (and UBC9 from *A. thaliana*). Thus, this might represent another example of bacteria manipulating the UPS system for the targeted degradation of host proteins, benefiting bacterial colonization and spread (Strohmayer *et al.*, 2019).

SWP12 is a potential effector protein secreted by '*Ca.* P. tritici,' the causal agent of wheat blue dwarf disease. The expression of SWP12 has a negative impact on plant resistance, as it destabilizes TaWRKY74, a transcription factor associated with plant resistance. This destabilization occurs through a 26S proteasome-dependent process, weakening plant immunity and promoting phytoplasma colonization (Bai *et al.*, 2022).

UPS and Plant Defense Mechanisms

The UPS system represents a prominent target for plant pathogens due to its key vital role in regulating plant immunity, including pathogen recognition and activation of defense responses (Jones and Dangl, 2006; Vierstra, 2009; Trujillo and Shirasu, 2010; Marino *et al.*, 2012; Sharma *et al.*, 2016; Langin *et al.*, 2023). As previously described, phytoplasmas produce various effector proteins that interfere with the UPS and its components. It is remarkable that, despite the reduced size of the phytoplasma genome, these effector proteins exhibit significant functional diversity, emphasizing their importance in modulating pathogenicity. The observation that effector proteins from different phytoplasma strains target the UPS highlights the significance of this specific system for both the plant and the phytoplasma (Bai *et al.*, 2009, 2022; Oshima *et al.*, 2013; MacLean *et al.*, 2014; Strohmayer *et al.*, 2019;

Iwabuchi *et al.*, 2020; Huang *et al.*, 2021; Kitazawa *et al.*, 2022).

By targeting the UPS, phytoplasmas can disrupt metabolic pathways that are not regularly linked. For example, SAP05 links SPLs and GATAs to RPN10 (MacLean *et al.*, 2014; Huang *et al.*, 2021), thereby bridging transcription factors and a proteasomal component. Interestingly, certain effectors such as SAP54, phyllogen, and SAP05 exhibit partial independence from ubiquitin (MacLean *et al.*, 2014; Huang *et al.*, 2021; Kitazawa *et al.*, 2022), while others, like PM19_00185, demonstrate E3 ubiquitin ligase activity (Strohmayer *et al.*, 2019). These findings suggest a highly refined co-evolutionary process, wherein phytoplasma effectors have adapted to exploit the UPS system for their survival and pathogenicity. The phytoplasma's strategy for attacking the UPS is a delicate balance. It disrupts the system to gain a pathogenic advantage while still relying on an operational UPS for the execution of its strategy. Since the UPS regulates numerous critical pathways, including growth, hormonal signaling, and stress responses, alongside plant immunity, phytoplasmas leverage their central role to exert a broad influence on host physiology (Trujillo and Shirasu, 2010; Sharma *et al.*, 2016; Langin *et al.*, 2023).

Understanding these intricate interactions is essential for developing strategies to counteract phytoplasma infections. Targeting the effector molecules offers a promising approach to mitigate infection and manage the associated symptoms effectively.

Transcription Factors and Phytoplasma Infection

TCP transcription factor family

The TCP gene family encodes plant-specific transcription factors across green plants. The acronym "TCP" originates from its first identified members: *Teosinte Branched 1* (*TB1*) from corn (maize, *Zea mays*) (Doebley *et al.*, 1995), *Cycloidea* (*CYC*) from snapdragon (*Antirrhinum majus*) (Luo *et al.*, 1996), and *Proliferating Cell Factors 1* and *2* (*PCF1* and *PCF2*) from rice (*Oryza sativa*) (Kosugi and Ohashi, 1997; Cubas *et al.*, 1999). TCP genes have expanded through duplication, with species like apple and corn containing 40 and 46 genes, respectively (Martín-Trillo and Cubas, 2010; Li, 2015; Ding *et al.*, 2019; Tabarelli *et al.*, 2022).

All TCP proteins share a unique N-terminal non-canonical basic-helix-loop-helix (bHLH) domain, known as the TCP domain (58–62 amino acids), that facilitates nuclear localization, DNA binding, and protein–protein interactions (Aggarwal *et al.*, 2010). TCPs are divided into class I and II, with class II further split into the Cincinnata (CIN) and CYC subclasses. A subgroup of the CIN clade is regulated by miR319, which modulates their activity post-transcriptionally (Palatnik *et al.*, 2003).

Class I and II TCPs bind distinct yet overlapping DNA motifs, showing specific differences between the members of the two classes (Kosugi and Ohashi, 2002; Nicolas and Cubas, 2016). Nonetheless, these motifs are not entirely exclusive to each class, implying a potential targeting overlap by both class I and class II TCP proteins, although an antagonistic interplay between the two classes has been documented in a very few instances. Like other bHLH proteins, TCPs require dimerization to bind DNA forming both homo- and heterodimers within the same class, affecting their DNA-binding affinity and specificity (Kosugi and Ohashi, 2002; Danisman *et al.*, 2013; Danisman, 2016; Nicolas and Cubas, 2016). Besides regulating key developmental processes such as flower and leaf development, branching, and axillary meristematic activity, TCPs also influence plant hormone biosynthesis and responses to abiotic and biotic stresses (Martín-Trillo and Cubas, 2010; Mukhtar *et al.*, 2011; Danisman *et al.*, 2012; Danisman, 2016; Nicolas and Cubas, 2016; Viola and Gonzalez, 2023).

In *A. thaliana*, CIN-TCPs control leaf morphology by repressing leaf marginal meristem activity (Sarvepalli and Nath, 2018). The expression of some members of the CIN subgroup is tightly regulated by the miRNA miR319. *A. thaliana* plants overexpressing miR319 display a decrease in miR319-targeted CIN-TCP transcripts, and a peculiar phenotype, characterized by crinkly leaves, cotyledon epinasty, and flowering delay (Palatnik *et al.*, 2003).

TCP transcription factors as phytoplasma targets

TCPs were among the first plant genes identified as being linked to phytoplasma-associated symptoms. This discovery followed early research showing that the effector SAP11 from the '*Ca.* P. asteris' AY-WB strain ($SAP11_{AYWB}$) binds to and destabilizes certain TCP members in *A. thaliana* (Sugio *et al.*, 2011a). In this first report on host–phytoplasma protein interactions, Sugio and collaborators observed a reduction in all eight CIN-TCPs of *A. thaliana* in the presence of $SAP11_{AYWB}$, along with a decrease in the level of jasmonic acid (JA) in the leaves of SAP11-overexpressing plants. The findings led to a proposed model in which the deactivation of one CIN-TCP, AtTCP4, downregulates *LOX2*, the gene encoding lipoxygenase, an enzyme that catalyzes the first step in JA biosynthesis. Consequently, the reduced expression of *LOX2* leads to diminished JA production. This model was further supported by studies on the SAP11 ortholog from '*Ca.* P. mali,' $SAP11_{CaPm}$, which was observed to alter JA levels both in *Nicotiana benthamiana* (Tan *et al.*, 2016) and in apple through the destabilization of MdTCP25, the AtTCP4 ortholog (Janik *et al.*, 2017). It is well documented that various pathogenic and symbiotic microorganisms can suppress JA levels to facilitate host colonization (Yan and Xie, 2015). Other studies have consistently demonstrated a connection between JA pathway disruption and the plant response to phytoplasma (Musetti *et al.*, 2013; Luge *et al.*, 2014; Dermastia *et al.*, 2015; Mardi *et al.*, 2015; Paolacci *et al.*, 2017). However, the regulation of JA response to phytoplasmas appears to be intricate and context-dependent, varying across different plant species and pathosystems. This complexity extends beyond the destabilization of TCPs, highlighting the broader role of JA in plant–phytoplasma interactions (Dermastia, 2019).

Another phytoplasma-associated symptom correlated to TCP deactivation is the alteration of leaf morphology, particularly the development of crinkled leaves. This phenotypic effect is linked to the destabilization of several members of the CIN-TCP subgroup and has been observed in *A. thaliana* (Sugio *et al.*, 2011a, 2011b; Chang *et al.*, 2018;), *N. benthamiana* (Tan *et al.*, 2016) as well as *M. domestica* (Janik *et al.*, 2017). Studies involving the expression of several SAP11-like genes in *A. thaliana* revealed differing effects on TCP stability and leaf phenotype. For example, $SAP11_{AYWB}$ destabilizes all CIN-TCPs, resulting in strongly crinkly and serrated leaves. In contrast, $SAP11_{CaPm}$ selectively degrades TCP2 and TCP10, producing relatively less pronounced effects on leaf morphology. $SAP11_{PnWB}$ and $SAP11_{OYM}$, from the '*Ca.* P. aurantifolia' strain peanut witches'-broom (PnWB) and '*Ca.* P. asteris' strain onion yellows mild (OY-M), respectively, exhibit a weak destabilization of TCP2, with minimal impacts on the leaf phenotype (Chang *et al.*, 2018). These findings underscore the variation in TCP-targeting efficiency among SAP11 homologs and their corresponding phenotypic effects.

Several effectors interact with CYC/TB1-TCPs as well. Interaction screenings have confirmed the association between SAP11 and members of the CYC/TB1 subclass, although the ability of SAP11 to destabilize this group of TCPs was only confirmed later. For example, in *A. thaliana*, $SAP11_{AYWB}$ destabilizes TCP18 (Pecher *et al.*, 2019). Similarly, the effector Zaofeng6 from '*Ca.* P. ziziphi' (the causative agent of jujube witches'-broom) destabilizes a TCP12-like protein in jujube (Chen *et al.*, 2022), while $SAP11_{CaPm}$ from '*Ca.* P. mali' destabilizes TCP18-like proteins in *M. domestica* (Mittelberger *et al.*, 2022). Additionally, SWP1, a SAP11-like effector of '*Ca.* P. tritici,' destabilizes TCP18 in *A. thaliana* (Wang *et al.*, 2018). $SAP11_{WBDL}$, isolated from '*Ca.* P. aurantifolia,' the strain responsible for witches'-broom disease of lime (WBDL), interacts strongly with two CYC-TCP proteins in *Citrus aurantifolia* (Al-Subhi *et al.*, 2021). Since TCP18 in *A. thaliana* acts as the major suppressor of axillary bud growth (Aguilar-Martínez *et al.*, 2007; Finlayson, 2007; González-Grandío *et al.*, 2013), its deactivation has been correlated with increased shoot branching and development of witches'-brooms, which have been observed in all the aforementioned studies.

Interestingly, two effectors of '*Ca.* P. ziziphi' – SJP1 and SJP2 – exhibit contrasting behavior towards a TCP compared with other effectors. In a 2024 study, Ma and colleagues demonstrated that overexpressing SJP1 and SJP2 in *A. thaliana* resulted in reduced leaf size compared with the wild type (Col-0). This phenotype was driven by the stabilization of ZjTCP2 mediated by the effectors. In transgenic plants, the accumulation of ZjTCP2 in the leaves was accompanied by a significant reduction in miR319, the miRNA responsible for its negative regulation. The phenotype could be reversed by inducing the constitutive expression of this miRNA (Ma *et al.*, 2024).

A study by Pecher and collaborators (2019) compared two SAP11-like proteins and revealed how the different behavior of effectors targeting TCPs directly affects the development of symptoms. $SAP11_{MBSP}$, from a '*Ca.* P. asteris'-related strain, maize bushy stunt phytoplasma (MBSP), interacts exclusively with CYC/TB1 TCPs and induces explicitly the development of witches'-broom in corn, without causing leaf malformations. However, corn plants expressing $SAP11_{AYWB}$, which are able to destabilize all class II TCPs, exhibit both phenotypes. Furthermore, results indicate that the overexpression of $SAP11_{MBSP}$ in *A. thaliana* prompted the formation of fertile flowers, whereas in corn it resulted in the absence of female inflorescences. This finding aligns with previous observations that TCP18 primarily impacts floral transition rather than floral structure of *A. thaliana* (Niwa *et al.*, 2013), while the corn ortholog TB1 controls the architecture of female inflorescences acting as a key regulator within a MADS-box transcription factor cascade (Studer *et al.*, 2017). Moreover, the study demonstrated that the deactivation of CIN and CYC/TB1 members TCP12 and TCP18, mediated by the $SAP11_{AYWB}$ effector, suppresses plant defense genes against insect vectors, ultimately increasing vector fitness. Conversely, this suppression was not observed with the $SAP11_{MBSP}$ effector, suggesting that class II TCPs may not play a role in regulating the defense response in corn. In conclusion, Pecher and colleagues put forward an intriguing hypothesis: the evolutionary trajectory of the SAP11 effectors across phytoplasma species is likely to be shaped by the varied functions of TCPs in governing both plant developmental processes and defense mechanisms.

Besides the well-documented interactions between SAP11-like effectors and class II TCPs, emerging evidence suggests that SAP11-like proteins may also target members of the class I TCP subgroup, and that non-SAP11-like effectors can target TCPs as well. For example, $SAP11_{PM19}$, secreted by '*Ca.* P. mali' strain PM19 binds to members of class I *A. thaliana* TCPs *in vitro*. Although the interaction does not seem to interfere with the DNA-binding capacity of the TCPs (Strohmayer *et al.*, 2021), it induces their relocation to different subcellular compartments. The functional consequences of this relocation remain unclear and warrant further exploration (Boonrod *et al.*, 2022). In another study, Correa Marrero and collaborators (2024) performed a yeast two hybrid (Y2H) screening involving 21 putative and confirmed phytoplasma effectors and a library of *A. thaliana* transcription-related proteins. The analysis indicated that 15 of these effectors showed enrichment in interactions with 20 *A. thaliana* TCPs, indicating a strong targeting preference for this transcription factor family. Interestingly, the structure and sequence analyses of the different effector proteins suggested that these effectors are likely unrelated phylogenetically. This finding suggests functional redundancy among effectors, implying that TCP targeting plays a critical role in the development of the phytoplasma infections. The potential universality and significance of TCP targeting by diverse phytoplasma effectors make this an interesting area for further investigation.

MADS-Box Transcription Factors and Phytoplasma Infection

MADS-box transcription factors

MADS-box transcription factors are characterized by a conserved 58 amino acid MADS-box domain that determines DNA binding and dimerization (Gramzow and Theissen, 2010). Within the protein, two distinct functional regions have been identified: The N-terminal portion primarily determines the specificity of DNA binding, while the C-terminal portion is essential for dimer formation. This arrangement enables the potential creation of a variety of proteins with slightly varying DNA-binding specificities from a restricted set of genes through the heterodimerization of different MADS-box proteins. These proteins are found in eukaryotic groups, like plants, animals, and fungi, with large MADS-box gene families. In plants, these transcription factors regulate nearly every fundamental aspect of life (Saedler *et al.*, 2001; Becker and Theissen, 2003; Gramzow and Theissen, 2010; Zhang *et al.*, 2024).

MADS-box transcription factors play crucial roles in plant development, controlling processes like flowering, seed development, and fruit maturation. They bind DNA as dimers or tetramers, using the N-terminus for DNA binding and the C-terminus for dimerization and nuclear localization. Plant-specific MIKC-type MADS proteins have four domains: M (DNA-binding), I (intervening), K (keratin-like), and C (C-terminal). Their combinatorial interactions enable precise regulation of developmental pathways. These transcription factors are also important targets for manipulation during pathogen attacks. Phytoplasmas actively target MTFs to manipulate plant metabolism for their own benefit (Hogenhout *et al.*, 2008).

Alterations in MADS-Box Transcription Factors during Phytoplasma Infection

SAP54

Given the important role of MADS-box transcription factors in plants, especially in flower development, the discovery that phytoplasmas actively target these developmental regulators is intriguing. The effector SAP54 from AY-WB phytoplasma interacts with MADS-box proteins, leading to the degradation of the respective transcription factors. This degradation is mediated by the interaction between SAP54 proteins of the RADIATION SENSITIVE 23 (RAD23) family, which are responsible for shuttling substrates to the proteasome for degradation. It is suggested that SAP54 specifically destabilizes MADS-box proteins, including the flower development regulators SEPALLATA3 and APETALA1, thus affecting flower development. This mechanism, involving the interaction of $SAP54_{AYWB}$ with RAD23 proteins and MTFs, allows phytoplasmas to manipulate plant development, leading to the formation of leaf-like flower structures in infected plants. These structural changes increase the attractiveness of the plants to insect vectors, thereby promoting phytoplasma proliferation and spread (MacLean *et al.*, 2011, 2014; Orlovskis and Hogenhout 2016). It has been furthermore shown that SAP54 promotes the degradation of the MADS-box transcription factor SVP, and the absence of SVP is linked to improved mating of the phytoplasma-transmitting leafhopper vector *Macrosteles quadrilineatus* (Orlovskis *et al.*, 2025).

Phyllogens

Effectors that induce phyllody (i.e. the abnormal development of floral parts into leafy structures) have been collectively designated phyllogens (Maejima *et al.*, 2014). This group comprises 25 phyllogens from nine different phytoplasma species, comprising SAP54 from AY-WB phytoplasma, $PHYL1_{OY}$ from OY phytoplasma, $PHYL1_{PnWB}$ from PnWB phytoplasma, $PHYL_{RYD}$ from rice yellow dwarf (RYD) phytoplasma, and $PHYL_{SY}$ from strawberry yellows (SY) phytoplasma (MacLean *et al.*, 2011; Maejima *et al.*, 2014, 2015; Liu *et al.*, 2015; Yang *et al.*, 2015; Iwabuchi *et al.*, 2019; Iwabuchi *et al.*, 2020; Suzuki *et al.*, 2024). Phyllogen effectors target and degrade different MADS domain proteins via the ubiquitin-proteasomal pathway (Maejima *et al.*, 2014). This degradation of transcription factors enables phyllogen to induce phyllody phenotypes in many plant species (Kitazawa *et al.*, 2017). Interestingly, some members of phyllogen are non-phyllody-inducing, and, for example, $PHYL_{RYD}$ and $PHYL_{SY}$ bind floral MTFs but do not degrade all of those they bind to (Iwabuchi *et al.*, 2020; Suzuki *et al.*, 2024).

Phytoplasmas attack MTFs for their own benefit and use them as a hub for host manipulation. It can thus be assumed that manipulating these transcription factors serves as a strategic mechanism for orchestrating metabolic alterations and symptom manifestation within the plant. This interaction and exploitation underscores the sophisticated interplay between these plant pathogens and host regulatory networks, shedding light on novel avenues for understanding the associated plant diseases.

Other Transcription Factors in Phytoplasma Infection

MYB transcription factors and their role in phytoplasma-related plant responses

The MYB transcription factor family is a large and functionally diverse group of proteins found in all eukaryotes. In plants, MYB proteins play vital roles in regulatory networks, including development, metabolism, and responses to a range of biotic and abiotic stresses. The first plant MYB gene, *COLORED1* in corn, was identified over 35 years ago for its role in anthocyanin synthesis (Paz-Ares *et al.*, 1987). Advances from the *Arabidopsis* genome project later revealed the extensive diversity and functions of MYB genes across species like corn, rice, petunia, snapdragon, grapevine, poplar, and apple (Stracke *et al.*, 2001; Dubos *et al.*, 2010). MYB proteins are characterized by their N-terminal highly conserved DNA-binding domain, known as the MYB domain. This domain consists of up to four imperfect amino acid sequence repeats, each forming a helix-turn-helix (HTH) structure. The third helix, referred to as the recognition helix, makes direct contact with DNA (Dubos *et al.*, 2010).

Based on the number of amino acid repeats, MYB transcription factors are categorized into 1R-MYB, 2R-MYB (R2R3-MYB), 3R-MYB (R1R2R3-MYB), and 4R-MYB, with R2R3-MYB being the most abundant in plants (Biswas *et al.*, 2023). This class of MYB genes has experienced significant plant expansion, with some species possessing over 100 members (Dubos *et al.*, 2010). This expansion reflects the essential functions that these transcription factors serve in various plant-specific processes.

Phytoplasma infections alter MYB transcription factor expression, impacting plant defense responses. In grapevine cv. Chardonnay, several MYB transcription factors were repressed during bois noir (BN) phytoplasma infection, impairing lignin biosynthesis and weakening shoot lignification (Hren *et al.*, 2009). Conversely, in grapevine cv. Manzoni, MYB transcription factors were induced under similar BN infection conditions, suggesting cultivar-specific defense strategies (Albertazzi *et al.*, 2009).

MYB transcription factors also regulate the phenylpropanoid biosynthesis pathway, which produces structural and defensive compounds, including lignin (Albertazzi *et al.*, 2009). In the grapevine cv. Chardonnay, BN phytoplasma infection induced the MYB repression, leading to the downregulation of key genes involved in the phenylpropanoid pathway, including phenylalanine ammonia lyase (PAL), cinnamate 4-hydroxylase (C4H), and 4-coumarate: CoA ligase (4CL). This repression reduces lignin biosynthesis and compromises cell wall integrity, making cv. Chardonnay more susceptible to BN infection than cv. Manzoni.

In addition to their individual functions, MYB proteins also form transcriptional complexes with bHLH transcription factors, further diversifying their regulatory roles (Pireyre and Burow, 2015).

bHLH transcription factors

The bHLH family, named for its conserved alkaline/helix-loop-helix domain, ranks as the second largest transcription factor family in animals and plants. This versatile family of transcription factors plays a fundamental role in almost every aspect of plant development, stress resilience, pathogen defense, and responses to the environment (Pireyre and Burow, 2015). Notably, bHLH proteins are crucial in interpreting light cues/pathways and engaging in intricate interactions with photoreceptors (Buti *et al.*, 2020).

Phytoplasma infections lead to alterations in the expression levels of genes encoding PIFs and MYC2. For instance, in *Paulownia* seedlings infected by paulownia witches'-broom (PaWB) phytoplasma, *PIF3* was significantly upregulated compared with healthy *Paulownia* seedlings (Dong *et al.*, 2018). *PIF3* plays an important role in orchestrating the initial stages of light-triggered chloroplast development by regulating photoresponsive nuclear genes. The increased expression of *PIF3* in PaWB phytoplasma-infected plants influences chloroplast development, resulting in a changed leaf color (Dong *et al.*, 2018). Also, the FPKM (fragments per kilobase of transcript per million fragments mapped) value for genes encoding MYC2 was higher in healthy *Paulownia* compared with PaWB-infected plants, which implies a different regulation of JA signaling. In phytoplasma-infected sweet cherry leaves, MYC2 was also upregulated, indicating an upregulated JA biosynthesis as part of plant responses to infection (Tan *et al.*, 2019).

WRKY transcription factors

WRKY transcription factors form a major family of regulatory proteins in plants, with more than

70 members in *Arabidopsis* and over 100 in rice and soybean (Bakshi and Oelmüller, 2014). WRKY transcription factors are characterized by their highly conserved 60 amino acid WRKY domains. These domains feature a consistent heptapeptide sequence, WRKYGQK, at the N-terminus and a distinctive zinc-finger-like motif at the C-terminus. Both the heptapeptide sequence and the zinc-finger-like motif are crucial for the strong binding capability of WRKY transcription factors to the consensus *cis*-acting elements named W-box (TTGACT/C) (Bakshi and Oelmüller, 2014).

Phytoplasma infections significantly alter the expression of WRKY transcription factors, affecting the plant's ability to mount effective defense responses. Specific WRKYs are targeted by phytoplasma effector proteins, disrupting normal regulatory pathways. A key example is the potential effector SWP12, secreted by '*Ca.* P. tritici,' responsible for the wheat blue dwarf disease. SWP12 has been shown to weaken plant immunity and promote phytoplasma colonization by targeting host WRKY transcription factors, specifically *Triticum aestivum* WRKY74 (TaWRKY74) in wheat and *N. benthamiana* WRKY17 (NbWRKY17) (Bai *et al.*, 2022). The expression of TaWRKY74 increased resistance to '*Ca.* P. tritici,' reducing symptoms such as yellowing. In contrast, silencing TaWRKY74 enhanced plant susceptibility to infection. The results indicate that SWP12 may contribute to the accumulation of phytoplasma by promoting the degradation of TaWRKY74, thereby weakening the plant's defense response. This degradation likely involves exploitation of the plant's UPS (see earlier in this chapter).

bZIP transcription factors

The basic leucine zipper (bZIP) transcription factor family is an evolutionarily conserved group in eukaryotic organisms and is characterized by their highly conserved bZIP domain, spanning *c.*40–80 amino acids (Agarwal *et al.*, 2019). bZIPs are key regulators of plant stress responses, particularly in the unfolded protein response (UPR) pathway. bZIP17, bZIP28, and bZIP60 are central to this process, which maintains endoplasmic reticulum homeostasis (Dröge-Laser *et al.*, 2018).

Phytoplasma infections disrupt the UPR, altering bZIP activity to facilitate colonization. In *Paulownia* plants infected with PaWB phytoplasma, genes encoding UPR-related proteins were notably altered upon phytoplasma infection (Mou *et al.*, 2013). Also, in *Arabidopsis* plants infected with '*Ca.* P. asteris,' elevated transcription levels of both spliced and unspliced forms of bZIP60 were observed, along with increased expression of bZIP17, and the two genes encoding S1P and S2P, which activate bZIP28 (Musetti *et al.*, 2023). These changes were accompanied by observed alterations in endoplasmic reticulum architecture. However, it remains unclear whether phytoplasma effectors can hijack the UPR and associated genes to promote their colonization and survival, and further research is needed to elucidate this aspect.

Conclusion and Outlook

Despite the noticeable differences among various species, it is exciting to observe how phytoplasmas have developed unique and sophisticated strategies to manipulate their plant hosts. A primary method involves interacting with several families of transcription factors, which not only aids phytoplasma survival but also facilitates their spread and colonization within plants. A remarkable example is the case of phyllogens, which hijack the UPS to induce the degradation of MADS-box transcription factors, whose inactivation causes phyllody (leaf-like flowers) and suppression of the plant immune -response. However, there is an opportunity for future research, as some aspects of the effects of effector–MADS binding remain unclear. Understanding the mechanisms underlying this degradation can enhance our knowledge of plant–pathogen interactions and potentially inform strategies for plant protection in the future.

Additionally, the SWP12 effector directly undermines plant immunity by targeting a WRKY transcription factor. Concurrently, other transcription factors such as MYBs, PIFs, and MYC2 show altered expression levels, further affecting plant susceptibility. Similarly, there is a direct correlation between the development of symptoms such as witches'-brooms, crinkled leaves, and JA alteration and the well-described proteasome-independent destabilization of TCPs by SAP11-like proteins. In the current state of knowledge, the deactivation of several members of TCPs appears to be one of the few general principles of plant–phytoplasma interaction. Recent data seem to show that not only SAP11-like, but also other effectors, may target this family of transcription factors. Likewise, it appears that some effectors can relocalize TCPs to different subcellular compartments, with direct effects on the phenotype, although further data will be needed to confirm these interesting findings. It is therefore clear that TCPs are emerging as one of the main target hubs for phytoplasmas, and although many of the effects of TCP deactivation have been unraveled, it is difficult to understand the full extent to which this favors phytoplasma infection.

It is tempting to speculate that transcription factors, in their role as major targets for phytoplasma, might confer susceptibility to phytoplasma infection, and that genetic varieties might thus be involved in high susceptibility or tolerance against these pathogens. In general, although resistance and susceptibility genes have been identified and characterized for a wide range of plant pathosystems, knowledge in the context of phytoplasma diseases is limited. Specific resistance or susceptibility genes for phytoplasma infections in plant hosts have not yet been identified, but there is evidence suggesting that a genetic basis exists. The fact that it is possible to rank plant host genotypes according to their susceptibility to phytoplasma infection is a promising start in revealing genetic traits for resistance. This would be very important not only for a better understanding of molecular mechanisms that underlie susceptibility to phytoplasma but also for breeding new tolerant varieties. Unraveling the role of transcription factors in plant susceptibility and tolerance to phytoplasma infection is thus an exciting path, not to be neglected in resistance breeding approaches.

References

Abramovitch, R.B. and Martin, G.B. (2005) AvrPtoB: A bacterial type III effector that both elicits and suppresses programmed cell death associated with plant immunity. *FEMS Microbiology Letters* 245, 1–8. DOI: 10.1016/j.femsle.2005.02.025.

Agarwal, P., Baranwal, V.K., and Khurana, P. (2019) Genome-wide analysis of bZIP transcription factors in wheat and functional characterization of a TabZIP under abiotic stress. *Scientific Reports* 9, 4608. DOI: 10.1038/s41598-019-40659-7.

Aggarwal, P., Das Gupta, M., Joseph, A.P., Chatterjee, N., Srinivasan, N., and Nath, U. (2010) Identification of specific DNA binding residues in the TCP family of transcription factors in Arabidopsis. *Plant Cell* 22, 1174–1189. DOI: 10.1105/tpc.109.066647.

Aguilar-Martínez, J.A., Poza-Carrión, C., and Cubas, P. (2007) Arabidopsis BRANCHED1 acts as an integrator of branching signals within axillary buds. *Plant Cell* 19, 458–472. DOI: 10.1105/tpc.106.048934.

Albertazzi, G., Milc, J., Caffagni, A., Francia, E., Roncaglia, E. *et al.* (2009) Gene expression in grapevine cultivars in response to bois noir phytoplasma infection. *Plant Science* 176, 792–804. DOI: 10.1016/j.plantsci.2009.03.001.

Al-Subhi, A.M., Al-Sadi, A.M., Al-Yahyai, R.A., Chen, Y., Mathers, T. *et al.* (2021) Witches' broom disease of lime contributes to phytoplasma epidemics and attracts insect vectors. *Plant Disease* 105, 2637–2648. DOI: 10.1094/PDIS-10-20-2112-RE.

Bai, B., Zhang, G., Li, Y., Wang, Y., Sujata, S. *et al.* (2022) The 'Candidatus Phytoplasma tritici' effector SWP12 degrades the transcription factor TaWRKY74 to suppress wheat resistance. *Plant Journal* 112, 1473–1488. DOI: 10.1111/tpj.16029.

Bai, X., Correa, V.R., Toruño, T.Y., Ammar, E.-D., Kamoun, S., and Hogenhout, S.A. (2009) AY-WB phytoplasma secretes a protein that targets plant cell nuclei. *Molecular Plant-Microbe Interactions* 22, 18–30. DOI: 10.1094/MPMI-22-1-0018.

Bakshi, M. and Oelmüller, R. (2014) WRKY transcription factors: Jack of many trades in plants. *Plant Signaling and Behavior*, 9, e27700. DOI: 10.4161/psb.27700.

Becker, A. and Theissen, G. (2003) The major clades of MADS-box genes and their role in the development and evolution of flowering plants. *Molecular Phylogenetics and Evolution* 29, 464–489. DOI: 10.1016/S1055-7903(03)00207-0.

Biswas, D., Gain, H., and Mandal, A. (2023) MYB transcription factor: A new weapon for biotic stress tolerance in plants. *Plant Stress* 10, 100252. DOI: 10.1016/j.stress.2023.100252.

Boller, T. and He, S.Y. (2009) Innate immunity in plants: An arms race between pattern recognition receptors in plants and effectors in microbial pathogens. *Science* 324, 742–744. DOI: 10.1126/science.1171647.

Boonrod, K., Strohmayer, A., Schwarz, T., Braun, M., Tropf, T., and Krczal, G. (2022) Beyond destabilizing activity of SAP11-like effector of *Candidatus* Phytoplasma mali strain PM19. *Microorganisms* 10. DOI: 10.3390/microorganisms10071406.

Buti, S., Hayes, S., and Pierik, R. (2020) The bHLH network underlying plant shade-avoidance. *Physiologia Plantarum* 169, 312–324. DOI: 10.1111/ppl.13074.

Büttner, D. (2016) Behind the lines-actions of bacterial type III effector proteins in plant cells. *FEMS Microbiology Reviews* 40, 894–937. DOI: 10.1093/femsre/fuw026.

Carreón-Anguiano, K.G., Vila-Luna, S.E., Sáenz-Carbonell, L., and Canto-Canché, B. (2023) Novel insights into phytoplasma effectors. *Horticulturae* 9, 1228. DOI: 10.3390/horticulturae9111228.

Chang, S.H., Tan, C.M., Wu, C.-T., Lin, T.-H., Jiang, S.-Y. *et al.* (2018) Alterations of plant architecture and phase transition by the phytoplasma virulence factor SAP11. *Journal of Experimental Botany* 69, 5389–5401. DOI: 10.1093/jxb/ery318.

Chen, H., Chen, J., Li, M., Chang, M., Xu, K. *et al.* (2017) A bacterial type III effector targets the master regulator of salicylic acid signaling, NPR1, to subvert plant immunity. *Cell Host and Microbe* 22, 777-788.e7. DOI: 10.1016/j.chom.2017.10.019.

Chen, P., Chen, L., Ye, X., Tan, B., Zheng, X. *et al.* (2022) Phytoplasma effector Zaofeng6 induces shoot proliferation by decreasing the expression of ZjTCP7 in *Ziziphus jujuba*. *Horticulture Research* 9, 1–10. DOI: 10.1093/hr/uhab032/6510945.

Correa Marrero, M., Capdevielle, S., Huang, W., Al-Subhi, A.M., Busscher, M. *et al.* (2024) Protein interaction mapping reveals widespread targeting of development-related host transcription factors by phytoplasma effectors. *Plant Journal* 117, 1281–1297. DOI: 10.1111/tpj.16546.

Couto, D. and Zipfel, C. (2016) Regulation of pattern recognition receptor signalling in plants. *Nature Reviews Immunology* 16, 537–552. DOI: 10.1038/nri.2016.77.

Cubas, P., Lauter, N., Doebley, J., and Coen, E. (1999) The TCP domain: A motif found in proteins regulating plant growth and development. *Plant Journal* 18, 215–222. DOI: 10.1046/j.1365-313x.1999.00444.x.

Danisman, S. (2016) TCP transcription factors at the interface between environmental challenges and the plant's growth responses. *Frontiers in Plant Science* 7, 1930. DOI: 10.3389/fpls.2016.01930.

Danisman, S., van der Wal, F., Dhondt, S., Waites, R., Folter, S. de *et al.* (2012) Arabidopsis class I and class II TCP transcription factors regulate jasmonic acid metabolism and leaf development antagonistically. *Plant Physiology* 159, 1511–1523. DOI: 10.1104/pp.112.200303.

Danisman, S., van Dijk, A.D.J., Bimbo, A., van der Wal, F., Hennig, L. *et al.* (2013) Analysis of functional redundancies within the Arabidopsis TCP transcription factor family. *Journal of Experimental Botany* 64, 5673–5685. DOI: 10.1093/jxb/ert337.

Dermastia, M. (2019) Plant hormones in phytoplasma infected plants. *Frontiers in Plant Science* 10, 477. DOI: 10.3389/fpls.2019.00477.

Dermastia, M., Nikolic, P., Chersicola, M., and Gruden, K. (2015) Transcriptional profiling in infected and recovered grapevine plant responses to 'Candidatus Phytoplasma solani'. *Phytopathogenic Mollicutes* 5, S123. DOI: 10.5958/2249-4677.2015.00053.5.

Ding, S., Cai, Z., Du, H., and Wang, H. (2019) Genome-wide analysis of TCP family genes in Zea mays L. identified a role for ZmTCP42 in drought tolerance. *International Journal of Molecular Sciences* 20. DOI: 10.3390/ijms20112762.

Doebley, J., Stec, A., and Gustus, C. (1995) Teosinte branched1 and the origin of maize: evidence for epistasis and the evolution of dominance. *Genetics* 141, 333–346. DOI: 10.1093/genetics/141.1.333.

Dong, Y., Zhang, H., Fan, G., Zhai, X., Wang, Z., and Cao, Y. (2018) Comparative transcriptomics analysis of phytohormone-related genes and alternative splicing events related to witches' broom in Paulownia. *Forests* 9, 318. DOI: 10.3390/f9060318.

Dröge-Laser, W., Snoek, B.L., Snel, B., and Weiste, C. (2018) The Arabidopsis bZIP transcription factor family – An update. *Current Opinion in Plant Biology* 45, 36–49. DOI: 10.1016/j.pbi.2018.05.001.

Dubos, C., Stracke, R., Grotewold, E., Weisshaar, B., Martin, C., and Lepiniec, L. (2010) MYB transcription factors in Arabidopsis. *Trends in Plant Science* 15, 573–581. DOI: 10.1016/j.tplants.2010.06.005.

Finlayson, S.A. (2007) Arabidopsis Teosinte Branched1-like 1 regulates axillary bud outgrowth and is homologous to monocot Teosinte Branched1. *Plant and Cell Physiology* 48, 667–677. DOI: 10.1093/pcp/pcm044.

González-Grandío, E., Poza-Carrión, C., Sorzano, C.O.S., and Cubas, P. (2013) BRANCHED1 promotes axillary bud dormancy in response to shade in Arabidopsis. *Plant Cell* 25, 834–850. DOI: 10.1105/tpc.112.108480.

Gramzow, L. and Theissen, G. (2010) A hitchhiker's guide to the MADS world of plants. *Genome Biology* 11, 214. DOI: 10.1186/gb-2010-11-6-214.

Groll, M., Schellenberg, B., Bachmann, A.S., Archer, C.R., Huber, R. *et al.* (2008) A plant pathogen virulence factor inhibits the eukaryotic proteasome by a novel mechanism. *Nature* 452, 755–758. DOI: 10.1038/nature06782.

Hogenhout, S.A., Oshima, K., Ammar, E.-D., Kakizawa, S., Kingdom, H.N., and Namba, S. (2008) Phytoplasmas: Bacteria that manipulate plants and insects. *Molecular Plant Pathology* 9, 403–423. DOI: 10.1111/j.1364-3703.2008.00472.x.

Hren, M., Nikolić, P., Rotter, A., Blejec, A., Terrier, N. *et al.* (2009) 'Bois noir' phytoplasma induces significant reprogramming of the leaf transcriptome in the field grown grapevine. *BMC Genomics* 10, 460. DOI: 10.1186/1471-2164-10-460.

Huang, W., MacLean, A.M., Sugio, A., Maqbool, A., Busscher, M. *et al.* (2021) Parasitic modulation of host development by ubiquitin-independent protein degradation. *Cell* 184, 5201–5214.e12. DOI: 10.1016/j.cell.2021.08.029.

Iwabuchi, N., Maejima, K., Kitazawa, Y., Miyatake, H., Nishikawa, M. *et al.* (2019) Crystal structure of phyllogen, a phyllody-inducing effector protein of phytoplasma. *Biochemical and Biophysical Research Communications* 513, 952–957. DOI: 10.1016/j.bbrc.2019.04.060.

Iwabuchi, N., Kitazawa, Y., Maejima, K., Koinuma, H., Miyazaki, A. *et al.* (2020) Functional variation in phyllogen, a phyllody-inducing phytoplasma effector family, attributable to a single amino acid polymorphism. *Molecular Plant Pathology* 21, 1322–1336. DOI: 10.1111/mpp.12981.

Janik, K., Mithöfer, A., Raffeiner, M., Stellmach, H., Hause, B., Schlink, K., and Mithofer, A. (2017) An effector of apple proliferation phytoplasma targets TCP transcription factors – A generalized virulence strategy of phytoplasma? *Molecular Plant Pathology* 18, 435–442. DOI: 10.1111/mpp.12409.

Jones, J.D.G. and Dangl, J.L. (2006) The plant immune system. *Nature* 444, 323–329. DOI: 10.1038/nature05286.

Kitazawa, Y., Iwabuchi, N., Himeno, M., Sasano, M., Koinuma, H. *et al.* (2017) Phytoplasma-conserved phyllogen proteins induce phyllody across the Plantae by degrading floral MADS domain proteins. *Journal of Experimental Botany* 68, 2799–2811. DOI: 10.1093/jxb/erx158.

Kitazawa, Y., Iwabuchi, N., Maejima, K., Sasano, M., Matsumoto, O. *et al.* (2022) A phytoplasma effector acts as a ubiquitin-like mediator between floral MADS-box proteins and proteasome shuttle proteins. *Plant Cell* 34, 1709–1723. DOI: 10.1093/plcell/koac062.

Kosugi, S. and Ohashi, Y. (1997) PCF1 and PCF2 specifically bind to cis elements in the rice proliferating cell nuclear antigen gene. *Plant Cell* 9, 1607–1619. DOI: 10.1105/tpc.9.9.1607.

Kosugi, S. and Ohashi, Y. (2002) DNA binding and dimerization specificity and potential targets for the TCP protein family. *Plant Journal* 30, 337–348. DOI: 10.1046/j.1365-313x.2002.01294.x.

Langin, G., González-Fuente, M., and Üstün, S. (2023) The plant ubiquitin-proteasome system as a target for microbial manipulation. *Annual Review of Phytopathology* 61, 351–375. DOI: 10.1146/annurev-phyto-021622-110443

Lei, L., Stevens, D.M., and Coaker, G. (2020) Phosphorylation of the Pseudomonas effector AvrPtoB by Arabidopsis SnRK2.8 is required for bacterial virulence. *Molecular Plant* 13, 1513–1522. DOI: 10.1016/j.molp.2020.08.018.

Leong, J.X., Raffeiner, M., Spinti, D., Langin, G., Franz-Wachtel, M. *et al.* (2022) A bacterial effector counteracts host autophagy by promoting degradation of an autophagy component. *EMBO Journal* 41, e110352. DOI: 10.15252/embj.2021110352.

Li, S. (2015) The *Arabidopsis thaliana* TCP transcription factors: A broadening horizon beyond development. *Plant Signaling and Behavior* 10, e1044192-1–e1044192-12. DOI: 10.1080/15592324.2015.1044192.

Liu, C.-T., Huang, H.-M., Hong, S.-F., Kuo-Huang, L.-L., Yang, C.-Y. *et al.* (2015) Peanut witches' broom (PnWB) phytoplasma-mediated leafy flower symptoms and abnormal vascular bundles development. *Plant Signaling and Behavior* 10, e1107690. DOI: 10.1080/15592324.2015.1107690.

Luge, T., Kube, M., Freiwald, A., Meierhofer, D., Seemüller, E., and Sauer, S. (2014) Transcriptomics assisted proteomic analysis of *Nicotiana occidentalis* infected by *Candidatus* Phytoplasma mali strain AT. *Proteomics* 14, 1882–1889. DOI: 10.1002/pmic.201300551.

Luo, D., Carpenter, R., Vincent, C., Copsey, L., and Coen, E. (1996) Origin of floral asymmetry in Antirrhinum. *Nature* 383, 794–799. DOI: 10.1038/383794a0.

Ma, F., Zheng, Y., Zhang, N., Deng, M., Zhao, M. *et al.* (2024) The 'Candidatus Phytoplasma ziziphi' effectors SJP1/2 negatively control leaf size by stabilizing the transcription factor ZjTCP2 in jujube. *Journal of Experimental Botany* 75, 3054–3069. DOI: 10.1093/jxb/erae042.

MacLean, A.M., Sugio, A., Makarova, O.V., Findlay, K.C., Grieve, V.M. *et al.* (2011) Phytoplasma effector SAP54 induces indeterminate leaf-like flower development in Arabidopsis plants. *Plant Physiology* 157, 831–841. DOI: 10.1104/pp.111.181586.

MacLean, A.M., Orlovskis, Z., Kowitwanich, K., Zdziarska, A.M., Angenent, G.C., Immink, R.G.H., and Hogenhout, S.A. (2014) Phytoplasma effector SAP54 hijacks plant reproduction by degrading MADS-box proteins and promotes insect colonization in a RAD23-dependent manner. *PLoS Biology* 12, e1001835. DOI: 10.1371/journal.pbio.1001835.

Maejima, K., Iwai, R., Himeno, M., Komatsu, K., Kitazawa, Y. *et al.* (2014) Recognition of floral homeotic MADS domain transcription factors by a phytoplasmal effector, phyllogen, induces phyllody. *Plant Journal* 78, 541–554. DOI: 10.1111/tpj.12495.

Maejima, K., Kitazawa, Y., Tomomitsu, T., Yusa, A., Neriya, Y. *et al.* (2015) Degradation of class E MADS-domain transcription factors in Arabidopsis by a phytoplasmal effector, phyllogen. *Plant Signaling and Behavior* 10, e1042635. DOI: 10.1080/15592324.2015.1042635.

Mardi, M., Karimi Farsad, L., Gharechahi, J., and Salekdeh, G.H. (2015) In-depth transcriptome sequencing of Mexican lime trees infected with Candidatus Phytoplasma aurantifolia. *PloS One* 10, e0130425. DOI: 10.1371/journal.pone.0130425.

Marino, D., Peeters, N., and Rivas, S. (2012) Ubiquitination during plant immune signaling. *Plant Physiology* 160, 15–27. DOI: 10.1104/pp.112.199281.

Marshall, R.S. and Vierstra, R.D. (2019) Dynamic regulation of the 26S proteasome: From synthesis to degradation. *Frontiers in Molecular Biosciences* 6, 40. DOI: 10.3389/fmolb.2019.00040.

Martín-Trillo, M. and Cubas, P. (2010) TCP genes: A family snapshot ten years later. *Trends in Plant Science* 15, 31–39. DOI: 10.1016/j.tplants.2009.11.003.

Misas-Villamil, J.C., Kolodziejek, I., Crabill, E., Kaschani, F., Niessen, S. *et al.* (2013) Pseudomonas syringae pv. syringae uses proteasome inhibitor syringolin A to colonize from wound infection sites. *PLoS Pathogens* 9, e1003281. DOI: 10.1371/journal.ppat.1003281.

Mittelberger, C., Hause, B., and Janik, K. (2022) The '*Candidatus* Phytoplasma mali' effector protein SAP-11_{CaPm} interacts with MdTCP16, a class II CYC/TB1 transcription factor that is highly expressed during phytoplasma infection. *PloS One* 17, e0272467. DOI: 10.1371/journal.pone.0272467.

Mou, H.-Q., Lu, J., Zhu, S.-F., Lin, C.-L., Tian, G.-Z., Xu, X., and Zhao, W.-J. (2013) Transcriptomic analysis of *Paulownia* infected by Paulownia witches'-broom phytoplasma. *PloS One* 8, e77217. DOI: 10.1371/journal.pone.0077217.

Mukhtar, M.S., Carvunis, A.-R., Dreze, M., Epple, P., Steinbrenner, J. *et al.* (2011) Independently evolved virulence effectors converge onto hubs in a plant immune system network. *Science* 333, 596–601. DOI: 10.1126/science.1203659.

Musetti, R., Farhan, K., Marco, F. de, Polizzotto, R., Paolacci, A. *et al.* (2013) Differentially-regulated defence genes in *Malus domestica* during phytoplasma infection and recovery. *European Journal of Plant Pathology* 136, 13–19. DOI: 10.1007/s10658-012-0147-6.

Musetti, R., Pagliari, L., Mian, G., Oliveira Cantao, F.R. de, Bernardini, C., Santi, S., and van Bel, A.J.E. (2023) The sieve-element endoplasmic reticulum: A focal point of phytoplasma-host plant interaction? *Frontiers in Microbiology* 14, 1030414. DOI: 10.3389/fmicb.2023.1030414.

Nicolas, M. and Cubas, P. (2016) TCP factors: New kids on the signaling block. *Current Opinion in Plant Biology* 33, 33–41. DOI: 10.1016/j.pbi.2016.05.006.

Niwa, M., Daimon, Y., Kurotani, K., Higo, A., Pruneda-Paz, J.L. *et al.* (2013) BRANCHED1 interacts with FLOWERING LOCUS T to repress the floral transition of the axillary meristems in Arabidopsis. *Plant Cell* 25, 1228–1242. DOI: 10.1105/tpc.112.109090.

Orlovskis, Z. and Hogenhout, S.A. (2016) A bacterial parasite effector mediates insect vector attraction in host plants independently of developmental changes. *Frontiers in Plant Science* 7, 885. DOI: 10.3389/fpls.2016.00885.

Orlovskis, Z., Singh, A., Kliot, A., Huang, W., and Hogenhout, S.A. (2025) The phytoplasma SAP54 effector acts as a molecular matchmaker for leafhopper vectors by targeting plant MADS-box factor SVP. *eLife* 13, RP98992. DOI: 10.7554/eLife.98992.

Oshima, K., Kakizawa, S., Nishigawa, H., Jung, H.-Y., Wei, W. *et al.* (2004) Reductive evolution suggested from the complete genome sequence of a plant-pathogenic phytoplasma. *Nature Genetics* 36, 27–29. DOI: 10.1038/ng1277.

Oshima, K., Maejima, K., and Namba, S. (2013) Genomic and evolutionary aspects of phytoplasmas. *Frontiers in Microbiology* 4, 230. DOI: 10.3389/fmicb.2013.00230.

Palatnik, J.F., Allen, E., Wu, X., Schommer, C., Schwab, R., Carrington, J.C., and Weigel, D. (2003) Control of leaf morphogenesis by microRNAs. *Nature* 425, 257–263. DOI: 10.1038/nature01958.

Paolacci, A.R., Catarcione, G., Ederli, L., Zadra, C., Pasqualini, S. *et al.* (2017) Jasmonate-mediated defence responses, unlike salicylate-mediated responses, are involved in the recovery of grapevine from bois noir disease. *BMC Plant Biology* 17, 118. DOI: 10.1186/s12870-017-1069-4

Paz-Ares, J., Ghosal, D., Wienand, U., Peterson, P.A., and Saedler, H. (1987) The regulatory c1 locus of Zea mays encodes a protein with homology to myb proto-oncogene products and with structural similarities to transcriptional activators. *EMBO Journal* 6, 3553–3558. DOI: 10.1002/j.1460-2075.1987.tb02684.x.

Pecher, P., Moro, G., Canale, M.C., Capdevielle, S., Singh, A. *et al.* (2019) Phytoplasma SAP11 effector destabilization of TCP transcription factors differentially impact development and defence of Arabidopsis versus maize. *PLoS Pathogens* 15, 1-27. DOI: 10.1371/journal.ppat.1008035.

Pireyre, M. and Burow, M. (2015) Regulation of MYB and bHLH transcription factors: A glance at the protein level. *Molecular Plant* 8, 378–388. DOI: 10.1016/j.molp.2014.11.022.

Poot, S.A.H. de, Tian, G., and Finley, D. (2017) Meddling with fate: The proteasomal deubiquitinating enzymes. *Journal of Molecular Biology* 429, 3525–3545. DOI: 10.1016/j.jmb.2017.09.015.

Ramachandran, P., J, B.J., Maupin-Furlow, J.A., and Uthandi, S. (2021) Bacterial effectors mimicking ubiquitin-proteasome pathway tweak plant immunity. *Microbiological Research* 250, 126810. DOI: 10.1016/j.micres.2021.126810.

Rosebrock, T.R., Zeng, L., Brady, J.J., Abramovitch, R.B., Xiao, F., and Martin, G.B. (2007) A bacterial E3 ubiquitin ligase targets a host protein kinase to disrupt plant immunity. *Nature* 448, 370–374. DOI: 10.1038/nature05966.

Saedler, H., Becker, A., Winter, K.U., Kirchner, C., and Theissen, G. (2001) MADS-box genes are involved in floral development and evolution. *Acta Biochimica Polonica* 48, 351–358.

Sarvepalli, K. and Nath, U. (2018) CIN-TCP transcription factors: Transiting cell proliferation in plants. *IUBMB Life* 70, 718–731. DOI: 10.1002/iub.1874.

Sharma, B., Joshi, D., Yadav, P.K., Gupta, A.K., and Bhatt, T.K. (2016) Role of ubiquitin-mediated degradation system in plant biology. *Frontiers in Plant Science* 7, 806. DOI: 10.3389/fpls.2016.00806.

Singer, A.U., Schulze, S., Skarina, T., Xu, X., Cui, H. *et al.* (2013) A pathogen type III effector with a novel E3 ubiquitin ligase architecture. *PLOS Pathogens* 9, e1003121. DOI: 10.1371/journal.ppat.1003121.

Stracke, R., Werber, M., and Weisshaar, B. (2001) The R2R3-MYB gene family in Arabidopsis thaliana. *Current Opinion in Plant Biology* 4, 447–456. DOI: 10.1016/s1369-5266(00)00199-0.

Strohmayer, A., Moser, M., Si-Ammour, A., Krczal, G., and Boonrod, K. (2019) '*Candidatus* Phytoplasma mali' genome encodes a protein that functions as a E3 ubiquitin ligase and could inhibit plant basal defense. *Molecular Plant-Microbe Interactions* DOI: 10.1094/MPMI-04-19-0107-R.

Strohmayer, A., Schwarz, T., Braun, M., Krczal, G., and Boonrod, K. (2021) The effect of the anticipated nuclear localization sequence of '*Candidatus* Phytoplasma mali' SAP11-like protein on localization of the protein and destabilization of TCP transcription factor. *Microorganisms* 9, 1–17. DOI: 10.3390/microorganisms9081756.

Studer, A.J., Wang, H., and Doebley, J.F. (2017) Selection during maize domestication targeted a gene network controlling plant and inflorescence architecture. *Genetics* 207, 755–765. DOI: 10.1534/genetics.117.300071.

Su, T., Yang, M., Wang, P., Zhao, Y., and Ma, C. (2020) Interplay between the ubiquitin proteasome system and ubiquitin-mediated autophagy in plants. *Cells*, 9. DOI: 10.3390/cells9102219.

Sugio, A., Kingdom, H.N., MacLean, A.M., Grieve, V.M., and Hogenhout, S.A. (2011a) Phytoplasma protein effector SAP11 enhances insect vector reproduction by manipulating plant development and defense hormone biosynthesis. *Proceedings of the National Academy of Sciences of the USA* 108, 1254–1263. DOI: 10.1073/pnas.1105664108.

Sugio, A., MacLean, A.M., Kingdom, H.N., Grieve, V.M., Manimekalai, R., and Hogenhout, S.A. (2011b) Diverse targets of phytoplasma effectors: From plant development to defense against insects. *Annual Review of Phytopathology* 49, 175–195. DOI: 10.1146/annurev-phyto-072910-095323.

Suzuki, M., Kitazawa, Y., Iwabuchi, N., Maejima, K., Matsuyama, J. *et al.* (2024) Target degradation specificity of phytoplasma effector phyllogen is regulated by the recruitment of host proteasome shuttle protein. *Molecular Plant Pathology* 25, e13410. DOI: 10.1111/mpp.13410.

Tabarelli, M., Malnoy, M., and Janik, K. (2022) Chasing consistency: An update of the TCP gene family of Malus × Domestica. *Genes* 13. DOI: 10.3390/genes13101696.

Tan, C.M., Li, C.-H., Tsao, N.-W., Su, L.-W., Lu, Y.-T. *et al.* (2016) Phytoplasma SAP11 alters 3-isobutyl-2-methoxypyrazine biosynthesis in *Nicotiana benthamiana* by suppressing *NbOMT1*. *Journal of Experimental Botany* 67, 4415–4425. DOI: 10.1093/jxb/erw225.

Tan, Y., Wang, J., Davis, R.E., Wei, H., Zong, X. *et al.* (2019) Transcriptome analysis reveals a complex array of differentially expressed genes accompanying a source-to-sink change in phytoplasma-infected sweet cherry leaves. *Annals of Applied Biology*, 175, 69–82. DOI: 10.1111/aab.12511.

Trujillo, M. and Shirasu, K. (2010) Ubiquitination in plant immunity. *Current Opinion in Plant Biology* 13, 402–408. DOI: 10.1016/j.pbi.2010.04.002.

Vierstra, R.D. (2009) The ubiquitin-26S proteasome system at the nexus of plant biology. *Nature Reviews Molecular Cell Biology* 10, 385–397. DOI: 10.1038/nrm2688.

Viola, I.L. and Gonzalez, D.H. (2023) TCP transcription factors in plant reproductive development: Juggling multiple roles. *Biomolecules* 13. DOI: 10.3390/biom13050750.

Wang, N., Yang, H., Yin, Z., Liu, W., Sun, L., and Wu, Y. (2018) Phytoplasma effector SWP1 induces witches' broom symptom by destabilizing the TCP transcription factor BRANCHED1. *Molecular Plant Pathology* 19, 2623–2634. DOI: 10.1111/mpp.12733.

Wang, W., Liu, N., Gao, C., Rui, L., and Tang, D. (2019) The Pseudomonas syringae effector AvrPtoB associates with and ubiquitinates Arabidopsis exocyst subunit EXO70B1. *Frontiers in Plant Science* 10, 1027. DOI: 10.3389/fpls.2019.01027.

Xu, F.-Q. and Xue, H.-W. (2019) The ubiquitin-proteasome system in plant responses to environments. *Plant, Cell and Environment* 42, 2931–2944. DOI: 10.1111/pce.13633.

Yan, C. and Xie, D. (2015) Jasmonate in plant defence: Sentinel or double agent? *Plant Biotechnology Journal* 13, 1233–1240. DOI: 10.1111/pbi.12417.

Yang, C.-Y., Huang, Y.-H., Lin, C.-P., Lin, Y.-Y., Hsu, H.-C. *et al.* (2015) MicroRNA396-targeted SHORT VEGETATIVE PHASE is required to repress flowering and is related to the development of abnormal flower symptoms by the Phyllody Symptoms1 Effector. *Plant Physiology* 168, 1702–1716. DOI: 10.1104/pp.15.00307.

Yuan, M., Ngou, B.P.M., Ding, P., and Xin, X.-F. (2021) PTI-ETI crosstalk: An integrative view of plant immunity. *Current Opinion in Plant Biology* 62, 102030. DOI: 10.1016/j.pbi.2021.102030.

Zhang, Z., Zou, W., Lin, P., Wang, Z., Chen, Y. *et al.* (2024) Evolution and function of MADS-box transcription factors in plants. *International Journal of Molecular Sciences* 25. DOI: 10.3390/ijms252413278.

Zhou, B. and Zeng, L. (2017) Conventional and unconventional ubiquitination in plant immunity. *Molecular Plant Pathology* 18, 1313–1330. DOI: 10.1111/mpp.12521.

10 Molecular Interactions of '*Candidatus* Phytoplasma solani' with its Hosts: Insights into Pathogenicity Mechanisms

Marina Dermastia,[1]* Marina Drčelić,[2] and Martina Šeruga Musić[2]
[1]*National Institute of Biology, Ljubljana, Slovenia;* [2]*Department of Biology, Faculty of Science, University of Zagreb, Zagreb, Croatia*

Abstract

'*Candidatus* Phytoplasma solani' (ribosomal group 16SrXII-A) is an insect vector-borne phytopathogen that causes devastating diseases in economically important crops, including grapevine. It colonizes the phloem of infected plants and the cells of insect vectors by secreting effector proteins that manipulate host immune responses and facilitate nutrient acquisition. Studies on infected plants reveal significant alterations in the expression of genes, proteins, and metabolites associated with defense mechanisms, photosynthesis, and nutrient transport. In addition, '*Ca.* P. solani' disrupts the phytohormone balance of the host, which further contributes to the development of symptoms and creates a favorable environment for pathogen proliferation. This chapter addresses the key mechanisms of '*Ca.* P. solani' pathogenicity and provides insights into the development of novel strategies to mitigate the impact of phytoplasma-induced diseases.

> ... whilst this planet has gone circling on according to the fixed law of gravity, from so simple a beginning endless forms most beautiful and most wonderful have been, and are being evolved.
>
> Charles Darwin, *The Origin of Species*, 6th edn

Introduction

The genus '*Candidatus* Phytoplasma' encompasses a very diverse group of endocellular pathogens affecting numerous plant species worldwide, including economically important crops. They are transmitted and hosted by phloem-feeding insects of the suborder Auchenorrhyncha and induce drastic symptoms such as yellowing or reddening of leaves, witches'-broom symptoms, abnormal flower formation, stunting, and dieback, reflecting deep disturbance of metabolic and developmental processes in infected plants (Namba, 2019; Wei and Zhao, 2022).

In contrast to some phytoplasma species that are transmitted by a specific insect vector and specialized to only a few or one plant species, '*Ca.* P. solani' from the ribosomal group 16SrXII has a wide plant host range, including many wild species, ornamentals, and crops, and is transmitted by different polyphagous insect vectors (Quaglino *et al.*, 2013, 2019). It is responsible for severe agricultural losses of important affected crops, including grapevine, corn (maize), potato, tomato, pepper, and strawberry, primarily in the Euro-Mediterranean region.

This chapter explores the molecular mechanisms underpinning the interactions between

*Corresponding author; marina.dermastia@nib.si

 Phytoplasmas: Genomes, Plant Hosts and Vectors, 2nd Edition. (Ed. W. Wei)
DOI: 10.1079/9781800627031.0010

'*Ca.* P. solani' and its host plants, focusing on genomic features, effector functions, and host responses revealed by several omics approaches.

Insights into the Genome of '*Candidatus* Phytoplasma solani': General Characteristics Shared with Phytoplasmas yet Demonstrating Unique Specificities

The class of Mollicutes, which includes mycoplasmas, acholeplasmas, spiroplasmas, and phytoplasmas, exhibits remarkable differences in genome size within a given order, family, or genus (Razin *et al.*, 1998). Accordingly, a wide range of genome sizes from 530 to 1350 kbp, including the largest phytoplasma chromosome reported to date, has been identified in the strains of the genus '*Ca.* P. solani' by pulsed field gel electrophoresis (PFGE) (Marcone *et al.*, 1999).

Despite the ongoing challenge of culturing phytoplasmas in pure *in vitro* cultures, the combined application of metagenomic approaches and phytoplasma DNA enrichment, together with advances in sequencing technologies such as Illumina and Oxford Nanopore Technologies, offers promising opportunities to obtain comprehensive phytoplasma genome assemblies. Currently, 18 fully assembled phytoplasma genomes are deposited in GenBank, in addition to a considerable number of draft phytoplasma genomes (Wei and Zhao, 2022; https://www.ncbi.nlm.nih.gov/genome/browse#!/overview/phytoplasma; last accessed June 2025). Most fully assembled genomes belong to subclade I, particularly '*Ca.* P. asteris'-related strains. The inability to assemble some phytoplasma genome drafts into a single contig may be attributed to the prevalence of repetitive sequences hindering assembly and the lack of closely related reference genomes.

For '*Ca.* P. solani'-related strains, only three genome drafts have been published to date. These include strains 284/9 and 231/09, which were obtained from tobacco and parsley, respectively (Mitrović *et al.*, 2014), and strain SA-1, which was sequenced from micropropagated periwinkle (*Catharanthus roseus* L. Don.) infected with phytoplasmas from grapevine (Šeruga Music *et al.*, 2019). Among these, the draft genome of strain SA-1 is the most complete, with a length of 821,322 bp assembled in 19 contigs, covering about 81% of the total estimated genome size. In contrast, strains 284/09 and 231/09 have a total genome assembly of 557,538 kb in 128 contigs and 515,758 kb in 298 contigs, respectively. Detailed analyses of the SA-1 draft genome, including comparisons with other Mollicute genomes, have been performed (Šeruga Musić *et al.*, 2019).

Several common features of phytoplasma genomes were identified, such as a low GC content, slightly higher than that of more distantly related phytoplasmas such as '*Ca.* P. mali' or flavescence dorée phytoplasma (FDp), but comparable to closely related subclade I phytoplasmas, including '*Ca.* P. australiense' and '*Ca.* P. asteris' (Table 10.1). In addition, the irregular GC skew pattern observed in the genome sequence of the SA-1 strain, which is similar to the genomes of strains aster yellows witches'-broom (AY-WB) and onion yellows mild (OY-M) of '*Ca.* P. asteris,' emphasizes the high plasticity of the genomes in comparison with the FDp, which has low genome plasticity (Debonneville *et al.*, 2022). The SA-1 strain genome also shows a remarkable presence of repetitive sequences and multicopy genes, accounting for 18.5% of its genome, more than other phytoplasma genomes (Table 10.1), which probably hindered the complete assembly. On the other hand, the SA-1 draft genome contains 709 coding DNA sequences (CDS), comparable to the number found in fully assembled, closely related phytoplasma genomes (Table 10.1), confirming the completeness of the genome. Whole-genome alignment of '*Ca.* P. solani' SA-1 genome with closely related genomes shows no conserved synteny, suggesting substantial genome rearrangements leading to biological divergence between these related phytoplasmas.

Comparative analysis of '*Ca.* P. solani' SA-1 strain with other fully sequenced phytoplasma genomes revealed a core set of 259 orthologous gene clusters shared by all analyzed phytoplasmas, with about 10% being unique and presumably phytoplasma-specific. '*Ca.* P. solani' SA-1 has 23.84% of its gene clusters that are unique to this strain, a feature also observed in other phytoplasmas, such as peanut witches'-broom (PnWB) phytoplasma. This suggests that these genome-specific genes may play an important role in adaptation to different plant and insect hosts and phytoplasma differentiation.

Table 10.1. General characteristics of '*Ca.* P. solani' SA-1 genome and comparison with selected representative phytoplasma genomes.

	'*Ca.* P. solani' SA-1	'*Ca.* P. australiense' PAa	'*Ca.* P. asteris' AY-WB	'*Ca.* P. asteris' OY	'*Ca.* P. mali'	FDp strain CH
16S rRNA group	16SrXII	16SrXII	16SrI	16SrI	16SrX	16SrV
No. of contigs	19	1	1	1	1	1
Chromosome size/combined length (bp)	821,322	879,324	706,569	853,092	601,943	654,223
GC content (%)	28.3	27	26.9	27.8	21.4	21.7
No. of CDS:	709	839	671	754	497	506
CDS with assigned function	452	502	450	446	338	350
CDS annotated as hypothetical proteins	257	337	221	308	159	156
No. of tRNA genes	32	35	32	32	32	32
No. of rRNA operons	2	2	2	2	2	2
GenBank accession number and reference paper	MPBG01000000 (Šeruga Musić *et al.*, 2019)	AM422018 (Tran-Nguyen *et al.*, 2008)	CP000061 (Bai *et al.*, 2006)	AP006628 (Oshima *et al.*, 2004)	CU469464 (Kube *et al.*, 2008)	CP097583 (Debonneville *et al.*, 2022)

CDS, coding DNA sequences.

Nevertheless, most genes involved in the major metabolic pathways were common across phytoplasmas. Interestingly, the *sucP* gene, which encodes sucrose phosphorylase, is absent in several phytoplasma genomes. However, it is present as a pseudogene in '*Ca.* P. solani' SA-1, and is also found in a similar form in '*Ca.* P. asteris' OY-M strain, while the *sucP* gene in '*Ca.* P. australiense' encodes the complete SucP protein (Šeruga Musić *et al.*, 2019). It has been suggested that the problem of entry into glycolysis is circumvented through the use of phosphorylated hexose (Kube *et al.*, 2012). The presence of alternative pathways in '*Ca.* P. solani' SA-1 was supported by the discovery of several glycolytic genes involved in initial stages of glycolysis as well as those related to the malate pathway (Šeruga Musić *et al.*, 2019).

In addition, two duplications were found in this draft genome, comprising mainly metabolic genes, as previously found in the '*Ca.* P. asteris' OY-W strain. Genes involved in replication, DNA repair, and modification were similar to those found in other phytoplasma genomes, including the excision repair complex *uvrABC*, five subunits of DNA polymerase, and *ruvX*, but lacking *ruvA* and *ruvB* genes. Interestingly, the '*Ca.* P. solani' SA-1 genome was found to contain a *recA* gene, which is absent in most sequenced phytoplasma genomes. Finally, '*Ca.* P. solani' SA-1 possesses all genes of the Sec-dependent secretion system, which encode the protein export machinery and are essential for the secretion of effectors that interact with host plants (Šeruga Musić *et al.*, 2019).

Transposon-like genetic elements, designated as potential mobile units (PMUs), were first described for the '*Ca.* P. asteris' strain AY-WB (Bai *et al.*, 2006). These putative replicative elements, measuring up to 20 kbp in size, contained genes involved in DNA replication and transposition, such as *dnaB*, *dnaG*, *ssb*, *hflB*, *himS*, *fliA*, *tmk*, *dam*, *uvrB*, and *tra5*. They also frequently encoded membrane-bound or secreted proteins such as effectors. In the '*Ca.* P. solani' SA-1 genome, several PMU or PMU-like elements were found and described, providing further evidence that these genomic regions are associated with genome instability and plasticity. The most complete PMU element was about 18 kbp in size, while others ranged from 10.3 to 14.3 kbp. Degenerated or truncated elements, which are 2–3 kbp in size were also observed. As these repetitive regions are difficult to assemble and are related to the unsuccessful completion of the '*Ca.* P. solani' SA-1 genome, the number of PMU-like elements could be even larger. However, the PMU-like elements found in '*Ca.* P. solani' exhibited a very high degree of heterogeneity and mosaicism in terms of their organization and content. Moreover, the molecular phylogeny of PMU-related genes revealed intermixing, large rearrangements, and horizontal gene transfer of PMUs from both closely related phytoplasmas such as '*Ca.* P. australiense' or the more distant '*Ca.* P. asteris' and even '*Ca.* P. mali.' Some of the PMU-associated genes, in particular several of the annotated *tra5* genes, were also found to be unique to '*Ca.* P. solani' strains, as revealed by a BLAST (basic local alignment search tool) search of the corresponding amino acid and nucleotide sequences. Of particular interest was the discovery of previously characterized domains found in alternative sigma factors involved in gene regulation in other pathogenic bacteria. All of these features were found in PMU-like regions of the '*Ca.* P. solani' SA-1 genome, giving this phytoplasma an evolutionary advantage and contributing to its high dynamics that allow successful adaptation and infection of various plant and insect hosts (Šeruga Musić *et al.*, 2019).

New OMICs Approaches Illuminate '*Candidatus* Phytoplasma solani' Infection in Plant Hosts

Advances in molecular biology, particularly high-throughput techniques, have elucidated the '*Ca.* P. solani' infection of host plants (Dermastia, 2017; Dermastia *et al.*, 2019). While in-depth studies of known phytoplasma genome sequences, including those of '*Ca.* P. solani' (Mitrović *et al.*, 2014; Šeruga Musić *et al.*, 2019), have shed light on the broad spectrum of possible interactions with host plants, many of these have been clarified by functional analyses of transcriptomic data (Albertazzi *et al.*, 2009; Hren *et al.*, 2009; Dermastia *et al.*, 2021) and metabolomic data (Prezelj *et al.*, 2016) on host plants infected with '*Ca.* P. solani.'

The collective findings from these studies on '*Ca.* P. solani' indicate that it affects both the primary and secondary metabolism of host plants by influencing hormonal, nutritional, and stress signaling pathways (Dermastia, 2017, 2019). These processes are intricately regulated by various factors, including transcription factors and small RNAs such as microRNAs and small-interfering RNAs (phasiRNAs) (Dermastia *et al.*, 2021). Transcriptomic analyses revealed dynamic responses in infected grapevines during the early growth phase, indicating significant transcriptional reprogramming even before the onset of symptoms. Interestingly, this reprogramming was not reflected in the levels of small RNAs during the initial phase of infection. However, as the disease progressed, the number of differentially expressed small RNAs increased significantly, particularly those associated with biotic stress processes. These results emphasize the importance of considering both mRNA and sRNA dynamics to understand grapevine responses to '*Ca.* P. solani' infection (Dermastia *et al.*, 2021).

In view of the close relationship between photosynthesis and the development of symptoms later in the growing season, particular attention was paid to this aspect when analyzing plants infected with phytoplasmas. Transcriptomic analysis of leaf veins, in which phytoplasmas thrive, from symptomatic '*Ca.* P. solani'-infected grapevines showed the suppression of several photosynthesis-related genes (Albertazzi *et al.*, 2009; Hren *et al.*, 2009). The genes coding for chlorophyll a/b-binding proteins in both photosystem I and photosystem II are significantly downregulated (Hren *et al.*, 2009). This suppression also extends to the *rbcL* gene, indicating the loss of the ribulose bisphosphate carboxylase/oxygenase (Rubisco) large subunit (Bertamini *et al.*, 2002). In addition, downregulation of the genes for the cytochrome b6f complex and ATP synthase was observed (Hren *et al.*, 2009). These molecular changes corresponded with a decrease in chlorophyll, carotenoids, and proteins, as well as activities of Rubisco and nitrate reductase (Bertamini and Nedunchezhian, 2001a, 2001b; Bertamini *et al.*, 2002; Endeshaw *et al.*, 2012; Rusjan *et al.*, 2012a). The upregulation of phasiRNAs in '*Ca.* P. solani'-infected grapevines correlated with genes involved in photosystem II and the Calvin cycle, which is associated with a significant reduction in photosynthesis-related gene transcripts (Dermastia *et al.*, 2021, 2023). As symptoms progressed, significant changes in secondary metabolism also became evident. In agreement with previous observations on genes involved in flavonoid synthesis (Rusjan *et al.*, 2012b; Rotter *et al.*, 2018), similar trends were observed in cv. Zweigelt (Dermastia *et al.*, 2021). Comparison of responses between early and late growing seasons revealed that anthocyanins, chalcones, and dihydroflavonols were more strongly upregulated in the early growing season, whereas isoprenoids were predominantly upregulated in the late growing season (Dermastia *et al.*, 2021).

Phytoplasma infection in grapevine leads to changes in glycolysis and oxidative stress pathways, resulting in upregulation of aldolase genes (Dermastia *et al.*, 2021). These genes facilitate the reversible conversion of fructose 1,6-bisphosphate to dihydroxyacetone phosphate and glyceraldehyde-3-phosphate, consistent with the observed increase in enzyme activity. Interestingly, transcription of a gene encoding dehydroascorbate reductase was only slightly increased in the late growing season in response to infection, while its enzymatic activity was significantly higher at both the beginning and end of sampling, suggesting possible post-translational regulation. In addition, the expression of seven ascorbate peroxidase genes was found to be differentially regulated in response to infection (Dermastia *et al.*, 2021).

In the context of other plant–pathogen interactions, increased activities of carbohydrate transporters and sucrolytic enzymes, coupled with increased respiratory metabolism, play a crucial role in establishing an energetically favorable state for plant defense (Dermastia, 2017; Dermastia *et al.*, 2019). This induction of the energy-associated network is manifested by elevated levels of fructose 6-phosphate and various metabolites of the tricarboxylic acid cycle, such as malate and citrate, in grapevines infected with '*Ca.* P. solani' (Prezelj *et al.*, 2016; Dermastia, 2017). Notably, phytoplasmas lack enzymes of the membrane-bound phosphoenolpyruvate-dependent phosphotransferase system, which is commonly used by bacteria for efficient sugar import and phosphorylation. However, they possess genes required for the initial steps of the glycolytic pathway (Kube *et al.*, 2012; Mitrović *et al.*, 2014; Šeruga Musić *et al.*, 2019), suggesting a mechanism by which phytoplasmas utilize phosphorylated hexoses from the host. The availability of fructose 6-phosphate, possibly derived from the host, could serve as a key substrate for phytoplasma glycolysis. Furthermore, the observed increase in malate concentrations in infected grapevines could contribute to an alternative energy production pathway of glycolysis, possibly involving malate uptake by specific transporters encoded in the phytoplasma genome (Dermastia *et al.*, 2021; Prezelj *et al.*, 2016).

There is increasing evidence that the feedback inhibition of photosynthesis causes leaf chlorosis due to the accumulation of carbohydrates in the source leaves, possibly influenced by phytoplasmas (Dermastia, 2017; Dermastia *et al.*, 2019). This transition to carbohydrate sinks in the grapevine infected with '*Ca.* P. solani' is characterized by increased invertase activity, although transcript levels do not always match enzyme activity (Hren *et al.*, 2009; Covington Dunn *et al.*, 2016). In addition, the differential expression of sucrose transporter genes such as *SUC27*, *SUC11*, and *SUC12* during infection suggests that they are involved in facilitating the transition from source to sink in the leaf phloem (Santi *et al.*, 2013a, 2013b). Sucrose synthase is involved in carbohydrate accumulation and phloem obstruction and shows increased expression during infection of the grapevine with '*Ca.* P. solani' (Hren *et al.*, 2009; Santi *et al.*, 2013a, 2013b). In addition, genes associated with callose synthase and β-1-3-glucanase show increased transcript levels, which supports callose deposition and degradation (Hren *et al.*, 2009; Santi *et al.*, 2013a, 2013b; Dermastia *et al.*, 2015). There is an increased transcription of *AGPL*, which encodes the large subunit of ADP-glucose pyrophosphorylase, a key enzyme in starch biosynthesis (Hren *et al.*, 2009; Dermastia *et al.*, 2015; Rotter *et al.*, 2018).

In grapevines infected with '*Ca.* P. solani,' the importance of hormonal regulation in phytoplasma diseases was reconfirmed (Dermastia, 2019). Transcriptomic analyses revealed a remarkable upregulation of genes related to salicylic acid biosynthesis and signaling, especially in leaf vein-enriched samples and whole grapevine leaves infected with '*Ca.* P. solani.' This included the induction of transcripts encoding *S*-adenosyl-L-methionine: salicylic acid carboxyl methyltransferase, which is responsible for the formation of methyl salicylate. In addition, genes encoding pathogenesis-related proteins (PR-1, PR-2, PR-5) regulated by salicylic acid and the *DMR6* gene associated with systemic acquired resistance are upregulated (Dermastia, 2017; Dermastia *et al.*, 2019). Studies suggest that '*Ca.* P. solani' infection induces salicylic acid-dependent systemic acquired resistance in tomato and grapevine and delays the multiplication of phytoplasmas (Dermastia, 2017; Rotter *et al.*, 2018; Dermastia *et al.*, 2019). Although plants respond to phytoplasmas via salicylic acid-mediated signaling, these responses do not confer resistance to the disease (Paolacci *et al.*, 2017; Dermastia *et al.*, 2019). Furthermore, in grapevines interacting with '*Ca.* P. solani,' activation of the salicylic acid signaling pathway suppresses JA defense responses. However, in plants recovering from bois noir disease caused by this phytoplasma, the JA signaling pathway is activated, possibly preventing further disease development (Paolacci *et al.*, 2017). Changes in miRNAs and phasiRNAs targeting hormone metabolism genes in grapevine infected with '*Ca.* P. solani' suggest intricate regulatory networks that modulate hormone responses during phytoplasma infection (Dermastia *et al.*, 2021). These findings emphasize the complex interplay between phytoplasma infection and hormonal regulation and provide insights into potential targets for disease control strategies.

Under stress conditions, plants undergo metabolic changes that lead to the accumulation of various metabolites, especially amino acids. In grapevine vein samples, infection with '*Ca.* P. solani' leads to increased levels of various amino acids, including serine, glycine, valine, leucine, alanine, β-alanine, threonine, aspartate, pyroglutamate, and proline (Prezelj *et al.*, 2016). Proline, in particular, plays a central role in plant stress responses, acting as a metal chelator, antioxidant, and signaling molecule, thus contributing to stress mitigation and recovery (Hayat *et al.*, 2012). The substantial sixfold increase in proline levels observed in '*Ca.* P. solani'-infected plants aligns with the proposed role of proline in stress attenuation. Moreover, the upregulation of a gene encoding cysteine synthase in '*Ca.* P. solani'-infected cv. Chardonnay was further indicative of the plant's response to stress (Albertazzi *et al.*, 2009; Hren *et al.*, 2009).

In addition to the general omics approaches, the researchers constructed a statistical model of the disease triangle for bois noir disease in grapevine caused by '*Ca.* P. solani' using collected datasets (Rotter *et al.*, 2018). This model integrates the basic concepts of the disease triangle from plant pathology to explain the interactions between the host plant, the pathogen, and the environment. By using original experimental data, a statistical proof-of-concept model was developed using various statistical and data mining techniques. The focus was on '*Ca.* P. solani' infection in cv. Chardonnay grapevines within a given system. Individual plants in a vineyard were observed over a period of 6 years. The phytoplasma content, the expression patterns of 21 selected grapevine genes, and the prevailing environmental conditions were recorded, all of which correlated with disease severity. The developed statistical model predicts that disease severity depends on the expression of the grapevine gene *DMR6*, which plays a crucial role in mediating salicylic acid homeostasis and systemic acquired resistance (Dermastia, 2017; Dermastia *et al.*, 2019), as well as the amount of summer precipitation and the frequency of '*Ca.* P. solani.' Furthermore, the integration of heterogeneous network analyses with biological datasets (Dermastia *et al.*, 2021; Škrlj *et al.*, 2021) has revealed the diversity within biological networks and has assisted in reconstructing networks from RNA-seq data. Within this framework, the concept of community dissipation was introduced, measuring the temporal dispersion of gene communities. These innovative approaches have confirmed early hypotheses about how '*Ca.* P. solani' causes disease and have led to the identification of new genes – for instance those encoding expansins and cryptdin – that were not previously associated with '*Ca.* P. solani' pathogenicity (Dermastia *et al.*, 2021).

Pathogenicity Factors/Effector Proteins of '*Candidatus* Phytoplasma solani' and Their Impact on Host Plant Metabolism

The pathogenicity of intracellular plant pathogenic bacteria relies on specific factors known as effectors. In contrast to typical Gram-negative bacterial pathogens, phytoplasmas do not possess the type III secretion system and instead utilize the Sec-dependent pathway for protein export through membranes. These proteins are secreted via this system directly into the cytoplasm of the plant host's phloem sieve cells, where they act as pathogenicity factors or effectors that target key host proteins to influence cellular processes (Bai *et al.*, 2009; Hoshi *et al.*, 2009; Huang *et al.*, 2021; MacLean *et al.*, 2014). Identified effectors in aster yellows witches broom (AY-WB) phytoplasma, such as SAP05, SAP11, and SAP54, interact with transcription factors to induce symptoms such as witches'-broom or leaf-like flowers (Tomkins *et al.*, 2018; Yan *et al.*, 2024). Recent studies on SAP05 have revealed the first phytoplasma effector crystal structure, showing its interaction with transcription factors SPL and GATA in a ubiquitin-independent manner by hijacking the 26S proteasome subunit RPN10 of the plant host, thereby greatly altering host plant architecture and reproduction (Huang *et al.*, 2021; Yan *et al.*, 2024). TENGU, another effector identified in the OY-M strain, impacts auxin and JA signaling pathways (Minato *et al.*, 2014).

While much attention has focused on the effectors of '*Ca.* P. asteris,' similar effectors have also been identified in other phytoplasma genomes (Lu *et al.*, 2014; Siewert *et al.*, 2014; Anabestani *et al.*, 2017; Janik *et al.*, 2017; Pecher *et al.*, 2019; Zhou *et al.*, 2021; Bai *et al.*, 2022, 2023). However, the molecular mechanisms underlying many effectors and their targets remain elusive. Recent research suggests that phytoplasmas may employ unique infection strategies and that their effectors have unusual targets (Marrero *et al.*, 2023). In addition, membrane-associated proteins of phytoplasmas, such as the major antigenic membrane protein, may play a role in recognition or interaction with the insect vectors (Suzuki *et al.*, 2006; Galetto *et al.*, 2011), although their specific functions in plant cells are not known. These uncertainties extend to the mechanisms of effectors of '*Ca.* P. solani' as well. However, the combined results of earlier genomic and transcriptomic studies on plants infected with '*Ca.* P. solani' are highly complementary to recent effector studies on this phytoplasma.

In a 2019 study by the Šeruga Musić team, 38 putative genes for secreted effectors were identified in the genome of '*Ca.* P. solani' strain SA-1, 20 of which are located within PMU-like regions. Among the identified putative effectors, five were unique to the SA-1 strain with no similar nucleotide or protein sequences in other phytoplasma genomes. Further comparative analyses revealed that three putative secreted protein sequences (PSSA1 v1c1220, PSSA1 v1c4850, and PSSA1 v1c6880) were shared with '*Ca.* P. solani' strains 284/09 and 231/09 and two predicted putative secreted proteins were identical or similar to proteins secreted by '*Ca.* P. asteris' strain AY-WB (i.e. SAP11 and SAP21, respectively) (Šeruga Musić *et al.*, 2019).

In another study conducted by the Brader research group (Dermastia *et al.*, 2023), six potential pathogenicity factors/effectors (PoStoSP04, PoStoSP06, PoStoSP13, PoStoSP14, PoStoSP18, and PoStoSP28) of '*Ca.* P. solani' were revealed from the grapevine cv. Zweigelt, infected with a '*Ca.* P. solani' genotype CPsM4_At6 (Aryan *et al.*, 2014). Two of them have been previously annotated as functional genes – one (PoStoSP18) as a maltose-binding protein (malE) within the ABC sugar transport system, and the other (PoStoSP28) is 91–100% identical to an antigenic membrane protein StAMP from '*Ca.* P. solani' (Fabre *et al.*, 2011). The identification of possible species-specific and even strain-specific putative effectors, together with their presence in unstable genomic regions such as PMUs, indicates their importance in pathogenicity and their role in the maintenance of the cosmopolitan lifestyle of this phytoplasma.

The effects of '*Ca.* P. solani' effector candidates on the host cellular machinery were investigated through the transient transformation of *Nicotiana benthamiana* leaves using agrobacterium-mediated methods (Dermastia *et al.*, 2023). Some of the tested proteins localized to the nucleus and cytosol of *N. benthamiana* cells. Their involvement in carbohydrate metabolism was demonstrated, particularly in sugar phosphorylation pathways and through the confirmed interaction of PoStoSP28 (StAMP) with grapevine phosphoglucomutase enzymes. Dermastia *et al.* (2023) revealed, for the first time, the potential pathway used by phytoplasmas to import sugars and generate hexose-6-phosphate to support glycolysis (Tan *et al.*, 2021). This finding is significant as phytoplasmas lack the hexokinase and sugar-specific phosphotransferase system required for the entry of phosphorylated hexose into glycolysis (Kube *et al.*, 2012), which is the first step in glucose degradation to generate ATP.

When plants face stress or infections that overwhelm their defense systems, the balance between oxidants and antioxidants is disrupted, leading to excess reactive oxygen species (ROS) in their cells (Sies *et al.*, 2017). To counteract these metabolic disturbances, plants rely on robust ROS elimination systems. However, as ROS also plays a role in plant defense, it would be beneficial for pathogens to neutralize the effects of ROS. The ascorbate-glutathione cycle is an important antioxidant system for ROS elimination and the maintenance of cellular balance. Importantly, induction of the complete ascorbate-glutathione cycle in '*Ca.* P. solani'-infected grapevines late in the growing season positively correlates with increased enzyme activities of key antioxidants in the leaves of *N. benthamiana* after transient transformation with different effector candidates (Dermastia *et al.*, 2023).

Co-localization of '*Ca.* P. solani' PoStoSP28 (StAMP) with autophagosomes after transient transformation of *N. benthamiana* leaves (Dermastia *et al.*, 2023) is consistent with autophagosome-like structures in mesophyll tomatoes infected with purple top phytoplasma (Wei *et al.*, 2022; Junichi *et al.*, 2023). However, the role of these autophagosome-like structures in phytoplasma infection remains to be elucidated.

Conclusion

The rapid advancement of methods and data management has uncovered a plethora of previously undiscovered signaling pathways and molecules involved in the interaction of '*Ca.* P. solani' with its host plants. While the functions of many of these molecules remain elusive or have been inferred from comparisons with existing studies on other phytoplasmas, the point has now been reached where the intricate mechanisms underlying '*Ca.* P. solani' plant interactions can be unraveled. This newfound understanding promises to reveal new avenues for additional disease management strategies centerd on these key pathways and molecules.

Acknowledgments

The authors would like to thank numerous researchers whose contributions have been invaluable in collecting information on phytoplasmas especially '*Candidatus* Phytoplasma solani,' in order to draw possible conclusions regarding their interactions with host plants. This study incorporates data from projects P4-0165 J4-6459, J4-0813, J4-2544, J1-7151, J4-N2-0078, J1-60018, P4-0165, and P2-0103, funded by the Slovenian Agency for Research and Innovation (ARIS). In addition, it was funded by Croatian Science Foundation grants UIP-2014-09-9744 and IP 2019-04-2469.

References

Albertazzi, G., Milc, J., Caffagni, A., Francia, E., Roncaglia, E. *et al.* (2009) Gene expression in grapevine cultivars in response to bois noir phytoplasma infection. *Plant Science* 176, 792–804. DOI: 10.1016/j.plantsci.2009.03.001.

Anabestani, A., Izadpanah, K., Abbà, S., Galetto, L., Ghorbani, A. *et al.* (2017) Identification of putative effector genes and their transcripts in three strains related to '*Candidatus* Phytoplasma aurantifolia.' *Microbiological Research* 199, 57–66. DOI: 10.1016/j.micres.2017.03.001.

Aryan, A., Brader, G., Mörtel, J., Pastar, M., and Riedle-Bauer, M. (2014) An abundant '*Candidatus* Phytoplasma solani' tuf b strain is associated with grapevine, stinging nettle and Hyalesthes obsoletus. *European Journal of Plant Pathology* 140, 213–227. DOI: 10.1007/s10658-014-0455-0.

Bai, B., Zhang, G., Li, Y., Wang, Y., Sujata, S. *et al.* (2022) The "Candidatus Phytoplasma tritici" effector SWP12 degrades the transcription factor TaWRKY74 to suppress wheat resistance. *Plant Journal* 112, 1473–1488. DOI: 10.1111/tpj.16029.

Bai, B., Zhang, G., Pei, B., Song, Q., Hao, X. *et al.* (2023) The function of the phytoplasma effector SWP12 depends on the properties of two key amino acids. *Journal of Biological Chemistry* 299, 103052. DOI: 10.1016/J.JBC.2023.103052.

Bai, X., Zhang, J., Ewing, A., Miller, S.A., Jancso Radek, A. *et al.* (2006) Living with genome instability: The adaptation of phytoplasmas to diverse environments of their insect and plant hosts. *Journal of Bacteriology* 188, 3682–96. DOI: 10.1128/JB.188.10.3682-3696.2006.

Bai, X., Correa, V.R., Toruño, T.Y., Ammar, E.-D., Kamoun, S., and Hogenhout, S.A. (2009) AY-WB phytoplasma secretes a protein that targets plant cell nuclei. *Molecular Plant-Microbe Interactions* 22, 18–30. DOI: 10.1094/MPMI.

Bertamini, M. and Nedunchezhian, N. (2001a) Decline of photosynthetic pigments, ribulose-1,5-bisphosphate carboxylase and soluble protein contents, nitrate reductase and photosynthetic activities, and changes in tylakoid membrane protein pattern in canopy shade grapevine (*Vitis vinifera* L.). *Photosynthetica* 39, 529–537.

Bertamini, M. and Nedunchezhian, N. (2001b) Effects of phytoplasma [stolbur-subgroup (Bois noir-BN)] on photosynthetic pigments, saccharides, ribulose 1,5-bisphosphate carboxylase, nitrate and nitrite reductases, and photosynthetic activities in field-grown grapevine (*Vitis vinifera* L.) cv. *Photosynthetica* 39, 119–122. DOI: 10.1023/A:1012412406727.

Bertamini, M., Nedunchezhian, N., Tomasi, F., and Grando, M. (2002) Phytoplasma [Stolbur-subgroup (Bois Noir-BN)] infection inhibits photosynthetic pigments, ribulose-1,5-bisphosphate carboxylase and photosynthetic activities in field grown grapevine (*Vitis vinifera* L. cv. Chardonnay) leaves. *Physiological and Molecular Plant Pathology* 61, 357–366. DOI: 10.1006/pmpp.2003.0449.

Covington Dunn, E., Roitsch, T., Dermastia, M., Covington, E.D., Roitsch, T., and Dermastia, M. (2016) Determination of the activity signature of key carbohydrate metabolism enzymes in phenolic-rich grapevine tissues. *Acta Chimica Slovenica* 63, 757–762. DOI: 10.17344/acsi.2016.2484.

Debonneville, C., Mandelli, L., Brodard, J., Groux, R., Roquis, D., and Schumpp, O. (2022) The complete genome of the "flavescence dorée" phytoplasma reveals characteristics of low genome plasticity. *Biology* 11, 953. DOI: 10.3390/BIOLOGY11070953.

Dermastia, M. (2017) Interactions between grapevines and grapevine yellows phytoplasmas BN and FD. In: Dermastia, M., Bertaccini, A., Constable, F., and Mehle, N. (eds) *Grapevine Yellows Diseases and*

Their Phytoplasma Agents. Springer Briefs in Agriculture. Springer, Cham, Switzerland, pp. 47–67. DOI: 10.1007/978-3-319-50648-7_3.

Dermastia, M. (2019) Plant hormones in phytoplasma infected plants. *Frontiers in Plant Science* 10, 477. DOI: 10.3389/fpls.2019.00477.

Dermastia, M., Nikolic, P., Chersicola, M., and Gruden, K. (2015) Transcriptional profiling in infected and recovered grapevine plant responses to '*Candidatus* Phytoplasma solani'. *Phytopathogenic Mollicutes* 5, S123. DOI: 10.5958/2249-4677.2015.00053.5.

Dermastia, M., Kube, M., and Šeruga-Musić, M. (2019) Transcriptomic and proteomic studies of phytoplasma-infected plants. In: Bertaccini, A., Oshima, K., Kube, M., and Rao, G.P. (eds) *Phytoplasmas: Plant Pathogenic Bacteria – III*. Springer, Singapore, pp. 35–55. DOI: 10.1007/978-981-13-9632-8_3.

Dermastia, M., Škrlj, B., Strah, R., Anžič, B., Tomaž, Š. *et al.* (2021) Differential response of grapevine to infection with "*Candidatus* Phytoplasma solani" in early and late growing season through complex regulation of mRNA and small RNA transcriptomes. *International Journal of Molecular Science* 22, 3531. DOI: 10.3390/IJMS22073531.

Dermastia, M., Tomaž, Š., Strah, R., Lukan, T., Coll, A. *et al.* (2023). Candidate pathogenicity factor/effector proteins of 'Candidatus Phytoplasma solani' modulate plant carbohydrate metabolism, accelerate the ascorbate–glutathione cycle, and induce autophagosomes. *Frontiers in Plant Science* 14, 1232367. DOI: 10.3389/FPLS.2023.1232367/BIBTEX.

Endeshaw, S.T., Murolo, S., Romanazzi, G., and Neri, D. (2012) Effects of Bois noir on carbon assimilation, transpiration, stomatal conductance of leaves and yield of grapevine (Vitis vinifera) cv. Chardonnay. *Physiologica Plantarum* 145, 286–295. DOI: 10.1111/j.1399-3054.2012.01576.x.

Fabre, A., Danet, J.-L., and Foissac, X. (2011) The stolbur phytoplasma antigenic membrane protein gene *stamp* is submitted to diversifying positive selection. *Gene* 472, 37–41. DOI:10.1016/j.gene.2010.10.012.

Galetto, L., Bosco, D., Balestrini, R., Genre, A., Fletcher, J., and Marzach, C. (2011) The major antigenic membrane protein of "Candidatus Phytoplasma asteris" selectively interacts with ATP synthase and actin of leafhopper vectors. *PLoS One* 6, e22571. DOI:10.1371/journal.pone.0022571.

Hayat, S., Hayat, Q., Alyemeni, M.N., Wani, A.S., Pichtel, J., and Ahmad, A. (2012) Role of proline under changing environments: A review. *Plant Signaling and Behavior* 7, 1456–66. DOI: 10.4161/psb.21949.

Hoshi, A., Oshima, K., Kakizawa, S., Ishii, Y., Ozeki, J., and Hashimoto, M. (2009) A unique virulence factor for proliferation and dwarfism in plants identified from a phytopathogenic bacterium. *Proceedings of the National Academy of Science of the USA* 106, 6416–6421. DOI: 10.1073/pnas.0813038106.

Hren, M., Nikolić, P., Rotter, A., Blejec, A., Terrier, N. *et al.* (2009) "Bois noir" phytoplasma induces significant reprogramming of the leaf transcriptome in the field grown grapevine. *BMC Genomics* 10, 460. DOI:10.1186/1471-2164-10-460.

Huang, W., MacLean, A.M., Sugio, A., Maqbool, A., Busscher, M. *et al.* (2021) Parasitic modulation of host development by ubiquitin-independent protein degradation. *Cell* 184, 5201–5214.e12. DOI: 10.1016/J.CELL.2021.08.029.

Janik, K., Mithöfer, A., Raffeiner, M., Stellmach, H., Hause, B., and Schlink, K. (2017) An effector of apple proliferation phytoplasma targets TCP transcription factors-a generalized virulence strategy of phytoplasma? *Molecular Plant Pathology* 18, 435–442. DOI: 10.1111/mpp.12409.

Junichi, I., Bo Min, K., Yan, Z., and Wei, W. (2023) Phytoplasma infection alters polar lipid composition and triggers chloroplast autophagy in host plants. *Phytopathogenic Mollicutes* 13, 3–4. DOI: 10 5958/2249-4677.2023.00002.6.

Kube, M., Schneider, B., Kuhl, H., Dandekar, T., Heitmann, K. *et al.* (2008) The linear chromosome of the plant-pathogenic mycoplasma "*Candidatus* Phytoplasma mali". *BMC Genomics* 9, 1–14. DOI: 10.1186/1471-2164-9-306.

Kube, M., Mitrovic, J., Duduk, B., Rabus, R., and Seemüller, E. (2012) Current view on phytoplasma genomes and encoded metabolism. *Scientific World Journal* 2012, 1–25. DOI: 10.1100/2012/185942.

Lu, Y.-T., Li, M.-Y., Cheng, K.-T., Tan, C.M., Su, L.-W. *et al.* (2014) Transgenic plants that express the phytoplasma effector SAP11 show altered phosphate starvation and defense responses. *Plant Physiology* 164, 1456–69. DOI: 10.1104/pp.113.229740.

MacLean, A.M., Orlovskis, Z., Kowitwanich, K., Zdziarska, A.M., Angenent, G.C. *et al.* (2014) Phytoplasma effector SAP54 hijacks plant reproduction by degrading MADS-box proteins and promotes insect colonization in a RAD23-dependent manner. *PLoS Biology* 12, e1001835. DOI: 10.1371/journal.pbio.1001835.

Marcone, C., Neimark, H., Ragozzino, A., Lauer, U., and Seemüller, E. (1999) Chromosome sizes of phytoplasmas composing major phylogenetic groups and subgroups. *Phytopathology* 89, 805.

Marrero, M.C., Capdevielle, S., Huang, W., Al-Subhi, A.M., Busscher, M. *et al.* (2023) Protein interaction mapping reveals widespread targeting of development-related host transcription factors by phytoplasma effectors. *Plant Journal* 117, 1281–1297. DOI: 10.1111/TPJ.16546.

Minato, N., Himeno, M., Hoshi, A., Maejima, K., Komatsu, K. *et al.* (2014) The phytoplasmal virulence factor TENGU causes plant sterility by downregulating of the jasmonic acid and auxin pathways. *Scientific Reports* 4, 7399. DOI: 10.1038/srep07399.

Mitrović, J., Siewert, C., Duduk, B., Hecht, J., Molling, K. *et al.* (2014) Generation and analysis of draft sequences of "stolbur" phytoplasma from multiple displacement amplification templates. *Journal of Molecular Microbiology and Biotechnology* 24, 1–11. DOI: 10.1159/000353904.

Namba, S. (2010) Molecular and biological properties of phytoplasmas. *Proceedings of the Japan Academy, Series B, Physical and Biological Sciences* 95, 401–418. DOI: 10.2183/PJAB.95.028.

Oshima, K., Kakizawa, S., Nishigawa, H., Jung, H.Y., Wei, W. *et al.* (2004) Reductive evolution suggested from the complete genome sequence of a plant-pathogenic phytoplasma. *Nature Genetics* 36, 27–29. DOI: 10.1038/ng1277.

Paolacci, A.R., Catarcione, G., Ederli, L., Zadra, C., Pasqualini, S. *et al.* (2017) Jasmonate-mediated defence responses, unlike salicylate-mediated responses, are involved in the recovery of grapevine from bois noir disease. *BMC Plant Biology* 17, 118. DOI: 10.1186/s12870-017-1069-4.

Pecher, P., Moro, G., Canale, M.C., Capdevielle, S., Singh, A. *et al.* (2019) Phytoplasma SAP11 effector destabilization of TCP transcription factors differentially impact development and defence of Arabidopsis versus maize. *PLoS Pathogens* 15, e1008035. DOI: 10.1371/journal.ppat.1008035.

Prezelj, N., Fragener, L., Weckwerth, W., and Dermastia, M. (2016) Metabolome of grapevine leaf vein-enriched tissue infected with '*Candidatus* Phytoplasma solani'. *Mitteilungen Klosterneuburg Rebe und Wein, Obstbau und Früchteverwertung* 66, 74–78.

Quaglino, F., Zhao, Y., Casati, P., Bulgari, D., Bianco, P.A. *et al.* (2013). '*Candidatus* Phytoplasma solani', a novel taxon associated with stolbur- and bois noir-related diseases of plants. *International Journal of Systematic and Evolutionary Microbiology* 63, 2879–94. DOI: 10.1099/ijs.0.044750-0.

Quaglino, F., Sanna, F., Moussa, A., Faccincani, M., Passera, A. *et al.* (2019) Identification and ecology of alternative insect vectors of 'Candidatus Phytoplasma solani' to grapevine. *Scientific Reports* 9, 1–11. DOI: 10.1038/s41598-019-56076-9.

Razin, S., Yogev, D., and Naot, Y. (1998) Molecular biology and pathogenicity of mycoplasmas. *Microbiology and Molecular Biology Reviews* 62, 1094–1156. DOI: 10.1128/MMBR.62.4.1094-1156.1998.

Rotter, A., Nikolić, P., Turnšek, N., Kogovšek, P., Blejec, A. *et al.* (2018) Statistical modeling of long-term grapevine response to '*Candidatus* Phytoplasma solani' infection in the field. *European Journal of Plant Pathology* 150, 653–668. DOI: 10.1007/s10658-017-1310-x.

Rusjan, D., Halbwirth, H., Stich, K., Mikulič-Petkovšek, M., and Veberič, R. (2012a) Biochemical response of grapevine variety "Chardonnay" (Vitis vinifera L.) to infection with grapevine yellows (Bois noir). *European Journal of Plant Pathology* 134, 231–237. DOI: 10.1007/s10658-012-9988-2.

Rusjan, D., Veberič, R., and Mikulič-Petkovšek, M. (2012b) The response of phenolic compounds in grapes of the variety "Chardonnay" (*Vitis vinifera* L.) to the infection by phytoplasma Bois noir. *European Journal of Plant Pathology* 133, 965–974. DOI: 10.1007/s10658-012-9967-7.

Santi, S., De Marco, F., Polizzotto, R., Grisan, S., and Musetti, R. (2013a) Recovery from stolbur disease in grapevine involves changes in sugar transport and metabolism. *Frontiers in Plant Science* 4, 171. DOI: 10.3389/fpls.2013.00171.

Santi, S., Grisan, S., Pierasco, A., De Marco, F., and Musetti, R. (2013b) Laser microdissection of grapevine leaf phloem infected by stolbur reveals site-specific gene responses associated to sucrose transport and metabolism. *Plant, Cell and Environment* 36, 343–355. DOI: 10.1111/j.1365-3040.2012.02577.x.

Šeruga Musić, M., Samarzija, I., Hogenhout, S.A., Haryono, M., Cho, S.T., and Kuo, C.H. (2019) The genome of "Candidatus Phytoplasma solani" strain SA-1 is highly dynamic and prone to adopting foreign sequences. *Systematic and Applied Microbiology* 42, 117–127. DOI: 10.1016/J.SYAPM.2018.10.008.

Sies, H., Berndt, C., and Jones, D.P. (2017) Oxidative stress. *Annual Review of Biochemistry* 86, 715–748. DOI: 10.1146/ANNUREV-BIOCHEM-061516-045037.

Siewert, C., Luge, T., Duduk, B., Seemüller, E., Büttner, C. *et al.* (2014) Analysis of expressed genes of the bacterium "Candidatus phytoplasma mali" highlights key features of virulence and metabolism. *PLoS One* 9, e94391. DOI:10.1371/journal.pone.0094391.

Škrlj, B., Novak, M.P., Brader, G., Anžič, B., Ramšak, Ž. *et al.* (2021) New cross-talks between pathways involved in grapevine infection with "*Candidatus* Phytoplasma solani" revealed by temporal network modelling. *Plants (Basel)* 10, 646. DOI: 10.3390/PLANTS10040646.

Suzuki, S., Oshima, K., Kakizawa, S., Arashida, R., Jung, H.-Y. *et al.* (2006) Interaction between the membrane protein of a pathogen and insect microfilament complex determines insect-vector specificity. Proceedings of the National Academy of Science of the USA 103, 4252–4257.

Tan, Y., Li, Q., Zhao, Y., Wei, H., Wang, J. *et al.* (2021) Integration of metabolomics and existing omics data reveals new insights into phytoplasma-induced metabolic reprogramming in host plants. *PLoS One* 16, e0246203. DOI: 10.1371/journal.pone.0246203.

Tomkins, M., Kliot, A., Marée, A.F., and Hogenhout, S.A. (2018) A multi-layered mechanistic modelling approach to understand how effector genes extend beyond phytoplasma to modulate plant hosts, insect vectors and the environment. *Current Opinion in Plant Biology* 44, 39–48. DOI: 10.1016/J.PBI.2018.02.002.

Tran-Nguyen, L.T.T., Kube, M., Schneider, B., Reinhardt, R., and Gibb, K.S. (2008) Comparative genome analysis of "Candidatus Phytoplasma australiense" (subgroup tuf-Australia I; rp-A) and "Ca. Phytoplasma asteris" strains OY-M and AY-WB. *Journal of Bacteriology* 190, 3979–91. DOI: 10.1128/JB.01301-07.

Wei, W. and Zhao, Y. (2022) Phytoplasma taxonomy: Nomenclature, classification, and identification. *Biology (Basel)* 11, 1119. DOI: 10.3390/BIOLOGY11081119/S1.

Wei, W., Inaba, J., Zhao, Y., Mowery, J.D., and Hammond, R. (2022) Phytoplasma infection blocks starch breakdown and triggers chloroplast degradation, leading to premature leaf senescence, sucrose reallocation, and spatiotemporal redistribution of phytohormones. *International Journal of Molecular Sciene* 23, 1810. DOI: 10.3390/IJMS23031810/S1.

Yan, X., Yuan, X., Lv, J., Zhang, B., Huang, Y. *et al.* (2024) Molecular basis of SAP05-mediated ubiquitin-independent proteasomal degradation of transcription factors. *Nature Communications* 15, 1170. DOI: 10.1038/S41467-024-45521-7.

Zhou, J., Ma, F., Yao, Y., Deng, M., Chen, M. *et al.* (2021). Jujube witches' broom phytoplasma effectors SJP1 and SJP2 induce lateral bud outgrowth by repressing the ZjBRC1-controlled auxin efflux channel. *Plant, Cell and Environment* 44, 3257–3272. DOI: 10.1111/pce.14141.

11 Dissecting the Response of Plants to Phytoplasmas: Structural and Molecular Changes at the Site of Infection

Rita Musetti*
Department of Land, Environment, Agriculture and Forestry (TESAF), University of Padova, Legnaro, Italy

Abstract

Phytoplasmas greatly affect the ultrastructural organization and function of the sieve elements, with important effects on assimilate translocation. Huge callose deposition at the sieve plates and anomalous accumulation of proteinaceous material are some of the most common structural responses opposed by the host plants. In association with these modifications, genes encoding occlusion proteins and callose synthase are induced in phytoplasma-infected grapevine, apple trees, tomato, and *Arabidopsis*. As the phloem tissue is fundamental for systemic transport of energy-rich molecules, building blocks, mineral nutrients, and signals in the plant, the compromised phloem functionality may have disastrous consequences for plant health and development. This chapter focuses on some of the responses of the host plants to phytoplasma infection at the site of infection, to illuminate the mechanism(s) by which phytoplasmas overcome plant immune system.

Introduction

The presence and activity of pathogens and their recognition by host plants lead to many biochemical changes that activate the plant defense response. Phytoplasmas, plant-pathogenic prokaryotes, are restricted to the phloem sieve elements (SEs) throughout their entire life cycle within the host plant. The SEs are enucleated cells, separated by porous sieve plates that facilitate long-distance transport (van Bel, 2003). In addition to sugars, the phloem is a conduit for the (re)mobilization of carbon and nitrogen from source tissues to sink tissues, often in the form of amino acids (Jiang *et al.*, 2019; Koenig and Hoffmann-Benning, 2020). Moreover, the phloem transmits systemic signals, including fast electrical signals relayed by Ca^{2+} (van Bel *et al.*, 2011) and reactive oxygen species (ROS), as well as slower signals like proteins, mRNAs, and hormones (Turgeon and Wolf, 2009; Dinant *et al.*, 2010; Johns *et al.*, 2021).

Due to their reduced genomes, phytoplasmas have evolved to finely interact with the SEs in order to spread systemically within the plant host, an essential step in establishing a successful infection. Phytoplasma infection greatly affects the ultrastructural organization of SEs and induces severe impairment of the SE functions, with dramatic consequences for assimilate translocation, particularly sugar metabolism and transport (van Bel and Musetti, 2019). Even though phytoplasmas are economically relevant plant pathogens, as they induce huge crop

*Corresponding author: rita.musetti@unipd.it

 Phytoplasmas: Genomes, Plant Hosts and Vectors, 2nd Edition. (Ed. W. Wei)
DOI: 10.1079/9781800627031.0011

damage, the molecular events involved in plant response mechanisms to infection remain largely unexplored. For example, nothing is known about the mechanism(s) by which phloem cells perceive the phytoplasma presence and initiate the signaling processes to respond to the infection. Anomalous callose deposition at the sieve plates and accumulation of proteinaceous material are common structural responses opposed by the host plants at the site of infection (Pagliari *et al.*, 2017; Bernardini *et al.*, 2022) (Figs 11.1 and 11.2). In association with these modifications, genes encoding occlusion proteins and the sieve pore-specific callose synthase 7 were induced in many phytoplasma-infected plants – such as grapevine, apple, tomato, and *Arabidopsis* (Musetti *et al.*, 2010; Santi *et al.*, 2013a, 2013b; De Marco *et al.*, 2016; Pagliari *et al.*, 2017; Bernardini *et al.*, 2022).

As the phloem tissue is fundamental for the systemic transport of energy-rich molecules, building blocks, mineral nutrients, and signals, the compromised phloem functionality may have serious consequences for plant health and development. This chapter focuses on the responses of the host plants to phytoplasma infection at the site of infection, aiming to shed light on the mechanism(s) by which phytoplasmas overcome the plant immune system.

Phytoplasmas Cause Significant Alterations in Phloem Organization and Sieve Element Architecture

Phloem hyperplasia and cell wall modifications

Phloem hyperplasia is one of the histological modifications induced by phytoplasma infection in both woody and herbaceous host plants (Braun and Sinclair, 1976; Oshima *et al.*, 2001; Liu *et al.*, 2015). Measurements performed on thin sections of *Arabidopsis* midribs revealed that phloem cells were almost two times more numerous in phytoplasma-infected samples than in uninfected ones (Pagliari *et al.*, 2016). The stolbur phytoplasma '*Candidatus* Phytoplasma solani' induces a peculiar arrangement of the typical bicollateral vascular bundle of tomato plants (*Solanum lycopersicum* cv. Micro-Tom) (De Marco *et al.*, 2016), as both phloem strands along the two sides of xylem, show hyperplasia, and the xylem vessels are arranged in strings. Similar modifications are described in other phytoplasma-diseased *Solanaceae*, for example *S. malacoxylon* (Iriti *et al.*, 2008). Hyperplasia has been associated with phloem cell necrosis (Braun and Sinclair, 1979; Uehara *et al.*, 1999)

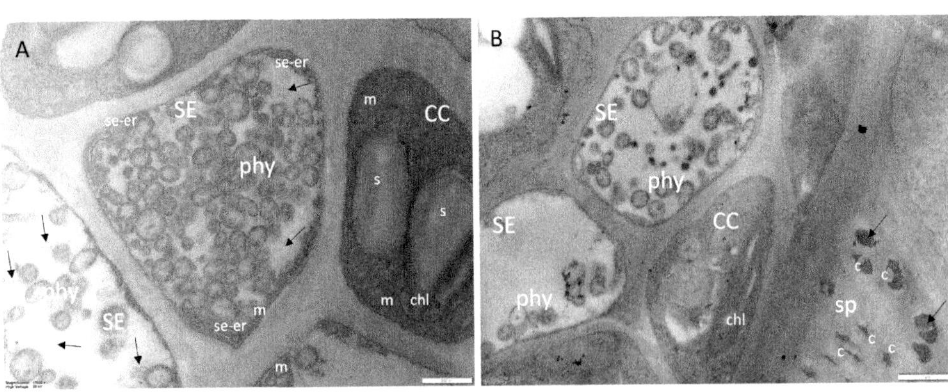

Fig. 11.1. Representative transmission electron micrographs of an *Arabidopsis thaliana* leaf midrib infected with chrysanthemum yellows (CY) phytoplasma. (A, B) In infected midribs, numerous phytoplasmas and agglomerates of filamentous phloem protein (arrows) are visible inside the sieve elements. In companion cells, chloroplasts are deformed and may contain big starch granules. (B) The sieve plates show the most pores constricted by callose depositions. c, callose; CC, companion cell; chl, chloroplast; m, mitochondrion; phy, phytoplasmas; s, starch; SE, sieve element; se-er, sieve element endoplasmic reticulum; sp, sieve plate. Scale bars: (A) 500 nm; (B) 1 μm.

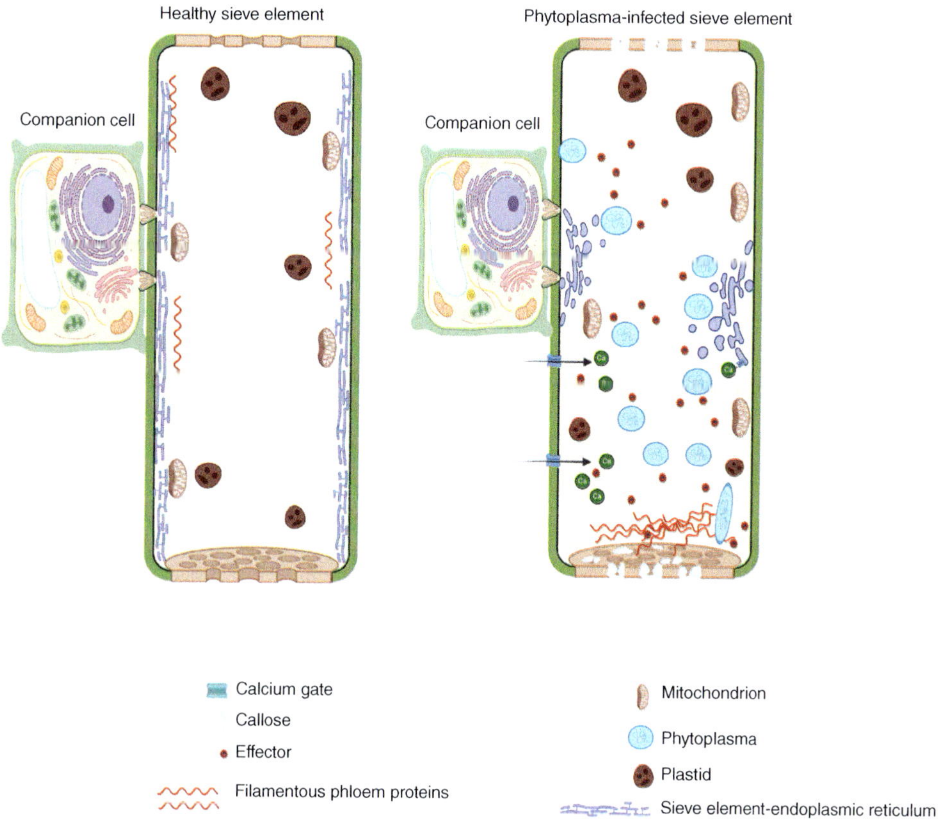

Fig. 11.2. Graphic representation of normal (left) and phytoplasma-infected (right) sieve element–companion cell complexes in *Arabidopsis thaliana*. Phytoplasma infection causes gating of Ca^{2+} channels in the sieve elements and the sieve element endoplasmic reticulum (SE-ER), leading to the sieve tube occlusion by phloem protein filament aggregation and callose deposition at the sieve plates, with relevant effects on phytoplasma spread and photoassimilate distribution. The SE-ER displays ultrastructural changes, which may function as anchor sites for phytoplasmas. (Created with BioRender.com.)

and to the formation of new cells due to cambium hyperactivity (Braun and Sinclair, 1976). Hyperplasia has been considered as a plant response to the reduced phloem functionality in infected tissue (Oshima *et al.*, 2001) or as a possible consequence of phytoplasma effector activity (Bai *et al.*, 2009; Sugio *et al.*, 2011). Interestingly, mild phytoplasma strains do not induce phloem necrosis or hyperplasia in host plants (Oshima *et al.*, 2001). Gallinger *et al.* (2021) suggested that the diverse anatomical (and consequently physiological) changes in both phloem and xylem tissues of fruit trees following phytoplasma infection may be related to different degrees of adaptation, reflecting the plant's ability to survive phytoplasma disease.

At the ultrastructural level, phytoplasmas induce profound modification in the SE architecture (Musetti *et al.*, 2013a, 2023; Buxa *et al.*, 2015; Zimmermann *et al.*, 2015; Bernardini *et al.*, 2022; Musetti *et al.*, 2023). Regardless of the cell types (i.e. SEs, companion cells (CCs), phloem parenchyma cells (PPCs)), microscopical observations revealed cell wall thickening and distortion in the phloem of host plants (Musetti *et al.*, 2013a; Santi *et al.*, 2013a; De Marco *et al.*, 2016; Pagliari *et al.*, 2016; Buoso *et al.*, 2022), likely due to changes in wall composition caused by defensive depositions of phenolic compounds (Messiaen *et al.*, 1993) and/or homogalacturonans (Cosgrove, 2005). Julius *et al.* (2021) demonstrated that alterations in the SE wall integrity and physical strength impair sugar translocation and change carbohydrate partitioning across the plant. These physiological modifications are observed in numerous host plant–phytoplasma interactions (Braun and Sinclair, 1979; De Marco *et al.*, 2021; Bernardini *et al.*, 2022), where reduced phloem transport leads to sugar accumulation, along with other polysaccharides such as cellulose and starch in source organs (Lepka *et al.*, 1999; Musetti *et al.*, 2007; Santi *et al.*, 2013a, 2013b; De Marco *et al.*, 2021).

Changes in sieve element substructural components

Restricted to the SEs of the host plants, phytoplasmas must exploit these cells to obtain the nutrients necessary for their metabolism, growth, and multiplication. The first and fundamental step in the establishment of a successful infection by a pathogen, is adhesion to the host structures (Agrios, 2005). The discovery of genes encoding putative membranc proteins with an adhesion motif on the surface of phytoplasmas supports their adhesion to SE substructures (Neriya *et al.*, 2014). Buxa *et al.* (2015) and Pagliari *et al.* (2016) reported that in infected tomatoes and *Arabidopsis* the SE plasma membrane undergoes invagination or undulation, coming into close contact with the phytoplasma membrane, to form a structure that anchors the phytoplasma body to the SE plasma membrane (Musetti *et al.*, 2016). Under transmission electron microscope, this phytoplasma "adhesion structure" appears tubular in shape, 30–40 nm in width, perpendicularly placed to the SE plasma membrane, and connected to the plant actin filaments (Musetti *et al.*, 2016). Phytoplasmas are, in fact, able to interact with the plant cytoskeleton by means of their immunodominant membrane proteins (Boonrod *et al.*, 2012), exposed on the outer part of the cell (Konnerth *et al.*, 2016). Buxa *et al.* (2015) reported that phytoplasma infection may impose actin severing and reorganization in the SEs, likely for promoting phytoplasma motility within SEs and through the sieve plates (Musetti *et al.*, 2016).

The SE endoplasmic reticulum (SE-ER) also exhibits morphological changes in the presence of phytoplasma (Buxa *et al.*, 2015; Pagliari *et al.*, 2016; Musetti *et al.*, 2023). Stacks are detached from the SE plasma membrane, lobated, and fragmented in vesicles (Buxa *et al.*, 2015). The SE-ER may function as a "docking berth" for phytoplasmas, which often appear anchored to the malformed SE-ER membranes (Musetti *et al.*, 2023). The SE-ER restructuring is accompanied by the modulation of genes encoding proteins involved in shaping and anchoring the SE-ER. A selective change in the expression of genes associated with the unfolded protein response (UPR) is also reported, which likely reflects a tradeoff between activating plant defense responses, sustaining host cells needed for the phytoplasma's biotrophic lifestyle, and supporting the survival of phytoplasmas (Musetti *et al.*, 2023). Interestingly, Inaba and co-workers (2023) reported that treatments using UPR chemical inducers led to the inhibition of phytoplasma multiplication in tomato plants affected by potato purple top phytoplasma. Mittelberger *et al.*, (2019) hypothesized that phytoplasma effectors may primarily target plant cell membrane systems, reporting significant structural alterations and loss of viability of *Nicotiana benthamiana* protoplasts expressing the '*Ca.* P. mali' effector $PME2_{ST}$. Effectors might also induce the formation of contact sites between phytoplasmas and SE-ER, similar to how numerous intracellular bacteria dock onto diverse host organelles (Dumox *et al.*, 2012). Such junctions may serve as bridges for effector trafficking and pathways for the supply of host resources (Tilney *et al.*, 2001).

Formation and role of sieve element occlusions in phytoplasma-infected plants

It has been widely reported that phytoplasma infection is associated with the formation of massive occlusions in the SEs, which impair photosynthate translocation in the infected plant (Braun and Sinclair, 1978; Lherminier *et al.*, 2003). The slowdown of phloem transport provokes sugar accumulation in source organs (Santi *et al.*, 2013b), reflecting the phytoplasma's high need for readily available high-energy oligosaccharides (Oshima *et al.*, 2007). The surplus of photosynthates in infected plants may also promote the synthesis of polysaccharides such as cellulose or starch (Musetti *et al.*, 2007; Giorno *et al.*, 2013; Santi *et al.*, 2013a, 2013b; De Marco *et al.*, 2021). As the reduction in sugar transport capability was also observed in plants hosting a low phytoplasma titre (Christensen *et al.*, 2004), it is argued that the SE occlusions are not exclusively due to the high abundance of phytoplasma, which may physically obstruct the SE lumen, but also to the plant response to the stress imposed by the pathogen. Abnormal callose deposition, caused by *de novo* synthesis of callose and the conformational changes in SE structural proteins, mainly observed at the sieve plates (Musetti *et al.*, 2010, 2013a), are two characteristic reactions of the SEs to mechanical injury (Knoblauch and van Bel, 1998; Furch *et al.*, 2007) and microbial attack (Santi *et al.*, 2013b; Pagliari *et al.*, 2017; Bernardini *et al.*, 2022) to prevent sugar losses and impede pathogen spread. The involvement of callose as well as proteinaceous substances in the SE defense reactions has been hard to demonstrate, given the native occlusion reactions in response to sectioning (Knoblauch and van Bel, 1998). An *in vivo* observation method was set up (Knoblauch and van Bel, 1998) and applied to broad bean plants (*Vicia faba* L.) infected with flavescence dorée phytoplasma (Musetti *et al.*, 2013a). *In vivo* experiments demonstrated that phytoplasma infection causes Ca^{2+} influx into SEs, leading to the occlusion of the sieve plates by callose deposition or protein plugging (Musetti *et al.*, 2013a). As described, Ca^{2+} influx may also confer cell wall thickening of the conducting elements (Hepler, 2005; Wolf, 2022).

The role of callose and phloem proteins seems to extend beyond merely obstructing SEs to prevent the systemic spread of pathogens. They also appear to play a role in the regulation of resistance mechanisms at the molecular level. In *Arabidopsis*, the expression of *AtCALS7*, the gene encoding the enzyme responsible for callose deposition around the sieve pores (Barratt *et al.*, 2011; Xie *et al.*, 2011), is strongly enhanced during phytoplasma infection, corresponding with increased callose deposition at the sieve pores (Bernardini *et al.*, 2022). In line with this, the expression of the sucrose synthase genes *AtSUS5* and *AtSUS6*, which function together in a callose-spinning apparatus, is upregulated. Thus, precursors are provided for the synthesis of callose (De Marco *et al.*, 2021) and the carbon skeletons of defense-related compounds (Bolouri-Moghaddam *et al.*, 2010). Conversely, the loss of callose synthase 7 in the *Atcals7ko* line increases susceptibility to phytoplasma infection, affecting sugar transport and metabolism, which supports pathogen multiplication (Bernardini *et al.*, 2022).

The presence of structural proteins in the sieve tubes has been observed for a long time (Cronshaw and Sabnis, 1990). In Fabaceae, the so-called sieve element occlusion (SEO) proteins (Pélissier *et al.*, 2008) are aggregated in giant protein bodies, the forisomes (Knoblauch *et al.*, 2001). In response to different stresses, including phytoplasma infection (Musetti *et al.*, 2013a), forisomes undergo a conformational change from a condensed to a dispersed state, which plugs the sieve plates and prevents the loss of photoassimilates and pathogen spread (Knoblauch *et al.*, 2001; Furch *et al.*, 2007; Musetti *et al.*, 2013a). Genes encoding SEO protein components not only occur in Fabaceae, but are widespread among dicotyledonous plants (Rüping *et al.*, 2010). In *Arabidopsis thaliana*, the formation of filamentous SE proteins requires the expression of both sieve element occlusion-related (SEOR) AtSEOR1 or AtSEOR2 proteins (Anstead *et al.*, 2012). Knocking-out of *AtSEOR1* and *AtSEOR2* results in the inability to seal sieve tubes in phytoplasma-infected mutant lines (Pagliari *et al.*, 2017), despite the phytoplasma-imposed stress. Interestingly, *Arabidopsis Atseor1ko* mutants, which express AtSEOR2 protein free from its bond with AtSEOR1 (Anstead *et al.*, 2012), exhibit reduced phytoplasma titer compared with wild-type plants with early-stage infection. Recent studies suggest that the AtSEOR2 acts as an "effector hub" involved in plant immunity (González-Fuente *et al.*, 2020; Ceulemans *et al.*, 2021). *Atseor1ko* mutant plants display a significant increase in abscisic acid (ABA) and indole 3-acetic acid (IAA) levels even at the beginning of phytoplasma infection compared with the wild type, suggesting that faster hormone-mediated signaling may be induced in these mutant plants (Bernardini *et al.*, 2020).

Phytoplasmas Generate Ca^{2+}-Mediated Alarm Signals in the Sieve Elements, Responsible for Phloem Structural and Molecular Changes

Phytoplasmas manipulate their host plants and insect vectors by the synthesis and secretion of effectors (Furch *et al.*, 2021; Oshima *et al.*, 2023 and literature within). In plants, effectors can migrate from the SEs to the adjacent cells and systemically via the SEs (Huang *et al.*, 2021). Plants have evolved receptor proteins to rapidly recognize the pathogen effectors, triggering an immune response to prevent pathogen colonization in the host (Han, 2019). It remains unknown what proteins trigger the previously described SE changes during phytoplasma–SE interaction. Phytoplasmas provoke Ca^{2+} influx in the SEs, leading to sieve pore occlusion (Musetti *et al.*, 2013a), with effects on phytoplasma spread, carbohydrate distribution, and the whole phloem physiology (Bernardini *et al.*, 2022) (Fig. 11.2). Ca^{2+} influx into SEs also activates downstream signaling molecules, generating various intracellular cascades leading to changes in metabolism and gene expression (Kudla *et al.*, 2010; van Bel *et al.*, 2014).

The presence of glutamate receptor-like calcium sensors (Toyota *et al.*, 2018; Sanden and Schulz, 2021) and Ca^{2+}-permeable channels on the SE plasma membrane (Furch *et al.*, 2009; Hafke *et al.*, 2009) appears important in this network, even if the precise role(s) of these proteins during phytoplasma infection is not elucidated yet. In establishing an effective systemic response to pathogens, it is obvious that signaling molecules are rapidly produced prior to SE occlusion. Increased cytosolic Ca^{2+} levels in SEs can activate respiratory burst oxidase (RBOH) proteins through binding to the N-terminal EF hands on these enzymes, increasing the level of ROS and leading to further ion channel gating and wave propagation (Hilleary and Gilroy, 2018). Among RBOH proteins, RBOHD plays a critical role in the generation of systemic Ca^{2+} and ROS waves, which are at the base of the mechanisms leading to the acquisition of systemic resistance (Miller *et al.*, 2009). Miller and co-workers (2009) also proposed that the ROS that primarily mediates the rapid systemic signal through the SEs is hydrogen peroxide (H_2O_2).

Interestingly, using cytochemical methods, H_2O_2 was detected in the SEs of grapevines and apple plants, which were spontaneously recovered from flavescence dorée and apple proliferation diseases, respectively (Musetti *et al.*, 2004, 2007). H_2O_2 was not found in the SEs of phytoplasma-symptomatic apple and grapevine leaves, nor in those of healthy leaves (Musetti *et al.*, 2004, 2007). This suggests that a discrete H_2O_2 accumulation in the SEs is peculiar to recovered plants, in which the induction of resistance against phytoplasmas is established (Musetti *et al.*, 2013b; Osler *et al.*, 2014). In contrast, biotrophic phytoplasmas may limit the accumulation of H_2O_2 in the SEs during disease symptom development to prevent cell death. This is consistent with the fact that phytoplasmas must maintain host cell viability to ensure their own survival and propagation.

Many other phloem-mobile signaling molecules, such as sugars (Li and Sheen, 2016), hormones, microRNA, lipids, and proteins, are involved in the SE responses to pathogens (Koenig and Hoffmann-Benning, 2020). Investigations have examined the levels of these molecules in phytoplasma-infected plants, but analyses have been performed on diverse plant–phytoplasma pathosystems and using different plant parts, for example whole leaves (Gai *et al.*, 2014a, 2014b; Dermastia, 2019), leaf midribs (Margaria *et al.*, 2013; Chitarra *et al.*, 2018) or plants grown in different conditions (*in vitro* micropropagated, in greenhouses, or fields). As a result, the findings have not been consistent.

The use of phloem sap samples, collected through spontaneous exudation or facilitated by the use of ethylenediaminetetraacetic acid (EDTA) (Gaupels *et al.*, 2008; Zimmerman *et al.*, 2019), allows for a reduction in contamination and more precise results about the roles of the phloem-mobile molecules during phytoplasma infection. Gai *et al.* (2018) reported that many transcripts associated with the signaling and metabolism of diverse phytohormones (IAA, ABA, jasmonates, and cytokinins), transcription factors, and proteins were differentially abundant in the phloem sap of phytoplasma-infected mulberry plants. They hypothesized that multiple hormone pathways and the expression of hormone-sensitive proteins may work cooperatively in the response to phytoplasma infection and be responsible for symptom development in the infected plants (Gai *et al.*, 2018).

Conclusion

Phytoplasmas are associated with economically relevant plant diseases, and no curative methods are available for their control. Phytoplasma recognition by the host plant induces many biochemical changes, leading to the activation of defense responses that are effective when triggered early. Increasing knowledge on plant responses at the site of the phytoplasma infection, particularly in the phloem, is fundamental to understanding phytoplasma biology and their pathogenic behavior and in developing novel biotechnology-assisted strategies to counteract pathogen spread in host plants.

References

Agrios, G. (2005) *Plant Pathology*, 5th edn. Elsevier Academic Press, Amsterdam.

Anstead, J.A., Froelich, D.R., Knoblauch, M., and Thompson, G.A. (2012) Arabidopsis P-protein filament formation requires both AtSEOR1 and AtSEOR2. *Plant and Cell Physiology* 53, 1033–1042. DOI: 10.1093/pcp/pcs046.

Bai, X., Correa, V.R., Toruño, T.Y., Ammar, E.D., Kamoun, S., and Hogenhout, S.A. (2009) AY-WB phytoplasma secretes a protein that targets plant cell nuclei. *Molecular Plant-Microbe Interaction* 22, 18–30. DOI: 10.1094/MPMI-22-1-0018.

Barratt, D.H.P., Kölling, K., Graf, A., Pike, M., Calder, G. *et al.* (2011) Callose synthase GSL7 is necessary for normal phloem transport and inflorescence growth in *Arabidopsis*. *Plant Physiology* 155, 328–341.

Bernardini, C., Pagliari, L., De Rosa, V. *et al.* (2020) Pre-symptomatic modified phytohormone profile is associated with lower phytoplasma titres in an Arabidopsis *seor1ko* line. *Scientific Reports* 10, 14770. DOI: 10.1038/s41598-020-71660-0.

Bernardini, C., Santi, S., Mian, G., Levy, A., Buoso, S. *et al.* (2022) Increased susceptibility to Chrysanthemum Yellows phytoplasma infection in *Atcals7ko* plants is accompanied by enhanced expression of carbohydrate transporters. *Planta* 256, 43. DOI: 10.1007/s00425-022-03954-8.

Bolouri-Moghaddam, M.R., Le Roy, K., Xiang, L., Rolland, F., and van den Ende, W. (2010) Sugar signaling and antioxidant network connections in plant cells. *FEBS Journal* 277, 2022–2037. DOI: 10.1111/j.1742-4658.2010.07633.x.

Boonrod, K., Munteanu, B., Jarausch, B., Jarausch, W., and Krczal, G. (2012) An immunodominant membrane protein (Imp) of '*Candidatus* Phytoplasma mali' binds to plant actin. *Molecular Plant-Microbe Interaction* 25, 889–895. DOI: 10.1094/MPMI-11-11-0303.

Braun, E.J. and Sinclair, W.A. (1976) Histopathology of phloem necrosis in *Ulmus americana*. *Phytopathology* 66, 598–607.

Braun, E.J. and Sinclair, W.A. (1978) Translocation in phloem necrosis-diseased American elm seedlings. *Phytopathology* 68, 1733–1737.

Braun, E.J. and Sinclair, W.A. (1979) Phloem necrosis of elms: Symptoms and histopathological observations in tolerant host. *Phytopathology* 69, 354–358.

Buoso, S., Musetti, R., Marroni, F., Calderan, A., Schmidt, W., and Santi, S. (2022) Infection by phloem-limited phytoplasma affects mineral nutrient homeostasis in tomato leaf tissues. *Journal of Plant Physiology* 271, 153659. DOI: 10.1016/j.jplph.2022.153659.

Buxa, S.V., Degola, F., Polizzotto, R., De Marco, F., Loschi, A. *et al.* (2015) Phytoplasma infection in tomato is associated with re-organization of plasma membrane, ER stacks, and actin filaments in sieve elements. *Frontiers in Plant Science* 6, 650. DOI: 10.3389/fpls.2015.00650.

Ceulemans, E., Ibrahim, H.M., De Coninck, B., and Goossens, A. (2021) Pathogen effectors: Exploiting the promiscuity of plant signaling hubs. *Trends in Plant Science* 26, 780–795.

Chitarra, W., Pagliarani, C., Abbà, S., Boccacci, P., Birello, G. *et al.* (2018) miRVIT: A novel miRNA database and its application to uncover *Vitis* responses to Flavescence dorée infection. *Frontiers in Plant Science* 9, 1034.

Christensen, N.M., Nicolaisen, M., Hansen, M., and Schulz, A. (2004) Distribution of phytoplasmas in infected plants as revealed by real-time PCR and bioimaging. *Molecular Plant-Microbe Interaction* 17, 1175–1184.

Cosgrove, D.J. (2005) Growth of the plant cell wall. *Nature Review Molecular Cell Biology* 6, 850–861. DOI: 10.1038/nrm1746.

Cronshaw, J. and Sabnis, D.D. (1990) Phloem proteins. In: Behnke, H.D. and Sjolund, R.D. (eds) *Sieve Elements. Comparative Structure, Induction, and Development*. Springer, Berlin, pp. 257–283.

De Marco, F., Pagliari, L., Degola, F., Buxa, S.V., Loschi, A. *et al.* (2016) Combined microscopy and molecular analyses show phloem occlusions and cell wall modifications in tomato leaves in response to '*Candidatus* Phytoplasma solani'. *Journal of Microscopy* 263, 212–225. DOI: 10.1111/jmi.12426.

De Marco, F., Batailler, B., Thorpe, M.R., Razan, F., Le Hir, R. *et al.* (2021) Involvement of SUT1 and SUT2 sugar transporters in the impairment of sugar transport and changes in phloem exudate contents in phytoplasma-infected plants. *International Journal of Molecular Sciences* 22, 745. DOI: 10.3390/ijms22020745.

Dermastia, M. (2019) Plant hormones in phytoplasma infected plants. *Frontiers in Plant Science* 10, 477. DOI: 10.3389/fpls.2019.00477.

Dinant, S., Bonnemain, J.L., Girousse, C. and Kehr, J. (2010) Phloem sap intricacy and interplay with aphid feeding. *Comptes Rendus Biologies* 333, 504–515.

Dumox, M., Clare, D.K., Saibil, H.R., and Hayward, R.D. (2012) *Chlamydiae* assemble a pathogen synapse to hijack the host endoplasmic reticulum. *Traffic* 13, 1612–1627. DOI: 10.1111/tra.12002.

Furch, A.C., Hafke, J.B., Schulz, A., and van Bel, A.J.E. (2007) Ca^{2+}-mediated remote control of reversible sieve tube occlusion in *Vicia faba*. *Journal of Experimental Botany* 58, 2827–2838.

Furch, A.C.U., van Bel, A.J.E., Fricker, M.D., Felle, H.H., Fuchs M., and Hafke, J.B. (2009) Sieve element Ca^{2+} channels as relay stations between remote stimulus and sieve tube occlusion. *Plant Cell* 21, 2118–2132.

Furch, A.C., Theißen, G., and Zimmermann, M.R. (2021) New phytoplasma effector: 50 shades of green. *Cell Host and Microbe* 29, 1601–1603.

Gai, Y.P., Han, X.J., Li, Y.Q., Yuan, C.Z., Mo, Y.Y. *et al.* (2014a) Metabolomic analysis reveals the potential metabolites and pathogenesis involved in mulberry yellow dwarf disease. *Plant, Cell and Environment* 37, 1474–1490. DOI: 10.1111/pce.12255.

Gai, Y.P., Li, Y.Q., Guo, F.Y., Yuan, C.Z., Mo, Y.Y. *et al.* (2014b) Analysis of phytoplasma-responsive sRNAs provide insight into the pathogenic mechanisms of mulberry yellow dwarf disease. *Scientific Reports* 4, 5378. DOI: 10.1038/srep05378.

Gai, Y.P., Yuan, S.S., Liu, Z.Y., Zhao, H.N., Liu, Q. *et al.* (2018) Integrated phloem sap mRNA and protein expression analysis reveals phytoplasma-infection responses in mulberry. *Molecular Cell Proteomics* 17, 1702–1719. DOI: 10.1074/mcp.RA118.000670.

Gallinger, J., Zikeli, K., Zimmermann, M.R., Görg, L.M., Mithöfer, A. *et al.* (2021) Specialized 16SrX phytoplasmas induce diverse morphological and physiological changes in their respective fruit crops. *PLoS Pathogens* 17, e1009459. DOI: 10.1371/journal. ppat.1009459.

Gaupels, F., Knauer, T., and van Bel, A.J.E. (2008) A combinatory approach for analysis of protein sets in barley sieve-tube samples using EDTA-facilitated exudation and aphid stylectomy. *Journal of Plant Physiology* 165, 95–103. DOI: 10.1016/j.jplph.2007.07.023.

Giorno, F., Guerriero, G., Biagetti, M., Ciccotti, A.M., and Baric, S. (2013) Gene expression and biochemical changes of carbohydrate metabolism in *in vitro* micro-propagated apple plantlets infected by '*Candidatus* Phytoplasma mali'. *Plant Physiology and Biochemistry*, 70, 311–317. DOI: 10.1016/j.plaphy.2013.05.040.

González-Fuente, M., Carrère, S., Monachello, D., Marsella, B.G., Cazalé, A.C. *et al.* (2020) EffectorK, a comprehensive resource to mine for *Ralstonia, Xanthomonas,* and other published effector interactors in the *Arabidopsis* proteome. *Molecular Plant Pathology* 21, 1257–1270. DOI: /10.1111/mpp.12965.

Hafke, J.B., Furch, A.C., Fricker, M.D., and van Bel A.J.E. (2009) Forisome dispersion in *Vicia faba* is triggered by Ca^{2+} hotspots created by concerted action of diverse Ca^{2+} channels in sieve elements. *Plant Signaling and Behavior* 4, 968–972.

Han, G.Z. (2019) Origin and evolution of the plant immune system. *New Phytologist* 222, 70–83.

Hepler, P.K. (2005) Calcium: A central regulator of plant growth and development. *Plant Cell* 17, 2142–55. DOI: 10.1105/tpc.105.032508.

Hilleary, R. and Gilroy, S. (2018). Systemic signaling in response to wounding and pathogens. *Current Opinion in Plant Biology* 43, 57–62.

Huang, W., MacLean, A.M., Sugio, A., Maqbool, A., Busscher, M. *et al.* (2021) Parasitic modulation of host development by ubiquitin-independent protein degradation. *Cell* 184, 5201–5214.

Inaba, J., Kim, B.M., Zhao, Y., Jansen, A.M., and Wei, W. (2023) The endoplasmic reticulum is a key battleground between phytoplasma aggression and host plant defense. *Cells* 12, 2110. DOI: 10.3390/cells12162110.

Iriti, M., Quaglino, F., Maffi, D., Casati, P., Bianco, P.A., and Faoro, F. (2008) *Solanum malacoxylon*, a new natural host for Stolbur phytoplasma. *Journal of Phytopathology* 156, 8–14.

Jiang, Y., Zhang, C.X., Chen, R., and He, S.Y. (2019) Challenging battles of plants with phloem-feeding insects and prokaryotic pathogens. *Proceedings of the National Academy of Sciences of the USA* 116, 23390–23397.

Johns, S., Hagihara, T., Toyota, M., and Gilroy, S. (2021) The fast and the furious: Rapid long-range signaling in plants. *Plant Physiology* 185, 694–706. DOI: 10.1093/plphys/kiaa098.

Julius, B.T., McCubbin, T.J., Mertz, R.A., Baert, N., Knoblauch, J. *et al.* (2021) Maize *Brittle Stalk2-Like3*, encoding a COBRA protein, functions in cell wall formation and carbohydrate partitioning. *Plant Cell* 33, 3348–3366. DOI: 10.1093/plcell/koab193.

Knoblauch, M. and van Bel, A.J.E. (1998) Sieve tubes in action. *Plant Cell* 10, 35–50.

Knoblauch, M., Peters, W.S., Ehlers, K., and van Bel, A.J.E. (2001) Reversible calcium regulated stopcocks in legume sieve tubes. *Plant Cell*, 13, 1221–1230. DOI: /10.1105/tpc.13.5.1221.

Koenig, A.M. and Hoffmann-Benning, S. (2020) The interplay of phloem-mobile signals in plant development and stress response. *Bioscience Reports* 40, BSR20193329.

Konnerth, A., Krczal, G., and Boonrod, K. (2016) Immunodominant membrane proteins of phytoplasmas. *Microbiology* 162, 1267–1273. DOI: 10.1099/mic.0.000331.

Kudla, J., Batistic, O., and Hashimoto, K. (2010) Calcium signals: The lead currency of plant information processing. *Plant Cell* 22, 541–63. DOI: 10.1105/tpc.109.072686.

Lepka, P., Stitt, M., Moll, E., and Seemüller, E. (1999) Effect of phytoplasmal infection on concentration and translocation of carbohydrates and amino acids in periwinkle and tobacco. *Physiological and Molecular Plant Pathology* 55, 59–68.

Lherminier, J., Benhamou, N., Larrue, J., Milat, M.L., Boudon-Padieu, E., Nicole, M., and Blein, J.P. (2003) Cytological characterization of elicitin-induced protection in tobacco plants infected by *Phytophthora parasitica* or phytoplasma. *Phytopathology* 93, 1308–1319.

Li, L. and Sheen, J. (2016) Dynamic and diverse sugar signaling. *Current Opinion in Plant Biology* 33, 116–125. DOI: 10.1016/j.pbi.2016.06.018.

Liu, C.T., Huang, H.M., Hong, S.F., Kuo Huang, L.L., Yan, C.Y. *et al.* (2015) Peanut witches' broom (PnWB) phytoplasma-mediated leafy flower symptoms and abnormal vascular bundles development. *Plant Signaling and Behavior* 10, e1107690.

Margaria, P., Abbà, S., and Palmano, S. (2013) Novel aspects of grapevine response to phytoplasma infection investigated by a proteomic and phospho-proteomic approach with data integration into functional networks. *BMC Genomics* 14, 38. DOI: 10.1186/1471-2164-14-38.

Messiaen, J., Read, N.D., van Cutsem, P., and Trewavas, A.J. (1993) Cell wall oligogalacturonides increase cytosolic free calcium in carrot protoplasts. *Journal of Cell Science* 104, 365–371.

Miller, G., Schlauch, K., Tam, R., Cortes, D., Torres, M.A. *et al.* (2009) The plant NADPH oxidase RBOHD mediates rapid systemic signaling in response to diverse stimuli. *Science Signaling* 2, ra45.

Mittelberger, C., Stellmach, H., Hause, B., Kerschbamer, C., Schlink, K. *et al.* (2019) A novel effector protein of apple proliferation phytoplasma disrupts cell integrity of nicotiana spp. protoplasts. *International Journal of Molecular Science* 20, 4613. DOI: 10.3390/ijms20184613.

Musetti, R., Sanità di Toppi, L., Ermacora, P., and Favali, M.A. (2004) Recovery in apple trees infected with the apple proliferation phytoplasma: An ultrastructural and biochemical study. *Phytopathology* 94, 203–208.

Musetti, R., Marabottini, R., Badiani, M., Martini, M., Sanità di Toppi, L. *et al.* (2007) On the role of H_2O_2 in the recovery of grapevine (*Vitis vinifera*, cv. Prosecco) from Flavescence dorée disease. *Functional Plant Biology* 34, 750–758.

Musetti, R., Paolacci, A., Ciaffi, M., Tanzarella, O.A., Polizzotto, R. *et al.* (2010) Phloem cytochemical modification and gene expression following the recovery of apple plants from apple proliferation disease. *Phytopathology* 100, 390–399.

Musetti, R., Buxa, S.V., De Marco, F., Loschi, A., Polizzotto, R., Kogel, K.-H., and van Bel, A.J.E. (2013a) Phytoplasma-triggered Ca($^{2+}$) influx is involved in sieve-tube blockage. *Molecular Plant-Microbe Interaction* 26, 379–86. DOI: 10.1094/MPMI-08-12-0207-R.

Musetti, R., Farhan, K., De Marco, F., Polizzotto, R., Paolacci, A. *et al.* (2013b) Differentially-regulated defence genes in Malus domestica during phytoplasma infection and recovery. *European Journal of Plant Pathology* 136, 13–19.

Musetti, R., Pagliari, L., Buxa, S.V., Degola, F., De Marco, F. *et al.* (2016) OHMS**: Phytoplasmas dictate changes in sieve-element ultrastructure to accommodate their requirements for nutrition, multiplication and translocation. *Plant Signaling and Behavior* 11, e1138191. DOI: 10.1080/15592324.2016.1138191.

Musetti, R., Pagliari, L., Mian, G., De Oliveira Cantao, F.R., Bernardini, C., Santi, S., and van Bel, A.J.E. (2023) The sieve-element endoplasmic reticulum: A focal point of phytoplasma-host plant interaction? *Frontiers in Microbiology* 14, 1030414. DOI: 10.3389/fmicb.2023.1030414.

Neriya, Y., Maejima, K., Nijo, T., Tomomitsu, T., Yusa, A. *et al.* (2014) Onion yellow phytoplasma P38 protein plays a role in adhesion to the hosts. *FEMS Microbiology Letters* 361, 115–122.

Osler, R., Borselli, S., Ermacora, P., Loschi, A., Martini, M., Musetti, R., and Loi, N. (2014) Acquired tolerance in apricot plants that stably recovered from European stone fruit yellows. *Plant Disease* 98, 492–496.

Oshima, K., Shiomi, T., Kuboyama, T., Sawayanagi, T., Nishigawa, H. *et al.* (2001). Isolation and characterization of derivative lines of the onion yellows phytoplasma that do not cause stunting or phloem hyperplasia. *Phytopathology* 91, 1024–1029.

Oshima, K., Kakizawa, S., Arashida, R., Ishii, Y., Hoshi, A., and Hayashi, Y. (2007). Presence of two glycolytic gene clusters in a severe pathogenic line of '*Candidatus* Phytoplasma asteris'. *Molecular Plant Pathology* 8, 481–489.

Oshima, K., Maejima, K., Isobe, Y., Endo, A., Namba, S. *et al.* (2023) Molecular mechanisms of plant manipulation by secreting effectors of phytoplasmas. *Physiological and Molecular Plant Pathology* 125, 102009.

Pagliari, L., Martini, M., Loschi, A., and Musetti, R. (2016) Looking inside phytoplasma-infected sieve elements: A combined microscopy approach using *Arabidopsis thaliana* as a model plant. *Micron* 89, 87–97. DOI: 10.1016/j.micron.2016.07.007.

Pagliari, L., Buoso, S., Santi, S., Furch, A.C.U., Martini, M. *et al.* (2017) Filamentous sieve element proteins are able to limit phloem mass flow, but not phytoplasma spread. *Journal of Experimental Botany* 68, 3673–3688. DOI: 10.1093/jxb/erx199.

Pélissier, H.C., Peters, W.S., Collier, R., van Bel A.J.E., and Knoblauch, M. (2008) GFP tagging of sieve element occlusion (SEO) proteins results in green fluorescent forisomes. *Plant and Cell Physiology* 49, 1699–1710.

Rüping, B., Ernst, A.M., Jekat, S.B., Nordzieke, S., Reineke, A.R. *et al.* (2010) Molecular and phylogenetic characterization of the sieve element occlusion gene family in *Fabaceae* and non-*Fabaceae* plants. *BMC Plant Biology* 10, 219. DOI: 10.1186/1471-2229-10-219.

Sanden, N.C. and Schulz, A. (2021) Stationary sieve element proteins. *Journal of Plant Physiology* 266, 153511. DOI: 10.1016/j.jplph.2021.153511.

Santi, S., De Marco, F., Polizzotto, R., Grisan, S., and Musetti, R. (2013a) Recovery from stolbur disease in grapevine involves changes in sugar transport and metabolism. *Frontiers in Plant Science* 4, 171–182.

Santi, S., Grisan, S., Pierasco, A., De Marco, F., and Musetti, R. (2013b) Laser microdissection of grapevine leaf phloem infected by stolbur reveals site-specific gene responses associated to sucrose transport and metabolism. *Plant, Cell and Environment* 36, 343–355.

Sugio, A., MacLean, A.M. Kingdom, H.N., Grieve, V.M., Manimekalai, R., and Hogenhout, S.A. (2011) Diverse targets of phytoplasma effectors: From plant development to defense against insects. *Annual Review of Phytopathology* 49, 175–195.

Tilney, L.G., Harb, O.S., Connelly, P.S., Robinson, C.G., and Roy, C.R. (2001) How the parasitic bacterium *Legionella pneumophila* modifies its phagosome and transforms it into rough ER: implications for conversion of plasma membrane to the ER membrane. *Journal of Cell Science* 114, 4637–4650. DOI: 10.1242/jcs.114.24.4637.

Toyota, M., Spencer, D., Sawai-Toyota, S., Jiaqi, W., Zhang, T. *et al.* (2018) Glutamate triggers long-distance, calcium-based plant defense signaling. *Science* 361, 1112–1115.

Turgeon, R. and Wolf, S. (2009) Phloem transport: Cellular pathways and molecular trafficking. *Annual Review of Plant Biology* 60, 207–221.

Uehara, T., Tanaka, M., Shiomi, T., Namba, S., Tsuchizaki, T., and Matsuda, I. (1999). Histopathological studies on two symptom types of phytoplasma associated with lettuce yellows. *Japanese Journal of Phytopathology* 65, 465–469.

van Bel, A.J.E. (2003) The phloem, a miracle of ingenuity. *Plant, Cell and Environment* 26, 125–150.

van Bel, A.J.E., Furch, A.C., Hafke, J.B., Knoblauch, M., and Patrick, J.W. (2011) (Questions)*n* on phloem biology. 2. Mass flow, molecular hopping, distribution patterns and macromolecular signalling. *Plant Science* 181, 325–330.

van Bel, A.J.E., Furch, A.C.U., Will, T., Buxa, S.V., Musetti, R., and Hafke, J.B. (2014) Spread the news: Systemic dissemination and local impact of Ca^{2+} signals along the phloem pathway. *Journal of Experimental Botany* 65, 1761–1787. DOI: 10.1093/jxb/ert425.

van Bel, A.J.E. and Musetti, R. (2019) Sieve element biology provides leads for research on phytoplasma lifestyle in plant hosts. *Journal of Experimental Botany* 70, 3737–3755. DOI: 10.1093/jxb/erz172.

Wolf, S. (2022) Cell wall signaling in plant development and defense. *Annual Review of Plant Biology* 73, 323–353.

Xie, B., Wang, X., Zhu, M., Zhang, Z., and Hong, Z. (2011) *CalS7* encodes a callose synthase responsible for callose deposition in the phloem. *Plant Journal* 65, 1–14. DOI: 10.1111/j.1365-313X.2010. 04399.x.

Zimmermann, M.R., Schneider, B., Mithöfer A., and Reichelt, M. (2015) Implications of '*Candidatus* Phytoplasma mali' infection on phloem function of apple trees. *Endocytobiosis Cell Research* 26, 67–75.

Zimmermann, M.R., Knauer, T., and Furch, A.C.U. (2019) Collection of phloem sap in phytoplasma-infected plants. In: Musetti, R., Pagliari, L. (eds) *Phytoplasmas. Methods in Molecular Biology*, Vol. 1875. Humana Press, New York, pp. 291–299. DOI: 10.1007/978-1-4939-8837-2_21.

12 Psyllid Vectors and Their Control

Barbara Jarausch[1]* and Wolfgang Jarausch[2]
[1]*Julius Kuehn-Institute, Siebeldingen, Germany;* [2]*RLP AgroScience, Neustadt an der Weintraße, Germany*

Abstract
Jumping plant lice or psyllids are small plant-feeding insects that have garnered much attention as direct pests or as vectors of '*Candidatus* Phytoplasma' species. All recognized psyllid vectors belong to the genus *Cacopsylla* of the family Psyllidae. They transmit a narrow range of 16SrX group phytoplasmas that cause diseases on fruit trees of great economic importance. Thus, the impact of the psyllids on fruit production is high. As infected trees cannot be cured and resistant plant material is not yet available to the growers, preventive vector control is of paramount importance to limit the disease spread and is discussed. For this, knowledge about the biology of the vector species, as well as about the transmission parameters, is crucial. This chapter provides an up-to-date overview of the multidisciplinary approaches that have been adopted to deepen the understanding of the different features of phytoplasma transmission by psyllid vectors.

Diseases Associated with Psyllid-Transmitted Phytoplasmas

Since the beginning of the 21st century, the importance of psyllids as phytoplasma vectors has become the focus of scientific attention, although the first reports are from the 1960s. All recognized psyllid vectors today belong to a single genus of the family Psyllidae (Sternorrhyncha: Psylloidea). They transmit a narrow range of phytoplasmas belonging to the 16Sr group X fruit tree phytoplasmas. These diseases, however, are of great and increasing economic importance, and thus the impact of the psyllids on fruit production is high. The diseases transmitted by psyllid vectors are pear decline (PD) of *Pyrus*, apple proliferation (AP) of *Malus*, European stone fruit yellows (ESFY), and peach yellow leaf roll (PYLR) of *Prunus*.

There are two other reports about psyllids involved in the transmission of aster yellows or stolbur-type phytoplasmas to carrots. Leclant *et al.* (1974) found *Trioza nigricornis* (Förster) as a vector, and Font *et al.* (1999) obtained transmission with *Bactericera trigonica* Hodk. As these reports were not confirmed by other research groups, these putative psyllid vectors will not be the scope of the present chapter.

All four phytoplasma diseases of *Pyrus*, *Malus*, and *Prunus* are associated with distinct but closely related phytoplasmas. They all belong to the 16Sr group X, the AP cluster (Seemüller *et al.*, 1998). This phylogenetic clade comprises the AP agent, '*Candidatus* Phytoplasma mali,' the PD

*Corresponding author: barbara.jarausch@julius-kuehn.de

 Phytoplasmas: Genomes, Plant Hosts and Vectors, 2nd Edition. (Ed. W. Wei)
DOI: 10.1079/9781800627031.0012

agent, '*Ca.* P. pyri,' and the ESFY agent, '*Ca.* P. prunorum,' as well as the PYLR agent and the newly identified PD agent from Taiwan (PDTW) (Seemüller *et al.*, 1998; Seemüller and Schneider, 2004; Liu *et al.*, 2007). All these phytoplasmas are phylogenetically closely related to each other and share 16S rDNA sequence similarities of between 98.6% and 99.6%.

A survey of the disease distribution and prevalence of confirmed and putative insect vectors throughout Europe has been performed in the framework of COST action FA0807. Based on these results, a database consisting of maps and detailed tables from 28 European and Middle Eastern countries was compiled (Tedeschi *et al.*, 2013; http://costphytoplasma.ipwgnet.org/InsectVectors.htm; last accessed June 2025). Table 12.1 gives an introductive overview of the most important phytoplasma diseases and their agents, which are transmitted by psyllid vectors.

Pear decline

Pear decline is one of the most important diseases of pear and induces a more-or-less quick decline of the tree. The quick decline leads to the sudden death of the entire tree within a few weeks, whereas a slow decline is characterized by continuous degradation and death after some years (Seemüller, 1989). Premature leaf reddening is an unspecific but highly indicative symptom of a PD infection of pear (Jarausch *et al.*, 2023c). A decline-like disorder called "moria del pero" had already been reported in Italy at the beginning of the 20th century (Mader, 1908; Refatti, 1964). In North America, PD was first described in British Columbia (McLarty, 1948) and then spread along the Pacific Coast (Woodbridge *et al.*, 1957). The disease presumably originated from Europe and was introduced into North America as along with its vector (Seemüller, 1989). Devastating epidemics of PD were observed in the 1950s and 1960s along the Pacific Coast of North America and in Italy (Refatti, 1964). Today, PD probably occurs wherever pear is grown in Europe and North America. PD has also been reported from the Middle East in Asia: from Turkey (Sertkaya *et al.*, 2008), Iran (Salehi *et al.*, 2008), and recently from Jordan and Israel (Alloush *et al.*, 2024; Sharon *et al.*, 2025). A similar decline in pear has been found in Taiwan (Liu *et al.*, 2007). New outbreaks of PD are currently observed across Europe.

Apple proliferation

Apple proliferation induces specific symptoms such as witches'-brooms, enlarged stipules, or small-sized fruits with poor taste (Seemüller, 1990). These typical symptoms can be best seen in late summer or fall. Early break of leaf buds in spring and reddening in late summer can be the first indication of the disease. AP-infected trees are not usually killed by the phytoplasma infection. However, AP is one of the economically most important phytoplasma diseases in Europe and

Table 12.1. Phytoplasma diseases of fruit crops, its agent, its psyllid vector, and the vector's host plant.

Psyllid species	Phytoplasma	Disease	Reproduction host plant
Cacopsylla picta	'*Ca.* P. mali'	Apple proliferation	*Malus*
Cacopsylla melanoneura	'*Ca.* P. mali'	Apple proliferation	*Crataegus*, *Malus*
Cacopsylla pruni	'*Ca.* P. prunorum'	European stone fruit yellows	*Prunus* species
Cacopsylla pyri	'*Ca.* P. pyri'	Pear decline	*Pyrus*
Cacopsylla pyricola	'*Ca.* P. pyri'	Pear decline	*Pyrus*
Cacopsylla pyrisuga	'*Ca.* P. pyri'	Pear decline	*Pyrus*
Cacopsylla bidens	'*Ca.* P. pyri'	Pear decline	*Pyrus*
Cacopsylla pyri	'*Ca.* P. pyri'	Peach yellow leaf roll	*Prunus persica*
Cacopsylla pyricola	'*Ca.* P. pyri'	Peach, yellow leaf roll	*Prunus persica*
*Cacopsylla qianli**	PDTW phytoplasma	Pear decline-Taiwan	*Pyrus*
Cacopsylla chinensis	PDTW phytoplasma	Pear decline-Taiwan	*Pyrus*

*Presumed vectors.

seriously impairs fruit size, quality, yield, and vigor of the trees (Kunze, 1989). Fruits of infected trees are poor in taste, and their size and weight are reduced by 30–60%, rendering them unmarketable. AP was first reported in northern Italy (Rui *et al.*, 1950). Since then, AP has been detected in most European countries where apple is grown commercially and has been recorded in Austria, Belgium, Bosnia and Herzegovina, Bulgaria, Croatia, Czech Republic, Finland, France, Germany, Hungary, Italy, Norway, Poland, Romania, Serbia, Slovenia, Spain, Switzerland, the Netherlands, and Turkey (Tedeschi *et al.*, 2013). The highest incidences of AP were found in the past in temperate climatic zones of middle, western, southern, and eastern Europe (Kunze, 1989), but nowadays the disease has spread further northwards and southwards of this zone. The geographic distribution of AP is restricted to Europe and neighboring regions like the Asian part of Turkey (Canik and Ertunc, 2007).

European stone fruit yellows

European stone fruit yellows is the common name of several economically important decline diseases of stone fruits (*Prunus* spp.) in Europe. A decline in apoplexy was first reported for apricots in France (Chabrolin, 1924) and Japanese plums in Italy (Goidànich, 1933). Since then, several diseases have been described as apricot chlorotic leaf roll of apricot (*P. armeniaca*), leptonecrosis and decline of Japanese plum (*P. salicina*), peach yellows, and decline of peach (*P. persica*). Similar disorders have been found in European plum (*P. domestica*), almond (*P. amygdalus*), and flowering cherry (*P. serrulata*) (Lorenz *et al.*, 1994). The most characteristic symptoms that are observed in most of the infected *Prunus* species are off-season growth, premature break of leaf buds before flowering, and leaf yellowing or reddening in combination with leaf roll in summer (Marcone *et al.*, 2010). ESFY is the limiting factor in the production of apricot and Japanese plum in several major stone fruit-growing areas of Europe and neighboring Mediterranean regions, where it causes considerable economic losses due to the high mortality of infected trees. Its economic incidence is less important in peach, almond, and European plum. Disorders in cherry are mostly unrelated to the ESFY agent but are caused by other types of phytoplasmas (Marcone *et al.*, 2010). Molecular characterization of the pathogen revealed that all these diseases are caused by a similar organism, the ESFY phytoplasma (Lorenz *et al.*, 1994; Jarausch *et al.*, 1998). Currently, the disease is reported in 16 out of the 28 EU countries, as well as in Switzerland, the Balkan States (Serbia, Bosnia and Herzegovina, and Albania), Turkey, and Ukraine. Further records outside Europe are available from northern Africa (Egypt and Tunisia) and western Asia (Azerbaijan) (Steffek *et al.*, 2012).

Peach yellow leaf roll disease

Whereas ESFY disease of *Prunus* is limited to Europe, PYLR is a disease of stone fruits in North America caused by a similar phytoplasma. It is distinct from western X-disease, which also occurs in *Prunus* species in North America. It differs from X-disease by symptomatology, speed of disease spread, and by the insect vector (Blomquist and Kirkpatrick, 2002a). While X-disease is transmitted exclusively by leafhoppers, the main vector of PYLR is a pear psyllid. PYLR was first described in the middle of the 20th century and caused major losses in peaches in the late 1970s (Purcell *et al.*, 1981). PYLR was probably introduced to North America and is predominantly found in peach orchards in the neighborhood of pear orchards (Purcell *et al.*, 1981). Only recently, Sabaté *et al.* (2014) detected peach orchards in Spain with PYLR-like symptoms. PCR tests revealed infection with '*Ca.* P. pyri' but not with '*Ca.* P. prunorum,' and they identified *Cacopsylla pyri* as a vector of PYLR in Spain (Sabaté *et al.*, 2018). Molecular characterization of the pathogen revealed that PYLR phytoplasma is a genetic variant of '*Ca.* P. pyri' (Sabaté *et al.*, 2018).

Taxonomy of Psyllid Vectors

The Hemiptera (= Rhynchota) is divided into three suborders: Sternorrhyncha, Auchenorrhyncha, and Heteroptera. Among the

Sternorrhyncha, the jumping plant-lice or psyllids form the well-defined superfamily Psylloidea. About 4000 species of this superfamily are described worldwide, including about 400 species in Europe (Ouvrard, 2015). The taxonomy of this superfamily has recently undergone far-reaching revisions. This first revision was still based only on morphological reinterpretation (Burckhardt and Ouvrard, 2012), leading to eight families: Aphalaridae, Carsidaridae, Calophyidae, Homotomidae, Liviidae, Phacopteronidae, Psyllidae, and Triozidae, of which Aphalaridae, Liviidae and Psyllidae were newly redefined. Applying comprehensive molecular analyses, an important update of this classification of Psylloidea has been presented in more recent years (Burckhardt *et al.*, 2021). This second revised classification reflects the phylogenies presented by Percy *et al.* (2018) and Cho *et al.* (2019). Percy *et al.* (2018) studied more than 400 psyllid species using a next-generation sequencing approach. They obtained near-complete mitogenomes for most species and partial nuclear genomes (data from *c.*373 genes) for a representative subset of taxa. Cho *et al.* (2019) applied a phylogenetic approach using mitochondrial and nuclear gene fragments to reconstruct phylogenetic relationships to elucidate the evolution of Psylloidea. The results of both studies are primarily congruent and contribute to a consensual hypothesis on the phylogeny of Psylloidea. All recognized phytoplasma vectors are found in one subfamily of Psyllidae. Within this subfamily, Psyllinae, the genus *Cacopsylla* is the largest genus and includes all important vector species for fruit tree phytoplasmas (Ouvrard, 2015; Jarausch *et al.*, 2019a).

Biology of Psyllid Vectors

Psyllids are phloem feeders, and nymphs, as well as adults, feed on plant sap. However, they can also feed on the xylem, as demonstrated for *Cacopsylla pyri* (Civolani *et al.*, 2011), *C. pruni* (Gallinger and Gross, 2018), or *C. picta* (Görg and Gross, 2021). Psyllids have highly specialized host requirements. Most psyllids are monophagous, having only one host plant; others are oligophagous with a few closely related host plants; and only a few are polyphagous (Ossiannilsson, 1992; Burckhardt, 1994). Adults are often able to exploit other food plants temporarily. Depending on the species, the eggs are laid on the new buds, bark crevices, or leaves, where they can produce pit-like deformations on the leaf blade. The egg is oblong with a basal pedicel, which is inserted into the plant tissue. The larval development passes through five instars, which are more-or-less strongly flattened dorsoventrally (Burckhardt, 1994).

Many tropical and southern temperate species are polyvoltine with overlapping generations. By contrast, northern temperate species tend to be univoltine or bivoltine. Species overwinter as eggs, young nymphal instars, or adults. These often migrate to overwintering or shelter plants, such as conifers, and return to their host plants in spring, where they complete their immature to adult life cycle (Burckhardt *et al.*, 2014). Important univoltine vector species with an obligate alternation between host plants for reproduction and plants for overwintering are *C. pruni*, *C. melanoneura*, *C. picta*, and *C. pyrisuga* (Mayer and Gross, 2007; Cermák and Lauterer, 2008; Thébaud *et al.*, 2009; Jarausch *et al.*, 2013; Jarausch and Jarausch, 2014, 2016). In these cases, overwintering was observed only on conifers at higher altitudes (Mayer and Gross, 2007; Thébaud *et al.*, 2009; Pizzinat *et al.*, 2011; Ulubas Serçe *et al.*, 2011). Migration to the respective reproduction or overwintering plant seems to be direct and may take place even over long distances, for example 27 km in the case of *C. pruni* (Thébaud *et al.*, 2009).

When feeding, psyllid nymphs may cause considerable damage to their host plants since their salivary injections may contain phytotoxins and thus produce serious necroses and malformations. Certain phytoplasma vector species, such as pear psyllids, may also occur as direct plant pests, which makes them economically particularly important (Ossiannilsson, 1992).

The basic geographic distribution of the most important psyllid vectors in Europe and neighboring regions has been assessed during the COST action FA0807 and is available online (Tedeschi *et al.*, 2013). An electronic determination key for the most important *Cacopsylla* species found on Rosaceae in Europe is available online at www.psyllidkey.info and www.dlr.rlp.de/Psylliden (both last accessed June 2025) (Burckhardt *et al.*, 2008). The database Psyl'list – The World Psylloidea Database provides an almost complete inventory of all Psylloidea worldwide (Ouvrard, 2015; https://data.nhm.ac.uk/dataset/psyl-list/resource/8746ceec-4846-4899-b607-9ba603002033; last accessed June 2025).

Vectors of pear decline

The pear psyllids (Hemiptera: Psylloidea: *Cacopsylla*) comprise 24 known species of small sap-feeding insects distributed in Europe, temperate Asia, and the Americas, that are limited to pear as their developmental hosts (Cho *et al.*, 2017; Civolani *et al.*, 2023b; Horton, 2024). Pear psyllids can cause direct damage to pear trees: the larvae affect plant growth by withdrawal of plant sap, and the secreted honeydew burns plant tissue and favors the growth of sooty mold; adults damage the plants by injection of salivary phytotoxins. In Europe, three confirmed vectors of PD live on pear: *C. pyri* (Linnaeus), *C. pyricola* (Foerster), and *C. pyrisuga* (Foerster). Using DNA barcoding, recent studies of Cho *et al.* (2019) showed that there is no evidence for the presence of any European pear psyllid in eastern Asia. No single life cycle describes the seasonal biology of all pear psyllids. Species may differ in the life stages that overwinter, the number of generations, dispersal tendencies, and the use of winter shelter plants.

Cho *et al.* (2017) classified the pear psyllids into three groups by morphology and life cycle: the pyri group, pyrisuga group, and nigella group. Life cycle categories are based upon generation numbers, the presence or absence of seasonal dimorphism, and the overwintering stage. Understanding of the annual cycle is far more extensive for species in the pyri-group than for the other two groups. The pyri-group consists of 16 species from the Western or Eastern Palearctic regions and includes important pests such as *C. pyri*, *C. pyricola*, *C. bidens*, and *C. chinensis*. Species in this group are multivoltine and overwinter in reproductive diapause as a seasonally distinctive form or morphotype. Detailed descriptions of the life cycle and an updated geographic distribution of all three groups are given by Civolani *et al.* (2023b). Regarding phytoplasma vectors, *C. pyri* is reported from Europe, the Caucasus, Central Asia, the Russian Far East, and China; *C. pyricola* naturally occurs in the Western Palearctics and was introduced into the USA and Canada in the early 19th century. The two species are oligophagous on *Pyrus* species, such as *P. communis*, *P. eleagrifolia*, *P. pyraster*, *P. amygdaliformis*, and *P. salicifolia* (Burckhardt, 1994). The biology of *C. pyri* and *C. pyricola* is similar: both are polyvoltine. Thus, *C. pyri* can produce 4–5 generations in Central Europe and up to 8 generations in southern France. Two morphologically distinct forms can be distinguished: a darkish winter form (*C. pyri* f. *pyri*) and a light summer form (*C. pyri* f. *pyrarboris*). *C. pyricola* has 4–5 generations in France and 3–4 generations in the USA, with the darker winter form (*C. pyricola* f. *simulans*) and the lighter summer forms (*C. pyricola* f. *pyricola*). The first oviposition of the winter form coincides with rising temperatures in early spring on the buds and midribs of the leaves (Burckhardt, 1994). Another polyvoltine member of the pyri-group that is of increasing importance is *C. bidens*. Its current distribution covers France, Armenia, Bulgaria, Estonia, Greece, Iran, Israel, Italy, Jordan, Kazakhstan, Kyrgyzstan, Latvia, Lebanon, Moldova, Uruguay, Mongolia, Romania, Slovakia, Slovenia, Turkmenistan, Ukraine, and Uzbekistan, Argentina, and Chile (Ouvrard, 2015). It has been recently reported from India (Akbar *et al.*, 2017). The most abundant psyllid reported from pear orchards in China was *C. chinensis*, while *C. quianli* was only found rarely (Chen *et al.*, 2018). Katoh *et al.* (2013) observed three distinct clusters, which comprised all Japanese *C. chinensis*, only Taiwanese *C. chinensis*, and Taiwanese and Chinese *C. chinensis*, respectively.

Pear psyllids of the pyrisuga-group have a substantially different life cycle. The pyrisuga-group consists of one Western Palearctic species and two Eastern Palearctic species (Civolani *et al.*, 2023b). The Western Palearctic *C. pyrisuga* is univoltine, overwinters on conifers, and remigrates to its host plant *Pyrus*, in early spring. A detailed description of the life cycle of *C. pyrisuga* is given by Civolani *et al.* (2023b). Egg deposition starts at the end of March when buds begin to open, and new adults appear in May and June after 6 weeks' larval development (Jarausch and Weintraub, 2014; Jarausch *et al.*, 2019a), followed by dispersal of the new adults to high-altitude conifers assisted by up-slope summer winds (Cermák and Lauterer, 2008).

Vectors of apple proliferation

Two psyllids, *C. picta* (Foerster) (syn. *C. costalis*) and *C. melanoneura* (Foerster) are recognized vectors of '*Ca.* P. mali' (Frisinghelli *et al.*, 2000; Tedeschi *et al.*, 2002; Jarausch *et al.*, 2003). In most of the studied areas, both species are present and occur sympatrically (Jarausch *et al.*, 2003; Delić *et al.*, 2005; Carraro *et al.*, 2008; Mattedi *et al.*, 2008; Miñarro *et al.*, 2016); in some others (north-western Italy and Norway), only *C. melanoneura* has been found (Tedeschi *et al.*, 2002; Hatteland *et al.*, 2022). The spatial distribution, natural infection rate, and transmission capacity of *C. picta* and *C. melanoneura* are heterogeneous among different geographic regions.

C. picta (Foerster, 1848) is distributed across Europe and Asia Minor and is monophagous on *Malus* spp. (Jarausch and Jarausch, 2010; Ouvrard, 2015). The insect completes one generation per year and overwinters as an adult on conifers, particularly of the genera *Picea* and *Pinus* (Mayer *et al.*, 2011; Jarausch and Jarausch, 2014). At the end of winter (March/April), *C. picta* remigrants move from the overwintering sites to apple trees for oviposition (Jarausch and Jarausch, 2014). In northern Italy, *C. picta's* reproduction takes place from mid-March until the beginning of August. During this period, the overwintered adults (remigrants) of *C. picta* can be detected in apple orchards until the beginning of June, while the young adults of the new generation (emigrants) occur from the end of June until the beginning of August (Mattedi *et al.*, 2008; Fischnaller *et al.*, 2017). The insects of the new generation (emigrants) feed on their reproductive hosts until the beginning of July when they leave the apple trees as adults (Mattedi *et al.*, 2008; Jarausch *et al.*, 2011). However, in the course of climate change, the migration and reproduction phases have shifted forward. In apple orchards in south-western Germany, remigrants have been seen in the second half of February (Jarausch and Jarausch, 2014; W. Jarausch, unpublished data). Jarausch and Jarausch (2014) succeeded for the first time to maintain a hibernating continuous colony of *C. picta* until the next year and hence could establish a permanent rearing of this vector species. The results suggested a direct migration of emigrants from the reproductive host apple to conifers as estivation and overwintering plants in summer and vice versa to apple orchards for remigrants in early spring. The study confirmed that *C. picta* is univoltine and cannot reproduce on conifers. However, *C. picta* adults can feed on conifers during hibernation. Barthel *et al.* (2020) studied the feeding behaviour of *C. picta* using an amplicon deep sequencing approach to identify plant DNA in the insects. They found a great diversity of DNA from different non-host plants, and among those, DNA from conifers.

C. melanoneura has a Palearctic distribution and is oligophagous on Rosaceae such as *Crataegus*, *Malus*, and *Pyrus* spp. (Ouvrard, 2015). The life cycle is similar to that of *C. picta*, but the overwintering adults appear earlier in the year on *Crataegus* or apple trees, and the new generation abandons the host plant earlier than *C. picta* to migrate to the overwintering plants (Tedeschi *et al.*, 2002; Mattedi *et al.*, 2008; Mayer *et al.*, 2011; Pizzinat *et al.*, 2011). Pizzinat *et al.* (2011) collected adults of *C. melanoneura* also on *Abies alba* (Miller) and *Larix decidua* (Miller) at a high-altitude range in the mountains in north-western Italy. Using sleeve cages, they were able to hibernate *C. melanoneura* in branch cages on different coniferous species at a high altitude. Tedeschi *et al.* (2012) analyzed in detail the population dynamics of *C. melanoneura* in north-east Italy. They observed a very variable population trend of overwintered *C. melanoneura* during the years of investigation. Barthel *et al.* (2020) evaluated the effect of daytime and tree canopy height on sampling of *C. melanoneura* in South Tyrol, Italy. There was a tendency to catch more emigrant adults of *C. melanoneura* in the morning and in the lower part of the canopy, while for *C. melanoneura* remigrants, no differences were observed. With regard to host plants, defined as plants where *C. melanoneura* completes its immature to adult life cycle, several studies attributed diverse roles to hawthorn or apple depending on different geographic areas. In north-western Italy, *C. melanoneura* is equally common in both apple orchards and on *Crataegus monogyna* plants; moreover, hawthorn plants have been found infected with '*Ca.* P. mali' entailing a possible role of this plant as a phytoplasma source of inoculum (Tedeschi *et al.*, 2009). Conversely, in Germany, this psyllid preferred hawthorn as a host plant, and it was not found infected with the phytoplasma (Mayer *et al.*, 2009). To assess the ecological differences, Malagnini *et al.* (2013) conducted host-switching experiments on *C. melanoneura* samples collected from apple and hawthorn and investigated genetic differences between these samples by genotyping seven specific microsatellite markers. Both ecological and genetic results indicated a differentiation between *C. melanoneura* samples associated with the two host plants. As *C. melanoneura* is often mistaken on hawthorn and conifers for another species, *C. affinis* (Löw), which is morphologically very similar, Tedeschi and Nardi (2010) explored a DNA-based discrimination in these two hawthorn-feeding species.

Vectors of European stone fruit yellows

So far, *C. pruni* (Scopoli, 1763) is the only described vector of '*Ca.* P. prunorum' (Carraro *et al.*, 1998b; Jarausch *et al.*, 2001, 2007a, 2008; Thébaud *et al.*, 2009; Marcone *et al.*, 2010). *C. pruni* is widespread in the Western Palearctic (Ossiannilsson, 1992; Ouvrard, 2015). A comprehensive overview of the distribution of '*Ca.* P. prunorum' and its vector *C. pruni* in European fruit-growing areas were given by Steffek *et al.* (2012). *C. pruni* can be divided into a complex of two genetic groups (A and B), which have similar biology and morphology, but there are some differences between the effectiveness of phytoplasma transmission (Sauvion *et al.*, 2007, 2021; Peccoud *et al.*, 2013). *C. pruni* is strictly oligophagous on *Prunus* spp. and overwinters as an adult on conifers (Jarausch and Jarausch, 2016; Gallinger and Gross, 2018). Jarausch and Jarausch (2016) succeeded in establishing a permanent rearing of *C. pruni* by completing the entire life cycle, including a host plant switch from *Prunus* to conifer and back to *Prunus*. At the end of winter/early spring, according to the climatic conditions and geographic zones, adult remigrants move from their overwintering plants, such as spruce, back to *Prunus* spp. for oviposition (Thébaud *et al.*, 2009; Jarausch and Jarausch, 2016; Gallinger and Gross, 2018). A mark, release, and recapture experiment conducted by Maier *et al.* (2013) in a model apricot orchard in Lower Austria showed that the released *C. pruni* easily migrated from tree to tree within 24 h after release but the majority remained within the model orchard after 8 days. Using electrical penetration graph (EPG) recordings and survival bioassays, Gallinger and Gross (2018) were able to demonstrate that adults of *C. pruni* feed on conifers while nymphs were only able to ingest phloem and xylem sap of coniferous trees but cannot develop on a conifer diet.

Many studies showed that wild *Prunus* spp. play an essential role in the epidemiology of the disease, not only as a reservoir of the psyllid vectors but also of the phytoplasma, above all, blackthorn (*P. spinosa*) and myrobalan plum (*P. cerasifera*) (Jarausch *et al.*, 2019a). Sabaté *et al.* (2016) pointed out the role of *P. mahaleb* as a potential reservoir of '*Ca.* P. prunorum' and its vector *C. pruni* in Spain. In contrast, low populations of *C. pruni* were found on cultivated *Prunus* spp., such as *P. armeniaca*, *P. persica*, *P. amygdalus*, and *P. domestica*, and much higher vector densities were reported from different wild *Prunus* spp. such as *P. spinosa*, *P. cerasifera*, *P. domestica*, and *P. salicina*. Interestingly, the infected wild *P. spinosa* and *P. cerasifera* rarely showed any typical symptoms (Carraro *et al.*, 2002; Jarausch *et al.*, 2008, 2019b). Nationwide monitoring of *C. pruni* and the presence of '*Ca.* P. prunorum' in the host plants of the insect in Germany and France showed that both are widespread in wild autochthonous European *Prunus* such as *P. spinosa* and *P. cerasifera* (Jarausch *et al.*, 2019b; Marie-Jeanne *et al.*, 2020). The results supported the hypothesis that ESFY is an endemic European disease with a natural cycle including wild *Prunus* as a reservoir. Thus, *P. spinosa* represents a major source of infection in stone fruit orchards. This was also supported by the analysis of the genetic variants of '*Ca.* P. prunorum' in plants and insects (Riedle-Bauer *et al.*, 2019) and by a multiscale study of the dispersal of the pathogen and its vector between cultivated and wild *Prunus*-growing compartments in France (Marie-Jeanne *et al.*, 2020).

In a recent survey conducted in Hungary (Koncz *et al.*, 2023), high population densities of *C. crataegi* were found in apricot orchards, and 2.6% out of 231 tested specimens were positive for '*Ca.* P. prunorum.' No transmission trials have been conducted so far, but *C. crataegi* has already been found infected with '*Ca.* P. pyri' in north-western Italy (Tedeschi *et al.*, 2009) and with '*Ca.* P. mali' in Spain (Miñarro *et al.*, 2016). Phylogenetic analyses showed that Hungarian and Italian populations of *C. crataegi* form two distinct clades (Koncz *et al.*, 2023), which led to the hypothesis that *C. crataegi* might contribute to the observed disease spread of ESFY in Hungary.

Acquisition and Transmission Characteristics of Psyllid Vectors

The spread of phytoplasmas by their insect vectors is based on specific acquisition and transmission characteristics, which *a priori* exclude certain insect species as competent vectors. Insect vectors acquire the phytoplasma by feeding on the phloem of infected plants. Solely acquiring phytoplasmas by an insect does not imply that the insect is a vector, since phytoplasmas may be acquired but not reinjected by feeding (Jarausch and Weintraub, 2013). Therefore, the natural infection rate is not necessarily correlated with its transmission capacity (Jarausch and Jarausch, 2010). The acquisition and transmission process consists of three phases: acquisition (AAP), latency (LP), and inoculation (IAP). After acquisition, the phytoplasma must overcome several barriers inside the insect's body. Details are described in Weintraub and Beanland (2006) and Bosco and D'Amelio (2010). Only if the pathogen is able to arrive at the critical part of the salivary gland and replicate to a minimum titer for transmission, can an insect become a competent vector (Jarausch and Weintraub, 2013).

The length of time needed for an individual to become infectious, the latent period or incubation period differs among the psyllid vectors. For example, the univoltine vector *C. pruni* has a long latent phase and predominantly transmits the phytoplasma after overwintering when remigrating into the orchards (Thébaud *et al.*, 2009; Marie-Jeanne *et al.*, 2020). In this case, the disease spread is monocyclic on a regional scale. In contrast, vectors with short latent periods, like *C. picta*, transmit the phytoplasma before migration as well as after overwintering, leading to a polycyclic disease spread on a local and a regional scale (Jarausch *et al.*, 2011, 2013).

Vectors of pear decline

First reports of pear psyllids as vectors of phytoplasmas came from the Pacific Coast of North America. Jensen *et al.* (1964) identified *C. pyricola* as the vector of '*Ca.* P. pyri' at a time when the disease was thought to be virus-borne. In California, Blomquist and Kirkpatrick (2002a) detected the pathogen in both the winter form and summer form of *C. pyricola* but without a clear seasonal trend. Since then, no further vector has been described for the USA (Jarausch *et al.*, 2019a). In Europe and the whole Palearctic region, the distribution of the putative vectors of '*Ca.* P. pyri' is diverse: while for Great Britain only *C. pyricola* has been described as a vector (Davies *et al.*, 1992), *C. pyri* was identified as the main vector in France (Lemoine, 1984), Italy (Carraro *et al.*, 1998a), and Spain (Garcia-Chapa *et al.*, 2005). In the Czech Republic (Ludvikova *et al.*, 2011) and in Lower Austria (Lethmayer *et al.*, 2011) all three pear psyllid species were found naturally infected with '*Ca.* P. pyri.' In Turkey, only *C. pyri* carried the phytoplasma, albeit all three species were present in the investigated regions (Kaya *et al.*, 2016). In Germany, only recently, all three known pear psyllid vectors were captured in various population densities. Winter forms of *C. pyri* had infection rates between 0.4% and 4.4%, while winter forms of *C. pyricola* had infection rates of 2.9–9.4%. The univoltine species *C. pyrisuga* was scarcely captured, with only one infected individual (Jarausch *et al.*, 2023b). In contrast, infection rates of 9.6% have recently been reported for *C. pyrisuga* remigrants from Austria (Riedle-Bauer *et al.*, 2022).

A fourth pear psyllid, the polyvoltine species *C. bidens* (Šulc, 1907), has been found with '*Ca.* P. pyri' in Bulgaria (Etropolska *et al.*, 2015), Jordan (Alloush *et al.*, 2023, 2024), and Israel (Sharon *et al.*, 2025). Only recently, Tomer *et al.* (2025) confirmed the vector capacity of *C. bidens* in transmission trials. PD has also been found in Taiwan (PDTW), where the European species *C. pyri* and *C. pyricola* are not present. Liu *et al.* (2007) found two other *Cacopsylla* species, *C. qianli* and *C. chinensis*, infected with the PDTW phytoplasma, and Liu *et al.* (2011) proved the transmission capacity of *C. chinensis* during transmission trials in Taiwan. Based on PCR detection and symptom development, they showed that pear trees were either infected by PDTW or PDTWII phytoplasma, or co-infected by both, when exposed to *C. chinensis* specimens.

The proof of the vector capacity via transmission trials was undertaken in the UK, where transmission trials carried out with field-collected *C. pyricola* yielded transmission rates between

3% and 61 %, depending on the collection site of the psyllids (Davies *et al.*, 1992). Acquisition of '*Ca.* P. pyri' by *C. pyricola* from experimentally infected pear seedlings was highest in August and lowest in winter. In Italy, Carraro *et al.* (1998a, 2001a) detected '*Ca.* P. pyri' in 55% of groups of *C. pyri* collected from March to October in the orchards, and 30% of the inoculated test plants became infected. They could furthermore show that *C. pyri* retained infectivity during winter but could not transmit PD to dormant plants. In Spain, the infection rate of *C. pyri* reached almost 100% in September, coinciding with the maximum phytoplasma titer in the aerial plant parts. The highest transmission rate to an artificial sucrose medium was obtained in August and in October. Although the percentage of infected psyllids was similar for both genders, '*Ca.* P. pyri' transmission by females was significantly higher than by males (Garcia-Chapa *et al.*, 2005). Çağlayan *et al.* (2010) showed the capability of *C. pyri* to transmit PD from infected pears to healthy periwinkles under controlled conditions and confirmed it as a vector of '*Ca.* P. pyri' in Turkey. In Austria, winter-form individuals of *C. pyri and C. pyricola* transmitted the pathogen to 19.2% and 4.9% of pear seedlings as test plants, respectively (Riedle-Bauer *et al.*, 2022). The same authors proved for the first time the vectoring ability of *C. pyrisuga* in transmission experiments with remigrants that infected 9.5% of pear seedlings. Also, in Germany, *C. pyri* and *C. pyricola* transmitted the phytoplasma to healthy test plants, but quantitative PCR data indicated that only about 40% of infected individuals of both species had sufficiently high phytoplasma loads to be considered as infectious (Jarausch *et al.*, 2023b). The data for phytoplasma concentration were normalized to the number of insect cells. The authors estimated that individuals with less than one phytoplasma per insect cell had just acquired phytoplasma but were not yet infectious, while potentially infectious individuals had phytoplasma loads of 1–200 phytoplasmas per insect cell or 1×10^5 to 1.7×10^7 phytoplasmas per *C. pyri* individual and 1×10^5 to 7.5×10^7 phytoplasmas per *C. pyricola* individual. Quantitative data on the phytoplasma concentration in pear psyllids were also reported by Blomquist and Kirkpatrick (2002b). They estimated the number of phytoplasmas per psyllid within a range from 1×10^6 to 8.2×10^7 with a higher titer in the winter form. They concluded that psyllid-mediated spring infections could happen well before '*Ca.* P. pyri' would normally recolonize the upper part of the tree from the roots. Raddadi *et al.* (2011) tested *C. pyri* collected in Italian pear orchards, detecting '*Ca.* P. pyri' in 26% of the specimens (in 27% of males and in 25% of females) with nested PCR and in 51% of the individuals through real-time PCR with a number of '*Ca.* P. pyri' cells ranging from 1.43×10^1 to 8.5×10^5 per *C. pyri* individual. Moreover, fluorescent *in situ* hybridization (FISH) allowed the detection of '*Ca.* P. pyri' in Malphigian tubules and salivary glands of *C. pyri*.

Vectors of apple proliferation

Two psyllids are recognized vectors of '*Ca.* P. mali': *C. picta* and *C. melanoneura*. Both psyllids transmit the pathogen in a persistent manner, and the presence of phytoplasma-infected and infective psyllids among the first remigrants collected in apple orchards suggests winter retention of the pathogen (Tedeschi *et al.*, 2003; Jarausch *et al.*, 2004, 2011; Mattedi *et al.*, 2008).

Findings of phytoplasma-infected individuals of *C. picta* have been reported from many different countries: Italy, Germany, Bosnia-Herzegovina, France, Switzerland, Finland, Czech Republic, Bulgaria, Spain, Turkey, and Croatia (Frisinghelli *et al.*, 2000; Jarausch *et al.*, 2003, 2011; Carraro *et al.*, 2008; Delić *et al.*, 2008; Lemmetty *et al.*, 2011; Ludvikova *et al.*, 2011; Etropolska *et al.*, 2015; Kaya *et al.*, 2016; Miñarro *et al.*, 2016; Križanac *et al.*, 2017). In northeastern Italy, the natural infection rate of *C. picta* was found to be between 9% and 13%, respectively, for overwintering and offspring adults (Carraro *et al.*, 2008). In the main apple-growing regions of Italy in Trentino and South Tyrol, mean infection rates of 6% (Mattedi *et al.*, 2008), 11% (Baric *et al.*, 2010), and 21% (Fischnaller *et al.*, 2017) for remigrants were reported. In Germany, about 10% of overwintered *C. picta* were found naturally infected with the pathogen every year (Jarausch *et al.*, 2004, 2007b, 2011). Carraro *et al.* (2008) proved that *C. picta* adults are already highly infective when moving from shelter plants to apple trees, and Jarausch *et al.*

(2011) showed that the vectors retain infectivity during their entire presence in apple orchards. Data for northern Switzerland (10%; Jarausch *et al.*, 2011), western France (Alsace: 14%; Jarausch *et al.*, 2011), Finland (11%; Lemmetty *et al.*, 2011), and Spain (13%; Miñarro *et al.*, 2016) were in a similar range. A lower infection rate of roughly 3% was reported in Bulgaria (Etropolska *et al.*, 2015) and in Turkey at 4.36% (Kaya *et al.*, 2016). Jarausch *et al.* (2011) found no significant relationship between the infestation status of the orchard and the infection rate of *C. picta* remigrants captured within these orchards, although a tendency for higher infestation in abandoned orchards was observed. These data can be explained by a regional dispersal of remigrant individuals and by a certain percentage of transovarial transmission of the phytoplasma, as detected by Mittelberger *et al.* (2017). During experimental transmission trials in different laboratories in Germany and Italy, it was confirmed that both emigrants and remigrants of *C. picta* can transmit the agent efficiently (Jarausch *et al.*, 2004, 2007b, 2011; Carraro *et al.*, 2008; Mattedi *et al.*, 2008). Recently, Oppedisano *et al.* (2020) found transmission rates of 10% for *C. picta* versus 1.5% for *C. melanoneura*, and they demonstrated that all life stages of *C. picta* (remigrants, emigrants, and nymphs) are able to acquire and transmit '*Ca.* P. mali' to a higher extent than any life stage of *C. melanoneura*. Jarausch *et al.* (2011) found a usually higher transmission efficiency with naturally infected remigrants than with emigrants in the field. Individuals experimentally born on infected plants acquired the phytoplasma to nearly 100% while new adults emerging on healthy plants and feeding thereafter on infected plants had an acquisition rate of only about 10% (Jarausch *et al.*, 2010a). Mattedi *et al.* (2008) described that natural transmission in the orchards in Trentino takes place during the migration period of the new generation of *C. picta*. Transmission by emigrants before leaving the orchard is also supported by quantitative real-time data, which indicate that '*Ca.* P. mali' multiplies very quickly to high concentrations after experimental acquisition (Pedrazzoli *et al.*, 2007). This was recently confirmed by Oppedisano *et al.* (2020), who analyzed a higher acquisition ability of the F1 generation from infected plants, and they concluded that the probability of infection is mainly influenced by the presence of F1 nymphs and the phytoplasma load. Mittelberger *et al.* (2017) showed that *C. picta* vertically transmits '*Ca.* P. mali' and that a critical phytoplasma concentration threshold is necessary for successful maternal transmission of the phytoplasma. Moreover, the phytoplasma titer in newly emerged F1 adults was similar to that in infected parental individuals, indicating that transovarially infected F1 adults are as infective as remigrants.

The phytoplasma concentration in *C. picta* remigrants was constantly high, ranging between 10^6 and 10^8 phytoplasmas per insect, and did not differ significantly after arrival in the orchards and emergence of the first offspring (Mayer *et al.*, 2009; Jarausch *et al.*, 2011). This indicates that remigrants are infective during their whole lifespan on apple after overwintering. Jarausch *et al.* (2010a) also reported differences in acquisition, multiplication, and transmission efficiencies among different strains of '*Ca.* P. mali.' In Germany, *C. picta* transmits all three '*Ca.* P. mali' subtypes AT1, AT2, and AP (Jarausch *et al.*, 2000), while in South Tyrol, the spread of AP seems to be predominantly linked to the transmission of subtype AT2 by *C. picta* (Baric *et al.*, 2011). All three subtypes were also identified in *C. picta* of north-eastern Italy (Martini *et al.*, 2008). *C. picta* individuals captured in Finland were infected with subtypes AT2 and AP (Lemmetty *et al.*, 2011).

In apple-growing regions where *C. melanoneura* and *C. picta* coexist, the latter most often has a major vector role (Frisinghelli *et al.*, 2000; Jarausch *et al.*, 2003, 2004, 2007b; Mattedi *et al.*, 2008; Kaya *et al.*, 2016, Miñarro *et al.*, 2016; Oppedissano *et al.*, 2020). Otherwise, in areas where *C. picta* does not occur, *C. melanoneura* has an important role in spreading the disease (Tedeschi and Alma, 2004; Tedeschi *et al.*, 2012; Hatteland *et al.*, 2022). It is considered the main AP vector in north-western Italy (Tedeschi *et al.*, 2002, 2003, 2012) but plays a minor role in north-eastern Italy and other European regions (Mayer *et al.*, 2009; Fischnaller *et al.*, 2017; Oppedisano *et al.*, 2020). Tedeschi *et al.* (2003) evidenced the importance of the overwintered adults of *C. melanoneura* in vectoring '*Ca.* P. mali' due to the longer period spent in the apple orchards and a higher proportion of '*Ca.* P. mali'-infected specimens compared with newly emerged adults. Thus, natural infection rates of *C. melanoneura* in apple orchards are less than 1% in Germany, northern Switzerland, and eastern France (Mayer *et al.*, 2009). In contrast, in Turkey, Kaya *et al.* (2016) found 2.8% *C. melanoneura*/*C. affinis* infected with '*Ca.* P. mali,' and in Spain, Miñarro *et al.* (2016) analyzed a considerably higher infection rate of *C. melanoneura* of 10% compared with 13% infected *C. picta* in the same region. No transmission of '*Ca.* P. mali' by *C. melanoneura* could be obtained in Germany (Mayer *et al.*, 2009), and only 0.36% of infected plants were obtained after transmission trials with *C. melanoneura* in the Trentino region (north-eastern Italy) (Mattedi *et al.*, 2008). In contrast, 29.4% of plants in north-western Italy inoculated with naturally infected overwintered adults tested positive for '*Ca.* P. mali' (Tedeschi and Alma, 2004). These high variations in the vector competence of *C. melanoneura* led to the hypothesis of genetic differences among distinct populations, which can be associated with different subtypes of '*Ca.* P. mali.' A strict co-occurrence of AT1 subtypes and *C. melanoneura* has been found in several regions of north-west Italy (Casati *et al.*, 2010), while in other regions where *C. picta* is the principal vector, there is the dominance of subtype AT2 (Jarausch *et al.*, 2000; Cainelli *et al.*, 2004; Baric *et al.*, 2011) or subtype AP (Jarausch *et al.*, 2000, 2004; Martini *et al.*, 2008). Natural infection rates in overwintered adults of *C. melanoneura* were 6.98% for populations from north-west Italy and 3.18% for populations from north-east Italy, with an average concentration of around 10^6 phytoplasmas per insect for both populations (Corretto *et al.*, 2023). These results are in line with previous studies (Tedeschi *et al.*, 2012; Oppedisano *et al.*, 2020). Monti *et al.* (2013) determined a concentration of 10^4–10^6 copies per individual. In contrast, in Germany, the maximum phytoplasma concentration never exceeded 40,000 copies per individual in field-collected *C. melanoneura*, far below the minimum titer threshold determined for effective transmission by *C. picta* (10^6–10^8 phytoplasma DNA copies per insect) (Jarausch *et al.*, 2007b, 2011; Mayer *et al.*, 2009). Only recently, Candian *et al.* (2020) investigated the temporal dynamics of '*Ca.* P. mali' load in *C. melanoneura* in the different developmental stages in north-western Italy. After the controlled acquisition, they evaluated the phytoplasma concentration in all developmental stages from the first instar to newly hatched adults. While the highest phytoplasma concentration of naturally infected overwintered adults as reference was 2.04×10^3 genomic units (GU)/pg insect 18S rDNA, they determined mean phytoplasma titers ranging around 1.23×10^2 GU/pg insect in newly emerged adults, with individual titers higher than 1×10^3 GU/pg insect 18S rDNA. '*Ca.* P. mali' was detected in all *C. melanoneura* developmental stages, and a higher acquisition efficiency was observed in the third to fifth instar nymphs and newly emerged adults. The phytoplasma titer increased during the period spent on conifers for estivation and overwintering, demonstrating a long latency period. They concluded that even though overwintered adults retained higher quantities of the agent, several nymphs and newly emerged adults may also reach a sufficient phytoplasma load quantity to become infective and hence may pose a certain risk for the disease spread.

Vectors of European stone fruit yellows

Infected individuals of *C. pruni* were found in several European countries. The natural infection rates in *C. pruni* specimens collected from cultivated and/or wild hosts showed high variations: lower percentages are reported from, for example, Germany with 0.8–3% (Jarausch *et al.*, 2019b), 2.3% from Bulgaria (Etropolska *et al.*, 2015), 1.5–4.2% from Poland (Warabieda *et al.*, 2018) or between 3% (Yvon *et al.*, 2004) and 4.8% in France (Marie-Jeanne *et al.*, 2020). In Austria, infection rates ranged between 11.5% for the remigrants and 3.4% for the emigrant insect generation (Maier *et al.*, 2013). In Spain, Sabaté *et al.* (2016) determined infection rates varying between 4% and 40% in different regions. Higher infection rates were found in Hungary at around 15% (Mergenthaler *et al.*, 2017), in Turkey at 23% (Ulubas Serçe *et al.*, 2011), and in Italy, where the infection rate of the remigrants increased from 56% up to 80% after arrival on infected *Prunus* spp. in the orchards (Ermacora *et al.*, 2011).

The transmission of '*Ca.* P. prunorum' by different developmental stages of *C. pruni* was studied in detail under controlled conditions (Carraro *et al.*, 2001b, 2004). The overwintered adults, as well as the adults of the new generation of *C. pruni*, were able to transmit the phytoplasma to healthy test plants. However, transmission rates were consistently lower with emigrants than with remigrants. In Italy, remigrants showed transmission rates between 20% and 100% to different micropropagated test plants, in Germany, field-captured remigrants transmitted the pathogen to 23% of micropropagated test plants (Jarausch *et al.*, 2007a, 2008); and in Austria, *C. pruni* remigrants infected 15.7% of *P. armeniaca* plants, but no transmission occurred with field-captured emigrants (Riedle-Bauer *et al.*, 2019). This is consistent with previous reports indicating a very low transmission rate by young adults before overwintering (Carraro *et al.*, 2004; Jarausch *et al.*, 2007a, 2008; Thébaud *et al.*, 2009; Poggi Pollini *et al.*, 2010). Thébaud *et al.* (2009) demonstrated that the population of *C. pruni* has an extremely long "effective latency" period, which lasts the overwintering period. During this time, the phytoplasma concentration within the insects continuously rises, reaching a maximum of 5×10^7 phytoplasmas per insect at remigration. They concluded that only overwintered adults can efficiently transmit the disease and, thus, the disease spread is monocyclic. Martini *et al.* (2007) and Jarausch *et al.* (2010b) described similar phytoplasma concentrations in remigrants of *C. pruni*. A vertical (or transovarial) transmission of '*Ca.* P. prunorum' was not observed by Carraro *et al.* (1998b) and Thébaud *et al.* (2009), whilst Tedeschi *et al.* (2006) found indications for this passage in *C. pruni*, and Poggi Pollini *et al.* (2010) were also able to detect the ESFY pathogen in some eggs of *C. pruni*.

Vectors of peach yellow leaf roll

Insect vectors for PYLR phytoplasma were searched for intensively in California in the 1980s and 1990s. In field surveys for leafhoppers and psyllids in diseased peach orchards, only *C. pyricola* proved to be infected with PYLR phytoplasma. Infected psyllids were captured from peaches as well as pears grown in the neighborhood (Blomquist and Kirkpatrick, 2002a). In Europe, Sabaté *et al.* (2018) found that *C. pyri* was abundant in pear orchards and was also captured in peach orchards with a high infection rate (49%) with '*Ca.* P. pyri.' Transmission trials showed high transmission rates (up to 100%) of '*Ca.* P. pyri' by *C. pyri* to peach.

Genetics

Molecular discrimination of psyllid vectors

Traditionally, psyllid vector species were identified by morphological means. However, this can be problematic due to the extensive resemblance of some psyllid species, especially among females, and is error-prone for nymphs (Oettl and Schlink, 2015). DNA-based techniques offer new possibilities not only for the identification of difficult-to-identify species but also for the identification of insects that are damaged by the removal from sticky traps. As cytochrome c oxidase subunit I (COI) is widely used as a barcoding marker for species determination, it has also been used in a PCR–restriction fragment length polymorphism (RFLP)-based approach to discriminate *C. melanoneura* and *C. picta* from eight other *Cacopsylla* species (Oettl and Schlink, 2015). This rapid and cost-effective approach also allowed a reliable identification of nymphal stages.

Furthermore, molecular approaches revealed genetic variants and different genetic populations within a vector species. Tedeschi and Nardi (2010) selected specific PCR primers in the mitochondrial control region (CR) to distinguish the two species, *C. melanoneura* and *C. affinis*, which live together on hawthorn. While screening *C. melanoneura* populations with the specific primers, they discovered two genetic variants, which differed by the presence (WI, with indel) or the absence (WOI, without indel) of a 56 bp indel. Whereas the WOI type was dominant in all tested regions in northern Italy, the WI type was consistently found at low percentages, ranging from 5% to 31%. Etropolska *et al.* (2016) detected WI and WOI variants in different populations of *C. melanoneura* in Bulgaria, but the WI type accounted for less than 1% of the samples.

As *C. melanoneura* develops on different host plants (e.g. apple and hawthorn), populations from different geographic regions exhibit striking differences in '*Ca.* P. mali' acquisition and transmission capacities and microsatellite markers (SSR) were developed to therefore characterize the different populations (Malagnini *et al.*, 2007). Using these SSR markers, populations from apple and hawthorn could be differentiated in different geographic regions, indicating the existence of two different host races of *C. melanoneura* (Malagnini *et al.*, 2013). Whether these genetic differences are linked to the observed difference in phytoplasma vectoring ability is not clear. Corretto *et al.* (2023) found a correlation between the phytoplasma subtype and its concentration in infected trees, rather than with the psyllid genetic lineage.

Microsatellite genotyping was also applied to the analysis of the population structure of *C. pruni* (Sauvion *et al.*, 2007, 2009). Two distinct populations, A and B, of *C. pruni* were identified based on the analysis of nine microsatellite loci. Both groups can be phylogenetically separated by their ITS2 sequences and are considered two cryptic species (Peccoud *et al.*, 2013, 2018) as they cannot be distinguished morphologically. The development of specific PCR primers for each group (Peccoud *et al.*, 2013) enabled the establishment of the geographic distribution of both biotypes of *C. pruni* (Sauvion *et al.*, 2021). Both populations occur sympatrically in France (Sauvion *et al.*, 2021). While biotype A is more prevalent in south-western Europe, biotype B is dominant in Central and eastern Europe (Etropolska *et al.*, 2016; Jarausch *et al.*, 2019b). All *C. pruni* individuals in Hungary could be assigned to the genetic group B (Koncz *et al.*, 2023). As mitogenomes were recently used to study the phylogenetic relationships among psyllid species (Percy *et al.*, 2018; Cho *et al.*, 2019), sequencing of the mitogenomes was also applied to characterize the two AP vectors *C. picta* and *C. melanoneura* (Šafářová *et al.*, 2023). The obtained results show a high genetic homogeneity of the mitochondrial DNA of both psyllid species, which share common genetic characteristics with other members of the genus *Cacopsylla* described to date. Both mitogenomes are phylogenetically closely related to the univoltine Asian pear psyllid *C. burkhardti* but not to the polyvoltine European pear psyllid *C. pyri*. The intraspecies genetic variability among individuals of different localities and years of capture is very low for both species (Šafářová *et al.*, 2023).

The next step in genetic psyllid vector characterization will be the assembly of psyllid genomes. A first step in this regard has been accomplished by the first *de novo* transcriptome assembly of *C. melanoneura* (Weil *et al.*, 2020).

Molecular interactions of psyllid vectors

Hemipteran insects are well known for their ancient associations with beneficial bacterial endosymbionts, particularly nutritional symbionts that provide hosts with essential nutrients such as amino acids or vitamins lacking in their diets (Dittmer *et al.*, 2023). All investigated psyllids harbor a bacteriocyte-associated primary endosymbiont ('*Candidatus* Carsonella ruddii'), which supplies essential amino acids to the host and exhibits strict host–symbiont co-divergence (Thao *et al.*, 2000; Schuler *et al.*, 2022). Metagenome sequencing of *C. picta*, *C. melanoneura*, *C. pyri*, and *C. pyricola* enabled the establishment of the complete genomes of their '*Ca.* C. rudii' primary endosymbionts, which are closely related to each other (Šafářová *et al.*, 2023). In addition, a new psyllid-associated co-occurring endosymbiont was characterized for all four psyllids, which perfectly complements the tryptophan biosynthesis pathway that is incomplete in *Carsonella*. The name '*Candidatus* Psyllophila symbiotica' is proposed for this new enterobacterium. Quantitative analyses of the titers of both endosymbionts throughout the year in individuals of *C. pyricola* showed seasonal dynamics among the vegetation periods linked to the different metabolic roles of both endosymbionts at different life stages of the psyllid host (Štarhová Serbina *et al.*, 2023).

So far, there are no indications that these essential endosymbionts influence phytoplasma acquisition, replication, or transmission by the respective psyllid vector. This might differ for other endosymbionts (Gonella *et al.*, 2019). Therefore, Schuler *et al.* (2022) analyzed the microbiomes of four *Cacopsylla* vector species and six *Cacopsylla* non-vector species by 16S rRNA metabarcoding. They found stable differences in the microbiomes of the different species, with *Wolbachia* and *Rickettsia* being the most abundant facultative endosymbionts. Both are known for their possible protective effect on other insect species. While *Wolbachia* was almost absent in

C. picta and *C. melanoneura*, it was dominant in *C. pruni* (Schuler *et al.*, 2022). A similar study conducted with the three pear psyllids *C. pyri*, *C. pyricola*, and *C. pyrisuga* yielded three distinct microbiomes, which were more variable in *C. pyri* and *C. pyricola* compared with *C. pyrisuga*. *Wolbachia* was found in high relative abundance only in *C. pyri*, particularly in the adult stage (Štarhová Serbina *et al.*, 2022). Considering the major biological roles that *Wolbachia* plays in other insect taxa, its impact on biology and vector competence of phytoplasma transmitting *Cacopsylla* is of great interest (Štarhová Serbina *et al.*, 2022) and needs further investigation. Another potentially beneficial endosymbiont, '*Candidatus* Arsenophonus,' has been detected in several psyllid species, including *C. pyri*, *C. pyricola*, and *C. pyrisuga* (Schuler *et al.*, 2022; Štarhová Serbina *et al.*, 2022). In particular, it has been found in high relative abundance in overwintered adults of *C. pyrisuga* (Štarhová Serbina *et al.*, 2022). As phytoplasmas colonize niches in the insect that are occupied by bacterial symbionts and therefore might interact with them (Gonella *et al.*, 2019), their presence might have a direct impact on the microbial community of the host. However, Schuler *et al.* (2022) found no significant effect of phytoplasma infection on the microbial community of *C. melanoneura* and *C. picta*. A further insect-transmitted bacterium, '*Candidatus* Liberibacter europeaus,' was discovered in high prevalence in Italian populations of *C. pyri* (Raddadi *et al.*, 2011) and also in *C. pyricola* and *C. pyrisuga* (Camerota *et al.*, 2012). However, the liberibacter does not induce disease symptoms in psyllid-inoculated pear plants and is thus considered an endophyte (Raddadi *et al.*, 2011).

The vector competence of certain psyllid species seems to be related to specific phytoplasma genetic variants and insect haplotypes. This has been intensively studied for *C. melanoneura*, which is considered the main AP vector in north-west Italy (Aosta Valley and Piedmont) (Tedeschi *et al.*, 2012), but is not a vector in Germany and other regions (Mayer *et al.*, 2009). It has been speculated that this difference is due to the existence of different populations as revealed by Malagnini *et al.* (2013). However, a relation of subtype AT1 of '*Ca.* P. mali' to *C. melanoneura* has been observed (Casati *et al.*, 2010). Corretto *et al.* (2023) recently demonstrated through a multiple factor analysis that the primary factors influencing the acquisition of '*Ca.* P. mali' by *C. melanoneura* were the phytoplasma subtype, its concentration in the leaves of infected trees, and the psyllid lineage. They detected a new subtype, AT1-AO, in north-west Italy that was acquired much more efficiently by certain local *C. melanoneura* lineages.

The direct effect of the phytoplasma on the fitness of the psyllid vector has also been studied in the *C. melanoneura* system. Malagnini *et al.* (2010) showed that '*Ca.* P. mali' does not affect the development and the survival of *C. melanoneura* but causes a lower number of laid eggs on infected plants and a lower percentage of egg hatching rate. A transcriptomic study comparing infected and non-infected adults of *C. melanoneura* showed that the presence of '*Ca.* P. mali' influences several metabolic pathways (i.e. circadian clock, immune and nervous system, and carbohydrate metabolism), highlighting its broad impact on the psyllid's physiology (Weil *et al.*, 2020).

Multitrophic Interactions

In the interplays between host plant, pathogen, and insect vector, the latter plays a crucial role as they are responsible for the spread of the pathogen. Psyllids communicate with their host plant by volatile and gustative cues that can be manipulated by the phytoplasma in its own favor. For the psyllid vector *C. picta*, it was shown that emigrants were attracted to '*Ca.* P. mali'-infected apple plants because the phytoplasma induces apple trees to produce more β-caryophyllene (Mayer *et al.*, 2008a, 2008b). By feeding on infected plants, the probability of acquiring the phytoplasma increases. After overwintering on conifers, the psyllids return to apple plants but now prefer to lay their eggs on uninfected plants due to the deleterious effects of the pathogen on the developing offspring, thereby increasing the opportunity to transmit the phytoplasma (Gross, 2016). This behavior might be attributed to phloem metabolites perceived during probing and feeding, recognizing infected plants as less suitable hosts (Mayer *et al.*, 2011). This preference is correlated with differences in the phloem composition between infected and non-infected apple trees (Görg *et al.*, 2021). In addition, the pattern of plant volatiles emitted by the infected apple plant is modified by the virulence of the phytoplasma (Rid *et al.*, 2016). In pear trees infected with '*Ca.* P. pyri,' the expression of ethylbenzoate differed significantly between infected and healthy trees (Rid *et al.*, 2016).

The European psyllid species *C. pruni* exhibits a clear host plant preference for its natural host *P. spinosa* (Jarausch *et al.*, 2019b) as well as to *Prunus insititia*, while it is found only in low abundance on species of non-European origin, like peach and apricot (Gallinger *et al.*, 2019). The results of Gallinger and Gross (2020) suggest that *C. pruni* avoids *P. persica* because nymphs that fed on peach exhibited a prolonged developmental time and reduced developmental success compared with those observed on *P. insititia*. The phloem sap composition of *P. insititia* seemed to be more adapted to *C. pruni* (Gallinger and Gross, 2020). An infection with '*Ca.* P. prunorum' did not alter the phloem sap composition or affect the nymphal development.

Young adults of univoltine psyllid vectors migrate in early summer to conifers and spend most of their lifespan there. The results of Gallinger and Gross (2018) showed that adults of *C. pruni* can feed and survive on conifers, albeit the plant sap composition between *Prunus* and conifers differed strongly for organic acids and sugars. Nymphs can feed on, but cannot develop on, spruce and fir.

Signals triggering interactions between psyllids, phytoplasma, and plants and within psyllid species are not restricted to volatile and non-volatile infochemicals, as the existence of acoustic communication in courtship behavior of *C. pyri* has been shown (Eben *et al.*, 2015). The vibrational sequences of males and females are different, and females call only in the presence of calling males. Similar vibrational signals have been described for *C. pyricola* (Jocson, 2020). Females of *C. picta* initiate communication on the host plant by emitting trains of vibrational pulses, followed by a duet consisting of a male call and a female reply (Oppedisano *et al.*, 2019). In the same study, the existence of male calls by *C. melanoneura* was also demonstrated. Methods exploiting vibrational signals to disrupt mating may effectively control psyllid pests in the field (Liao *et al.*, 2022).

Control of Psyllid Vectors

As there are no applicable means to cure a phytoplasma-infected fruit tree, insecticide treatments were the first measures to control the spread of fruit tree phytoplasma diseases whenever a vector species was identified. However, classic pest management is hampered for various reasons: insecticide resistance of polyvoltine psyllid species, low abundance of univoltine vectors, missing or incomplete homologation of appropriate products, and, not least, environmental protection considerations. Therefore, alternative strategies of pest control, as well as long-term solutions based on the natural resistance of the plant, are currently under investigation.

The control of psyllid vectors has two different objectives: (i) preventing the transmission of the phytoplasma; and (ii) reducing the vector population. In particular, polyvoltine pear psyllids are pests themselves; they have been treated for a long time with insecticides aiming to reduce their population density. However, insecticide treatments do not necessarily prevent phytoplasma transmission either due to their slow efficacy or by increasing the probing. Therefore, push-and-pull or push-and-kill strategies are being developed in the context of searching for more environmentally friendly control measures.

The optimal timing of vector control is essential for the efficacy of repellent strategies to prevent phytoplasma transmission. A prerequisite is perfect knowledge of the biology and the transmission characteristics of the different psyllid vectors. The highest risk for phytoplasma transmission by univoltine vectors is the period of immigration into the orchards after overwintering (see earlier). Although polyvoltine pear psyllids have the highest phytoplasma infection rates in fall, the risk for phytoplasma transmission is highest after overwintering in late winter/early spring. This is likely due to the physiology of the pear trees: infections in the fall cannot efficiently establish in the tree because of winter degeneration of the phloem of the aerial parts (Schaper and Seemüller, 1982), while spring infections into the newly formed phloem are able to multiply throughout the vegetative season and are, thus, able to infect the tree efficiently. Therefore, vector control of non-migrating pear psyllids is also most important in late winter/early spring.

Timing of vector control

Insect behavior is commonly triggered by temperature and day length. Both factors may vary considerably between geographic regions and climatic zones. Global warming dramatically influences the factor "temperature" and, thus, insect behavior and development (W. Jarausch, personal observation, south-west Germany). This makes the timing of vector control more and more a lottery game. Univoltine vectors that overwinter on conifers at higher altitudes are affected by the temperatures in these altitudes and they probably migrate to their host plants with the help of downdrafts (Cermák and Lauterer, 2008). *C. melanoneura* is the earliest migrating univoltine vector psyllid, followed by *C. picta* and *C. pruni*, which seem to have a similar migration time, and *C. pyrisuga* as likely the latest migrating psyllid vector (W. Jarausch, personal observation, south-west Germany).

Tedeschi *et al.* (2012) defined an empirical immigration index for *C. melanoneura* based on the average maximum temperatures registered in apple orchards within 7 days before the overwintered adults arrive. In practice, in the mountainous region of Valsugana in north-eastern Italy, where this study was carried out, psyllids start to reach the apple orchards when the threshold is 9.5°C. In the analyzed period between 2002 and 2010, the first arrival of the psyllids in apple orchards ranged from calendar weeks 4 to 11, indicating that the day length was not an important factor influencing psyllid migration. In addition, apple phenology, which is important for egg deposition, was not correlated with migration. The temperature threshold should be adjusted for other regions either according to historical collection data or by programming periodical field collections. Moreover, other geographic factors associated with the winter site locations (e.g. the regional orography, the main air streams, and distance from apple orchards) may differently affect the psyllid migration process and influence its presence or absence, both in terms of time and quantity, in any given apple orchard (Tedeschi *et al.*, 2012).

Based on Tedeschi *et al.* (2012), a temperature-based immigration model has been developed for *C. melanoneura* and *C. picta* to predict the first appearance in apple orchards in South Tyrol (Panassiti *et al.*, 2019). The temperature threshold ranged from 5.65°C to 9.77°C for *C. melanoneura* and 12.73°C to 13.82°C for *C. picta*, confirming field monitoring data that the immigration of *C. melanoneura* in apple orchards was observed as early as late January, while *C. picta* remigrants arrive only in March. Panassiti *et al.* (2019) showed strong differences in a small-scale geographic comparison, indicating that additional data and further validations are necessary to generate regionally customized, reliable thresholds useful in insect vector management strategies.

Also, the polyvoltine pear psyllids may migrate and re-enter the orchards in late winter after completion of the diapause. As reviewed in Civolani *et al.* (2023b), winter forms migrate in late fall to shelter plants or other orchards for dispersal under certain climatic conditions. Thus, phytoplasma-infected individuals might enter new orchards in late winter and spread PD disease. In order to apply well-timed control strategies, Schaub *et al.* (2005) developed a phenological model of first- and second-generations of *C. pyri*. The model included a time-distributed delay and was based on data regarding the termination of reproductive diapause, oviposition period length, and developmental duration of juvenile stages depending on temperature. An updated temperature-driven model forecasting the dynamics and population structure (eggs/larvae/adults) of pear psylla was used by Belien *et al.* (2017) to optimise the timing of control sprays targeting a specific life stage of the pest.

Chemical control

Polyvoltine pear psyllids can increase to high population densities, causing direct damage to pear fruits and weakening the trees. In particular, immature stages of the psyllids produce large quantities of honeydew and probably inject toxins into the tree with their saliva (reviewed by Civolani *et al.*, 2023b). Therefore, synthetic insecticides have been applied since the

1940s to reduce the population of pear psyllids (for more details, see Civolani *et al.*, 2023b and Belien *et al.*, 2013). Prebloom treatments targeted the overwintering adults in order to reduce the development of the damaging second generation during fruit development. Postbloom treatments target the development of the different summer generations. However, pear psyllids have a pronounced capability to produce resistance against insecticides due to their host specificity and high reproductivity. A survey initiated by the Insecticide Resistance Action Committee (IRAC) showed that *C. pyri* developed resistance against several organophosphates and pyrethroids in many European and American countries (Sterck and Highwood, 1992). Nowadays, a limited number of specific pesticides, such as abamectin and spirotetramat, are used in integrated pest management (IPMs) programs, leading to reduced damage by pear psyllids. Unfortunately, the EU has confirmed the expiration of approval for spirotetramat in 2024 (Regulation EU 2022/489), while the approval expiration date for abamectin is under review (Regulation EU 2023/515) (Civolani *et al.*, 2023a). Resistance development, as well as the actual ban of synthetic insecticides due to environmental reasons, led to the conclusion that traditional control strategies are not sufficient for the management of pear psyllids. Thus, alternative control mechanisms, such as the release of predators and the use of repellents, were integrated into the new pest management programs and will be discussed here. Civolani *et al.* (2023a) recently showed that potassium bicarbonate, known as an inorganic fungicide, also has insecticidal activity against the juvenile forms of *C. pyri*.

Considering the phenology of the two AP vector species, it appears that their control is possible only when the insects are present in cultivated plants. In Trentino, since 1999, field trials have been carried out in order to find efficient insecticides to control *C. picta* and *C. melanoneura* and to determine the timing of the treatments. Etofenprox was found to be the most efficient product for controlling the overwintering adults of both species before blossoming (Mattedi *et al.*, 2007; Baldessari *et al.*, 2010). *C. melanoneura* was also efficiently depleted with organophosphates (Baldessari *et al.*, 2010). Aiming to find new insecticides with lower ecotoxicological impact, Baldessari *et al.* (2017) also showed that fosmet was effective against both psyllids. The control strategy aimed to prevent the reproduction of both species on apples. A particular problem arose for the control of overwintered adults of *C. picta* in years when oviposition coincided with the period of blossom, when standard insecticides cannot be applied. Another problem with the control of the highly efficient AP vector *C. picta* is its low abundance in the orchards, which necessitates highly efficient products. These are not homologated in many AP-infected countries (Jarausch and Torres, 2014). As both generations of *C. picta* can transmit phytoplasma, both the remigrants and the new generation must be controlled. Organophosphates, as well as neonicotinoids (thiametoxan, thiacloprid), were found to be appropriate products to control the larval development of *C. picta* (Mattedi *et al.*, 2007).

Only very few attempts have been undertaken to control the vector of ESFY by classic means of spraying. Poggi Pollini *et al.* (2007) conducted a field trial for vector control after a severe disease outbreak in the Province of Trentino (Italy). However, most of the applications had no efficacy in controlling the disease. Paleskić *et al.* (2017) tested the efficacy and the instant effect of 13 insecticides and antifeedants on *C. pruni* in cage studies and subsequent field trials. Application of cypermethrin before bloom and thiacloprid after bloom showed the best effects, both on the reduction of the *C. pruni* population and the pathogen transmission.

Biological control

Biological control is rarely efficient when applying only one method, instead it necessitates a sophisticated combination of different strategies well adapted to the pathosystem. For example, *C. pruni* has well-defined host plant preferences (Gallinger *et al.*, 2019) which have to be considered in the agricultural practices of apricot and peach growing. Both fruit tree species are not preferred host plants of *C. pruni* but the suckers of their rootstocks, which are often *P. domestica* and *P. cerasifera*, are highly attractive to the psyllids. Prompt elimination of these rootstock suckers reduces the vector population and the phytoplasma transmission.

Mineral clays have been assessed to repel psyllids from moving and sucking on the plant and thus from transmitting the phytoplasma. Processed kaolin was first shown to prevent *C. pyri* from oviposition and egg development (Daniel *et al.*, 2005; Erler and Cetin, 2007; Pasqualini *et al.*, 2007). Its efficacy against *C. melanoneura* has also been demonstrated (Tedeschi *et al.*, 2007). Late winter treatments with kaolin were able to reduce the number of laid eggs and, consequently, of nymphs. In contrast, kaolin treatments did not significantly affect *C. pruni* remigrants in cage experiments conducted by Paleskić *et al.* (2017). Since kaolin is not considered an insecticide and its impact on beneficial and phytotoxic effects is low, it can also be used in organic farming. It has already been integrated into modern pest management strategies against *C. pyri* (Belien *et al.*, 2013). While the kaolin particle film has a white color and the clay is processed, a totally natural, trunk-colored clay was recently assessed in double-choice cage experiments for its repellent effect against pear psyllids as well as *C. pruni* and *C. picta* (Jarausch *et al.*, 2023a). In addition to kaolin treatments, Nottingham *et al.* (2022) successfully tested reflective mulch (aluminum-metalized polyethylene sheets laid beneath trees) to deter *C. pyricola* by reflecting light back to the canopy of the trees.

A large community of predators and parasitoids attack pear psyllids both within their native Old World range and in regions where the psyllids have been introduced (Civolani *et al.*, 2023b). In particular, the predatory bug *Anthocoris nemoralis* was integrated into selective insecticide programs in Italy and the USA (Civolani and Pasqualini, 2003; Daugherty *et al.*, 2007; Horton, 2024). Phenology data for natural enemies have now been incorporated into psyllid control programs (Nottingham *et al.*, 2022). As insecticide use destroys the population of natural enemies within the orchard, the conservation of natural habitats in the surrounding areas may play an important role as a reservoir for predators and parasites of psyllids (Horton, 2024).

Biological control can be integrated into a sustainable modern pest management program (Belien *et al.*, 2013): the population buildup early in the season can be prevented by multiple sprayings of kaolin, further suppression of the pear psyllid population in spring can be achieved either by multiple mineral oil applications or selective insecticides with no side effects on beneficials, so that during summer, predation by predatory bugs will be sufficient to reduce the following pear psyllid generations until the end of September. Reducing the population density of pear psyllids, namely *C. pyri*, will also help decrease the spread of PD disease, as the infection rates of *C. pyri* are usually low.

Push-and-pull strategies

The outcome of research on multitrophic interactions is the key to developing innovative and sustainable applications such as attract-and-kill, push-and-pull, or push–pull–kill strategies (Gross and Gündermann, 2016). Attractants can be plant-emitted volatiles, specific colors used in traps, or even acoustic signals. Repellents can be plant-emitted volatiles applied in specific blends, natural extracts, or physical barriers like clay minerals or mineral oils. Kill components may be trapped with attractive volatiles, such as β-caryophyllene used to attract *C. picta* (Mayer *et al.*, 2008a), or colored traps, such as green-colored traps for attracting of *C. pyri* (Czarnobai de Jorge *et al.*, 2023), or even entomopathogenic fungi, for example *Pandora* sp. used against *Cacopsylla* species (Görg *et al.*, 2021).

The push component is most important in repelling the psyllids from probing the host plant and thus preventing the transmission of phytoplasmas. As univoltine vectors remigrate in early spring from their overwintering shelter plants, conifers, into the orchards, changes in the composition of conifer volatiles may play a major role in the induction of the migration. Thus, conifer volatiles are thought to have a repellent effect when applied to the reproduction host plant in this period. Ermacora *et al.* (2012) demonstrated in semi-field trials that conifer extracts have a repellent effect on *C. pruni* remigrants. The repellent effect of conifer plants on *C. pruni* was confirmed in field trials conducted in Germany (W. Jarausch, unpublished data). Gallinger *et al.* (2019) found that a synthetic blend of conifer volatiles exhibits a repellent effect on *C. pruni* in the field when applied using a large number of dispensers. Jarausch and Jarausch (2019) showed in double-choice cage experiments that extracts of spruce needles sampled during the migration phase had a repellent effect on *C. picta*.

Essential oils (EOs) are known to have a repellent effect on arthropods and, thus, have also been studied for psyllids. Czarnobai de Jorge *et al.* (2022) tested the major synthetic compounds of clove EO in olfactometer experiments and demonstrated a repellent effect of a specific mixture of these compounds on *C. pyri* and *C. pyricola*. The repellent mixture was loaded on biodegradable dispensers produced by electrospinning biocompatible polymer fibers. However, field trials were inconclusive and suggested further improvement of the dispensers. Recently, Czarnobai de Jorge *et al.* (2024) demonstrated the repellent activities of EOs extracted from *Juniperus mexicana* (cedarwood oil) and *Cinnamomum zeylanicum* (cinnamon bark oil) and their nanoformulations against *C. pyri* in the laboratory. As an attractant pull component, the authors tested a mixture of the five most abundant volatile organic compounds of pear plants against *C. pyri* and *C. pyrisuga* females. In field trials, they tested the selected attractive and repellent volatiles, formulated in biocompatible nanofibers or marker pen dispensers, by counting the psyllids on attractive green sticky traps. In a push-and-count design, they demonstrated the repellent effect of the EOs on *C. pyri* under field conditions.

Outlook of vector control

The ban on classic insecticides and the search for environmentally sustainable control strategies resulted in intensified research on new technologies for insect control. RNA interference (RNAi) is a powerful post-transcriptional gene-silencing mechanism that is increasingly studied for its use in pest control. The application of double-stranded RNA (dsRNA) targeting a gene of interest can specifically target a pest species by inducing mortality or reducing the fitness of the insect. However, the efficiency of RNAi can vary significantly among different insect orders and target genes (Lu and Shen, 2024). For example, Coleoptera are highly susceptible to RNAi, and the effect of a dsRNA treatment is strong and long-lasting. Psyllids also seem to be susceptible to RNAi, as several methods for RNAi-based control of the Asian citrus psyllid (*Diaphorina citri*), the vector of the devastating citrus Huanglongbing disease, have been developed (Taning *et al.*, 2016; Andrade *et al.*, 2017; Galdeano *et al.*, 2017). Many genes are currently evaluated for their efficiency in establishing an RNA pesticide. In addition, efficient dsRNA delivery methods are being studied (Lu and Shen, 2024). However, dsRNA delivery will be particularly challenging for phloem-sucking insects like psyllids.

As psyllids use vibrational signals for mate finding and courtship, they might be susceptible to vibrational manipulation. Thus, a future vibrational management technique could be developed as a control strategy, as proposed for the potato psyllid *Bactericera cockerelli* (Avosani *et al.*, 2020).

References

Akbar, S.A., Dar, M.A., Mahendiran, G., and Wachkoo, A.A. (2017) The first record of pear psylla *Cacopsylla bidens* (Hemiptera: Psyllidae) from India along with notes on seasonal occurrence and some elements of its biology. *Oriental Insects* 52, 101–111. DOI: 10.1080/00305316.2017.1378598.

Alloush, A.H.A., Bianco, P.A., Busato, E., AlMahasneh, A., Alma, A., Tedeschi, R., and Quaglino, F. (2023) Association of seven '*Candidatus* Phytoplasma' species to an almond disease complex in Jordan, and preliminary information on their putative insect vectors. *Crop Protection* 164, 106147. DOI: 10.1016/j.cropro.2022.106147.

Alloush, A.H.A., Bianco, P.A., Alma, A., Tedeschi, R., and Quaglino, F. (2024) Phytoplasma identification in pome fruit trees and Cacopsylla bidens (Hemiptera: Psyllidae) in Jordan. *Europran Journal of Plant Pathology* 169, 65–71. DOI: 10.1007/s10658-023-02808-7.

Andrade, E.C. and Hunter, W. B. (2017) RNAi feeding bioassay: Development of a non-transgenic approach to control Asian citrus psyllid and other hemipterans. *Entomologia Experimentalis et Applicata* 162, 389–396. DOI: 10.1111/eea.12544.

Avosani, S., Sullivan, T.E.S., Ciolli, M., Mazzoni, V., and Suckling, D.M. (2020) Vibrational communication and evidence for vibrational behavioural manipulation of the tomato potato psyllid, Bactericera cockerelli. *Entomologia Generalis* 40, 351–363. DOI: 10.1127/entomologia/2020/0984.

Baldessari, M., Trona, F., Angeli, G., and Ioriatti, C. (2010) Effectiveness of five insecticides for the control of adults and young stages of *Cacopsylla melanoneura* (Förster) (Hemiptera: Psyllidae) in a semi-field trial. *Pest Management Science* 66, 308–312. DOI: 10.1002/ps.1876. PMID: 19937976.

Baldessari, M., Angeli, G., and Oppedisano, T. (2017) Nuove strategiecontro le psille vettori degli scopazzi del melo. *L'Informatore Agrario* 9, 47–51.

Baric, S., Oettl, S., and Dalla Via, J. (2010) Infection rates of natural psyllid populations with '*Candidatus* Phytoplasma mali' in South Tyrol (northern Italy). *Julius-Kühn-Archiv* 427, 189–192.

Baric, S., Berger, J., Cainelli, C., Kerschbamer, C., and Dalla Via, J. (2011) Molecular typing of '*Candidatus* Phytoplasma mali' and epidemic history tracing by a combined T-RFLP/VNTR analysis approach. *European Journal of Plant Pathology* 131, 573–584.

Barthel, D., Schuler, H., Galli, J., Borruso, L., Geier, J. *et al.* (2020) Identification of plant DNA in adults of the phytoplasma vector *Cacopsylla picta* helps understanding its feeding behavior. *Insects* 11, 835. DOI: 10.3390/insects11120835.

Belien, T., Bangels, E., and Peusens, G. (2013) Integrated control of psyllid vectors of European fruit tree phytoplasmas. *Phytopathogenic Mollicutes* 3, 31–36.

Belien, T., Bangels, E., Brenard, N., Reijniers, J., Leirs, H., and Bylemans, D. (2017) Optimized timing of IPM treatments against pear psylla (*Cacopsylla pyri*) based on a temperature driven population dynamics model. *Integrated Protection of Fruit Crops IOBC-WPRS Bulletin* 123, 96–100.

Blomquist, C.L. and Kirkpatrick, B.C. (2002a) Identification of phytoplasma taxa and insect vectors of peach yellow leaf roll disease in California. *Plant Disease* 86, 759–763.

Blomquist, C.L. and Kirkpatrick, B.C. (2002b) Frequency and seasonal distribution of pear psylla infected with the pear decline phytoplasma in California pear orchards. *Phytopathology* 92, 1218–1226.

Bosco, D. and D'Amelio, R. (2010) Transmission specificity and competition of multiple phytoplasmas in the insect vector. In: Weintraub, P.G. and Jones, P. (eds) *Phytoplasmas – Genomes, Plant Hosts and Vectors*. CAB International, Wallingford, UK, pp. 250–271.

Burckhardt, D. (1994) Psylloid pests of temperate and subtropical crop and ornamental plants (Hemiptera, Psylloidea): A review. *Trends in Agricultural Sciences* 2, 173–186.

Burckhardt, D. and Ouvrard, D. (2012) A revised classification of the jumping plant-lice (Hemiptera: Psylloidea). *Zootaxa* 3509, 1–34.

Burckhardt, D., Jarausch, B., and Jarausch, W. (2008) www.psyllidkey.info: ein elektronischer Schlüssel zur Bestimmung der Psylliden auf Rosaceen in Mitteleuropa. *Mitteilungen aus dem Julius-Kühn-Institut* 417, 177.

Burckhardt, D., Ouvrard, D., Queiroz, D., and Percy, D. (2014) Psyllid host-plants (Hemiptera: Psylloidea): Resolving a semantic problem. *Florida Entomologist* 97, 242–246.

Burckhardt, D., Ouvrard, D., and Percy, D.M. (2021) An updated classification of the jumping plant-lice (Hemiptera: Psylloidea) integrating molecular and morphological evidence. *European Journal of Taxonomy* 736, 137–182. DOI: 10.5852/ejt.2021.736.1257.

Çağlayan, K., Gazel, M., Ulubaş Serçe, Ç., and Can, F. (2010) Experimental transmission trials by *Cacopsylla pyri*, collected from Pear Decline infected orchards in Turkey. *Julius-Kühn-Archiv* 427, 403–406.

Cainelli, C., Bisognin, C., Vindimian, M.E., and Grando, M.S. (2004) Genetic variability of AP phytoplasmas detected in the apple growing area of Trentino (North Italy). *Acta Horticulturae* 657, 425–430.

Camerota, C., Raddadi, N., Pizzinat, A., Gonella, E., Crotti, E. *et al.* (2012) Incidence of "*Candidatus* Liberibacter europaeus" and phytoplasmas in *Cacopsylla* species (Hemiptera: Psyllidae) and their host/shelter plants. *Phytoparasitica* 40, 213–221. DOI: 10.1007/s12600-012-0225-5.

Candian, V., Monti, M., and Tedeschi, R. (2020) Temporal dynamics of '*Ca.* Phytoplasma mali' load in the insect vector *Cacopsylla melanoneura*. *Insects* 11, 592. DOI: 10.3390/INSECTS11090592.

Canik, D. and Ertunc, F. (2007) Distribution and molecular characterization of apple proliferation phytoplasma in Turkey. *Bulletin of Insectology* 60, 335–336.

Carraro, L., Loi, N., Ermacora, P., Gregoris, A., and Osler, R. (1998a) Transmission of pear decline by using naturally infected *Cacopsylla pyri*. *Acta Horticulturae* 472, 665–668.

Carraro, L., Osler, R., Loi, N., Ermacora, P., and Refatti, E. (1998b) Transmission of European stone fruit yellows phytoplasma by *Cacopsylla pruni*. *Journal of Plant Pathology* 80, 233–239.

Carraro, L., Loi, N., and Ermacora, P. (2001a) The life cycle of pear decline phytoplasma in the vector *Cacopsylla pyri*. *Journal of Plant Pathology* 83, 87–90.

Carraro, L., Loi, N., and Ermacora, P. (2001b) Transmission characteristics of the European stone fruit yellows phytoplasma and its vector *Cacopsylla pruni*. *European Journal of Plant Pathology* 107, 695–700.

Carraro, L., Ferrini, F., Ermacora, P., and Loi, N. (2002) Role of wild *Prunus* species in the epidemiology of European stone fruit yellows. *Plant Pathology* 51, 513–517.

Carraro, L., Ferrini, F., Labonne, G., Ermacora, P., and Loi, N. (2004) Seasonal infectivity of *Cacopsylla pruni*, the vector of European stone fruit yellows phytoplasma. *Annals of Applied Biology* 144, 191–195.

Carraro, L., Ferrini, F., Labonne, G., Ermacora, P., and Loi, N. (2008) Infectivity of *Cacopsylla picta* (syn. *Cacopsylla costalis*), vector of '*Candidatus* Phytoplasma mali' in north east Italy. *Acta Horticulturae* 781, 403–407.

Casati, P., Quaglino, F., Tedeschi, R., Spiga, F.M., Alma, A., Spadone, P., and Bianco, P.A. (2010) Identification and molecular characterization of '*Candidatus* Phytoplasma mali' isolates in North-western Italy. *Journal of Phytopathology* 158, 81–87.

Cermák, V. and Lauterer, P. (2008) Overwintering of psyllids in South Moravia (Czech Republic) with respect to the vectors of the apple proliferation cluster phytoplasmas. *Bulletin of Insectology* 61, 147–148.

Chabrolin, C. (1924) Quelques maladies des arbres fruitiers de la Vallée du Rhône. *Annales Epiphytes* 10, 265–333.

Chen, P., Liu, Q., Qiao, X., Wang, J., and Zhang, T. (2018) Identification and phylogenetic analysis of pear psyllids (Hemiptera: Psyllidae) in Chinese pear orchards. *Journal of Economic Entomology* 111, 2908–2913. DOI: 10.1093/jee/toy230.

Cho, G., Burckhardt, D., Inoue, H., Luo, X., and Lee, S. (2017) Systematics of the east Palaearctic pear psyllids (Hemiptera: Psylloidea) with particular focus on the Japanese and Korean fauna. *Zootaxa* 4362, 75–98. DOI: 10.11646/zootaxa.4362.1.4.

Cho, G., Malenovsky, I., and Lee, S. (2019) Higher-level molecular phylogeny of jumping plant lice (Hemiptera: Sternorrhyncha: Psylloidea). *Systematic Entomology* 44, 638–651. DOI: 10.1111/syen.12345.

Civolani, S. and Pasqualini, E. (2003) *Cacopsylla pyri* L. (Hom. Psyllidae) and its predators relationship in Italy's Emilia-Romagna region. *Journal of Applied Entomology* 127, 214–220.

Civolani, S., Leis, M., Grandi, G., Gonzalez, E., Pasqualini, E., Chicca, M., and Tjallingii, W.F. (2011) Plant penetration by *Cacopsylla pyri*, an electrical penetration graph (EPG) study. *Journal of Insect Physiology* 57, 1407–1419. DOI: 10.1016/j.jinsphys.2011.07.008.

Civolani, S., Boselli, M., Radicetti, E., and Bernacchia, G. (2023a) Efficacy and selectivity of potassium bicarbonate salts against *Cacopsylla pyri* on pears. *Insects* 14, 491. DOI: 10.3390/insects14060491.

Civolani, S., Soroker, V., Cooper, W.R., and Horton, D.R. (2023b) Diversity, biology, and management of the pear psyllids: A global look, *Annals of the Entomological Society of America* 116, 331–357. DOI: 10.1093/aesa/saad025.

Corretto, E., Trenti, M., Štarhová Serbina, L., Howle, J.M., Dittmer, J. *et al.* (2023) Multiple factors driving the acquisition efficiency of apple proliferation phytoplasma in *Cacopsylla melanoneura*. *Journal of Pest Science* 97, 1299–1314. DOI: 10.1007/s10340-023-01699-1.

Czarnobai de Jorge, B., Hummel, H.E., and Gross, J. (2022) Repellent activity of clove essential oil volatiles and development of nanofiber-based dispensers against pear psyllids (Hemiptera: Psyllidae). *Insects* 13, 743. DOI: 10.3390/insects13080743.

Czarnobai de Jorge, B., Meyhöfer, R., Jürgens, A., and Gross, J. (2023) Preference of pear psyllid (*Cacopsylla pyri*) for specific colour inspires new application in plant protection. *Journal of Applied Entomology* 147, 976–989. DOI: 10.1111/jen.13194.

Czarnobai de Jorge, B., Kossmann, A., Hummel, H.E., and Gross, J. (2024) Evaluation of a push-and-pull strategy using volatiles of host and non-host plants for the management of pear psyllids in organic farming. *Frontiers in Plant Science* 15, 1375495. DOI: 10.3389/fpls.2024.1375495.

Daniel, C., Pfammatter, W., Kehrli, P, and Wyss, E. (2005) Processed kaolin as an alternative insecticide against the European pear sucker, *Cacopsylla pyri* (L). *Journal of Applied Entomology* 129, 363–367. DOI: 10.1111/j.1439-0418.2005.00981.x.

Daugherty, M.P., Briggs, C. J., and Welter, S.C. (2007) Bottom-up and top-down control of pear psylla (*Cacopsylla pyricola*): Fertilization, plant quality, and the efficacy of the predator *Anthocoris nemoralis*. *Biological Control* 43, 257–264.

Davies, D.L., Guise, C.M., and Adams, A.N. (1992) Parry's disease is similar to pear decline and is associated with mycoplasma-like organisms transmitted by *Cacopsylla pyricola*. *Plant Pathology* 41, 195–203.

Delić, D., Martini, M., Ermacora, P., Carraro, L., and Myrta, A. (2005) First report of fruit tree phytoplasmas and their psyllid vectors in Bosnia and Herzegovina. *Journal of Plant Patholoigy* 87, 150.

Delić, D., Martini, M., Ermacora, P., Carraro, L., and Myrta, A. (2008) Identification of fruit tree phytoplasmas and their vectors in Bosnia and Herzegovina. *Acta Horticulturae* 781, 429–434.

Dittmer, J., Corretto, E., Štarhová Serbina, L., Michalik, A., Nováková, E., and Schuler H. (2023) Division of labor within psyllids: Metagenomics reveals an ancient dual endosymbiosis with metabolic complementarity in the genus Cacopsylla. *mSystems* 8, e00578–23. DOI: 10.1128/msystems.00578-23.

Eben, A., Mühlethaler, R., Gross, J., and Hoch, H. (2015) First evidence of acoustic communication in the pear psyllid *Cacopsylla pyri* L. (Hemiptera: Psyllidae). *Journal of Pest Science* 88, 87–95. DOI: 10.1007/s10340-014-0588-0.

Erler, F. and Cetin, H. (2007) Effect of kaolin particle film treatment on winterform oviposition of the pear psylla *Cacopsylla pyri*. *Phytoparasitica* 35, 466–473. DOI: 10.1007/BF03020605.

Ermacora, P., Ferrini, F., Loi, N., Martini, M., and Osler, R. (2011) Population dynamics of *Cacopsylla pruni* and '*Candidatus* Phytoplasma prunorum' infection in North-Eastern Italy. *Bulletin of Insectology* 64, 143–144.

Ermacora, P., Ferrini, F., Martini, M., and Loi, N. (2012) Conifer extracts as repellent for *Cacopsylla pruni* and their effects on the spread of ESFY in apricot. *Petria* 22, 388–392.

Etropolska, A., Jarausch, W., Jarausch, B., and Trenchev, G. (2015) Detection of European fruit tree phytoplasmas and their insect vectors in important fruit-growing regions in Bulgaria. *Bulgarian Journal of Agricultural Science* 21, 1248–1253.

Etropolska, A., Jarausch, W., Jarausch, B., and Trenchev G (2016) Molecular typing of Bulgarian specimen of the phytoplasma vectors *Cacopsylla pruni* Scopoli and *Cacopsylla melanoneura* (Foerster). *Bulgarian Journal of Agricultural Science* 22, 98–102.

Fischnaller, S., Parth, M., Messner, M., Stocker, R., Kerschbamer, C., Reyes-Dominguez, Y., and Janik, K. (2017) Occurrence of different *Cacopsylla* species in apple orchards in South Tyrol (Italy) and detection of apple proliferation phytoplasma in *Cacopsylla melanoneura* and *Cacopsylla picta* (Hemiptera: Psylloidea). *Cicadina* 17, 37–51.

Font, I., Abad, P., Albinana, M., Espino, A.I., Dally, E.L., Davies, R.E., and Jorda, C. (1999) Amarilleos y enrojecimientos en zanahoria: una enfermedad a diagnostico. *Boletin Sanidad Vegetal Plagas* 25, 405–415.

Frisinghelli, C., Delaiti, L., Grando, M.S., Forti, D., and Vindimian, M.E. (2000) *Cacopsylla costalis* (Flor, 1861), as a vector of apple proliferation in Trentino. *Journal of Phytopathology* 148, 425–431.

Galdeano, D.M., Breton, M.C., Lopes, J.R.S., Falk, B.W., and Machado, M.A. (2017) Oral delivery of double-stranded RNAs induces mortality in nymphs and adults of the Asian citrus psyllid, *Diaphorina citri*. *PLoS One* 12, e0171847. DOI: 10.1371/journal.pone.0171847.

Gallinger, J. and Gross, J. (2018) Unraveling the host plant alternation of *Cacopsylla pruni* – Adults but not nymphs can survive on conifers due to phloem/xylem composition. *Frontiers in Plant Science* 9, 484. DOI: 10.3389/fpls.2018.00484.

Gallinger, J. and Gross, J. (2020) Phloem metabolites of *Prunus* sp. rather than infection with *Candidatus* Phytoplasma prunorum influence feeding behavior of *Cacopsylla pruni* nymphs. *Journal of Chemical Ecology* 46, 756–770. DOI: 10.1007/s10886-020-01148-8.

Gallinger, J., Dippel, C., and Gross, J. (2019) Interfering host location of *Cacopsylla pruni* with repellent plant volatiles. *IOBC WPRS Bulletin* 146, 10–12.

Garcia-Chapa, M., Sabaté, J., Lavina, A., and Batlle, A. (2005) Role of *Cacopsylla pyri* in the epidemiology of pear decline in Spain. *European Journal of Plant Pathology* 111, 9–17.

Goidànich, G. (1933). Un deperimento dei susini. *Bolletino Stazione Patologia Vegetale Roma* 13, 160–173.

Gonella, E., Tedeschi, R., Crotti, E., and Alma, A. (2019) Multiple guests in a single host: Interactions across symbiotic and phytopathogenic bacteria in phloem-feeding vectors – A review. *Entomologia Experimentalis et Applicata* 167, 171–185.

Görg, L.M. and Gross, J. (2021) Influence of ontogenetic and migration stage on feeding behavior of *Cacopsylla picta* on '*Candidatus* Phytoplasma mali' infected and non-infected apple plants *Journal of Insect Physiology* 131, 104229. DOI: 10.1016/j.jinsphys.2021.104229.

Görg, L.M., Eilenberg, J., Jensen, A.B., Jensen, A.H., and Gross, J. (2021) Pathogenicity against hemipteran vector insects of a novel insect pathogenic fungus from Entomophthorales (Pandora sp. nov. inedit.) with potential for biological control. *Journal of Invertebrate Pathology* 183, 107621. DOI: 10.1016/j.jip.2021.107621.

Gross, J. (2016) Chemical communication between phytopathogens, their host plants and vector insects and eavesdropping by natural enemies. *Frontiers in Ecology and Evolution* 4, 104. DOI: 10.3389/fevo.2016.00104.

Gross, J. and Gündermann, G. (2016) Principles of IPM in cultivated crops and implementation of innovative strategies for sustainable plant protection, In: Horowitz, R.A. and Ishaaya, I. (eds) *Advances in Insect Control and Resistance Management.* Springer, Dordrecht, the Netherlands, pp. 9–26.

Hatteland, B.A., Brede, Å., Brurberg, M.B., Tedeschi, R., Jarausch, B., Jarausch, W., and Blystad, D.-R. (2022) *Cacopsylla melanoneura* as a vector of apple proliferation in Norway. In: *PheroFruits 2022: Joint Meeting of the IOBC/WPRS Working Groups; Pheromones and Other Semiochemicals in IP and Integrated Protection of Fruit Crops; IPM in the XXI Century: New Tools, Tactics and Strategies to Improve Sustainability from Old and New Pests and Diseases; 25–29 September 2022; Girona, Spain.* Abstract book, p. 156. Available at: https://www.openagrar.de/receive/openagrar_mods_00083411 (last accessed June 2025).

Horton, D.R. (2024) Psyllids in natural habitats as alternative resources for key natural enemies of the pear psyllids (Hemiptera: Psylloidea). *Insects* 15, 37. DOI: 10.3390/insects15010037.

Jarausch, B. and Jarausch, W. (2010) Psyllid vectors and their control. In: Weintraub, P.G. and Jones, P. (eds) *Phytoplasmas – Genomes, Plant Hosts and Vectors.* CAB International, Wallingford, UK, pp. 250–271.

Jarausch, B. and Jarausch, W. (2014) Establishment of a permanent rearing of *Cacopsylla picta* (*Hemiptera: Psylloidea*), the main vector of '*Candidatus* Phytoplasma mali' in Germany. *Journal of Pest Science* 87, 459–467.

Jarausch, B. and Weintraub, P.G. (2013) Spread of phytoplasmas by insect vectors. *Phytopathogenic Mollicutes* 3, 5–8.

Jarausch, B. and Weintraub, P.G. (2014) Psyllid vectors of plant pathogens: Collection, rearing and use in laboratory research studies. In: Maramorosch, K. and Mahmood, F. (eds) *Rearing Animal and Plant Pathogen Vectors.* CRC Press, Boca Raton, FL, pp. 305–325.

Jarausch, B., Schwind, N., Jarausch, W., Krczal, G., Seemüller, E., and Dickler, E. (2003) First report of *Cacopsylla picta* as a vector for apple proliferation phytoplasma in Germany. *Plant Disease* 87, 101.

Jarausch, B., Schwind, N., Jarausch, W., and Krczal, G. (2004) Overwintering adults and springtime generation of *Cacopsylla picta* (synonym *C. costalis*) can transmit apple proliferation phytoplasmas. *Acta Horticulturae* 657, 409–413.

Jarausch, B., Fuchs, A., Mühlenz, I., Lampe, I., Harzer, U., and Jarausch, W. (2007a) Research on European stone fruit yellows (ESFY) in Germany. *Bulletin of Insectology* 60, 389–390.

Jarausch, B., Fuchs, A., Schwind, N., Krczal, G., and Jarausch, W. (2007b) *Cacopsylla picta* as most important vector for "*Candidatus* Phytoplasma mali" in Germany and neighbouring regions. *Bulletin of Insectology* 60, 189–190.

Jarausch, B., Mühlenz, I., Beck, A., Lampe, I., Harzer, U., and Jarausch, W. (2008) Epidemiology of European stone fruit yellows in Germany. *Acta Horticulturae* 781, 417–422.

Jarausch, B., Fuchs, A., König, D., Krzcal, G., and Jarausch, W. (2010a) Analysis of the acquisition and multiplication efficiency of different strains of *Ca.* Phytoplasma mali by the vector *Cacopsylla picta. Julius-Kühn-Archiv* 427, 175–177.

Jarausch, B., Schwind, N., Fuchs, A., and Jarausch, W. (2011) Characteristics of the spread of apple proliferation by its vector *Cacopsylla picta. Phytopathology* 101, 1471–1480.

Jarausch, B., Sauvion, N., and Jarausch, W. (2013) Spread of European fruit tree phytoplasma diseases. *Phytopathogenic Mollicutes* 3, 25–30.

Jarausch, B., Tedeschi, R., Sauvion, N., Gross, J., and Jarausch, W. (2019a) Psyllid vectors. In: Bertaccini, A., Weintraub, P.G., Rao, G.P., and Mori, N. (eds) *Phytoplasmas: Plant Pathogenic Bacteria – II. Transmission and Management of Phytoplasma – Associated Diseases.* Springer, Singapore, pp. 53–78. DOI: 10.1007/978-981-13-2832-9.

Jarausch, W. and Jarausch, B. (2016) A permanent rearing system for *Cacopsylla pruni*, the vector of '*Candidatus* Phytoplasma prunorum'. *Entomologia Experimentalis et Applicata* 159, 112–116.

Jarausch, W. and Jarausch, B. (2019) Repellent strategy to avoid transmission of '*Candidatus* Phytoplasma mali' by its vector *Cacopsylla picta. Phytopathogenic Mollicutes* 9, 177–178.

Jarausch, W. and Torres, E. (2014) Management of phytoplasma-associated diseases. In: Bertaccini, A. (ed.) *Phytoplasmas and Phytoplasma Disease Management: How to Reduce Their Economic Impact.* IPWG-International Phytoplasmologist Working Group, Bologna, Italy, pp. 199–208.

Jarausch, W., Lansac, M., Saillard, C., Broquaire, J.M., and Dosba, F. (1998) PCR assay for specific detection of European stone fruit yellows phytoplasmas and its use for epidemiological studies in France. *European Journal of Plant Pathology* 104, 17–27.

Jarausch, W., Saillard, C., Helliot, B., Garnier, M., and Dosba, F. (2000) Genetic variability of apple proliferation phytoplasmas as determined by PCR-RFLP and sequencing of a non-ribosomal fragment. *Molecular and Cellular Probes* 14, 17–24.

Jarausch, W., Danet, J.L., Labonne, G., Dosba, F., Broquaire, J.M., Saillard, C., and Garnier, M. (2001) Mapping the spread of apricot chlorotic leaf roll (ACLR) in southern France and implication of *Cacopsylla pruni* as a vector of European stone fruit yellows (ESFY) phytoplasmas. *Plant Pathology* 50, 782–790.

Jarausch, W., Fuchs, A., and Jarausch, B. (2010b) Establishment of a quantitative real-time PCR assay for the specific quantification of *Ca.* Phytoplasma prunorum in plants and insects. *Julius-Kühn-Archiv* 427, 392–394.

Jarausch, W., Jarausch, B., Fritz, M., Runne, M., Etropolska, A., and Pfeilstetter, E. (2019b) Epidemiology of European stone fruit yellows in Germany: The role of wild *P. spinosa*. *European Journal of Plant Pathology* 154, 463–476. DOI: 10.1007/s10658-019-01669-3.

Jarausch, W., Runne, M., Jarausch, B., Alexander, S., Häußer, S., and Diedel, R. (2023a) Entwicklung eines Tonminerals zur Vergrämung von Psylliden aus Obstanlagen. *Julius-Kühn-Archiv* 475, 402.

Jarausch, W., Runne, M., Schwind, N., Alexander, S., and Jarausch, B. (2023b) Epidemiology of pear decline in orchards without vector control in Southwest Germany. *Phytopathogenic Mollicutes* 13, 47–48. DOI: 10.5958/2249-4677.2023.00024.5.

Jarausch, W., Runne, M., Schwind, N., and Jarausch, B. (2023c) Leaf reddening as suitable symptom of Pear decline for remote sensing. *Phytopathogenic Mollicutes* 13, 137–138. DOI: 10.5958/2249-4677.2023.00069.5.

Jensen, D.D., Griggs, W.H., Gonzales, C.Q., and Schneider, H. (1964) Pear decline virus transmission by pear psylla. *Phytopathology* 54, 1346–1351.

Jocson D. (2020) *Good Vibrations: A Potential IPM Strategy for Pear Psylla (Cacopsylla pyricola).* WSU Tree Fruit, Washington State University Tree Fruit Extension. https://treefruit.wsu.edu/article/good-vibrations-a-potential-ipm-strategy-for-pear-psylla-cacopsylla-pyricola/ (last accessed June 2025).

Katoh, H., Inoue, H., Kuchiki, F., Ide, Y., Uechi, N., and Iwanami, T. (2013) Identification of a distinct lineage of *Cacopsylla chinensis* (Hemiptera: Psyllidae) in Japan on the basis of two mitochondrial DNA sequences. *Journal of Economic Entomology* 106, 536–542.

Kaya, K., Ulubas Serce, C., Gazel, M., Caglayan, K., and Sauvion, N. (2016) Potential psyllid vectors of *Candidatus* Phytoplasma mali and *Candidatus* Phytoplasma pyri in Turkey. *Pakistan Journal of Agricultural Sciences* 53, 383–392.

Koncz, L.S., Petróczy, M., Pénzes, B., Ladányi, M., Palkovics, L. *et al.* (2023) Detection of '*Candidatus* Phythoplasma prunorum' in apricot trees and its associated psyllid samples. *Agronomy* 13, 199. DOI: 10.3390/agronomy13010199.

Križanac, I., Plavec, J., Budinšcak, Ž., Ivic, D., Škoric, D., and Šeruga Music, M. (2017) Apple proliferation disease in Croatian orchards: A molecular characterization of '*Candidatus* Phytoplasma mali'. *Journal of Plant Pathology* 99, 95–101.

Kunze, L. (1989) Apple proliferation. In: Fridlund, P.R. (ed.) *Virus and Viruslike Diseases of Pome Fruits and Simulating Noninfectious Disorders*. Cooperative Extension College of Agriculture and Home Economics, Washington State University, Pullmann, WA, pp. 99–113.

Leclant, F., Marchoux, G., and Gianotti, J. (1974) Demonstration of the vector role of the psyllid *Trioza nigricornis* Foerst. (Homopter. insect) in the transmission of a proliferation disease of *Daucus carota* L. *Comptes Rendus Herbdomadaires des Séances de l'académie de Sciences D* 278, 57–59.

Lemmetty, A., Tuovinen, T., and Kemppainen R (2011) '*Candidatus* Phytoplasma mali' infected *Cacopsylla picta* found in apple orchards in South-Western Finnland. *Bulletin of Insectology* 64 (Suppl.), S257–S258.

Lemoine, J. (1984) Is *Psylla pyri* a vector for pear decline in France? *Bulletin-SROP* 7, 245–251.

Lethmayer, C., Hausdorf, H., Suarez-Mahecha, B., and Reisenzein, H. (2011) The importance of psyllids (Hemiptera Psyllidae) as vectors of phytoplasmas in pome and stone fruit trees in Austria. *Bulletin of Insectology* 64, S255–256.

Liao, Y.-C., Percy, D.M., and Yang, M.-M. (2022) Biotremology: Vibrational communication of Psylloidea. *Arthropod Structure and Development* 66, 101138. DOI: 10.1016/j.asd.2021.101138.

Liu, H.-L., Chen, C.-C., and Lin, C.-P. (2007) Detection and identification of the phytoplasma associated with pear decline in Taiwan. *European Journal of Plant Pathology* 117, 281–291.

Liu, S.-L., Liu, H.-L., Chang, S.-C., and Lin, C.-P. (2011) Phytoplasmas of two 16S rDNA groups are associated with pear decline in Taiwan. *Botanical Studies* 52, 313–320.

Lorenz, K.-H., Dosba, F., Poggi Pollini, C., Llácer, G., and Seemüller, E. (1994) Phytoplasma diseases of *Prunus* species in Europe are caused by genetically similar organisms. *Zeitschrift für Pflanzenkrankheiten und Pflanzenschutz* 101, 567–575.

Lu, J. and Shen, J. (2024) Target genes for RNAi in pest control: A comprehensive overview. *Entomologia Generalis* 44, 95–114. DOI: 10.1127/entomologia/2023/2207.

Ludvikova, H., Lauterer, P., Sucha, J., and Franova, J. (2011) Monitoring of psyllid species (Hemiptera, Psylloidea) in apple and pear orchards in East Bohemia. *Bulletin of Insectology* 64, 121–122.

Mader, C. (1908) La mortalità dei peri nella plaga di Bolzano-Gries. *Almanacco Agrario Sezione Trento Consiglio Provinciale d'Agricoltura per il Tirolo* 26, 347–350.

Maier, C., Bachinger, K., Mörtel, J., Engel, C., Czipin, L., and Riedle-Bauer, M. (2013) European stone fruit yellows: A mark, release and recapture experiment tracking the dispersal of its vector *Cacopsylla pruni* (Hemiptera: Psyllidae) in a model apricot orchard and epidemiological studies in Lower Austria. *Journal of Phytopathology* 161, 713–722.

Malagnini, V., Pedrazzoli, F., Forno, F., Komjanc, M., and Ioriatti, C. (2007) Characterization of microsatellite loci in *Cacopsylla melanoneura* Föster (Homoptera: Psyllidae). *Molecular Ecology Notes* 7, 495–497.

Malagnini, V., Pedrazzoli, F., Gualandri, V., Forno, F., Zasso, R., Pozzebon, A., and Ioriatti, C. (2010) A study of the effects of '*Candidatus* Phytoplasma mali' on the psyllid *Cacopsylla melanoneura* (Hemiptera: Psyllidae). *Journal of Invertebrate Pathology* 103, 65–67.

Malagnini, V., Pedrazzoli, F., Papetti, C., Cainelli, C., Zasso, R. *et al.* (2013) Ecological and genetic differences between *Cacopsylla melanoneura* (Hemiptera, Psyllidae) populations reveal species host plant preference. *PLoS One* 8, e69663. DOI: 10.1371/journal.pone.0069663.

Marcone, C., Jarausch, B., and Jarausch, W. (2010) *Candidatus* Phytoplasma prunorum, the causal agent of European stone fruit yellows: An overview. *Journal of Plant Pathology* 92, 19–34.

Marie-Jeanne, V., Bonnot, F., Thébaud, G., Peccoud, J., Labonne, G., and Sauvion, N. (2020) Multi-scale spatial genetic structure of the vector-borne pathogen '*Candidatus* Phytoplasma prunorum' in orchards and in wild habitats. *Scientific Reports* 10, 5002. DOI: 10.1038/s41598-020-61908-0.

Martini, M., Loi, N., Ermacora, P., Carraro, L., and Pastore, M., (2007) A real-time PCR method for detection and quantification of '*Candidatus* Phytoplasma prunorum' in its natural hosts. *Bulletin of Insectology* 60, 251–252.

Martini, M., Ermacora, P., Falginella, L., Loi, N., and Carraro, L. (2008) Molecular differentiation of '*Candidatus* Phytoplasma mali' and its spreading in Friuli Venezia Giulia region (North-East Italy). *Acta Horticulturae* 781, 395–402.

Mattedi, L., Forno, F., Cainelli, C., Grando, S., and Jarausch, W. (2007) Transmission of *Candidatus* Phytoplasma mali by psyllid vectors in Trentino. *IOBC/WPRS Bulletin* 30, 39.

Mattedi, L., Forno, F., Cainelli, C., Grando, M.S., and Jarausch, W. (2008) Research on *Candidatus* Phytoplasma mali transmission by insect vectors in Trentino. *Acta Horticulturae* 781, 369–374.

Mayer, C.J. and Gross, J. (2007) Different host plant odours influence migration behaviour of *Cacopsylla melanoneura* (Förster), an insect vector of the apple proliferation phytoplasma. *IOBC WPRS Bulletin* 30, 177–184.

Mayer, C.J., Vilcinskas, A., and Gross, J. (2008a) Phytopathogen lures its insect vector by altering host plant odor. *Journal of Chemical Ecology* 34, 1045–1049.

Mayer, C.J., Vilcinskas, A., and Gross, J. (2008b) Pathogen-induced release of plant allomone manipulates vector insect behaviour. *Journal of Chemical Ecology* 34, 1518–1522.

Mayer, C.J., Jarausch, B., Jarausch, W., Jelkmann, W., Vilcinskas, A., and Gross, J. (2009) *Cacopsylla melanoneura* has no relevance as vector of apple proliferation in Germany. *Phytopathology* 99, 729–738.

Mayer, C.J., Vilcinskas, A., and Gross, J. (2011) Chemically mediated multitrophic interactions in a plant-insect vector-phytoplasma system compared with a partially nonvector species. *Agricultural and Forest Entomology* 13, 25–35.

McLarty, H.R. (1948) Killing of pear trees. *Annual Reports of the Canadian Plant Disease Survey* 28, 77.

Mergenthaler, E., Kiss, B., Kiss, E., and Viczián, O. (2017) Survey on the occurrence and infection status of *Cacopsylla pruni*, vector of European stone fruit yellows in Hungary. *Bulletin of Insectology* 70, 171–176.

Miñarro, M., Somoano, A., Moreno, A., and García, R.R. (2016) Candidate insect vectors of apple proliferation in Northwest Spain. *SpringerPlus* 5, 1240. DOI: 10.1186/s40064-016-2907-9.

Mittelberger, C., Obkircher, L., Oettl, S., Oppedisano, T., Pedrazzoli, F. *et al.* (2017) The insect vector *Cacopsylla picta* vertically transmits the bacterium '*Candidatus* Phytoplasma mali' to its progeny. *Plant Pathology* 66, 1015–1021.

Monti, M., Martini, M., and Tedeschi, R. (2013) EvaGreen Real-time PCR protocol for specific '*Candidatus* Phytoplasma mali' detection and quantification in insects. *Molecular and Cellular Probes* 27, 129–136.

Nottingham, L.B., Orpet, R.J., and Beers, E.H. (2022) Integrated pest management programs for pear psylla, *Cacopsylla pyricola* (Förster) (Hemiptera: Psyllidae), using kaolin clay and reflective plastic mulch. *Journal of Economic Entomology* 115, 1607–1619. DOI: 10.1093/jee/toac121.

Oettl, S. and Schlink, K. (2015) Molecular identification of two vector species, *Cacopsylla melanoneura* and *Cacopsylla picta* (Hemiptera: Psyllidae), of apple proliferation disease and further common psyllids of Northern Italy. *Journal of Economic Entomology* 108, 2174–2183. DOI: 10.1093/jee/tov204.

Oppedisano, T., Polajnar, J., Kostanjšek, R., De Cristofaro, A., Ioriatti, C., Virant Doberlet, M., and Mazzoni, V. (2019) Substrate-borne vibrational communication in the vector of apple proliferation disease *Cacopsylla picta* (Hemiptera: Psyllidae). *Journal of Economic Entomology* 113, 596–603.

Oppedisano, T., Panassiti, B., Pedrazzoli, F., Mittelberger, C., Bianchedi, P.L. *et al.* (2020) Importance of psyllids' life stage in the epidemiology of apple proliferation phytoplasma. *Journal of Pest Science* 93, 49–61. DOI: 10.1007/s10340-019-01130-8.

Ossiannilsson, F. (1992) The Psylloidea (Homoptera) of Fennoscandia and Denmark. *Fauna Entomologica Scandinavica* 26, 1–347.

Ouvrard, D. (2015) *Psyl'list – The World Psylloidea Database*. Natural History Museum. Available at: https://data.nhm.ac.uk/dataset/psyl-list/resource/8746ceec-4846-4899-b607-9ba603002033 (last accessed June 2025).

Paleskić, C., Bachinger, K., Brader, G., Kickenweiz, M., Engel, C., Wurm, L., Czipin, L., and Riedle-Bauer, M. (2017) Cage and field experiments as basis for the development of control strategies against Cacopsylla pruni, the vector of European stone fruit yellows. *Annals of Applied Biology* 170, 357–368.

Panassiti, B., Sander, N., Mazzoni, V., Fischnaller, S., Parth, M., Messner, M., Janik, K., and Hartig, F. (2019) A temperature-based model for predicting the immigration of *Cacopsylla melanoneura* and *C. picta*, vectors of the apple proliferation disease, in South Tyrol, Northern Italy (Hemiptera: Psylloidea). *Cicadina* 18, 1–11.

Pasqualini, E., Civolani, S., and Grappadelli, L.E. (2007) Particle film technology: Approach for a biorational control of *Cacopsylla pyri* (Rhynchota Psyllidae) in Northern Italy. *Bulletin of Insectology* 55, 39–42.

Peccoud, J., Labonne, G., and Sauvion, N. (2013) Molecular test to assign individuals within the *Cacopsylla pruni* complex. *PLoS One* 8, e72454. DOI: 10.1371/journal.pone.0072454.

Peccoud, J., Pleydell, D.R.J., and Sauvion, N. (2018) A framework for estimating the effects of sequential reproductive barriers: Implementation using Bayesian models with field data from cryptic species, *Evolution* 72, 2503–2512. DOI: 10.1111/evo.13595.

Pedrazzoli, F., Gualandri, V., Forno, F., Mattedi, L., Malagnini, V. *et al.* (2007) Acquisition capacities of the overwintering adults of the psyllid vectors of "*Candidatus* Phytoplasma mali". *Bulletin of Insectology* 60, 195–196.

Percy, D.M., Crampton-Platt, A., Sveinsson, S., Lemmon, A.R., Lemmon, E.M., Ouvrard, D., and Burckhardt, D. (2018) Resolving the psyllid tree of life: phylogenomic analyses of the superfamily Psylloidea (Hemiptera). *Systematic Entomology* 43, 762–776. DOI: 10.1111/syen.12302.

Pizzinat, A., Tedeschi, R., and Alma. A. (2011) *Cacopsylla melanoneura* (Foerster): Aestivation and overwintering habitats in Northwest Italy. *Bulletin of Insectology* 64, S135–S136.

Poggi Pollini, C., Bianchi L., Forno, F., Franchini, S., Giunchedi, L. *et al.* (2007) Investigation on European stone fruit yellows in experimental apricot orchards in the province of Trentino (Italy). *Bulletin of Insectology* 60, 323–324.

Poggi Pollini, C., Forno, F., Franchini, S., Gobber, M., Lanzoni, C. *et al.* (2010) Detection and distribution of European stone fruit yellows in apricot cv. Bergeron and epidemiological studies in the province of Trento (Italy). *Julius-Kühn-Archiv* 427, 383–385.

Purcell, A.H., Nyland, G., Raju, B.C., and Heringer, M.R. (1981) Peach yellow leaf roll epidemic in northern California: Effects of peach cultivar, tree age, and proximity to pear orchards. *Plant Disease* 65, 365–368.

Raddadi, N., Gonella, E., Camerota, C., Pizzinat, A., Tedeschi, R. *et al.* (2011) '*Candidatus* Liberibacter europaeus' sp. nov. that is associated with and transmitted by the psyllid *Cacopsylla pyri* apparently behaves as an endophyte rather than a pathogen. *Environmental Microbiology* 13, 414–426.

Refatti, E. (1964) Pear moria in Italy. *FAO Plant Protectin Bulletin* 12, 1–7.

Rid, M., Mesca, C., Ayasse, M., and Gross, J. (2016) Apple proliferation phytoplasma influences the pattern of plant volatiles emitted depending on pathogen virulence. *Frontiers in Ecology and Evolution* 3, 152. DOI: 10.3389/fevo.2015.00152.

Riedle-Bauer, M., Paleskić, C., Schwanzer, J., Kölber, M., Bachinger, K. *et al.* (2019) Epidemiological and molecular study on '*Candidatus* Phytoplasma prunorum' in Austria and Hungary. *Annals of Applied Biology* 175, 400–414. DOI: 10.1111/aab.12541.

Riedle-Bauer, M., Paleskić, C., Schönhuber, C., Staples, M., and Brader, G. (2022) Vector transmission and epidemiology of '*Candidatus* Phytoplasma pyri' in Austria and identification of *Cacopsylla pyrisuga* as new pathogen vector. *Journal of Plant Diseases and Protection* 129, 375–386. DOI: 10.1007/s41348-021-00526-y.

Rui, D., Ciferri, R., and Refatti, E. (1950) La virosi degli "scopazzi del melo" nel Veronese. *Noticiario sulle Malattie delle Piante* 13, 7–11.

Sabaté, J., Laviña, A., and Batlle, A. (2014) First report of "*Candidatus* Phytoplasma pyri" causing peach yellow leaf roll (PYLR) in Spain. *Plant Disease* 98, 989.

Sabaté, J., Laviña, A., and Batlle, A. (2016) Incidence and distribution of '*Candidatus* Phytoplasma prunorum' and its vector *Cacopsylla pruni* in Spain: An approach to the epidemiology of the disease and the role of wild *Prunus*. *Plant Pathology* 65, 837–846.

Sabaté, J., Rodon, J., Artigues, M., Lavina, A., and Batlle, A. (2018) Transmission of '*Candidatus* Phytoplasma pyri' by naturally infected *Cacopsylla pyri* to peach, an approach to the epidemiology of peach yellow leaf roll (PYLR) in Spain. *Plant Pathology* 67, 978–986.

Šafářová, D., Zrníková, E., Holušová, K., Ourednícková, J., Starý, M., and Navrátil, M. (2023) Molecular characterization of mitogenome of *Cacopsylla picta* and *Cacopsylla melanoneura*, two vector species of '*Candidatus* Phytoplasma mali'. *Agronomy* 13, 2210. DOI: 10.3390/agronomy13092210.

Salehi, M., Izadpanah, K., Taghavi, S.M., and Rahimian, H. (2008) Characterization of a phytoplasma associated with pear decline in Iran. *Journal of Phytopathology* 156, 493–495.

Sauvion, N., Lachenaud, O., Genson, G., Rasplus, J.-Y., and Labonne, G. (2007) Are there several biotypes of *Cacopsylla pruni*? *Bulletin of Insectology* 60, 185–186.

Sauvion, N., Lachenaud, O., Mondor-Genson, G., Rasplus, J.Y., and Labonne, G. (2009) Nine polymorphic microsatellite loci from the psyllid *Cacopsylla pruni* (Scopoli), the vector of European stone fruit yellows. *Molecular Ecology Resources* 9, 1196–1199.

Sauvion, N., Peccoud, J., Meynard, C.N., and Ouvrard, D. (2021) Occurrence data for the two cryptic species of *Cacopsylla pruni* (Hemiptera: Psylloidea). *Biodiversity Data Journal* 9: e68860. DOI: 10.3897/BDJ.9.e68860.

Schaper, U. and Seemüller, E. (1982) Condition of the phloem and the persistence of mycoplasmalike organisms associated with apple proliferation and pear decline. *Phytopathology* 72, 736–742.

Schaub, L., Graf, B., and Butturini, A. (2005) Phenological model of pear psylla *Cacopsylla pyri*. *Entomologia Experimentalis et Applicata* 117, 105–111. DOI: 10.1111/j.1570-7458.2005.00339.x.

Schuler, H., Dittmer, J., Borruso, L., Galli, J., Fischnaller, S. *et al.* (2022) Investigating the microbial community of Cacopsylla spp. as potential factor in vector competence of phytoplasma. *Environmental Microbiology* 10, 4771–4786. DOI: 10.1111/1462-2920.16138.

Seemüller, E. (1989) Pear decline. In: Fridlund, P.R. (ed.) *Virus and Viruslike Diseases of Pome Fruits and Simulating Noninfectious Disorders*. Cooperative Extension College of Agriculture and Home Economics, Washington State University, Pullmann, WA, pp. 188–201.

Seemüller, E. (1990) Apple proliferation. In: Jones, A.L. and Aldwinkle, H.S. (eds) *Compendium of Apple and Pear Diseases*. APS Press, St. Paul, MN, pp. 67–68.

Seemüller, E. and Schneider, B. (2004) '*Candidatus* Phytoplasma mali', '*Candidatus* Phytoplasma pyri' and '*Candidatus* Phytoplasma prunorum', the causal agents of apple proliferation, pear decline and European stone fruit yellows, respectively. *International Journal of Systematic and Evolutionary Microbiology* 54, 1217–1226.

Seemüller, E., Marcone, C., Lauer, U., Ragozzino, A., and Göschl, M. (1998) Current status of molecular classification of the phytoplasmas. *Journal of Plant Pathololgy* 80, 3–26.

Sertkaya, G., Sertkaya, E., and Kaya, K. (2008) Detection of pear decline disease in pear and quince in the eastern mediterranean region of Turkey. *Acta Horticulturae* 781, 511–515.

Sharon, R., Tomer, M., and Zisovich Charitt, A. (2025) Seasonal dispersal of *Cacopsylla bidens* and its role in the epidemiology of '*Candidatus* Phytoplasma pyri' in mixed-crop systems. *Phytopathogenic Mollicutes* 15, 53–54. DOI: 10.5958/2249-4677.2025.00031.2.

Štarhová Serbina, L., Gajski, D., Malenovský, I., Corretto, E., Schuler, H. and Dittmer, J. (2022) *Wolbachia* infection dynamics in a natural population of the pear psyllid *Cacopsylla pyri* (Hemiptera: Psylloidea) across its seasonal generations. *Scientific Reports* 12, 16502. DOI: 10.1038/s41598-022-20968-0.

Štarhová Serbina, L., Corretto, E., Enciso Garcia, J.S., Berta, M., Giovanelli, T., Dittmer, J., and Schuler, H. (2023) Seasonal wild dance of dual endosymbionts in the pear psyllid *Cacopsylla pyricola* (Hemiptera: Psylloidea). *Scientific Reports* 13, 16038. DOI: 10.1038/s41598-023-43130-w.

Steffek, R., Swen Follak, S., Sauvion, N., Labonne, G., and MacLeod, A. (2012) Distribution of '*Candidatus* Phytoplasma prunorum' and its vector *Cacopsylla pruni* in European fruit growing areas: a literature survey. *EPPO Bulletin* 42, 191–202.

Sterck, G. and Highwood, D.P. (1992) Implementation of the IRAC anti-resistance guidelines with IPM programs for Belgian apple and pear orchards. *Proceedings of the British Crop Protection Conference on Pests and Disease* 2, 517–520.

Taning, C.N.T., Andrade, E.C., Hunter, W.B., Christiaens, O., and Smagghe, G. (2016) Asian citrus psyllid RNAi pathway – RNAi evidence. *Scientific Reports* 6, 38082. DOI: 10.1038/srep38082.

Tedeschi, R. and Alma, A. (2004) Transmission of apple proliferation phytoplasma by *Cacopsylla melanoneura* (Homoptera: Psyllidae). *Journal of Economical Entomology* 97, 8–13.

Tedeschi, R. and Nardi, F. (2010) DNA-based discrimination and frequency of phytoplasma infection in the two hawthorn-feeding species, *Cacopsylla melanoneura* and *Cacopsylla affinis*, in northwestern Italy. *Bulletin of Entomological Research* 100, 741–747. DOI: 10.1017/S0007485310000118.

Tedeschi, R., Bosco, D., and Alma, A. (2002) Population dynamics of *Cacopsylla melanoneura* (Homoptera: Psyllidae), a vector of apple proliferation phytoplasma in northwestern Italy. *Journal of Economical Entomology* 95, 544–551.

Tedeschi, R., Visentin, C., Alma, A., and Bosco, D. (2003) Epidemiology of apple proliferation (AP) in northwestern Italy: Evaluation of the frequency of AP-positive psyllids in naturally infected populations of *Cacopsylla melanoneura* (Homoptera: Psyllidae). *Annals of Applied Biology* 142, 285–290.

Tedeschi, R., Ferrato, V., Rossi, J., and Alma, A. (2006) Possible phytoplasma transovarial transmission in the psyllids *Cacopsylla melanoneura* and *Cacopsylla pruni*. *Plant Pathology* 55, 18–24.

Tedeschi, R., Gamba, U., Pinna, M., Spagnolo, S., and Alma, A. (2007) Il caolino: Una nuova opportunità nella difesa biologica ed integrata. Un esempio di applicazione nel contenimento di Cacopsylla melanoneura (Forster) in meleti del Canavese affetti da AP. *Bollettino di Agricoltura Biologica a Cura del CRAB* 3, 9–21.

Tedeschi, R., Lauterer, P., Brusetti, L., Tota, F., and Alma A. (2009) Composition, abundance and phytoplasma infection in the hawthorn psyllid fauna of northwestern Italy. *European Journal of Plant Pathology* 123, 301–310.

Tedeschi, R., Baldessari, M., Mazzoni, V., Trona, F., and Angeli, G. (2012) Population dynamics of *Cacopsylla melanoneura* (Hemiptera: Psyllidae) in Northeast Italy and its role in the apple proliferation epidemiology in apple orchards. *Journal of Economic Entomology* 105, 322–328.

Tedeschi, R., Jarausch, B., Delic, D., and Weintraub, P.-G. (2013) Actual distribution of fruit tree and grapevine phytoplasma diseases and their vectors in Europe and neighboring regions. *Phytopathogenic Molicutes* 3, 3–4.

Thao, M.L., Clark, M.A., Baumann, L., Brennan, E.B., Moran, N.A., and Baumann, P. (2000) Secondary endosymbionts of psyllids have been acquired multiple times. *Current Microbiology* 41, 300–304. DOI: 10.1007/s002840010138.

Thébaud, G., Yvon, M., Alary, R., Sauvion, N., and Labonne, G. (2009) Efficient transmission of '*Candidatus* Phytoplasma prunorum' is delayed by eight months due to a long latency in its host-alternating vector. *Phytopathology* 99, 265–273.

Tomer, M., Zisovich Charitt, A., and Sharon, R. (2025) First identification of *Cacopsylla bidens* as a vector of '*Candidatus* Phytoplasma pyri', the agent associated with pear decline. *Phytopathogenic Mollicutes* 15, 55–56. DOI: 10.5958/2249-4677.2025.00032.9.

Ulubas Serce, C., Yvon, M., Kaya, K., Gazel, M., Cengiz, F.C., Caglayan, K., and Sauvion, N. (2011) Survey on the presence of *Cacopsylla pruni* in Turkey: Preliminary results. *Bulletin of Insectology* 64, 145–146.

Warabieda, W., Soika, G., and Cieślińska, M. (2018) *Cacopsylla pruni* in Poland and its significance as a vector of '*Candidatus* Phytoplasma prunorum'. *Zemdirbyste-Agriculture* 105, 177–182. DOI: 10.13080/z-a.2018.105.023.

Weil, T., Ometto, L., Esteve-Codina, A., Gómez-Garrido, J., Oppedisano, T. *et al.* (2020) Linking omics and ecology to dissect interactions between the apple proliferation phytoplasma and its psyllid vector *Cacopsylla melanoneura*. *Insect Biochemistry and Molecular Biology* 127, 103474. DOI: 10.1016/j.ibmb.2020.103474.

Weintraub, P.G. and Beanland L.A. (2006) Insect Vectors of Phytoplasmas. *Annual Review of Entomology* 51, 91–111.

Woodbridge, C.G., Blodgett, E.C., and Diener, T.O. (1957) Pear decline in the Pacific northwest. *Plant Disease Reporter* 41, 569–572.

Yvon, M., Labonne, G., and Thébaud, G. (2004) Survival of European stone fruit yellows phytoplasma outside fruit crop production areas: A case study in southeastern France. *Acta Horticulturae* 657, 477–481.

13 Evolutionary Insights into Hemipteran Insect–Phytoplasma Associations: Phylogenetic Conservatism and Historical Ecology as Tools to Predict Future Associations

Valeria Trivellone,[1]* Sabrina Araujo,[2] and Christopher H. Dietrich[1]
[1]*Illinois Natural History Survey, Prairie Research Institute, University of Illinois at Urbana-Champaign, Champaign, Illinois, USA;* [2]*Department of Physics, Universidade Federal do Paraná, Paraná, Brazil*

Abstract
Novel diseases often result from pathogens shifting onto new hosts in response to environmental change, host manipulation by the pathogen, and ecological processes increasing exposure to new potential hosts. Understanding the specific evolutionary, physiological, and ecological mechanisms influencing shifts among hosts by pathogens may help predict and mitigate future disease outbreaks. Proactively addressing phytoplasma diseases – impacting over 1000 plant species and causing billions of dollars in losses annually – requires the development of models that incorporate host–pathogen traits, ecology, and evolutionary history to simulate conditions for host switching. Fortunately, a wealth of prior experimental results and ecological observations is already available to inform such models. This chapter highlights key findings from studies on vector–phytoplasma associations, emerging trends, and future research directions to enhance predictions and control strategies, and safeguarding food security, rural economies, commercial farms, and investors.

Introduction

Host switches of potential pathogens from wild species to humans, livestock, or crop plants are thought to be the most prevalent mechanism by which new infectious diseases emerge (Brooks *et al.*, 2023a). Such host switches resulting inadvertently from environmental change and/or human activities have occurred throughout history and include the most recent SARS-CoV-2 pandemic as well as major outbreaks of plant diseases affecting agriculture (Boeger *et al.*, 2022; Massey, 2022; Trivellone *et al.*, 2022b). These outbreaks have led to devastating economic consequences, with the Covid pandemic expected to have caused a cumulative loss to global GDP of US$12.5 trillion in 2024 (MarketScreener, 2022) and plant diseases to have caused an estimated $220 billion in annual agricultural losses globally based on 2021 data (FAO, 2021). Predicting where and when such host switches from wild to domesticated plants

*Corresponding author: valeria3@illinois.edu

 Phytoplasmas: Genomes, Plant Hosts and Vectors, 2nd Edition. (Ed. W. Wei)
DOI: 10.1079/9781800627031.0013

are likely to occur could help prevent future outbreaks (Araujo *et al.*, 2015; Klimov *et al.*, 2022). Yet, most prior research on such systems has been reactive, in crisis response, focused on reducing the spread of disease after an outbreak has already occurred and mitigating its adverse effects (Ellwanger *et al.*, 2019; Ristaino *et al.*, 2021; Trivellone *et al.*, 2022b). Models that incorporate ecological and evolutionary trait data for potential pathogens and their hosts (including vectors) provide the possibility of predicting and preventing the emergence of new diseases (Gardner *et al.*, 2023). Prior research demonstrates that environmental (e.g. anthropogenic and climatic) perturbations may yield conditions conducive to host switches (reviewed by Brooks *et al.*, 2019). Additionally, the virulence of a particular pathosystem may be related to the relative age or historical compatibility between a pathogen and its host and the phylogenetic relatedness of alternative hosts. This chapter reviews significant results from prior ecological and evolutionary studies of the phytoplasma–plant–vector (PPV) association to identify important traits and processes relevant to the prediction of the emergence of new PPV associations and outbreaks of phytoplasma diseases of potential agricultural importance.

Phytoplasmas (phylum Mycoplasmatota, class Mollicutes) are a diverse lineage of obligate vector-borne bacteria associated with the phloem sieve elements of more than 1000 species of vascular plants and internal organs of more than 200 species of phloem-feeding hemipteran insects (Trivellone, 2018, 2019). To date, 37 groups and about 300 subgroups of these bacteria have been characterized using genetic data (i.e. restriction fragment length polymorphism (RFLP) profiles of the F2nR2 region of the 16S rRNA gene), each group designated by a Roman numeral, with subgroups designated by separate letters (e.g. 16SrII-A). This includes 48 formally described species with provisional '*Candidatus*' status denoting not-yet cultured bacteria (Lee *et al.*, 1998, 2011; Seemüller *et al.*, 1998; IRPCM Phytoplasma/Spiroplasma Working Team – Phytoplasma Taxonomy Group, 2004; Wei *et al.*, 2007; Zhao *et al.*, 2009, 2021; Nejat *et al.*, 2013; Safarova *et al.*, 2016; Naderali *et al.*, 2017; Kirdat *et al.*, 2021, 2023; Rodrigues Jardim *et al.*, 2021; Bertaccini *et al.*, 2022; Wei and Zhao, 2022), two of them recently flagged as chimeras (Tiwarekar *et al.*, 2023). The primary mode by which phytoplasmas move from plant to plant is via hemipteran insect vectors, which obtain them by feeding on phloem sap of infected plants and transmit them to other plants through dispersal and additional feeding (Weintraub and Beanland, 2006; Orlovskis *et al.*, 2015).

According to a recent molecular divergence time analysis (Cao *et al.*, 2020), phytoplasmas diverged from acholeplasma-like ancestors about 641 million years ago. For the next approximately 325 million years, phytoplasmas presumably retained a lifestyle similar to their closest relatives among the Mollicutes: free-living or commensal with animals. However, about 316 million years ago, approximately coinciding with the origin of seed plants and plant-feeding hemipteran insects, phytoplasmas acquired a key innovation: the ability to infect plants and to spread from plant to plant via association with sap-sucking hemipteran insect vectors. This association enabled phytoplasmas to diversify by manipulating the behavior of their insect vectors and the phenology of their plant hosts, thereby increasing the ability of phytoplasmas to spread across terrestrial environments and acquire new insect vectors and plant hosts. The phytoplasma–vector association has become fine-tuned over millions of years of evolution. As in many other vector-borne pathogens (Huang *et al.*, 2020), this intimate association with vectors is manifested in their complex transmission cycle and high turnover in different hosts. It may also be reflected at the genomic level, with phytoplasma genomes oscillating from complexity to simplification as observed in other organisms (Wolf and Koonin, 2013) with dramatic losses of certain genes (reductive evolution; Namba, 2019) counterbalanced by acquisition of new genes, such as membrane proteins and virulence factors, involved in host–pathogen interactions (Bertaccini *et al.*, 2019; Rossi *et al.*, 2019). Membrane protein–protein interactions between phytoplasmas and insect vectors (allowing phytoplasmas to migrate from the insect gut to the salivary glands) as well as ecological niche manipulation enabled the evolution of vector competence. Using a previous time tree reconstruction of hemipteroid insects based on multilocus transcriptome data (Johnson *et al.*, 2018), vector competence may be either an apomorphic (Trivellone *et al.*, 2019), homoplasious (Dietrich, 2013), or plesiomorphic (Zamharir and Shameli, 2023) trait depending on the particular hemipteran lineage. Known vectors of phytoplasmas are scattered across several major lineages of hemipteran insects but, in certain vector lineages, vector capacity has been retained over multiple speciation events (Dietrich, 2013).

Evolution of PPV Associations: Ecological Fitting and Host Manipulation by Pathogens

Prevailing theories of host–parasite evolution comprise two non-mutually exclusive paradigms. One emphasizes the role of specialization through a co-evolutionary process defined by the speciation of the ancestral parasite species as a consequence of the speciation of the host (May and Anderson, 1983; Alizon *et al.*, 2009). Although it was suggested that host–parasite co-evolution includes at least two components, co-speciation and co-accommodation (i.e. ongoing interaction of a parasite species and its host during the period in which the parasite exhibits no cladogenesis; Brooks, 1979), this theory assumes that the parasite is locked in a strict evolutionarily interdependent relationship with its host that ultimately evolves toward extreme specialization. For a long time, this theory was applied to plant–parasite systems under the unsupported assumption that co-speciation is the prevalent and least costly evolutionary process (for a review, see Trivellone, 2022). However, taking empirical evidence into account, the theory generates a paradox: if pathogens are tightly co-adapted to a restricted range of hosts, then the emergence of new infectious diseases should be rare (Agosta *et al.*, 2010). This is not what we observe in nature.

The alternative paradigm (Brooks *et al.*, 2019; Gardner *et al.*, 2023) considers parasites as drivers of their own evolution through host switches facilitated by ecological fitting (EF) or adaptation through pre-existing capacity. This is a mechanism by which phenotypically flexible genotypes enable organisms to persist in novel environments and, more specifically, allow pathogens to colonize new hosts, given the opportunity (Janzen, 1985; Agosta, 2006; Agosta and Klemens, 2008). Another mechanism known as host manipulation (HM) – by which parasites alter the phenotype (including behavior, physiology, and/or morphology) of the host in ways that increase the parasites' ability to spread and diversify (Holmes and Bethel, 1972) – may increase the opportunity for parasites to colonize new hosts. Both mechanisms facilitate the formation of new host–pathogen associations and, consequently, may lead to rapid emergence of novel diseases.

Previous studies showed that EF processes may influence evolutionary trajectories in host–pathogen interactions (Papaïx *et al.*, 2014; Hoberg and Brooks, 2015). Environmental variability may significantly affect the prevalence of genotypes and phenotypes in the field, and such variation may drive ecological processes at different spatiotemporal scales, ultimately providing opportunities for host switches (Burdon and Thrall, 2009; Longo *et al.*, 2014; Tomkins *et al.*, 2018). Overall, EF, HM, and related mechanisms may explain why phylogenetic congruence and strict co-speciation between hosts and parasites are rare and cannot explain the emergence of new associations (reviewed by Trivellone and Panassiti, 2023a, 2023b). If EF is the predominant mechanism, then host switches will occur more frequently, with parasites colonizing hosts that are historically compatible (i.e. sharing traits inherited from common ancestors), driving pathogen diversification across generations (e.g. as in the recent SARS-CoV-2 pandemic; Boeger *et al.*, 2022). Although environmental perturbations may expose parasites to many new potential hosts (providing ecological opportunity), host switches may be more likely to occur if the new potential host is historically compatible (i.e. phylogenetically related) to the old host. Thus, phylogenies of actual and potential hosts in a given environment may be important for identifying which new hosts are most likely to be colonized by a parasite. A few recent studies have begun to consider this framework to predict the emergence of diseases, given the retained capacity of pathogens to infect alternate hosts and a new opportunity to invade susceptible hosts (Apari *et al.*, 2019; McKee *et al.*, 2019; Gougherty and Davies, 2021; Trivellone, 2022; Janik *et al.*, 2023). D'Bastiani *et al.* (2023) showed that host switches at microevolutionary scales drive parasite diversification through EF mediated by historical compatibility among hosts. Historical compatibility (or phylogenetic conservatism) may partly explain the capacity for host switching by the pathogen based on traits conserved in closely related hosts. Nevertheless, the evolutionary and ecological processes of pathogen–vector–host interactions have been investigated only from the perspective of transmission dynamics (Shaw *et al.*, 2019), and more research is needed to assess host switch intensity and predict disease emergence (Mauck *et al.*, 2018; D'Bastiani *et al.*, 2023).

One case of new hemipteran vector–phytoplasma associations explained by EF and phylogenetic conservatism was documented 70 years ago with the discovery in Europe of the non-native Nearctic leafhopper *Scaphoideus titanus* (Ball), responsible for devastating outbreaks of flavescence dorée (FD) in grapevine, a disease associated with two phytoplasmas in the phylogenetic subgroups 16SrV-C and 16SrV-D (Trivellone and Dietrich, 2021; Gonella *et al.*, 2024). In its native area *S. titanus* has been recorded from various woody and herbaceous plant species (Delong, 1948; Barnett, 1977) but, once introduced into Europe, it became strictly associated with grapevine in agroecosystems likely as a result of a shift in its ecological niche (reviewed in Braga and Janz, 2021). The phylogenetic conservatism of this leafhopper's association with the FD phytoplasma strains was later documented by comparing protein–protein interaction affinity in the related Palearctic leafhopper species, *Euscelidius variegatus* and *Macrosteles quadripunctulatus* (Trivellone *et al.*, 2019).

Manipulation of host biological traits (e.g. behavior, physiology, morphology), resulting in an increase in parasite fitness, has been well documented among pathogenic bacteria including phytoplasmas (Poulin *et al.*, 1994; Thompson, 1998; Gandon *et al.*, 2002; Friesen *et al.*, 2011; Weinersmith and Faulkes, 2014; Penczykowski *et al.*, 2016; Ángeles-López *et al.*, 2018; Hernandez-Caballero *et al.*, 2022; Massey, 2022). The HM hypothesis suggests that such manipulation increases pathogen transmission and dispersal (Holmes and Bethel, 1972) with rapid switches to new hosts through EF and expansion of host species range (Agosta and Klemens, 2008). Tests of the HM hypothesis on vector-borne plant pathogens have shown that pathogens and vectors interact through their shared hosts (plants), resulting in increases in vector fitness and larger pathogen populations (Mauck *et al.*, 2014; Rajabaskar *et al.*, 2014; Orlovskis and Hogenhout, 2016; Ángeles-López *et al.*, 2018). HM may be phylogenetically constrained (Poulin, 1995) but few studies have investigated links between ecological and evolutionary processes driving diversification via host switches in pathogen–vector–host systems (Thomas *et al.*, 2012; Heil, 2016).

Previous studies of host–pathogen systems (Gandon, 2018) investigated the conditions under which different components of HM evolve but focused only on the epidemiological consequences at short time scales. Yet, pathogen-mediated changes in vector physiology and behavior may also operate at evolutionary time scales via phylogenetic conservatism (traits retained by specific lineages over long periods), co-option, and phenotypic plasticity, which ultimately affect the phylodynamics of rapidly evolving pathogens (Ingle *et al.*, 2021). Notably, expansion and contraction of host distribution (vectors in particular) are predicted to occur under current climate change models (Plante *et al.*, 2024) or landscape alteration (Hoberg and Brooks, 2015) and could expose pathogens to previously unencountered potential hosts, eliciting host switches through EF (pre-existing capacity) and HM.

Recent research showed that pathogens might use a suboptimal resource (or host) and invade new host species (EF) without the need for any genetic innovation (reviewed by Brooks *et al.*, 2019). Such host switches may, in turn, drive the rapid evolution of the pathogen populations (D'Bastiani *et al.*, 2023). Using a protein–protein interaction experiment, Trivellone *et al.* (2019) explained the switch to a new host by a phytoplasma through phylogenetic conservatism (retention of an ancestral trait). This mechanism may explain the rapid origin of novel PPV associations (i.e. colonization of novel hosts at ecological time scales) eventually followed by speciation (evolutionary time scale) (Malembic-Maher *et al.*, 2020).

Modeling Phytoplasma Host Switches and Disease Emergence

Theoretical simulation modeling provides tools to understand interacting species' ecological and evolutionary dynamics. Such models are useful for testing hypotheses about mechanisms driving empirically observed patterns of plant pathogen diversification (reviewed by Railsback and Grimm, 2019; Jeger, 2000; Yang *et al.*, 2023) and can be used both to predict real-world patterns and generate new hypotheses. Individual-based models (IBMs) are particularly useful for vector-borne pathogens, providing a more fine-grained, realistic representation of population dynamics and incorporating complex interactions at the individual level (Keeling and Grenfell, 2000). Using a bipartite host–pathogen system, Araujo *et al.* (2015) proposed an IBM to demonstrate how host switch events may occur by simulating the conditions under which pathogens are likely to disperse and successfully colonize a novel host. The model suggests that switching to a new host does not require prior evolutionary novelty. Pathogens may survive on suboptimal hosts, resulting in an increased chance of switching to more distantly related hosts. Higher values of behavioral and life-history parameters such as dispersal capability and reproductive rate facilitate host switching. Recent studies tested many elements of the new theoretical framework based on EF (Gardner *et al.*, 2023) using this model (Souza *et al.*, 2022). The original IBM for parasites was expanded to incorporate several customizable parameters to accommodate different types of bipartite consumer–host associations, an interactive visualization of the model, and an in-depth description of the parameters in a biological context (Trivellone *et al.*, 2023). By using published life-history trait data on host–parasite associations, this model, implemented as an R package, shows how varying the value of an individual trait affects the rate of host switching.

Using an IBM framework similar to that of Araujo *et al.* (2015), D'Bastiani *et al.* (2023) developed a new individual-based simulation model that calculates the likelihood of host switching by incorporating parameters that account for historical compatibility of the hosts (phylogenetic relatedness) and genetic differentiation by the pathogen. This model suggests that switching to a new host may occur through the retention of an ancestral trait, as Trivellone *et al.* (2019) demonstrated empirically. If such traits enable pathogens to survive on suboptimal hosts, the chance of switching to more distantly related hosts increases, and higher values of certain traits (e.g. dispersal capability and reproductive rate) facilitate host switching. Thus, even in a hypothetical scenario where genetic novelties (i.e. mutations) do not emerge, pathogen populations may achieve the same size regardless of how poorly the individual pathogens fit their hosts (Feronato, 2022). Braga *et al.* (2018) also observed that colonization of a host that is not a perfect fit may drive pathogen diversification.

These previous theoretical studies provided a framework for modeling pathogen evolution but did not explicitly incorporate the vector component. A new tripartite IBM incorporating the vector is currently under development. This model is intended to simulate pathogen diversification patterns through time and elucidate the roles of HM and EF. The new algorithm incorporates parameters to accommodate multiple vectors and plant hosts. Preliminary results suggest that because the infection is a product of opportunity (exposure to potential new hosts) and the pathogen's pre-existing capacity to colonize the new host, HM increases the infection rate due to the increase in opportunity. Also, under HM, the pathogen population does not need to accumulate variability in order to switch hosts. This suggests that increased opportunity compensates for limited capacity and, therefore, that HM may drive the evolutionary diversification of pathogens. The new model also evaluates the fit between the pathogen and the new host, and preliminary results suggest that HM favors misfitting colonization, i.e. infections of more distant hosts (Burginski Rosa *et al.*, in preparation). Moving forward from these IBM-based studies, it should be possible to build a modelling platform that integrates both IBM and predictive statistical models with the aim of: (i) testing the hypothesis that host switch frequency will increase with increased vector dispersal via HM and EF, yielding new associations and pathogen diversification over time; (ii) identifying the most important ecological, evolutionary, and environmental factors driving host switches and pathogen diversification; and (iii) creating a tool to predict risk of spillover and emerging phytoplasma diseases (Fig. 13.1). Further refinements are also needed to overcome limitations of the existing IBM algorithm, such as partitioning host-switch frequency by differentiating opportunity from capacity and explicitly considering host abundance and other biological parameters.

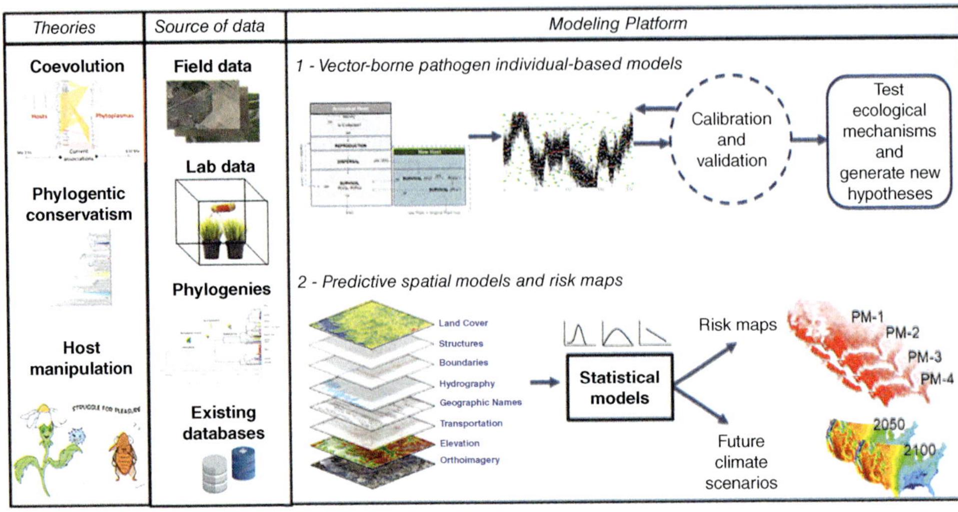

Fig. 13.1. Proposed modeling platform to test the roles of coevolution, phylogenetic conservatism, and host manipulation in driving the evolution of host–phytoplasma associations. Sources of data are a combination of empirical data from the field and laboratory experiments, phylogenies, and existing databases. Individual-based models are calibrated and validated with the data available. Predictive models are used to provide proactive tools to anticipate (re)emerging phytoplasma diseases. Map layers are adapted from the USGS Center of Excellence for Geospatial Information Science (CEGIS) via Wikimedia Commons.

Another algorithm, PACT (Phylogenetic Analysis for Comparing Trees) (Wojcicki and Brooks, 2004, 2005), has been used to estimate the frequency and relative prevalence of different processes driving host–parasite associations using phylogenies. This algorithm evaluates processes such as EF by using phylogenetic systematics directly to estimate and quantify the historical compatibility of potential host–parasite associations (Brooks, 1985). Using PACT for helminths parasitizing great apes, Brooks and Ferrao (2005) provided evidence that one-third of new associations representing emerging infectious diseases arise through host switches triggered by geographic dispersal and EF, constrained by historical compatibility. Incorporating phylogenetic information in this way will add a crucial historical component still missing from existing IBMs that aim to simulate phytoplasma host switches. Ultimately, multitrophic models will also be needed to evaluate the circulation and disease dynamics of vector-borne pathogens that use multiple hosts (Hoberg *et al.*, 2022).

New PPV Associations and Prediction of Emerging Phytoplasma Diseases

The basic biological characteristics of PPV associations are similar to those of many other vector-borne disease pathosystems affecting agriculture as well as those impacting veterinary and human health, including the potential roles played by HM, EF and, ultimately, host switches in the emergence of new diseases (Bosque-Pérez and Eigenbrode, 2011; Gandon, 2018; Cator *et al.*, 2020; Chala and Hamde, 2021; Mora *et al.*, 2022; Kostygov *et al.*, 2024). Thus, models developed for this specific system may be broadly applicable to understanding how new vector-borne diseases evolve (Trivellone and Panassiti, 2023a). Unfortunately, few attempts have been made so far to build such models for phytoplasmas and other plant pathogens, despite the wealth of published experimental results documenting genetic, physiological, and behavioral traits involved in HM and EF by phytoplasmas and the availability of detailed global climatic and landscape cover data.

The theoretical ecological and evolutionary framework proposed here and its policy extension, the document, assess, monitor, act (DAMA) protocol, were suggested as useful tools to cope with emerging infectious diseases, including plant diseases (Trivellone *et al.*, 2022b; Gardner *et al.*, 2023). The DAMA protocol was recently showcased using a case study from Bavaria (Germany) where data on silent infections were used to evaluate the potential for phytoplasma disease emergence through a phylogenetic triage of putative vector and plant hosts of aster yellows-related phytoplasma strains (Janik *et al.*, 2023). There are four important components that characterize the DAMA protocol; the first two are related to prediction and anticipation (document and assess), whereas the last two involve preparation, surveillance, and palliation (monitor and act) (Hoberg *et al.*, 2023).

Document

Further documentation of potential pathogens that are not currently causing problems for human society is needed to anticipate future problems through assessment and prediction. For phytoplasmas, an unexpectedly high diversity of new and previously documented strains was recently uncovered by screening insect specimens in a museum collection predominantly obtained from natural areas worldwide (Trivellone *et al.*, 2021, 2022a; Wei *et al.*, 2021). Such previously available resources, obtained through global and regional biodiversity inventories, may be leveraged to provide a broader understanding of the diversity and distribution of potential pathogens (Cook *et al.*, 2020; Janik *et al.*, 2023), facilitating the next steps of the DAMA protocol.

Assess

When a pathogen that is not currently causing an outbreak in a defined geographic area at a certain time is discovered, the next step for prediction is to evaluate its evolutionary history and its potential to switch hosts through the methods of phylogenetic triage and predictive modeling. Phylogenetic triage explicitly determines if the pathogen is a known strain or if it is closely related to one previously shown to cause disease outbreaks. For vector-borne pathogens, the same approach needs to be applied to the vector because vectors facilitate pathogen dispersal and exposure to more potential hosts. By assessing the risk associated with potential host switches and the potential impact of new associations between pathogens, vectors, and hosts, a subset of vectors and strains estimated to pose the highest risk can be selected for further monitoring (Janik *et al.*, 2023). An evolutionary-based risk heat map has been proposed to visualize the contribution of phylogenetic triage to the process of risk assessment (Brooks *et al.*, 2023b). Finally, data collected globally or regionally may be used to inform simulation and predictive models that evaluate potential outbreak risk under current and future climate change scenarios (Fig. 13.1).

Monitor

Monitoring high-risk pathogens identified in the previous step begins with targeted sampling strategies. Sampling schemes for documentation and monitoring should utilize tailored protocols designed to capture the full variability of pathogens and their distribution within a community (Molnár *et al.*, 2023a). Such schemes need to take into account landscape-scale variability to include wildlands, urban/peri-urban, and agricultural landscapes and interfaces between them (Trivellone *et al.*, 2022b). These efforts will produce detailed distribution maps that must be routinely updated and used periodically to refine predictive models.

Act

Ultimately, data obtained through monitoring, combined with historical and other relevant data, will be translated into actionable information, for instance by pinpointing pathogen/vector diversity hotspots with an elevated likelihood of future disease outbreaks. Interventions must be assessed collaboratively across sectors such as wildlife management, veterinary medicine, public health, education, and plant health. New policies and recommendations may be developed after thoroughly evaluating the pathogenic potential and dispersal dynamics of the target species. Effective prevention requires building interdisciplinary and transdisciplinary networks to identify optimal courses of action and ensure their coordinated implementation (Molnár *et al.*, 2023b).

Conclusion

Ecological interactions between phytoplasmas, their plant hosts, and insect vectors result in host expansion or contraction through ecological fitting and host manipulation that leads to the emergence of new associations and, potentially, new diseases. Previous research has identified several biological and environmental traits relevant for predicting where and when future outbreaks of phytoplasma diseases are most likely to occur. Future research should focus on incorporating such traits into predictive models and testing the models using real-world data. Ultimately, this will enable agronomists and other stakeholders to save time and resources through the development of surveillance and monitoring programs grounded in ecological and evolutionary theory and informed by the wealth of prior data on the ecology and evolution of PPV associations.

Acknowledgment

This work was supported by US National Science Foundation Grant DEB-2244871.

References

Agosta, S.J. (2006) On ecological fitting, plant–insect associations, herbivore host shifts, and host plant selection. *Oikos* 114, 556–565. DOI: 10.1111/j.2006.0030-1299.15025.x.

Agosta, S.J. and Klemens, J.A. (2008) Ecological fitting by phenotypically flexible genotypes: Implications for species associations, community assembly and evolution. *Ecology Letters* 11, 1123–1134. DOI: 10.1111/j.1461-0248.2008.01237.x.

Agosta, S.J., Janz, N., and Brooks, D.R. (2010) How specialists can be generalists: Resolving the "parasite paradox" and implications for emerging infectious disease. *Zoologia (Curitiba)* 27, 151–162. DOI: 10.1590/S1984-46702010000200001.

Alizon, S., Hurford, A., Mideo, N., and Van Baalen, M. (2009) Virulence evolution and the trade-off hypothesis: History, current state of affairs and the future. *Journal of Evolutionary Biology* 22, 245–259. DOI: 10.1111/j.1420-9101.2008.01658.x.

Ángeles-López, Y.I., Rivera-Bustamante, R., and Heil, M. (2018) Fatal attraction of non-vector impairs fitness of manipulating plant virus. *Journal of Ecology* 106, 391–400. DOI: 10.1111/1365-2745.12838.

Apari, P., Bajer, K., Brooks, D.R., and Molnár, O. (2019) Hiding in plain sight: An evolutionary approach to the South American Zika outbreak and its future consequences. *Zoologia* 36, 1–7. DOI: 10.3897/zoologia.36.e36272.

Araujo, S.B.L., Braga, M.P., Brooks, D.R., Agosta, S.J., Hoberg, E.P, Von Hartenthal, F.W., and Boeger, W.A. (2015) Understanding host-switching by ecological fitting. *PLoS One* 10, e0139225. DOI: 10.1371/journal.pone.0139225.

Barnett, D.E. (1977) A revision of the Nearctic species of the genus *Scaphoideus* (Homoptera: Cicadellidae). *Transactions of the American Entomological Society* 102, 485–593.

Bertaccini, A., Oshima, K., Maejima, K., and Namba, S. (2019) Phytoplasma effectors and pathogenicity factors. In: Bertaccini, A., Oshima, K., Kube, M., and Rao, G.P. (eds) *Phytoplasmas: Plant Pathogenic Bacteria – III: Genomics, Host Pathogen Interactions and Diagnosis*. Springer, Singapore, pp. 17–34. DOI: 10.1007/978-981-13-9632-8_2.

Bertaccini, A., Arocha-Rosete, Y., Contaldo, N., Duduk, B., Fiore, N. *et al.* (2022) Revision of the "*Candidatus* Phytoplasma" species description guidelines. *International Journal of Systematic and Evolutionary Microbiology* 72. DOI: 10.1099/ijsem.0.005353.

Boeger, W.A., Brooks, D.R., Trivellone, V., Agosta, S.J., and Hoberg, E.P. (2022) Ecological super-spreaders drive host–range oscillations: Omicron and risk space for emerging infectious disease. *Transboundary and Emerging Diseases* 60, e1280–e1288. DOI: 10.1111/tbed.14557.

Bosque-Pérez, N.A. and Eigenbrode, S.D. (2011) The influence of virus-induced changes in plants on aphid vectors: Insights from luteovirus pathosystems. *Virus Research* 159, 201–205. DOI: 10.1016/j.virusres.2011.04.020.

Braga, M.P. and Janz, N. (2021) Host repertoires and changing insect–plant interactions. *Ecological Entomology* 46, 1241–1253. DOI: 10.1111/een.13073.

Braga, M.P., Araujo, S.B., Agosta, S., Brooks, D., Hoberg, E., Nylin, S., Janz, N., and Boeger, W.A. (2018) Host use dynamics in a heterogeneous fitness landscape generates oscillations in host range and diversification. *Evolution* 72, 1773–1783. DOI:10.1111/evo.13557.

Brooks, D.R. (1979) Testing the context and extent of host-parasite coevolution. *Systematic Zoology* 28, 299–307. DOI: 10.2307/2412584.

Brooks, D.R. (1985) Historical ecology: A new approach to studying the evolution of ecological associations. *Annals of the Missouri Botanical Garden* 72, 660–680. DOI: 10.2307/2399219.

Brooks, D.R. and Ferrao, A.L. (2005) The historical biogeography of co-evolution: Emerging infectious diseases are evolutionary accidents waiting to happen: guest editorial. *Journal of Biogeography* 32, 1291–1299. DOI: 10.1111/j.1365-2699.2005.01315.x.

Brooks, D.R., Hoberg, E.P., and Boeger, W.A. (2019) *The Stockholm Paradigm: Climate Change and Emerging Disease*. University of Chicago Press, Chicago, IL.

Brooks, D.R., Hoberg, E.P., Boeger, W.A. and Gardner, S.L. (2023a) The emerging infectious disease crisis: changing a losing game. In: Gardner, S.L., Brooks, D.R., Boeger, W.A., and Hoberg, E.P. (eds) *An Evolutionary Pathway for Coping with Emerging Infectious Disease.* Zea Books, Lincoln, NB, pp. 14–27.

Brooks, D.R., Hoberg, E.P., Boeger, W.A., and Trivellone, V. (2023b) Assess: Using evolution to save time and resources. In: Gardner, S.L., Brooks, D.R., Boeger, W.A., and Hoberg, E.P. (eds) *An Evolutionary Pathway for Coping with Emerging Infectious Disease.* Zea Books, Lincoln, NB, pp. 146–157.

Burdon, J.J. and Thrall, P.H. (2009) Coevolution of rlants and their pathogens in natural habitats. *Science* 324, 755–756. DOI: 10.1126/science.1171663.

Burginski Rosa, J.L, Trivellone V., and Araujo, S. Host manipulation as a driver of pathogen evolution. In preparation.

Cao, Y., Trivellone, V., and Dietrich, C.H. (2020) A timetree for phytoplasmas (Mollicutes) with new insights on patterns of evolution and diversification. *Molecular Phylogenetics and Evolution* 149, 106826. DOI: 10.1016/j.ympev.2020.106826.

Cator, L.J., Johnson, L.R., Mordecai, E.A., El Moustaid, F., Smallwood, T.R. *et al.* (2020) The role of vector trait variation in vector-borne disease dynamics. *Frontiers in Ecology and Evolution* 8, 189. DOI: 10.3389/fevo.2020.00189.

Chala, B. and Hamde, F. (2021) Emerging and re-emerging vector-borne infectious diseases and the challenges for control: A review. *Frontiers in Public Health* 9, 715759. DOI: 10.3389/fpubh.2021.715759.

Cook, J.A., Arai, S., Armién, B., Bates, J., Bonilla, C.A.C. *et al.* (2020) Integrating biodiversity infrastructure into pathogen discovery and mitigation of emerging infectious diseases. *BioScience* 70, 531–534. DOI: 10.1093/biosci/biaa064.

D'Bastiani, E., Princepe, D., Marquitti, F.M., Boeger, W.A., Campião, K.M., and Araujo, S.B. (2023) Effect of host-switching on the ecological and evolutionary patterns of parasites. *Systematic Biology* 72, 912–924. DOI: 10.1093/sysbio/syad022.

Delong, D.M. (1948) The leafhoppers, or Cicadellidae, of Illinois. *Bulletin of the Illinois Natural History Survey* 24, 376.

Dietrich, C.H. (2013) Overview of the phylogeny, taxonomy and diversity of the leafhopper (Hemiptera: Auchenorrhyncha: Cicadomorpha: Membracoidea: Cicadellidae) vectors of plant pathogens. In: *Proceedings, International Symposium on Insect Vectors and Insect-Borne Diseases.* Taiwan Agricultural Research Institute, Taichung, Taiwan, pp. 47–70.

Ellwanger, J.H., Kaminski, V. de L., and Chies, J.A.B. (2019) Emerging infectious disease prevention: Where should we invest our resources and efforts? *Journal of Infection and Public Health* 12, 313–316. DOI: 10.1016/j.jiph.2019.03.010.

FAO (Food and Agriculture Organization) (2021) Scientific review of the impact of climate change on plant pests: A global challenge to prevent and mitigate plant-pest risks in agriculture, forestry and ecosystems. *ReliefWeb*. Available at: https://reliefweb.int/report/world/scientific-review-impact-climate-change-plant-pests-global-challenge-prevent-and (last accessed June 2025).

Feronato, S. G., Araujo, S., and Boeger, W. A. (2022) Accidents waiting to happen – Insights from a simple model on the emergence of infectious agents in new hosts. *Transboundary and Emerging Diseases* 69, 1727–1738. DOI: 10.1111/tbod.14146.

Friesen, M.L., Porter, S.S., Stark, S.C., Von Wettberg, E.J., Sachs, J.L., and Martinez-Romero, E. (2011) Microbially mediated plant functional traits. *Annual Review of Ecology, Evolution, and Systematics* 42, 23–46. DOI: 10.1146/annurev-ecolsys-102710-145039.

Gandon, S. (2018) Evolution and manipulation of vector host choice. *American Naturalist* 192, 23–34. DOI: 10.1086/697575.

Gandon, S., Agnew, P., and Michalakis, Y. (2002) Coevolution between parasite virulence and host life-history traits. *American Naturalist* 160, 374–388. DOI: 10.1086/341525.

Gardner, S., Brooks, D.R., Boeger, W.A., and Hoberg, E.P. (eds) (2023) *An Evolutionary Pathway for Coping with Emerging Infectious Disease.* Zea Books, Lincoln, NB.

Gonella, E., Benelli, G., Arricau-Bouvery, N., Bosco, D., Duso, C. *et al.* (2024) *Scaphoideus titanus* up-to-the-minute: Biology, ecology, and role as a vector. *Entomologia Generalis* 44, 481–496. DOI: 10.1127/entomologia/2023/2597.

Gougherty, A.V. and Davies, T.J. (2021) Towards a phylogenetic ecology of plant pests and pathogens. *Philosophical Transactions of the Royal Society B: Biological Sciences* 376, 20200359. DOI: 10.1098/rstb.2020.0359.

Heil, M. (2016) Host manipulation by parasites: Cases, patterns, and remaining doubts. *Frontiers in Ecology and Evolution* 4. DOI: 10.3389/fevo.2016.00080.

Hernandez-Caballero, I., Hernandez-Caballero, I., Garcia-Longoria, L., Gomez-Mestre, I., and Marzal, A. (2022) The Adaptive host manipulation hypothesis: Parasites modify the behaviour, morphology, and physiology of amphibians. *Diversity* 14, 739. DOI: 10.3390/d14090739.

Hoberg, E.P. and Brooks, D.R. (2015) Evolution in action: Climate change, biodiversity dynamics and emerging infectious disease. *Philosophical Transactions of the Royal Society B: Biological Sciences* 370, 20130553. DOI: 10.1098/rstb.2013.0553.

Hoberg, E.P., Trivellone, V., Cook, J.A., Dunnum, J.L., Boeger, W.A. *et al.* (2022) Knowing the biosphere: Documentation, specimens, archives, and names reveal environmental change and emerging pathogens. *MANTER: Journal of Parasite Biodiversity* 27. DOI: 10.32873/unl.dc.manter27.

Hoberg, E.P., Boeger, W.A., Molnár, O., Földvári, G., Gardner, S.L. *et al.* (2023) The DAMA Protocol: Anticipating to prevent and mitigate emerging infectious diseases. In: Gardner, S.L., Brooks, D.R., Boeger, W.A., and Hoberg, E.P. (eds) *An Evolutionary Pathway for Coping with Emerging Infectious Disease.* Zea Press, Lincoln, NB, pp. 112–121.

Holmes, J.C. and Bethel, W.M. (1972) Modification of intermediate host behavior by parasites. In: Canning, E.U. and Wright, C.A. (eds) *Behavioral Aspects of Parasite Transmission.* Academic Press, London, pp. 123–149.

Huang, W., Reyes-Caldas, P., Mann, M., Seifbarghi, S., Kahn, A. *et al.* (2020) Bacterial vector-borne plant diseases: Unanswered questions and future directions. *Molecular Plant* 13, 1379–1393. DOI: 10.1016/j.molp.2020.08.010.

Ingle, D.J., Howden, B.P., and Duchene, S. (2021) Development of phylodynamic methods for bacterial pathogens. *Trends in Microbiology* 29, 788–797. DOI: 10.1016/j.tim.2021.02.008.

IRPCM Phytoplasma/Spiroplasma Working Team – Phytoplasma Taxonomy Group (2004) "*Candidatus* Phytoplasma", a taxon for the wall-less, non-helical prokaryotes that colonize plant phloem and insects. *International Journal of Systematic and Evolutionary Microbiology* 54, 1243–1255. DOI: 10.1099/ijs.0.02854-0.

Janik, K., Panassiti, B., Kerschbamer, C., Burmeister, J., and Trivellone, V. (2023) Phylogenetic triage and risk assessment: how to predict emerging phytoplasma diseases. *Biology* 12, 732. DOI: 10.3390/biology12050732.

Janzen, D.H. (1985. On ecological fitting. *Oikos* 45, 308. DOI: 10.2307/3565565.

Jeger, M.J. (2000) Theory and plant epidemiology. *Plant Pathology* 49, 651–658.

Johnson, K.P., Dietrich, C.H., Friedrich, F., Beutel, R.G., Wipfler, B. *et al.* (2018) Phylogenomics and the evolution of hemipteroid insects. *Proceedings of the National Academy of Sciences of the USA* 115, 12775–12780. DOI: 10.1073/pnas.1815820115.

Keeling, M.J. and Grenfell, B.T. (2000) Individual-based perspectives on *R0*. *Journal of Theoretical Biology* 203, 51–61. DOI: 10.1006/jtbi.1999.1064.

Kirdat, K., Tiwarekar, B., Thorat, V., Sathe, S., Shouche, Y., and Yadav, A. (2021) "*Candidatus* Phytoplasma sacchari", a novel taxon – Associated with Sugarcane Grassy Shoot (SCGS) disease. *International Journal of Systematic and Evolutionary Microbiology* 71, 004591. DOI: 10.1099/ijsem.0.004591.

Kirdat, K., Kirdat, K., Tiwarekar, B., Sathe, S., and Yadav, A. (2023) From sequences to species: Charting the phytoplasma classification and taxonomy in the era of taxogenomics. *Frontiers in Microbiology* 14, 1123783. DOI: 10.3389/fmicb.2023.1123783.

Klimov, P.B., Pedroso, L.G.A., and He, Q. (2022) A probabilistic model predicting host specificity and host range expansion in mites parasitic on mammals. *Zoosymposia* 22, 48–48. DOI: 10.11646/zoosymposia.22.1.18.

Kostygov, A.Yu., Albanaz, A.T., Butenko, A., Gerasimov, E.S., Lukeš, J., and Yurchenko, V. (2024) Phylogenetic framework to explore trait evolution in Trypanosomatidae. *Trends in Parasitology* 40, 96–99. DOI: 10.1016/j.pt.2023.11.009.

Lee, I.-M., Gundersen-Rindal, D.E., and Bertaccini, A. (1998) Phytoplasma: Ecology and genomic diversity. *Phytopathology* 88, 1359–1366. DOI: 10.1094/PHYTO.1998.88.12.1359.

Lee, I.-M., Bottner-Parker, K.D., Zhao, Y., Villalobos, W., and Moreira, L. (2011) "*Candidatus* Phytoplasma costaricanum" a novel phytoplasma associated with an emerging disease in soybean (Glycine max). *International Journal of Systematic and Evolutionary Microbiology* 61, 2822–2826. DOI: 10.1099/ijs.0.029041-0.

Longo, A.V., Burrowes, P.A., and Zamudio, K.R. (2014) Genomic studies of disease-outcome in host–pathogen dynamics. *Integrative and Comparative Biology* 54, 427–438. DOI: 10.1093/icb/icu073.

Malembic-Maher, S., Desqué, D., Khalil, D., Salar, P., Bergey, B. *et al.* (2020) When a Palearctic bacterium meets a Nearctic insect vector: Genetic and ecological insights into the emergence of the grapevine Flavescence dorée epidemics in Europe. *PLoS Pathogens* 16, e1007967. DOI: 10.1371/journal.ppat.1007967.

MarketScreener (2022) IMF sees cost of COVID pandemic rising beyond $12.5 trillion. Available at: https://uk.marketscreener.com/news/latest/IMF-sees-cost-of-COVID-pandemic-rising-beyond-12-5-trillion-estimate-37600718/ (last accessed June 2025).

Massey, S.E. (2022) Host manipulation mechanisms of SARS-CoV-2. *Acta Biotheoretica* 70, 4. DOI: 10.1007/s10441-021-09425-z.

Mauck, K.E., De Moraes, C.M., and Mescher, M.C. (2014) Biochemical and physiological mechanisms underlying effects of Cucumber mosaic virus on host-plant traits that mediate transmission by aphid vectors. *Plant, Cell, and Environment* 37, 1427–1439. DOI: 10.1111/pce.12249.

Mauck, K.E., Chesnais, Q., and Shapiro, L.R. (2018) Evolutionary determinants of host and vector manipulation by plant viruses. In: Malmstrom, C.M. (ed.) *Advances in Virus Research*. Academic Press, Cambridge, MA, pp. 189–250. DOI: 10.1016/bs.aivir.2018.02.007.

May, R.M. and Anderson, R.M. (1983) Parasite-host coevolution. In: Futuyma, D.J. and Slatkin, M. (eds) *Coevolution*. Sinauer Associates, Sunderland, MA, pp.s 186–206.

McKee, C.D., Krawczyk, A.I., Sándor, A.D., Görföl, T., Földvári, M. *et al.* (2019) Host phylogeny, geographic overlap, and roost sharing shape parasite communities in European bats. *Frontiers in Ecology and Evolution* 7, 69. DOI: 10.3389/fevo.2019.00069.

Molnár, O., Hoberg, E.P., Trivellone, V., Földvári, G., and Brooks, D.R. (2023a) Prevent-Prepare-Palliate: the 3P framework – integrating the DAMA Protocol into global public health systems. In: Gardner, S.L., Brooks, D.R., Boeger, W.A., and Hoberg, E.P. (eds) *An Evolutionary Pathway for Coping with Emerging Infectious Disease*. Zea Books, Lincoln, NB, pp. 98-111.

Molnár, O., Knickel, M., and Marizzi, C. (2023b) All hands on deck: Turning evolutionary theory into preventive policies. In: Gardner, S.L., Brooks, D.R., Boeger, W.A., and Hoberg, E.P. (eds) *An Evolutionary Pathway for Coping with Emerging Infectious Disease*. Zea Books, Lincoln, NB, pp. 170-189.

Mora, C., McKenzie, T., Gaw, I.M., Dean, J.M., von Hammerstein, H. *et al.* (2022. Over half of known human pathogenic diseases can be aggravated by climate change. *Nature Climate Change* 12, 869–875. DOI: 10.1038/s41558-022-01426-1.

Naderali, N., Nejat, N., Vadamalai, G., Davis, R.E., Wei, W. *et al.* (2017) "*Candidatus* Phytoplasma wodyetiae", a new taxon associated with yellow decline disease of foxtail palm (Wodyetia bifurcata) in Malaysia. *International Journal of Systematic and Evolutionary Microbiology* 67, 3765–3772. DOI: 10.1099/ijsem.0.002187.

Namba, S. (2019) Molecular and biological properties of phytoplasmas. *Proceedings of the Japan Academy. Series B, Physical and Biological Sciences* 95, 401–418. DOI: 10.2183/pjab.95.028.

Nejat, N., Vadamalai, G., Davis, R.E., Harrison, N.A., Sijam, K. *et al.* (2013) "*Candidatus* Phytoplasma malaysianum", a novel taxon associated with virescence and phyllody of Madagascar periwinkle (Catharanthus roseus). *International Journal of Systematic and Evolutionary Microbiology* 63, 540–548. DOI: 10.1099/ijs.0.041467-0.

Orlovskis, Z. and Hogenhout, S.A. (2016) A bacterial parasite effector mediates insect vector attraction in host plants independently of developmental changes. *Frontiers in Plant Science* 7, 885. DOI: 10.3389/fpls.2016.00885.

Orlovskis, Z., Canale, M.C., Thole, V., Pecher, P., Lopes, J.R., and Hogenhout, S.A. (2015) Insect-borne plant pathogenic bacteria: Getting a ride goes beyond physical contact. *Current Opinion in Insect Science* 9, 16–23. DOI: 10.1016/j.cois.2015.04.007.

Papaïx, J., Burdon, J.J., Lannou, C., and Thrall, P.H. (2014) Evolution of pathogen specialisation in a host metapopulation: Joint effects of host and pathogen dispersal. *PLoS Computational Biology*10, e1003633. DOI: 10.1371/journal.pcbi.1003633.

Penczykowski, R.M., Laine, A.-L. and Koskella, B. (2016) Understanding the ecology and evolution of host–parasite interactions across scales. *Evolutionary Applications* 9, 37–52. DOI: 10.1111/eva.12294.

Plante, N., Durivage, J., Brochu, A.S., Dumonceaux, T., Santos, A.A. *et al.* (2024) Leafhoppers as markers of the impact of climate change on agriculture. *Cell Reports Sustainability* 1, 100029. DOI: 10.1016/j.crsus.2024.100029.

Poulin, R. (1995) "Adaptive" changes in the behaviour of parasitized animals: A critical review. *International Journal for Parasitology* 25, 1371–1383. DOI: 10.1016/0020-7519(95)00100-x.

Poulin, R., Brodeur, J., and Moore, J. (1994) Parasite manipulation of host behaviour: Should hosts always lose? *Oikos* 70, 479. DOI: 10.2307/3545788.

Railsback, S.F. and Grimm, V. (2019) *Agent-based and Individual-based Modeling: A Practical Introduction*. Princeton University Press, Princeton, NJ.

Rajabaskar, D., Bosque-Pérez, N.A., and Eigenbrode, S.D. (2014) Preference by a virus vector for infected plants is reversed after virus acquisition. *Virus Research* 186, 32–37. DOI: 10.1016/j.virusres.2013.11.005.

Ristaino, J.B., Anderson, P.K., Bebber, D.P., Brauman, K.A., Cunniffe, N.J. *et al.* (2021) The persistent threat of emerging plant disease pandemics to global food security. *Proceedings of the National Academy of Sciences of the USA* 118, e2022239118. DOI: 10.1073/pnas.2022239118.

Rodrigues Jardim, B., Kinoti, W.M., Tran-Nguyen, L.T., Gambley, C., Rodoni, B., and Constable, F.E. (2021) "*Candidatus* Phytoplasma stylosanthis", a novel taxon with a diverse host range in Australia, characterised using multilocus sequence analysis of 16S rRNA, secA, tuf, and rp genes. *International Journal of Systematic and Evolutionary Microbiology* 71, ijsem004589. DOI: 10.1099/ijsem.0.004589.

Rossi, M., Oshima, K., and Namba, S. (2019) Diversity and functional importance of phytoplasma Membrane Proteins. In: Bertaccini, A., Oshima, K., Kube, M., and Rao, G.P.(eds) *Phytoplasmas: Plant Pathogenic Bacteria – III: Genomics, Host Pathogen Interactions and Diagnosis*. Springer, Singapore, pp. 69–88. DOI: 10.1007/978-981-13-9632-8_5.

Safarova, D., Zemánek, T., Válová, P., and Navrátil, M. (2016) "*Candidatus* Phytoplasma cirsii", a novel taxon from creeping thistle [Cirsium arvense (L.) Scop.]. *International Journal of Systematic and Evolutionary Microbiology* 66, 1745–1753. DOI: 10.1099/ijsem.0.000937.

Seemüller, E., Marcone, C., Lauer, U., Ragozzino, A., and Göschl, M. (1998) Current status of molecular classification of the phytoplasmas. *Journal of Plant Pathology*, 80, 3–26.

Shaw, A.K., Igoe, M., Power, A.G., Bosque-Pérez, N.A., and Peace, A. (2019) Modeling approach influences dynamics of a vector-borne pathogen system. *Bulletin of Mathematical Biology* 81, 2011–2028. DOI: 10.1007/s11538-019-00595-z.

Souza, A.T.C., Araujo, S.B.L., and Boeger, W.A. (2022) The evolutionary dynamics of infectious diseases in an unstable planet: Insights from modeling the Stockholm Paradigm. *MANTER: Journal of Parasite Biodiversity*, 25. DOI: 10.10.32873/unl.dc.manter2.

Thomas, F., Rigaud, T., and Brodeur, J. (2012) Evolutionary routes leading to host manipulation by parasites. In: Hughes, D.P., Brodeur, J., and Thomas, F. (eds) *Host Manipulation by Parasites*. Oxford University Press, Oxford, UK, pp. 16–33.

Thompson, J.N. (1998) Rapid evolution as an ecological process. *Trends in Ecology and Evolution* 13, 329–332. DOI: 10.1016/S0169-5347(98)01378-0.

Tiwarekar, B., Kirdat, K., Sathe, S., Foissac, X., and Yadav, A. (2023) Chimera alert! The threat of chimeric sequences causing inaccurate taxonomic classification of phytoplasma strains. *bioRxiv*. DOI: 10.1101/2023.04.10.535501.

Tomkins, M., Kliot, A., Marée, A.F., and Hogenhout, S.A. (2018) A multi-layered mechanistic modelling approach to understand how effector genes extend beyond phytoplasma to modulate plant hosts, insect vectors and the environment. *Current Opinion in Plant Biology* 44, 39–48. DOI: 10.1016/j.pbi.2018.02.002.

Trivellone, V. (2018) *Hemiptera-Phytoplasma-Plant Biological Interaction Database*. Available at: http://trivellone.speciesfile.org/ (last accessed June 2025).

Trivellone, V. (2019) An online global database of Hemiptera-Phytoplasma-Plant biological interactions. *Biodiversity Data Journal* 7, e32910. DOI: 10.3897/BDJ.7.e32910.

Trivellone, V. (2022) Let emerging plant diseases be predictable. *MANTER: Journal of Parasite Biodiversity* 30, 1–8. DOI: 10.32873/unl.dc.manter30.

Trivellone, V. and Dietrich, C.H. (2021) Evolutionary diversification in insect vector-phytoplasma-plant associations. *Annals of the Entomological Society of America* 114, 137–150. DOI: 10.1093/aesa/saaa048.

Trivellone, V. and Panassiti, B. (2023a) A field synopsis, systematic seview, and meta-analyses of cophylogenetic studies: What is affecting congruence between phylogenies? *MANTER: Journal of Parasite Biodiversity*, 24. DOI: 10.32873/unl.dc.manter24.

Trivellone, V. and Panassiti, B. (2023b) Pathogen-host phylogenetic congruence varies with paradigmatic assumptions, analytical method, and type of association. In: Gardner, S.L., Brooks, D.R., Boeger, W.A., and Hoberg, E.P. (eds) *An Evolutionary Pathway for Coping with Emerging Infectious Disease.* Zea Books, Lincoln, NB, pp. 59–77.

Trivellone, V., Ripamonti, M., Angelini, E., Filippin, L., Rossi, M., Marzachí, C., and Galetto, L. (2019) Evidence suggesting interactions between immunodominant membrane protein Imp of Flavescence dorée phytoplasma and protein extracts from distantly related insect species. *Journal of Applied Microbiology* 127, 1801–1813. DOI: 10.1111/jam.14445.

Trivellone, V., Wei, W., Filippin, L., and Dietrich, C.H. (2021) Screening potential insect vectors in a museum biorepository reveals undiscovered diversity of plant pathogens in natural areas. *Ecology and Evolution* 11, 6493–6503. DOI: 10.1002/ece3.7502.

Trivellone, V., Cao, Y., and Dietrich, C.H. (2022a) Comparison of traditional and next-generation approaches for uncovering phytoplasma diversity, with discovery of new groups, subgroups and potential vectors. *Biology* 11, 977. DOI: 10.3390/biology11070977.

Trivellone, V., Hoberg, E.P., Boeger, W.A., and Brooks, D.R. (2022b) Food security and emerging infectious disease: Risk assessment and risk management. *Royal Society Open Science* 9, 211687. DOI: 10.1098/rsos.211687.

Trivellone, V., Araujo, S.B.L., and Panassiti, B. (2023) HostSwitch: An R package to simulate the extent of host-switching by a consumer. *R Journal* 14, 179–194. DOI: 10.32614/RJ-2023-005.

Wei, W. and Zhao, Y. (2022) Phytoplasma taxonomy: Nomenclature, classification, and identification. *Biology* 11, 1119. DOI: 10.3390/biology11081119.

Wei, W., Davis, R.E., Lee, I.M., and Zhao, Y. (2007) Computer-simulated RFLP analysis of 16S rRNA genes: Identification of ten new phytoplasma groups. *International Journal of Systematic and Evolutionary Microbiology* 57, 1855–1867. DOI: 10.1099/ijs.0.65000-0.

Wei, W., Trivellone, V., Dietrich, C.H., Zhao, Y., Bottner-Parker, K.D., and Ivanauskas, A. (2021) Identification of phytoplasmas representing multiple new genetic lineages from phloem-feeding leafhoppers highlights the diversity of phytoplasmas and their potential vectors. *Pathogens* 10, 352. DOI: 10.3390/pathogens10030352.

Weinersmith, K. and Faulkes, Z. (2014) Parasitic manipulation of hosts' phenotype, or how to make a zombie – An introduction to the symposium. *Integrative and Comparative Biology* 54, 93–100. DOI: 10.1093/icb/icu028.

Weintraub, P.G. and Beanland, L. (2006) Insect vectors of phytoplasmas. *Annual Review of Entomology* 51, 91–111. DOI: 10.1146/annurev.ento.51.110104.151039.

Wojcicki, M. and Brooks, D.R. (2004) Escaping the matrix: A new algorithm for phylogenetic comparative studies of co-evolution. *Cladistics* 20, 341–361. DOI: 10.1111/j.1096-0031.2004.00029.x.

Wojcicki, M. and Brooks, D.R. (2005) PACT: An efficient and powerful algorithm for generating area cladograms. *Journal of Biogeography* 32, 755–774. DOI: 10.1111/j.1365-2699.2004.01148.x.

Wolf, Y.I. and Koonin, E.V. (2013) Genome reduction as the dominant mode of evolution. *Bioessays* 35, 829–837. DOI: 10.1002/bies.201300037.

Yang, L.N., Ren, M., and Zhan, J. (2023) Modeling plant diseases under climate change: evolutionary perspectives. *Trends in Plant Science* 28, 519–526.

Zamharir, M.G. and Shameli, S. (2023) Study on soybean bud proliferation phytoplasma insect vector. *Phytopathogenic Mollicutes* 13(1), 37–38. DOI: 10.5958/2249-4677.2023.00019.1.

Zhao, Y., Wei, W., Lee, I.M., Shao, J., Suo, X., and Davis, R.E. (2009) Construction of an interactive online phytoplasma classification tool, iPhyClassifier, and its application in analysis of the peach X-disease phytoplasma group (16SrIII). *International Journal of Systematic and Evolutionary Microbiology* 59, 2582–2593. DOI: 10.1099/ijs.0.010249-0.

Zhao, Y., Wei, W., Davis, R.E., Lee, I.M., and Bottner-Parker, K.D. (2021) The agent associated with blue dwarf disease in wheat represents a new phytoplasma taxon, "*Candidatus* Phytoplasma tritici". *International Journal of Systematic and Evolutionary Microbiology* 71. DOI: 10.1099/ijsem.0.004604.

14 Exploring Plant Epigenetic Responses to Paulownia Witches'-Broom Phytoplasma Infection

Jidong Li,[1]* Assunta Bertaccini,[2] and Wei Wei[3]

[1]Henan Agricultural University, Zhengzhou, China; [2]Alma Mater Studiorum, University of Bologna – Plant Pathology, Bologna, Italy; [3]Molecular Plant Pathology Laboratory, ARS-USDA, Beltsville, Maryland, USA

Abstract

Phytoplasma infection induces epigenetic changes in plants, thereby affecting their growth and development. Paulownia trees, widely cultivated for timber, ornamental purposes, and agroforestry, are severely threatened by paulownia witches'-broom (PaWB) disease, associated with phytoplasma infection. Modifications to DNA, RNA, or protein (without altering DNA sequences) have been implicated in the development of PaWB disease. This chapter explores the epigenetic responses of *Paulownia* spp. to PaWB disease, highlighting the roles of DNA methylation, lysine acetylation and succinylation, histone methylation and acetylation, non-coding RNA regulation, mRNA modification, and protein modifications in disease development and resistance. These findings enhance our understanding of how plants respond to phytoplasma-induced stress and offer potential strategies for managing similar plant diseases through targeted epigenetic interventions.

Introduction

The genus *Paulownia*, within the Paulowniaceae family, encompasses 17 species of fast-growing, deciduous hardwood trees known for their significant ecological and economic contributions. Prominent species, including *P. fortunei*, *P. catalpifolia*, *P. tomentosa*, and *P. elongata*, are widely cultivated for their light-weight, durable wood, rapid growth, and remarkable adaptability to diverse environments (Dżugan *et al.*, 2021; Sławińska *et al.*, 2023). Native to South-East Asia, especially China, paulownia trees have been cultivated for over 2000 years for timber, ornamental purposes, and agroforestry. Nowadays, paulownia trees are widely planted across Asia, North America, Europe, and Australia, serving purposes such as ecological restoration, bioenergy production, and commercial forestry (Cao *et al.*, 2021; Jakubowski, 2022).

Paulownia trees are severely threatened by paulownia witches'-broom (PaWB) disease, the most destructive disease affecting *Paulownia* spp. PaWB disease is associated with the presence of phytoplasma identified as a strain of '*Candidatus* Phytoplasma asteris,' belonging to ribosomal subgroup 16SrI-D (Yue *et al.*, 2008). Infected trees exhibit symptoms such as shoot proliferation (witches'-broom), leafy flowers, smaller and thinner leaves, growth stunt,

*Corresponding author: lijidong@henau.edu.cn

DOI: 10.1079/9781800627031.0014

and die back, leading to substantial economic losses (Doi *et al.*, 1967).

Molecular analyses using simple sequence repeat (SSR) and amplified fragment length polymorphism (AFLP) markers have indicated that phytoplasma infection does not change the DNA sequences of paulownia (Cao *et al.*, 2012, 2014a, 2014b). This suggests that the disease exerts its effects through regulatory mechanisms rather than direct genetic mutations, pointing to the importance of studying epigenetic changes in understanding its impact.

Epigenetics, the study of heritable changes in gene function without altering DNA sequences, plays a pivotal role in regulating plant responses to biotic and abiotic stresses. These modifications include DNA and histone methylation, histone acetylation, non-coding RNA regulation, and mRNA modifications, all of which influence gene expression and plant physiology. Investigating the epigenetic responses of paulownia to PaWB disease provides an opportunity to uncover the mechanisms through which the disease affects growth and development, potentially offering avenues for improved management and resistance strategies.

While the epigenetic responses of paulownia to phytoplasma infection are still being elucidated, research on other plant species offers valuable insights. For instance, in grapevine (*Vitis vinifera*), flavescence dorée phytoplasma infection was associated with changes in DNA methylation, particularly in the promoters of genes tied to photosynthesis and secondary metabolism (Pagliarani *et al.*, 2020). Similarly, in mulberry (*Morus multicaulis*), differential methylation and expression of genes in mulberry dwarf phytoplasma-infected leaves were identified, suggesting a link between epigenetic changes and stress responses (Liu *et al.*, 2021). In sesame (*Sesamum indicum*), whole-genome bisulfite sequencing (WGBS) has revealed differential DNA methylation patterns associated with sesame phyllody phytoplasma infection (Verma *et al.*, 2022). These findings provide clues for exploring epigenetic mechanisms in paulownia.

This chapter investigates the epigenetic responses of paulownia to PaWB disease (Fig. 14.1), examining how DNA methylation, lysine acetylation and succinylation, histone methylation and acetylation, non-coding RNA regulation, and mRNA modification contribute to disease development and resistance. By investigating these mechanisms, this work aims to comprehensively understand how paulownia

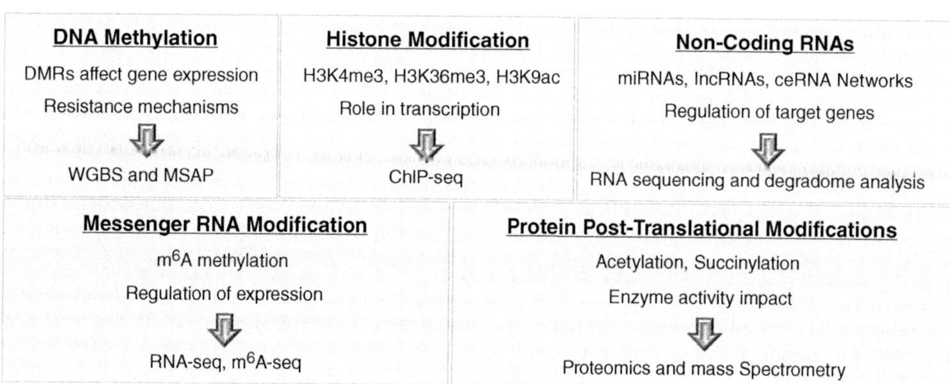

Fig. 14.1. A diagram illustrating the major epigenetic responses in paulownia trees induced by paulownia witches'-broom (PaWB) phytoplasma infection, and the methodologies used to study each category. DNA methylation: DMR, differentially methylated regions; WGBS, whole-genome bisulfite sequencing; MSAP, methylation-sensitive amplification polymorphism. Histone modification: H3K4me3, histone H3 at lysine 4; H3K36me3, trimethylation at lysine 36; H3K9ac, acetylation at lysine 9; ChIP-seq, chromatin immunoprecipitation sequencing. Non-coding RNAs: miRNAs, microRNAs; lncRNAs, long non-coding RNAs; ceRNA, competing endogenous RNA. Messenger RNA modification: m^{6}A, N^6-methyladenosine. Protein post-translational modifications.

trees respond to phytoplasma infection, offering potential strategies for breeding disease-resistant varieties and mitigating economic losses.

DNA Methylation Response to Phytoplasma

DNA methylation is a crucial epigenetic modification that is essential in various nuclear processes such as gene expression, DNA repair, and recombination. Multiple studies have indicated that DNA methylation at gene boundaries contributes to differential gene expression, particularly in response to pathogens. In paulownia, the relationship between DNA methylation and the response to PaWB phytoplasma infection has been explored since the early 2000s, revealing mechanisms that link methylation to disease progression and recovery.

A study conducted by Li *et al.* (2008) demonstrated that treatment of oxytetracycline (a derivative of tetracycline) effectively restores the normal growth and morphology of *P. tomentosa* × *P. fortunei* plantlets infected with PaWB phytoplasma. The treatment alleviates disease symptoms and eliminates phytoplasma, as confirmed by PCR, which shows no detectable phytoplasma DNA, demonstrating its effectiveness in symptom management and pathogen eradication. High-performance liquid chromatography (HPLC) analysis revealed that the global DNA methylation levels in healthy plantlets were nine times higher than those in PaWB-diseased plantlets. Moreover, the DNA methylation levels increase in oxytetracycline-treated plantlets in correlation with the concentration of oxytetracycline treatment, although they remained below the level observed in healthy plantlets (Li *et al.*, 2008).

Subsequent research revealed that exogenous methyl DNA adducts, such as methyl methanesulfonate (MMS) and dimethyl sulfonate (DMS), could recover PaWB-diseased *in vitro* plantlets of different *Paulownia* species to normal morphology. PCR detection of the 16S rRNA gene confirmed that PaWB phytoplasmas were eliminated following treatment with these adducts. SSR marker analysis showed no DNA sequence polymorphism among healthy, PaWB-diseased, and adduct-treated paulownia plantlets, suggesting that the observed changes were driven by epigenetic modifications rather than alterations in the genomic DNA (Zhai *et al.*, 2010; Fan *et al.*, 2011, 2012; Zhao *et al.*, 2011; Cao *et al.*, 2012).

Methylation-sensitive amplification polymorphism (MSAP) analysis also revealed that the DNA methylation levels in healthy paulownia plantlets were higher than those in PaWB-diseased ones. Unlike AFLP analysis, which showed no DNA sequence variation, MSAP data indicated that MMS treatment increased DNA methylation levels and facilitated the recovery of plantlets across different *Paulownia* species. This recovery was characterized by a restoration of healthy morphology (Cao *et al.*, 2014a, 2014b, 2014c).

In MMS-treated *P. tomentosa* × *P. fortunei* and *P. fortunei* plantlets, Cao *et al.* (2014a, 2014b) identified 81 and 72 MSAP fragments with altered methylation or demethylation patterns, respectively. Sequencing and alignment of these fragments revealed that homologous regions associated with annotated genes involved in diverse cellular functions, including substance metabolism, energy production, transcription, protein biosynthesis, pathogen defense, signal transduction, and transport (Cao *et al.*, 2014a, 2014b). Expression patterns of 12 key genes in these paulownia plantlets were also analyzed by quantitative real-time PCR (qRT-PCR). Among them, five were upregulated with increasing MMS concentrations, including those that encode chitin-inducible gibberellin-responsive protein and uncharacterized protein LOC 100796964 in *P. fortunei*, as well as glutamyl-tRNA amidotransferase subunit A, photosystem II 47 kDa protein, and nicotinamide adenine dinucleotide:ubiquinone oxidoreductase subunit 7 in *P. tomentosa* × *P. fortunei*. Conversely, seven genes were downregulated as MMS concentration decreased. The downregulated genes include those that encode leucyl aminopeptidase, cytochrome P450 76B6, ring finger protein, and β-hydroxyacyl-ACP dehydrase 1 in *P. fortunei*, and 26S protease regulatory subunit 6b, xanthine dehydrogenase/oxidase, and zinc finger protein ZAT5 (Cao *et al.*, 2014a, 2014b).

Cao *et al.* (2020) employed WGBS to construct a DNA cytosine methylation map at single base pair resolution to further explore the relationship between gene expression and DNA methylation. This approach was applied to both healthy and PaWB-diseased *P. fortunei* plantlets. A total of 422,662 differentially methylated regions (DMRs) were detected between the two groups. The combination of DNA methylation data and RNA-seq transcriptome analyses identified 436 genes with different methylation and transcript levels. KEGG (Kyoto Encyclopedia of Genes and Genomes) pathway analysis revealed that these genes mainly involve plant hormone signal transduction, carbon metabolism, and starch and sucrose metabolism. Validation using bisulfite-PCR and qRT-PCR confirmed that the DMR-associated genes suggest a repressive effect of DNA methylation on gene expression. Among these, two genes, *TPR1* and *R2R3-MYB*, were found potentially critical in the plant's response to PaWB phytoplasma infection (Cao *et al.*, 2020).

Histone Methylation and Acetylation Response to Phytoplasma Infection

In eukaryotes, DNA is organized into chromatin, with the nucleosome serving as its fundamental structural unit. Each nucleosome comprises a histone octamer, which consists of two subunits, each of the histone proteins H2A, H2B, H3, and H4. This octamer is wrapped by 147 bp of double-helix DNA, along with a linker histone H1. Histone modification, including methylation, acetylation, phosphorylation, ubiquitylation, and sumoylation, are important epigenetic mechanisms that modulate gene expression by altering chromatin structure and genome accessibility. In particular, histone methylation and acetylation play critical roles in promoting an open chromatin conformation, facilitating transcriptional activation and genome accessibility (Kouzarides, 2007; Jung and Kim, 2012).

Yan *et al.* (2019, 2020) further analyzed histone modifications in *P. fortunei* plantlets to investigate their response to PaWB phytoplasma infection. Samples included healthy plantlets, PaWB-diseased plantlets, and MMS-treated healthy and PaWB-diseased plantlets grown *in vitro*. Chromatin immunoprecipitation sequencing (ChIP-seq) was used to study three specific histone markers: methylation of histone H3 at lysine 4 (H3K4me3), methylation of histone H3 at lysine 36 (H3K36me3), and acetylation of histone H3 at lysine 9 (H3K9ac). These histone modifications are known to be associated with transcriptional activation and were chosen for their importance in regulating chromatin accessibility and gene expression (Yan *et al.*, 2019, 2020). The ChiP-seq analysis revealed that all three histone modifications were predominantly localized toward genic regions of the genome, such as promoters, untranslated regions (5' UTR and 3' UTR), coding sequences (CDS), and introns.

Significant differences in histone modification patterns were identified between healthy and PaWB-diseased paulownia plantlets. In diseased plantlets, genes with increased expression were associated with 1169, 5301, and 2688 DMGs marked by the H3K4me3, H3K36me3, and H3K9ac markers, respectively. In contrast, decreased expression was linked to 1265, 887, and 1228 DMGs with the same markers. MMS treatment of PaWB-diseased plantlets resulted in partial restoration of histone modification patterns, with 282, 87, and 165 upregulated DMGs, and 333, 507, and 436 downregulated DMGs for the three histone markers, respectively (Yan *et al.*, 2019, 2020).

Generally, the analysis showed that the number of histone modification regions was higher in diseased paulownia plantlets than in healthy plants. This suggests that phytoplasma infection induces widespread epigenetic alterations in the chromatin landscape. The combination of RNA-seq data and ChIP-qPCR analysis demonstrated that histone modifications positively regulate gene expression by enhancing chromatin accessibility. KEGG pathway analysis of differentially expressed genes/mRNAs (DEGs) associated with differential histone modification profiles identified their involvement in critical biological and metabolic processes. These pathways included metabolic pathways, biosynthesis of secondary metabolites, phenylpropanoid biosynthesis, plant–pathogen interactions, plant hormone signal transduction, and starch and sucrose metabolism. These findings highlight the pivotal role of histone modifications in modulating the transcriptional response of paulownia to PaWB phytoplasma infection, influencing both defense mechanisms and metabolic adjustments (Yan *et al.*, 2019, 2020).

Non-coding RNAs Regulation Response to Phytoplasma Infection

Non-coding RNAs (ncRNAs), such as microRNAs (miRNA), long non-coding RNAs (lncRNA), and circular RNAs (circRNAs), play important roles in regulating gene expression at the post-transcriptional level. These ncRNAs orchestrate cellular responses to biotic and abiotic stress, making them integral to plant defense mechanisms. miRNAs are small, single-stranded RNA molecules (21–23 nucleotides) that bind to complementary sequences on target mRNAs, leading to mRNA cleavage or destabilization, thereby suppressing gene expression (Voinnet, 2009). lncRNAs, transcripts exceeding 200 nucleotides, and circRNAs are single-stranded RNA molecules that form covalently closed continuous loops. Both lncRNAs and circRNAs are known to play regulatory roles in gene expression, although their mechanisms are not yet fully elucidated (Salmena *et al.*, 2011; Rinn and Chang, 2013).

MicroRNA regulation response to phytoplasma

Studies in *P. tomentosa* have revealed significant regulatory responses involving ncRNAs in the context of PaWB infection. High-throughput sequencing and degradome analysis of *in vitro* healthy plantlets, PaWB-diseased plantlets, and PaWB-diseased plantlets treated with 60 mg/l MMS identified 98,714 unigenes, 62 conserved miRNAs, and 35 novel miRNAs. Among them, 902 DEGs and 24 differentially expressed miRNAs (DERs) were detected, with 19 target genes of DERs being among the DEGs. Functional annotations linked these target genes to morphological changes, plant defense, and hormone signaling. Notable interactions include pau-miR156g-SPL, associated with morphological changes; pau-miR403-serine/threonine-protein kinase abkC, involved in stress signaling; and pau-miR166c-HD-ZIP, which is implicated in disease symptom development (Fan *et al.*, 2015a).

The effect of 20 mg/l MMS treatment was analyzed on *in vitro* PaWB-diseased, PaWB-diseased/treated, and healthy *P. tomentosa* plantlets, focusing on the expressions of miRNA and their target genes. A total of 20 conserved miRNAs were found to be differentially expressed among the samples. By degradome sequencing analysis, 55 unique target genes were targeted by ten of the 20 DERs. The functions of these target genes include acting as transcription factors and enhancing resistance to environmental stress (Fan *et al.*, 2017).

Similar investigations in *P. fortunei* demonstrated that ncRNAs also play key roles in this species' response to phytoplasma infection. High-throughput sequencing and degradome analysis of *in vitro* healthy, PaWB-diseased, and 20 mg/l MMS-treated PaWB-diseased plantlets uncovered 95 conserved and 122 novel miRNAs. These miRNAs were linked to 109 target genes, corresponding to 14 conserved miRNA families and 24 novel miRNAs. The expression levels of these miRNAs were validated by qRT-PCR, confirming their differential expression patterns. The majority of these miRNAs were responsive to PaWB disease, suggesting their pivotal roles in disease regulation and plant stress responses. Gene ontology and KEGG analyses annotated the target genes of these miRNAs, linking them to various biological functions and pathways, including cellular component, molecular function, and biological process (Niu *et al.*, 2016).

Another study conducted by Fan *et al.* (2016) on *in vitro* healthy, PaWB-diseased, and 60 mg/l MMS-treated PaWB-diseased *P. fortunei* plantlets identified 96 conserved miRNAs and 83 putative novel miRNAs. Among them, 17 conserved and 20 novel miRNAs were DMRs in response to PaWB phytoplasma infection. Degradome analysis identified 114 target genes associated with 18 conserved miRNA families and 33 target genes linked to 15 novel miRNAs. Functional annotation revealed these target genes are related to transcription regulation, stress response mechanisms, nitrogen metabolism, and other essential activities (Fan *et al.*, 2016).

In a hybrid of *P. tomentosa* × *P. fortunei*, studies on *in vitro* healthy, PaWB-diseased, and MMS-treated (15 and 30 mg/l) PaWB-diseased plantlets uncovered 125 conserved and 118 novel miRNAs, as well as 814 target genes linked to 23 conserved and 32 novel miRNAs. Among them, 33 DERs and 166 target genes were associated with PaWB disease. Functional analysis highlighted the role of these genes in plant–pathogen interactions and hormone signal transduction (Fan *et al.*, 2015b).

The miRNA-mediated regulation in response to phytoplasma infection involves a network of interactions. Genes involved in morphological changes, stress resistance, and hormone signaling pathways are among the key targets of differentially expressed miRNAs. These findings emphasize the role of ncRNAs, especially miRNAs, in modulating plant defenses against phytoplasma-induced diseases, offering valuable insights for developing resistant plant varieties through genetic and epigenetic approaches.

lncRNA regulation response to phytoplasma

To investigate the roles of lncRNAs in response to PaWB phytoplasma infection in *P. tomentosa*, RNA sequencing was conducted on various plantlets, including healthy, PaWB-diseased, and those treated with rifampicin and MMS. In one study, RNA sequencing of healthy, diseased, and rifampicin-treated plantlets identified 28,614 unique mRNAs and 3693 potential lncRNAs. Among them, 1063 unique mRNAs and 110 differentially expressed lncRNAs (DELs) were found. Comparations across the samples revealed 12 target genes of nine DELs that especially responded to PaWB phytoplasma infection. KEGG functional annotation of these genes indicated their involvement in eight pathways, including phenazine biosynthesis, RNA transport, cysteine and methionine metabolism, amino sugar and nucleotide sugar metabolism, glycerolipid metabolism, and monoterpenoid biosynthesis (Wang *et al.*, 2017).

Another study expanded the analysis to include MMS-treated plantlets, revealing 3689 lncRNAs, including 45 bidirectional, 563 antisense, 3012 intergenic, and 69 sense lncRNAs. A total of 112 DELs responsive to phytoplasma infection were identified, with computational predictions linking 89 DELs to 157 potential target genes. Among them, 27 target genes were differentially expressed between the healthy and PaWB-diseased plantlet samples. These 27 genes involve stress resistance, growth, metabolism, transport, and photosynthesis. In addition, 51 out of the 157 target genes (approximately 32.5%) underwent alternative splicing, resulting in 315 distinct transcript variants. This indicates that splicing plays a significant role in regulating gene expression within this system. The spliced transcript variants were mainly involved in photosynthesis and carbon metabolism. The observed changes in transcript diversity suggest that alternative splicing may be a regulatory mechanism to adapt or respond to perturbations caused by phytoplasma infection. Specifically, disruptions in primary metabolic pathways such as photosynthesis and carbon metabolism could indicate a reallocation of cellular resources or a stress-induced reprogramming of metabolic networks. These findings highlight the impact of phytoplasma infection on core physiological processes, with splicing potentially modulating the host response to the pathogen (Fan *et al.*, 2018a).

RNA-seq analysis conducted by Cao *et al.* (2018) on *P. tomentosa* plantlets, both healthy and PaWB-diseased, provided significant insights into the transcriptomic alterations caused by the infection. The study identified 2725 lncRNAs and 748 DELs. Among them, 1638 target genes associated with 514 DELs were characterized. Of these target genes, 1478 were identified as *cis*-acting targets of 468 DELs, suggesting that these lncRNAs regulate nearby genes on the same chromosome. Additionally, 448 target genes were found to be *trans*-acting targets of 447 DELs, indicating regulatory effects on distant genes or interchromosomal interactions. Further analysis uncovered that 224 of the 1638 target genes were DEGs between healthy and diseased samples. Of these, 99 genes exhibited increased expression levels upon phytoplasma infection, while 125 genes showed decreased expression. Functional enrichment analysis of the 224 DEGs demonstrated their involvement in 100 KEGG pathways. Among these pathways, plant hormone signal transduction emerged as the most significantly enriched category, emphasizing its critical role in the host's response to phytoplasma infection (Cao *et al.*, 2018).

Competing Endogenous RNA Response to Phytoplasma Infection

The theory of competing endogenous RNAs (ceRNAs) in plants provides a framework for understanding how RNA molecules regulate gene expression by competing for shared miRNAs. miRNAs bind to specific sequences on target RNAs, typically leading to their degradation or translation inhibition. In the ceRNA model, diverse RNA species, including mRNAs, lncRNAs, circRNAs, and even pseudogene-derived transcripts, compete for binding to miRNAs through common miRNA response elements (MREs). Because miRNAs are present in limited quantities, RNAs that share MREs can act as molecular sponges, sequestering miRNAs and reducing their availability to suppress other targets. Among these, lncRNAs play particularly significant regulatory roles by competing with other RNA molecules for miRNA binding, thus influencing gene expression (Salmena *et al.*, 2011).

To investigate the response of mRNAs, lncRNAs, circRNAs, and miRNAs to phytoplasma infection and to construct a ceRNA network elucidating the mechanisms of PaWB phytoplasma pathogenesis, RNA sequencing was performed using healthy and PaWB-diseased *P. fortunei in vitro* plantlets. A total of 550 miRNAs (395 conserved and 155 novel), 32,086 mRNAs, 3126 lncRNAs, and 1634 circRNAs were identified. Among them, 65 differentially expressed miRNAs (DEMs) were detected between the healthy and diseased samples, including 48 conserved and 17 novel miRNAs. Of these, 32 (25 conserved and 7 novel) were upregulated in the diseased sample, while 33 (23 conserved and 10 novel) were downregulated. The 65 DEMs were predicted to target 850 genes (Fan *et al.*, 2018b).

In addition to miRNA, Fan's team found a total of 1059 differentially expressed genes/mRNAs (DEGs), of which 639 were upregulated and 420 were downregulated. Functional annotations of these DEGs revealed their association with 129 KEGG pathways. The significantly enriched pathways include phenylpropanoid biosynthesis, carbon fixation, metabolism of xenobiotics by cytochrome P450, and isoquinoline alkaloid biosynthesis. Furthermore, analysis of lncRNAs identified a total of 229 DELs, including 81 upregulated and 148 downregulated, with 1048 target genes associated with these DELs, demonstrating their substantial regulatory potential in the host's response to the pathogen (Fan *et al.*, 2018b). The study also detected 65 differentially expressed circRNAs (DECs) between the healthy and PaWB-diseased samples. These 65 DECs were associated with 62 host genes. Out of these 62 host genes, nine showed differential expression and were linked to ten specific DECs. This suggests that some circRNAs may influence the expression of their host genes during phytoplasma infection (Fan *et al.*, 2018b).

To systematically explore the miRNA-mediated response mechanisms, 1059 predicted DEGs, 229 DELs, and 65 DECs were analyzed to construct ceRNA networks. The resulting networks, integrating lncRNA–miRNA–mRNA and circRNA–miRNA–mRNA interactions, revealed five miRNAs, four circRNAs, five lncRNAs, and 15 mRNAs. Visualization of the ceRNA networks indicated three primary hubs centered on specific miRNAs, including pf-miR156j, pf-miR172i, pf-miR8767a/b, and pf-miR172h/k, which play central roles in regulating host gene expression during infection (Fan *et al.*, 2018b).

Messenger RNA Modification Response to Phytoplasma Infection

Messenger RNA (mRNA) modification is an essential and widespread molecular mechanism critical for regulating key plant developmental processes, including embryo development, shoot stem cell fate, floral transition, trichome morphogenesis, leaf initiation, and root development. Among various post-transcriptional modifications, N^6-methyladenosine (m^6A) is the most prevalent. The m^6A modification plays a vital role in regulating mRNA expression, translation efficiency, alternative splicing, and degradation. Analogous to DNA methylation and histone modification, mRNA modification represents an additional layer of epigenetic regulation, referred to as the epitranscriptome (Meyer *et al.*, 2012; Schwartz *et al.*, 2013).

Xu *et al.* (2023) investigated the m^6A modification in response to PaWB phytoplasma infection. The m^6A transcriptome and RNA sequencing analyses of *in vitro* PaWB-diseased and MMS-treated PaWB-diseased *P. fortunei* plantlets revealed that MMS treatment caused reduced m^6A modification in PaWB-recovered paulownia plantlets compared with diseased samples, indicating a potential role of m^6A regulation in recovery. In PaWB-diseased samples, 20,568

m^6A peaks corresponding to 13,838 genes were detected, whereas MMS-treated samples exhibited 20,201 m^6A peaks across 13,505 genes (Xu *et al.*, 2023).

Comparison of the m^6A peaks between the two groups showed that 306 m^6A peaks corresponding to 319 genes were absent in the MMS-treated samples. The distribution of these absent peaks showed that 36.59% were in 3' UTRs, 17.41% in 5' UTRs, 21.58% in first exons, and 24.15% in other exons. These changes suggest a specific alteration in m^6A modification patterns during MMS-mediated recovery from phytoplasma infection. Combined with transcriptome sequencing analysis, 315 differentially methylated genes were also found to be significantly differentially expressed. Functional annotation using the KEGG pathway mapped these genes to critical biological processes, including ABC transport, diterpenoid biosynthesis, plant hormone signal transduction, and phenylpropanoid biosynthesis. Functional enrichment analysis identified two genes, *CLV2* and *STM*, involved in the maintenance of stem cells in the shoot apical meristem. These genes appear to play a significant role in PaWB phytoplasma–plant interactions, potentially influencing recovery and resistance mechanisms (Xu *et al.*, 2023).

Protein Modifications Response to Phytoplasma Infection

Protein post-translational modifications (PTMs) are crucial biological mechanisms that expand the functional diversity of the genetic code and regulate cellular physiology. These modifications alter protein structure, function, and interactions, allowing dynamic responses to environmental stimuli and metabolic changes. The complexity of PTM networks is highlighted by modifications at lysine residues, a critical amino acid for protein structure stability and function. Among these, lysine acetylation and lysine succinylation are two well-characterized PTMs that affect protein function, enzyme activity, and cellular metabolic regulation (Weinert *et al.*, 2013; Balparda *et al.*, 2021).

To study the lysine acetylation and succinylation responses to phytoplasma infection, Cao *et al.* (2019) conducted proteome, acetylome, and succinylome analyses using quantitative mass spectrometry. The study examined four sample groups: healthy, PaWB-diseased, MMS-treated healthy, and MMS-treated PaWB-diseased *P. tomentosa* plantlets grown *in vitro*. In total, 8963 proteins were identified across all samples, with 7826 quantified proteins. Comparative analysis revealed 276 differentially expressed proteins (DEPs) associated with PaWB phytoplasma infection. KEGG annotation showed significantly enriched pathways, including biosynthesis of secondary metabolites, porphyrin and chlorophyll metabolism, metabolic pathways, and phenylpropanoid biosynthesis (Cao *et al.*, 2019).

The study also identified 5558 unique modified lysine-acetylated sites on 2893 acetylated substrates, and 1970 lysine-succinylation sites on 1271 succinylated proteins. A total of 787 proteins were found to be both acetylated and succinylated, indicating significant cross-regulation between these PTMs. When comparing four groups of samples, 741 acetylation sites on 546 acetylated proteins and five succinylation sites on five succinylated proteins were specifically responsive to phytoplasma infection (Cao *et al.*, 2019).

In PaWB-diseased paulownia, reduced photosynthetic activity, along with higher starch and lower pigment levels, was observed compared with healthy plants. Several acetylation sites were detected on key enzymes involved in chlorophyll and starch synthesis, including Rubisco large subunit 1 (RubL1, acetylation site at K55), RuBisCO large subunit 2 (RubL2, acetylation site at K746), and protochlorophyllide reductase (POR, acetylation sites at K339 and K78). Site-directed mutagenesis of these acetylated residues significantly reduced the enzymatic activities of RuBisCO and POR. This reduction disrupted starch, pigment, and chlorophyll balance, resulting in the characteristic yellowing of leaves and stunted growth observed in *P. tomentosa* during PaWB phytoplasma infection. The findings provide insights into the role of lysine acetylation in regulating key metabolic enzymes and its impact on the physiological symptoms induced by phytoplasma infection (Cao *et al.*, 2019).

Conclusion

Epigenetics plays an important role in plant–pathogen interactions. This chapter explores the critical role of epigenetics in paulownia's response to PaWB phytoplasma infection. High-throughput sequencing reveals that phytoplasma infection disrupts DNA methylation, histone methylation and acetylation, non-coding RNA regulation, and mRNA and protein modifications, altering pathways related to hormone signaling, carbon metabolism, and plant defense. These epigenetic changes are associated with disease symptoms like witches'-broom (shoot proliferation) and little leaf. Treatments such as MMS effectively restore epigenetic balance, eliminate the pathogen, and reverse symptoms, demonstrating the therapeutic potential of targeting epigenetic mechanisms.

Decades of research into PaWB have provided valuable insights into plant–phytoplasma interactions. The release of genomic data for both paulownia and the PaWB phytoplasma has supported studies on epigenetic regulation during infection. These findings enhance understanding of how plants respond to phytoplasma-induced stress and offer potential strategies for managing similar plant diseases through targeted interventions.

References

Balparda, M., Elsässer, M., Badia, M.B., Giese, J., Bovdilova, A. *et al.* (2021) Acetylation of conserved lysines fine-tunes mitochondrial malate dehydrogenase activity in land plants. *Plant Journal* 109, 92–111. 10.1111/tpj.15556.

Cao X., Fan, G., and Zhai, X. (2012) Morphological changes of the witches' broom seedlings of *Paulownia tomentosa* treated with methyl methanesulphonate and SSR analysis. *Acta Phytopathologica Sinica* 42, 214–218. 10.13926/j.cnki.apps.2012.02.014.

Cao, X., Fan, G., Deng, M., Zhao, Z., and Dong, Y. (2014a) Identification of genes related to paulownia witches' broom by AFLP and MSAP. *International Journal of Molecular Science* 15, 14669–14683. DOI: 10.3390/ijms150814669.

Cao, X., Fan, G., Zhao, Z., Deng, M., and Dong, Y. (2014b) Morphological changes of paulownia seedlings infected phytoplasmas reveal the genes associated with witches' broom through AFLP and MSAP. *PLoS One* 9, e112533. DOI: 10.1371/journal.pone.0112533.

Cao, X., Zhao, Z., Fan, G., Deng, M., and Dong, Y. (2014c) Effect of methyl methanesulphonate on DNA methylation of witches' broom seedlings of *Paulownia tomentosa*. *Scientia Silvae Sinicae* 50, 99–108.

Cao, X., Zhai, X., Zhao, Z., Deng, M., Li, Y., and Fan, G. (2020) Genome-wide DNA methylation analysis of paulownia with phytoplasma infection. *Gene* 755, 144905. DOI: 10.1016/j.gene.2020.144905.

Cao, Y., Fan, G., Zhai, X., and Dong, Y. (2018) Genome-wide analysis of lncRNAs in *Paulownia tomentosa* infected with phytoplasmas. *Acta Physiologiae Plantarum* 40, 49. DOI: 10.1007/s11738-018-2627-6.

Cao, Y., Fan, G., Wang, Z., and Gu, Z. (2019) Phytoplasma-induced changes in the acetylome and succinylome of *Paulownia tomentosa* provide evidence for involvement of acetylated proteins in witches' broom disease. *Molecular and Cellular Proteomics* 18, 1210–1226. DOI: 10.1074/mcp.RA118.001104.

Cao, Y., Sun, G., Zhai, X., Xu, P., Ma, L. *et al.* (2021) Genomic insights into the fast growth of paulownias and the formation of *Paulownia* witches' broom. *Molecular Plant* 14, 1668–1682. DOI: 10.1016/j.molp.2021.06.021.

Doi, Y., Teranaka, M., Yora, K., and Asuyama, H. (1967) Mycoplasma or P. L. T. group-like microorganisms found in the phloem elements of plants infected with mulberry dwarf, potato witches'-broom, aster yellows or paulownia witches'-broom. *Annuals of Phytopathology Society of Japan* 33, 259–266. DOI: 10.3186/jjphytopath.33.259.

Dżugan, M., Miłek, M., Grabek-Lejko, D., Hęclik, J., Jacek, B., and Litwińczuk, W. (2021) Antioxidant activity, polyphenolic profiles and antibacterial properties of leaf extract of various *Paulownia* spp. clones. *Agronomy* 11, 2001. DOI: 10.3390/agronomy11102001.

Fan, G., Zhao, G., Zhai, X., and Cao, X. (2011) Morphological changes of *Paulownia tomentosa* × *P. fortunei* seedlings with witches' broom treated with dimethyl sulphonate and their SSR analysis. *Journal of Northeast Forestry University* 39, 30–33. DOI: 10.13759/j.cnki.dlxb.2010.12.016.

Fan, G., Zhao, G., Zhai, X., and Cao, X. (2012) Effect of dimethyl sulphate on phytoplasma of Paulownia tomentosa seedling infected by witches' broom and its DNA loci at SSR level. *Journal of Nanjing Forestry University* 36, 5–8. DOI: 10.3969/j.jssn.1000-2006.2012.03.002.

Fan, G., Cao, X., Niu, S., Deng, M., Zhao, Z., and Dong, Y. (2015a) Transcriptome, microRNA, and degradome analyses of the gene expression of paulownia with phytoplasma. *BMC Genomics* 16, 896. DOI: 10.1186/s12864-015-2074-3.

Fan, G., Niu, S., Xu, T., Deng, M., Zhao, Z. *et al.* (2015b) Plant-pathogen interaction-related micro-RNAs and their targets provide indicators of phytoplasma infection in *Paulownia tomentosa* × *Paulownia fortunei*. *PLoS One* 10, e0140590. DOI: 10.1371/journal.pone.0140590.

Fan, G., Niu, S., Zhao, Z., Dong, M., Xu, E., Wang, Y., and Yang, L. (2016) Identification of microRNAs and their targets in *Paulownia fortunei* plants free from phytoplasma pathogen after methyl methane sulfonate treatment. *Biochimie* 127, 271–280. DOI: 10.1016/j.biochi.2016.06.010.

Fan, G., Cao, Y., Deng, M., Zhai, X., Zhao, Z., Niu, S., and Ren, Y. (2017) Identification and dynamic expression profiling of microRNAs and target genes of *Paulownia tomentosa* in response to Paulownia witches' broom disease. *Acta Physiologiae Plantarum* 39, 28. DOI: 10.1007/s11738-016-2326-0.

Fan, G., Cao, Y., and Wang, Z. (2018a) Regulation of long noncoding RNAs responsive to phytoplasma infection in *Paulownia tomentosa*. *International Journal of Genomics* 2018, 3174352. DOI: 10.1155/2018/3174352.

Fan, G., Wang, Z., Zhai, X., and Cao, Y. (2018b) ceRNA cross-talk in paulownia witches' broom disease. *International Journal of Molecular Science* 19, 2463. DOI: 10.3390/ijms19082463.

Jakubowski, M. (2022) Cultivation potential and uses of paulownia wood: a review. *Forests* 13, 668. DOI: 10.3390/f13050668.

Jung, I. and Kim, D. (2012) Histone modification profiles characterize function-specific gene regulation. *Journal of Theoretical Biology* 310, 132–142. DOI: 10.1016/j.jtbi.2012.06.009.

Kouzarides, T. (2007) Chromatin modifications and their function. *Cell* 128, 693–705. DOI: 10.1016/j.cell.2007.02.005.

Li, M., Zhai, X., Fan, G., Zhang, B., and Liu, F. (2008) Effect of oxytetracycline on the morphology of seedlings with witches' broom and DNA methylation level of *Paulownia tomentosa* × *P. fortunei*. *Scientia Silvae Sinicae* 44, 152–156. DOI: 10.11707/j.1001-7488.20080927.

Liu, C., Dong, X., Xu, Y., Dong, Q., Wang, Y., Gai, Y., and Ji, X. (2021) Transcriptome and DNA methylome reveal insights into phytoplasma infection response to mulberry (*Morus multicaulis* Perr.). *Frontiers in Plant Science* 12, 697702. DOI: 10.3389/fpls.2021.697702.

Meyer, K.D., Saletore, Y., Zumbo, P., Elemento, O., Mason, C.E., and Jafferey, S.R. (2012) Comprehensive analysis of mRNA methylation reveals enrichment in 3' UTRs and near stop codons. *Cell* 149, 1635–1646. DOI: 10.1016/j.cell.2012.05.003.

Niu, S., Fan, G., Deng, M., Zhao, Z., Xu, E., and Cao, L. (2016) Discovery of microRNAs and transcript targets related to witches' broom disease in Paulownia fortunei by high-throughput sequencing and degradome approach. *Molecular Genetics and Genomics* 291, 181–191. DOI: 10.1007/s00438-015-1102-y.

Pagliarani, C., Gambino, G., Ferrandino, A., Chitarra, W., Vrhovsek, U. *et al.* (2020) Molecular memory of *Flavescence doree* phytoplasma in recovering grapevines. *Horticulture Research* 7, 126. DOI: 10.1038/s41438-020-00348-3.

Rinn, J.L. and Chang, H.Y. (2013) Genome regulation by long noncoding RNAs. *Annual Review of Biochemistry* 81, 145–166. DOI: 10.1146/annurev-biochem-051410-092902.

Salmena, L., Poliseno, L., Tay, Y., Kats, L., and Pandolfi, P.P. (2011) A *ceRNA* hypothesis: The rosetta stone of a hidden RNA language? *Cell* 146, 353–358. DOI: 10.1016/j.cell.2011.07.014.

Schwartz, S., Agarwala, S.D., Mumbach, M.R., Jovanovic, M., Mertins, P. *et al.* (2013) High-resolution mapping reveals a conserved, widespread, dynamic mRNA methylation program in yeast meiosis. *Cell* 155, 1409–1421. DOI: 10.1016/j.cell.2013.10.047.

Sławińska, N., Zając, J., and Olas, B. (2023) *Paulownia* organs as interesting new source of bioactive compounds. *International Journal of Molecular Science* 24, 1676. DOI: 10.3390/ijms24021676.

Verma, P., Singh, A., Purru, S., Bhat, K.V., and Lakhanpaul, S. (2022) Comparative DNA methylome of phytoplasma associated retrograde metamorphosis in sesame (*Sesamum indicum* L.). *Biology* 11, 954. DOI: 10.3390/biology11070954.

Voinnet, O. (2009) Origin, biogenesis, and activity of plant microRNAs. *Cell* 136, 669–687. DOI: 10.1016/j.cell.2009.01.046.

Wang, Z., Zhai, X., Cao, Y., Dong, Y., and Fan, G. (2017) Long non-coding RNAs responsive to witches' broom disease in *Paulownia tomentosa*. *Forests* 8, 348. DOI: 10.3390/f8090348.

Weinert, B.T., Schölz, C., Wagner, S.A., Iesmantavicius, V., Su, D., Daniel, J.A., and Choudhary, C. (2013) Lysine succinylation is a frequently occurring modification in prokaryotes and eukaryotes and extensively overlaps with acetylation. *Cell Reports* 4, 842–851. DOI: 10.1016/j.celrep.2013.07.024.

Xu, P., Huang, S., Zhai, X., Fan, Y., Li, X., Yang, H., Cao, Y., and Fan, G. (2023) N6-methyladenosine modification changes during the recovery processes of Paulownia witches' broom disease under the methyl methanesulfonate treatment. *Plant Direct* 7, e508. DOI: 10.1002/pld3.508.

Yan, L., Fan, G., and Li, X. (2019) Genome-wide analysis of three histone marks and gene expression in *Paulownia fortunei* with phytoplasma infection. *BMC Genomics* 20, 234. DOI: 10.1186/s12864-019-5609-1.

Yan, L., Zhai, X., Zhao, Z., and Fan, G. (2020) Whole-genome landscape of H3K4me3, H3K36me3 and H3K9ac and their association with gene expression during Paulownia witches' broom disease infection and recovery processes. *3 Biotech* 10, 336. DOI: 10.1007/s13205-020-02331-0.

Yue, H., Wu, Y., Shi, Y., Wu, K., and Li, Y. (2008) First report of paulownia witches' broom phytoplasma in China. *Plant Disease* 92, 1134. DOI: 10.1094/PDIS-92-7-1134A.

Zhai, X., Cao, X., and Fan, G. (2010) Growth of paulownia witches' broom seedlings treated with methylmethane sulphonate and SSR analysis. *Scientia Silvae Sinicae* 46, 176–181. DOI: 10.11707/j.1001-7488.20101229.

Zhao, G., Zhao, Z., Fan, G., and Cao, X. (2011) Effect of dimethyl sulphonate on the morphological changes of *Paulownia fortunei* seedlings with witches' broom and their DNA base sequences. *Journal of Henan Agricultural University* 45, 287–291. DOI: 10.16445/j.cnki.1000-2340.2011.03.009.

15 Unraveling Plant Responses to Jujube Witches'-Broom Phytoplasma Infection through Omics Analysis

Xianbo Zheng,[1] Wei Wei,[2] Jidong Li,[1]* and Jiancan Feng[1]
[1]Henan Agricultural University, Zhengzhou, China; [2]Molecular Plant Pathology Laboratory, ARS-USDA, Beltsville, Maryland, USA

Abstract
Phytoplasma infection disrupts plant gene expression at different levels, including transcription, translation, and metabolite production. Advances in high throughput analytical technologies have facilitated comprehensive studies of plant responses to phytoplasma infection from a multi-omics perspective. Jujube, a fruit tree in the Rhamnaceae family, is threatened by jujube witches'-broom (JWB) disease, caused by '*Candidatus* Phytoplasma ziziphi,' a strain belonging to the 16SrV-B subgroup. This chapter investigates jujube's responses to JWB phytoplasma infection through genomic, transcriptomic, proteomic, and metabolomic analyses. These approaches provide insights into the molecular mechanisms of disease progression and jujube's defense responses, deepening our understanding of plant–pathogen interactions and uncovering potential pathways to enhance disease resistance.

Introduction

In living organisms, genetic information within cells flows from DNA to the messenger RNA (mRNA), then to proteins, a process dictated by the central dogma of molecular biology. The study of DNA, mRNA, and proteins is broadly denoted as genomics, transcriptomics, and proteomics. Genomics examines the entire genetic blueprint of an organism by analyzing DNA sequences, enabling the identification of specific genes and genetic variations that influence cellular function. Transcriptomics studies the transcribed RNA, focusing on genes that are actively expressed to reveal dynamic changes in gene activity and provide insights into ongoing cellular processes. Proteomics studies the structure, function, and interactions of proteins, characterizing cellular mechanisms and pathways through protein networks and signaling cascades. Although metabolomics is not a direct part of the central dogma, it provides critical insights into cellular metabolism by analyzing metabolites and offers a comprehensive view of biochemical activities and physiological states within the cell (Reel *et al.*, 2021).

The Chinese jujube (*Ziziphus jujuba*), also called common jujube, Chinese date, red date, or "Zao" in Mandarin Chinese, is prized for its delicious and nutritious fruit (Qu & Wang, 1993; Liu *et al.* 2020a). The jujube fruit can be consumed fresh or dried and has long been used in traditional Chinese herbal medicine. Jujube has a long cultivation history of more than 7000 years and is grown in more than 40 countries worldwide (Yao, 2013; Liu *et al.*, 2020a). Despite

*Corresponding author: lijidong@henau.edu.cn

 Phytoplasmas: Genomes, Plant Hosts and Vectors, 2nd Edition. (Ed. W. Wei)
DOI: 10.1079/9781800627031.0015

its widespread cultivation and cultural importance, the jujube industry faces significant challenges, the most destructive of which is jujube witches'-broom (JWB) disease, also called "Zaofeng" disease in Chinese (Guo *et al.*, 2023). JWB disease is caused by JWB phytoplasma ('*Candidatus* Phytoplasma ziziphi'), a strain within the 16SrV-B subgroup (Jung *et al.*, 2003). JWB phytoplasma-infected jujube trees exhibit disrupted development, resulting in yield loss, reduced fruit quality, and gradual dieback within a few years, leading to substantial losses in the jujube industry (Guo *et al.*, 2023). To better understand and combat the effects of JWB disease, this chapter explores plant responses to JWB phytoplasma infection through omics analyses, including genomics, transcriptomics, proteomics, and metabolomics. The chapter uncovers the molecular mechanisms underlying jujube's defense responses, identifies potential biomarkers for early detection, and explores strategies for breeding resistant varieties.

Genomics of Jujube

The jujube genome is pivotal in fruit tree genomics, being one of the earliest sequenced fruit trees and the first in the Rhamnaceae family. The genome of fresh jujube cultivar Dongzao was first published in 2014 (Liu *et al.*, 2014). This assembly spans 437.65 Mb, covering 98.6% of the estimated genome size (JREP00000000). Of this, 321.45 Mb (73.56%) was anchored to 12 pseudo-chromosomes, with a total of 32,808 predicted genes. Another jujube genome is the dry cultivar Junzao, which was sequenced in 2016 (Huang *et al.*, 2016). The Junzao genome assembly spans 351 Mb, which is 86.5 Mb smaller than that of the Dongzao genome. Of the entire assembly, 293 Mb (83.6%) was anchored to 12 linkage groups and 27,443 protein-coding genes were predicted (LPXJ00000000).

Benefiting from the advance in long-read sequencing technologies, such as PacBio HiFi or ONT ultra-long, it is possible to assembly gap-free genomes containing highly repetitive regions, including centromere and telomeres. Recently, both genomes of the Dongzao and Junzao cultivars have been updated to telomere-to-telomere (T2T) resolution. The gapless genome of the Dongzao cultivar is 393.3 Mb (ASM3175591V1; Yang *et al.*, 2023b). Similarly, the T2T genome of the Junzao cultivar spans 385.8 Mb (LPXJ00000000.2; Li *et al.*, 2024).

Genome resequencing has become a critical tool in studying jujube germplasms, offering extensive insights into the species' population structure, evolutionary dynamics, domestication history, and interactions with JWB phytoplasma pathology. By analyzing genetic variations across a wide range of cultivated and wild accessions, researchers have generated comprehensive datasets that deepen the understanding of these processes. In 2016, Huang *et al.* initiated resequencing of 31 jujube accessions, including 21 jujube cultivars and ten wild jujube individuals. This effort generated 344 GB of sequences data, providing critical insights into the genetic structure and domestication patterns of jujube. Building on this, Guo *et al.* (2020) expanded the scope by resequencing 350 jujube accessions, consisting of 286 cultivars, 41 wild individuals, and 23 semi-wild accessions. This study achieved a sequencing depth greater than 15× coverage (Guo *et al.*, 2020), offering valuable information about the genetic basis of important traits, including those related to disease resistance, fruit quality, and environmental adaptation. In 2021, Guo *et al.* further advance this research by resequencing 493 jujube accessions, including 291 cultivated and 202 wild individuals, with an improved sequencing depth of more than 16× coverage. This additional dataset offered a more nuanced view of genetic diversity and evolutionary relationships. Recently, Li *et al.* (2024) conducted the most extensive resequencing project to date, analyzing 672 jujube individuals, which included 359 cultivated, 291 wild, and 22 semi-wild accessions. These resequencing data provide abundant genome information for jujube population structure, evolution, domestication research, and their relationship to JWB phytoplasma pathology.

Genomics of the JWB Phytoplasma

Currently, two JWB phytoplasma genomes have been published. The first published genome is from the '*Ca.* P. ziziphi' "nky" strain, which consists of a single circular chromosome measuring 750,803 bp, with a GC content of 23.3%. The genome of the "nky" strain is predicted to encode 694 protein-coding genes, along with two operons of rRNA genes, and 31 tRNA genes (Wang *et al.*, 2018b). Recently, the genome of another strain, '*Ca.* P. ziziphi' "Hebei-2018" was published in 2023. This genome is also circular, and spans 764,108 bp, with 735 predicted coding sequences. Compared with the "nky" strain, the "Hebei-2018" genome contains an additional 19,825 bp. These extra genes enrich the metabolic repertoire of the strain, particularly in the glycolysis pathway. The extra genes include those encoding pyruvate dehydrogenase subunits, such as *pdhA*, *pdhB*, *pdhC*, *pdhD*, acetate kinase (*ackA*), phosphate acetyltransferase (*pduL*), and lactate dehydrogenase (*LDH*), which may contribute to its pathogenicity and adaptability (Xue *et al.*, 2023).

Despite these additions, the metabolic pathways in the '*Ca.* P. ziziphi' genome remain limited, consistent with the minimalist genome characteristics typical of phytoplasmas. The JWB phytoplasma genome lacks genes encoding ATP synthase, the pentose phosphate pathway, and oxidative phosphorylation. Key pathways such as amino acid and fatty acid biosynthesis, as well as the tricarboxylic acid cycle, are also incomplete. This metabolic deficiency highlights the obligate parasite nature of phytoplasmas, which rely on their host plants for survival and reproduction, and lack genes involved in most amino acid synthesis.

A functional Sec-dependent protein translocation system has been identified in the JWB phytoplasma genome, implying an active mechanism for secreting virulence effectors into host cells (Wang *et al.*, 2018b; Xue *et al.*, 2023). These effectors likely play a critical role in manipulating host cellular functions to facilitate infection and proliferation.

Similar to the secreted protein in aster yellow phytoplasmas (SAPs), the JWB phytoplasma effectors also interact with certain target proteins in the host plants to induce JWB symptoms. Homologs of SAP11, JWB phytoplasma effectors SJP1, SJP2, and Zaofeng6 were determined to target the TEOSINTE BRANCHED 1/CYCLOIDEA/PROLIFERATING CELL FACTOR (TCP) transcription factor; these homologs promote accumulation of endogenous auxin, downregulate expression of genes in the strigolactone signal pathway, and induce shoot proliferation (Chen *et al.*, 2022; Zhou *et al.*, 2021).

Homologs of SAP54, JWB phytoplasma effector $PHYL1_{JWB}$, SJP3, and Zaofeng3, disrupt the expression of MADS-box transcription factors associated with floral organ identity and flowering time, inducing leafy flower, namely phyllody symptoms (Deng et al., 2021; Chen et al., 2024; Xue *et al.*, 2024). These findings demonstrated the interaction between JWB phytoplasma and their plant hosts on a molecular level, revealing the mechanisms of how virulence effectors induce JWB symptoms.

Transcriptional Analysis of Plant Response to JWB Phytoplasma

To better understand the plant's response mechanisms and identify the underlying pathways affected by JWB phytoplasmas, comprehensive transcriptional analyses have been performed, shedding light on differential gene expression patterns and key biological processes involved in disease progression and symptom manifestation. For example, a comparative transcriptome analysis was conducted to investigate the molecular responses of healthy and diseased jujube cultivar Langzao infected with JWB phytoplasmas (Fan *et al.*, 2017). RNA sequencing (RNA-seq) was performed using flowers and leaves collected from symptomatic and healthy plants. In the transcriptome libraries derived from the leaves, a total of 4266 differentially expressed genes (DEGs) were identified, comprising 2070 upregulated genes and 2196 downregulated genes. Similarly, an analysis of flower samples showed 3800 DEGs, including 1965 upregulated and 1835 downregulated genes. The expression profiles highlighted significant alterations in genes related to the amino acid metabolism and the carotenoid biosynthesis pathway. These pathways likely play roles in the manifestation of key symptoms of JWB disease, such as yellowing, witches'-brooming, and little leaf symptoms, suggesting nutrient deficiencies and disruptions in metabolic homeostasis as underlying factors (Fan *et al.*, 2017).

Transcriptome Analysis of Jujube during JWB Phytoplasma Infection Progression

To further elucidate the molecular mechanisms underpinning jujube's response to JWB phytoplasma infection, a detailed time-course transcriptome analysis was conducted (Ye *et al.*, 2017; Wang. *et al.*, 2018a). In this study, healthy jujube seedlings were graft inoculated with JWB phytoplasma-infected buds, and leaf samples were collected at six distinct stages of infection progression: 0 weeks after grafting (WAG), 2 WAG (no symptoms, PCR negative), 37 WAG (no symptoms, PCR positive), 39 WAG (symptoms appear), 48 WAG (whole-plant symptoms), and 52 WAG (starvation stage), during the infection progress for RNA-seq (Ye *et al.*, 2017; Wang. *et al.*, 2018a). RNA-seq identified a total of 58,060 jujube transcripts and 25,067 unigenes, with 8341 genes showing different expressions between infected and healthy seedlings. Among these DEGs, 3361 genes were upregulated, while 4980 genes were downregulated in infected plants compared with the healthy controls. Gene ontology (GO) analysis indicated that upregulated DEGs were mainly enriched in categories such as integral component of membrane, plasma membrane, transcription factor activity, sequence-specific DNA binding, defense response, and vacuole function. Conversely, downregulated DEGs were mainly over-represented in the categories of chloroplast, apoplast, integral component of membrane, extracellular region, and cell walls.

KEGG (Kyoto Encyclopedia of Genes and Genomes) pathway enrichment analysis also showed the functional divergence of DEGs. Upregulated genes were significantly associated with flavonoid biosynthesis, amino acid biosynthesis, phenylpropanoid biosynthesis and metabolism, α-linolenic acid metabolism, and amino sugar and nucleotide sugar metabolism. On the other hand, downregulated DEGs were mainly enriched in pathways related to ribosome function, carbon metabolism, porphyrin and chlorophyll metabolism, pentose and glucuronate interconversions, and photosynthesis. A notable shift in phytohormonal signaling was observed between 37 and 48 WAG. Genes associated with various phytohormones such as abscisic acid (ABA), brassinosteroid (BR), cytokinin (CTK), ethylene (ET), gibberellins (GA), auxin (IAA), jasmonic acid (JA), and salicylic acid (SA) exhibited differential expressions. All 37 JA-related DEGs were upregulated, with *LOX2* emerging as the most highly expressed gene, showing a 7.19-fold increase at 39 WAG, highlighting the role of JA in modulating defense responses during JWB phytoplasma infection (Ye *et al.*, 2017; Wang. *et al.*, 2018a).

Transcriptome Analysis of Jujube during JWB Phytoplasma Elimination Progression

Transcriptome analysis was also performed to investigate the molecular changes in jujube shoots during the elimination of JWB phytoplasma through tetracycline treatment (Wang *et al.*, 2019). This study utilized *in vitro* JWB phytoplasma-infected shoots subjected to tetracycline treatment over different time points. RNA-seq libraries were constructed using samples collected at three stages: untreated infected shoots at 0 months of tetracycline treatment (MTT) (with obvious symptoms), shoots at 3 MTT (PCR positive, exhibiting weaker symptoms such as witches'-broom), and shoots at 6 MTT (PCR negative, symptom-free).

High-throughput transcriptomic analysis identified 26,790 unigenes in the jujube shoot samples. Between the 3 and 6 MTT stages, 272 DEGs were identified, with 100 genes upregulated and 172 genes downregulated. GO enrichment analysis revealed that the top 20 enriched terms included key biological processes and molecular functions, such as metabolic process, cellular process, cell, cell part, catalytic activity, and binding. KEGG pathway analysis further showed the significant enrichment in pathways associated with the biosynthesis of secondary metabolites and general metabolic pathways, suggesting that the recovery process involves broad metabolic reprogramming.

Additionally, specific changes in the JA signaling pathway were observed between 3 and 6 MTT, reflecting the dynamic hormonal regulation during disease remission. Four JA-related genes that encode lipoxygenase (LOX; 107412659), oxophytodienoic acid reductase (OPR; 107432179), and phospholipase A1 (PLA1; 107416254 and 107416251) were downregulated. Conversely, gene encoding jasmonate *O*-methyltransferase (JMT; 107424658), a key enzyme involved in JA signaling and metabolism, were upregulated (Wang *et al.*, 2019). These findings indicate that JA-related pathways may be linked to symptom alleviation and recovery.

In a field experiment, oxytetracycline hydrochloride was applied to 20-year-old JWB diseased jujube trees (cultivar Muzao) by trunk injection to evaluate its potential for mitigating disease symptoms (Yang *et al.*, 2023a). Transcriptome analysis was performed on three groups of samples: healthy control trees (C group), JWB-diseased trees (D group), and tetracycline-treated JWB-diseased trees (T group). High-throughput RNA sequencing revealed 32,084 unigenes across the cDNA libraries. Differential gene expression analysis identified 755 DEGs in at least one pairwise comparison. Specifically, 488 DEGs were observed between healthy and diseased trees (group C vs group D), 345 DEGs between diseased and treated trees (group D vs group T), and 94 DEGs between healthy and treated trees (group C vs group T). Gene-enrichment analysis revealed that these DEGs were mainly involved in vital biological processes and pathways, including DNA and RNA metabolism, signaling, photosynthesis, and plant hormone metabolism and transduction, as well as primary and secondary metabolic processes (Yang *et al.*, 2023a). The findings revealed significant molecular disruptions in diseased trees compared with healthy controls, with tetracycline treatment partially restoring many affected processes. However, some pathways remained disrupted even after treatment, highlighting the complexity of disease recovery at the molecular level. These results provide insights into how tetracycline treatment alleviates the molecular effects of JWB phytoplasma, although full recovery at the transcriptomic level may not always be achieved.

Transcriptome Analysis of JWB-Resistant and -Susceptible Jujube Cultivars

To investigate natural resistance mechanisms against JWB phytoplasma, a resistant jujube cultivar, T13, which could recover from graft infection on diseased rootstocks, and a susceptible cultivar Pozao were selected to perform comparative transcriptome analysis (Wang *et al.*, 2022). Healthy scions of both Pozao and T13 were graft-inoculated onto JWB-diseased rootstocks to perform infection. Scions grafted on healthy rootstock were set as controls. Samples were collected at three different phases of disease progression: S1, when witches'-broom symptoms were first observed on both cultivars; S2, when severe witches'-broom developed on the susceptible cultivar Pozao but not on resistant T13; and S3, when T13 fully recovered to normal, while Pozao displayed more severe symptoms. High-throughput transcriptome sequencing identified 33,682 unigenes across all samples. In the susceptible cultivar Pozao, DEGs between the infected and healthy controls were detected at all three phases, reflecting a sustained transcriptional disturbance throughout disease progression. In contrast, in the resistant cultivar T13, DEGs were observed only at the S1 phase, suggesting an early and robust transcriptional response that successfully suppressed further disruptions. Among the 26 DEGs expressed in both cultivars, several genes were involved in auxin synthesis and efflux, cytokinin signaling, JA signaling, calcium signaling, mitogen-activated protein kinase (MAPK), and reactive oxygen species (ROS) signaling (Wang *et al.*, 2022). These pathways are critical for regulating plant defense, growth, and stress responses.

The resistant cultivar T13 demonstrated a well-coordinated early response to phytoplasma infection, which likely contributed to its ability to recover from symptoms. In contrast, the prolonged transcriptional disturbances observed in Pozao correlated with the worsening of symptoms. These findings offer valuable insights into the molecular mechanisms underlying resistance and susceptibility to JWB phytoplasma and may guide future strategies for breeding resistant jujube cultivars and improving disease management.

Transcriptome Analysis of JWB-Tolerant Indian Jujube

The Indian jujube (*Ziziphus mauritiana*) cultivar Cuimi has been demonstrated to exhibit high tolerance to JWB disease by graft infection tests conducted on JWB-diseased rootstocks (Xu *et al.*,

2023). To investigate the molecular basis of this tolerance, a comparative transcriptome analysis was performed on healthy and JWB phytoplasma-infected plants of the tolerant cultivar Cuimi and the susceptible cultivar Huping. Transcriptome analysis revealed significant differences in gene expression profiles between the healthy and JWB-infected states of the two cultivars. In the JWB-infected Huping plants, 1194 DEGs were upregulated, while 429 were downregulated, resulting in 1623 DEGs, including 1340 unique to Huping. In contrast, JWB-infected Cuimi plants exhibited 426 upregulated DEGs and 369 downregulated DEGs, a total of 795 DEGs, of which 512 were unique to Cuimi.

GO enrichment analysis revealed that the unique DEGs in Cuimi were significantly associated with biological processes and pathways that are potentially linked to JWB tolerance. The processes and pathways included protein ubiquitination, cell wall biogenesis, cell surface receptor signaling pathways, oxylipin biosynthetic process, and transcription factor activity. These pathways suggest that Cuimi employs mechanisms that enhance cellular defense, maintain structural integrity, and regulate stress-responsive signaling. Specific DEGs associated with ROS regulation, Ca^{2+} sensors, protein kinases, transcription factors, lignin biosynthesis, and plant hormone signaling pathways were identified as being related to the tolerance observed in Cuimi (Xu *et al.*, 2023). These findings highlight the importance of ROS detoxification, calcium-mediated signaling, and hormone-driven stress responses in mitigating the impact of JWB phytoplasma infection. Lignin biosynthesis, in particular, may contribute to strengthening the cell wall as a physical barrier against pathogen invasion.

The comparison of tolerant and susceptible cultivars underscores the distinct transcriptional reprogramming that occurs in response to JWB infection. While Huping exhibited a more extensive and possibly less targeted response, the transcriptional changes in Cuimi were more focused, involving pathways and genes that directly contribute to defense and stress tolerance. These results provide valuable insights into the molecular mechanisms underlying JWB tolerance in Indian jujube and offer potential targets for breeding and genetic improvement of JWB-resistant cultivars.

Identification of Gene Families and Expression Response to JWB Phytoplasma Infection

Although extensive jujube genome data have been released, detailed gene annotation remains limited. Using bioinformatics approaches, *Arabidopsis* protein sequences were typically employed as probes to perform BLAST (basic local alignment search tool) searches against the jujube genome data. This analysis enabled the identification of certain gene family members and a detailed exploration of their gene and protein structural features (Table 15.1). By mining the genome and transcriptome data, following quantitative real-time PCR (qRT-PCR) verification, many genes were significantly differentially expressed between healthy and JWB-diseased jujube samples. These genes are believed to be associated with JWB phytoplasma infection (Table 15.1).

Combined Proteome and Transcriptome Analysis of Jujube Response to Phytoplasma Infection

The integration of proteomic and transcriptomic analyses offers a comprehensive perspective on how plants respond to biotic stress by capturing changes at both the transcriptional and translational levels. High-throughput proteomic techniques, such as iTRAO labeling combined with mass spectrometry, have been employed to investigate jujube responses to JWB phytoplasma infection (Ye *et al.*, 2017). In the six stages of jujube infection progress – namely 0, 2, 37, 39, 48, and 52 WAG – the highest DEGs occurred at 37 WAG, while typical witches'-broom symptoms were observed at 48 WAG. Thus, iTRAQ proteome analysis was performed on samples at 37 and 48 WAG. The iTRAQ analysis produced 583,908 spectra, of which 41,465 were annotated as known peptides. Among them, 37,992 unique peptides were identified, corresponding to 6748 proteins. Building upon this comprehensive dataset, a more focused analysis was conducted to assess protein expression at specific stages of infection. At 37 WAG, 5378 proteins were identified, with 289 differentially expressed proteins (DEPs). Among them, 176 were upregulated in

Table 15.1. Identification and expression analysis of jujube gene families in response to Jujube witches'-broom (JWB) phytoplasma infection.

Gene family	No. of gene family members in jujube	Jujube genes responding to JWB phytoplasma infection	Reference
Auxin response factor (ARF)	16	*ZjARF1*, *-2*, *-3*, *-4*, and *-8*	Ma *et al.*, 2020
Basic helix-loop-helix (bHLH)	92	*ZjbHLH12*, *-18*, *-23*, *-24*, *-34*, *-49*, *-53*, *-62*, *-63*, *-70*, *-83*, and *-88*	Li *et al.*, 2019
Basic leucine zipper (bZIP)	45	*ZjbZIP3*, *-11*, *-12*, *-15*, *-17*, *-18*, *-19*, *-20*, and *-26*	Zhang *et al.*, 2020
Lipoxygenase (LOX)	13	*ZjLOX2*, *-5*, *-6*, and *-8*	Li *et al.*, 2022
Mitogen-activated protein kinase (MAPK)	10	*ZjMPK2*, *ZjMKK2* and *-4*	Liu *et al.*, 2017, 2019
Mitogen-activated protein kinase kinase (MAPKK)	5	*ZjMPK1, -3, -4,* and *-9*	Liu *et al.*, 2017
Mitogen-activated protein kinase kinase kinase (MAPKKK)	56	*ZjMAPKKK4*, *-7*, *-10*, *-17*, *-18*, *-25*, *-30*, *-34*, *-35*, *-37*, *-40*, *-41*, *-43*, *-44*, *-46*, *-52*, and *-53*	Liu *et al.*, 2020b
TEOSINTE BRANCHED 1/ CYCLOIDEA/ PROLIFERATING CELL FACTOR (TCP)	21	*ZjTCP6*, *-7*, *-12*, *-15*, *-16*, *-18*, and *-19*	Chen *et al.*, 2019
SQUAMOSA promoter-binding protein-like (SPL)	18	*ZjSPL6*, *-7*, *-9*, *-12*, *-13*, *-16*, *-17*, and *-18*	Shao *et al.*, 2017
WRKY	54	*ZjWRKY2*, *-3*, *-5*, *-6*, *-8*, *-9*, *-10*, *-15*, *-18*, *-22*, *-24*, *-26*, *-33*, *-34*, *-36*, *-37*, *-38*, *-42*, *-44*, *-45*, *-47*, and *-49*	Xue *et al.*, 2019

graft-infected seedlings, while 113 were downregulated, compared with the healthy controls. By 48 WAG, 5377 proteins were identified, with 753 DEPs detected. Of these, 358 were upregulated and 395 were downregulated. Among the most upregulated proteins, three of the top ten DEPs were L-gluconolactone oxidase-like proteins, enzymes known for vitamin C production. Additionally, two of the top ten upregulated proteins were JA-induced protein-like proteins, suggesting the involvement of JA in the defense response.

Integrating the proteome and transcriptome data, a hypothetical working model for the jujube response to phytoplasma infection has been proposed. The response progresses through three distinct stages. During the primary stage of disease (0–2 WAG), the phytoplasmas invade the plant, triggering initial recognition and signaling mechanisms. By the second stage (37–48 WAG), namely the defense activation stage, the plants mount a defense response, marked by increased levels of JA-related proteins and JA and a decrease in auxin-related proteins and auxin levels. Genes involved in plant–pathogen interactions are generally downregulated, potentially due to pathogen suppression mechanisms. In contrast, those involved in phenylpropanoid biosynthesis and metabolism are upregulated to strengthen the plant's structural and chemical defenses. In the third stage (48–52 WAG), as symptoms begin to manifest, plants exhibit noticeable symptoms, such as witches' broom, yellowing, and growth stunt. This stage is characterized by a decline in JA-related protein expression and JA content, as well as the reduced expression of photosynthesis-related genes, contributing to chlorosis and compromised energy production.

The combined analysis of proteome and transcriptome data emphasizes the complexity of the jujube response to phytoplasma infection. This approach highlights key pathways, such as hormonal crosstalk between JA and auxin, phenylpropanoid biosynthesis, and photosynthesis modulation, that are critical in the plant's defense and adaptation mechanisms.

Combined Proteome and Transcriptome Analysis during JWB Phytoplasma Elimination

Proteome and transcriptome analyses were also utilized to examine the JWB phytoplasma elimination process in the tetracycline-treated jujube plants (Wang *et al.*, 2019). This integrated approach provided insights into molecular changes during the recovery phase following treatment.

High-throughput iTRAQ analysis generated a total of 904,782 spectra, with 23,630 uniquely matched peptides. Among them, 6185 proteins were identified, enabling a detailed examination of DEPs and their corresponding transcripts. Comparison of samples at 3 and 6 MTT showed 20 DEPs were up- and downregulated based on the criteria (fold change >1.5 and Q-value <0.05). A cross-analysis of the 6185 identified proteins and 28,940 detected transcripts indicated a high correlation, with at least 5912 gene–protein pairs matching between datasets. Among the paired DEGs/DEPs, 14 of the 20 pairs demonstrated consistent expression patterns across time points at 3 and 6 MTT. All DEG/DEP pairs shared similar expression trends. These 14 correlated DEGs/DEPs shared a similar downregulation trend, with minor differences in fold changes, reflecting the close link between transcriptional and translational responses during the elimination process.

The analysis also highlighted significant trends in JA-related proteins during phytoplasma elimination. All seven jasmonate-induced protein-like (JIP) proteins were consistently downregulated in response to phytoplasma elimination. This trend contrasts sharply with their upregulation observed during JWB phytoplasma infection (Ye *et al.*, 2017), suggesting that JIP proteins may play a pivotal role in mediating the plant's stress response to the pathogen. Similarly, key DEPs associated with JA biosynthesis, such as phospholipase A (PLA) and lipoxygenase (LOX), were downregulated during phytoplasma elimination. These findings further prove the central role of JA-related proteins and pathways in the plant–phytoplasma interaction.

Metabolomics Analysis of Jujube Leaves Infected by Phytoplasma

Infection with phytoplasma changes the content of endogenous substances in jujube. For example, during the phytoplasma infection experiment performed by grafting diseased buds on healthy seedlings, the auxin content decreased while JA content increased by 37 WAG. At 48 WAG, there is no difference in auxin content between the infected seedling and the health control. Yet, the JA content is still much higher in the infected samples, which is consistent with the expression trends of JA-related genes and proteins (Ye *et al.*, 2017). During the JWB phytoplasma elimination experiment performed by tetracycline treatment on *in vitro* diseased shoots, the JA content slightly decreased at the sixth month compared with the third month after treatment (Wang *et al.*, 2019). These findings indicate that certain metabolic pathways, such as JA or auxin, respond to phytoplasma infection. Yet a wider analysis of endogenous substances is needed to understand jujube's response to phytoplasma infection.

With the ultra-performance liquid chromatography–tandem mass spectrometry (UPLC-MS/MS) technique, a widely targeted metabolomics analysis was performed to examine the metabolite composition of JWB-infected and healthy jujube plants (Liu *et al.*, 2023). Leaves from 2-year-old jujube plants grown in pots under controlled conditions, including healthy and JWB-infected leafhoppers (*Hishimonus hamatus*), were collected for UPLC-MS/MS analysis. In total, there were 1183 metabolites detected in the samples, including 82 alkaloids, 95 amino acids and derivatives, 266 flavonoids, 64 free fatty acids, 19 glycerol esters, 43 lignans and coumarins, 64 organic acids, 183 phenolic acids and derivatives, 29 phosphatidycholines (LPCs), 23 phosphatidylethanolamine (LPEs), 66 saccharides, 11 tannins, 128 terpenoids, 19 vitamins, and some other metabolites (Liu *et al.*, 2023).

Among these metabolites, there were 646 significantly differentially accumulated metabolites (DAMs) between the JWB-infected jujube and healthy controls, with 252 downregulated and 394 upregulated. Of the DAMs, 51 of 58 amino acids, 12 of 15 saccharides, 17 of 18 free fatty acids, 11 of 13 glycerol esters, 21 of 23 LPEs, and all 28 LPCs were up-accumulated in the JWB-infected samples. In contrast, 16 of 19 lignans, 11 of 13 coumarins, and 80 of 93 terpenoids were down-accumulated in the JWB-infected samples. The research also revealed that the JWB insect vector *H. hamatus* exhibited an increased frequency of phloem ingestion and spent longer in the phloem-feeding phase when feeding on JWB-infected leaves, compared with feeding on healthy leaves (Liu *et al.*, 2023).

Primary metabolites, such as amino acids fatty acids, glycerol esters, and saccharides, may serve as energy resources for phytoplasma and its insect vectors. Secondary metabolites like lignans and tannins are typical insect- and pathogen-resistant materials. The accumulation of primary metabolites and decrease of secondary metabolites revealed by metabolomics analysis explained the interaction mechanisms among plants–phytoplasmas–insects and their implications for the spread of phytoplasma-associated diseases on a physiological and biochemical level.

Conclusion

The difficulty in culturing phytoplasmas *in vitro* has limited the study of their pathogenic mechanisms. Recent advancements in DNA sequencing have provided sufficient genome information for both jujube and JWB phytoplasma, enabling the identification and annotation of genes related to JWB pathogenicity. High-throughput transcriptome and proteome sequencing facilitated the exploration of jujube's molecular response to phytoplasma infection, revealing the complicated mechanisms of JWB disease at both transcriptional and translational levels. Metabolomic analysis further explained the interaction mechanisms among plants–phytoplasmas–insects. These studies have significantly enhanced the understanding of plant–phytoplasma interactions from the perspectives of transcriptomics and proteomics.

The combination of transcriptome and proteome analyses has revealed the pivotal role of plant hormone-related pathways, especially the JA biosynthesis pathway, in plant–phytoplasma interactions. Transcription factors, such as members of the TCP family, and functional genes like *LOX* have been identified as key players in the response to phytoplasma infection and elimination. These genes and proteins were significantly differentially expressed, indicating their involvement in the development and mitigation of JWB disease. The study of JWB phytoplasma effectors has provided further insights into how JWB manipulates plant signaling pathways and immune responses, highlighting their role in disease progression. The new findings highlight the effectiveness of omics-based approaches in uncovering phytoplasma-pathogenic mechanisms, identifying key pathways and effectors, and informing strategies to manage diseases and enhance plant resilience.

References

Chen, P., Li, J., Ye, X., Tan, B., Zheng, X. *et al.* (2019) Genome-wide identification of Ziziphus jujuba TCP transcription factors and their expression in response to infection with jujube witches' broom phytoplasma. *Acta Physiologiae Plantarum* 41, 86. DOI: 10.1007/s11738-019-2879-9.

Chen, P., Chen, L., Ye, X., Tan, B., Zheng, X. *et al.*. (2022) Phytoplasma effector Zaofeng6 induces shoot proliferation by decreasing the expression of ZjTCP7 in *Ziziphus jujuba*. *Horticulture Research* 9, uhab032. DOI: 10.1093/hr/uhab032.

Chen, P., Zhang, Y., Li, Y., Yang, Q., Li, Q. *et al.* (2024) Jujube witches' broom phytoplasma effector Zaofeng3, a homologous effector of SAP54, induces abnormal floral organ development and shoot proliferation. *Phytopathology* 114, 200–210. DOI: 10.1094/PHYTO-10-21-0448-R.

Deng, M., Ma, Fu., Zhang, X., Huang, J., Yang, J. *et al.* (2021) Genome-wide identification of jujube witches' broom phytoplasma effectors revealed the role of SJP3 in inducing phyllody. *Scientia Horticulturae* 290, 110548. DOI: 10.1016/j.scienta.2021.110548.

Fan, X., Liu, W., Qiao, Y., Shang, Y., Wang, G. *et al.* (2017) Comparative transcriptome analysis of *Ziziphus jujuba* infected by jujube witches' broom phytoplasmas. *Scientia Horticulturae* 226, 50–58. DOI: 10.1016/j.scienta.2017.08.026.

Guo, M., Zhang, Z., Cheng, Y., Li, S., Shao, P. *et al.* (2020) Comparative population genomics dissects the genetic basis of seven domestication traits in jujube. *Horticulture Research* 7, 89. DOI: 10.1038/s41438-020-0312-6.

Guo, M., Zhang, Z., Li, S., Lian, Q., Fu, P. *et al.* (2021) Genomic analyses of diverse wild and cultivated accessions provides insights into the evolutionary history of jujube. *Plant Biotechnology Journal* 19, 517–531. DOI: 10.1111/pbi.13480.

Guo, S., Gu, L., Zhang, Y., Wu, Y., Tan, B. *et al.* (2023) Jujube witches' broom ('Zaofeng') disease: bacteria that drive the plants crazy. *Fruit Research* 3, 35. DOI: 10.48130/FruRes-2023-0035.

Huang, J., Zhang, C., Zhao, X., Fei, Z., Wan, K. *et al.* (2016) The jujube genome provides insights into genome evolution and the domestication of sweetness/acidity taste in fruit trees. *PLoS Genetics* 12, e1006433. DOI: 10.1371/journal.pgen.1006433.

Jung, H., Sawayanagi, T., Kakizawa, S., Nishigawa, H., Wei, W. *et al.* (2003) '*Candidatus* Phytoplasma ziziphi', a novel phytoplasma taxon associated with jujube witches'-broom disease. *International Journal of Systematic and Evolutionary Microbiology* 53, 1037–1041. DOI: 10.1099/ijs.0.02393-0.

Li, H., Gao, W., Xue, C., Zhang, Y., Liu, Z. *et al.* (2019) Genome-wide analysis of the bHLH gene family in Chinese jujube (*Ziziphus jujuba* Mill.) and wild jujube. *BMC Genomics* 20, 568. DOI: 10.1111/pbi.13480.

Li, J., Chen, L., Chen, P., Li, Q., Yang, Q. *et al.* (2022) Genome-wide identification and expression of the lipoxygenase gene family in jujube (*Ziziphus jujuba*) in response to phytoplasma infection. *Journal of Plant Biochemistry and Biotechnology* 31, 139–153. DOI: 10.1007/s13562-021-00670-4.

Li, K., Chen, R., Abudoukayoumu, A., Wei, Q., Ma, Z. *et al.* (2024) Two haplotype-resolved T2T reference genomes for *Ziziphus jujuba* and *Z. jujuba* var. *spinosa* provides new insights into the domestication of jujube. *Horticulture Research* 11, uhae071. DOI: 10.1093/hr/uhae071.

Liu, M., Zhao, J., Cai, Q., Liu, G., Wang, J. *et al.* (2014) The complex jujube genome provides insights into fruit tree biology. *Nature Communications* 5, 5315. DOI: 10.1038/ncomms6315.

Liu, M., Wang, J., Wang, L., Liu, P., Zhao, J. *et al.* (2020a) The historical and current research progress on jujube-a superfruit for the future. *Horticulture Research* 7, 119. DOI: 10.1038/s41438-020-00346-5.

Liu, R., Li, B., Chen, X., Liu, J., Luo, K., and Li, G. (2023) '*Candidatus* Phytoplasma ziziphi' changes the metabolite composition of jujube tree leaves and affects the feeding behavior of its insect vector *Hishimonus hamatus* Kuoh. *Insects* 14, 750. DOI: 10.3390/insects14090750.

Liu, Z., Wang, L., Xue, C., Chu, Y., Gao, W. *et al.* (2020b) Genome-wide identification of *MAPKKK* genes and their response to phytoplasma infection in Chinese jujube (*Ziziphus jujuba* Mill.) *BMC Genomics* 21, 142. DOI: 10.1186/s12864-020-6548-6.

Liu, Z., Zhang, L., Xue, C., Fang, H., Zhao, J., and Liu, M. (2017) Genome-wide identification and analysis of MAPK and MAPKK gene family in Chinese jujube (*Ziziphus jujuba* Mill.). *BMC Genomics* 18, 855. DOI: 10.1186/s12864-017-4259-4.

Liu, Z., Zhao, Z., Xue, C., Wang, L., Wang, L. *et al.* (2019) Three main genes in the MAPK cascade involved in the Chinese jujube-phytoplasma interaction. *Forests* 10, 392. DOI: 10.3390/f10050392.

Ma, F., Huang, J., Yang, J., Zhou, J., Sun, Q., and Sun, J. (2020) Identification, expression and miRNA targeting of auxin response factor genes, related to phyllody in the witches' broom disease of jujube. *Gene* 746, 144656. DOI: 10.1016/j.gene.2020.144656.

Qu, Z. and Wang, Y. (1993) *Chinese Fruit Tree Records – Chinese Jujube*. China Forestry Publishing House, Beijing, China.

Reel, P.S., Reel, S., Pearson, E., Trucco, E., and Jefferson, E. (2021) Using machine learning approaches for multi-omics data analysis: a review. *Biotechnology Advances* 49, 107739. DOI: 10.1016/j.biotechadv.2021.107739.

Shao, F., Lu, Q., Wilson, I., and Qiu, D. (2017) Genome-wide identification and characterization of the *SPL* gene family in *Ziziphus jujuba*. *Gene* 627, 315–321. DOI: 10.1016/j.gene.2017.06.044.

Wang, H., Ye, X., Li, J., Tan, B., Chen, P. *et al.* (2018a) Transcriptome profiling analysis revealed co-regulation of multiple pathways in jujube during infection by '*Candidatus* Phytoplasma ziziphi'. *Gene* 665, 82–95. DOI: 10.1016/j.gene.2018.04.070.

Wang, H., Ye, X., Li, J., Tan, B., Chen, P. *et al.* (2019) Combination of iTRAQ proteomics and RNA-seq transcriptomics reveals jasmonate-related-metabolisms central regulation during the process of jujube witches' broom recovery by tetracycline treatment. *Scientia Horticulturae* 243, 197–206. DOI: 10.1016/j.scienta.2018.08.015.

Wang, J., Song, L., Jiao, Q., Yang, S., Gao, R., Lu, X., and Zhou, G. (2018b) Comparative genome analysis of jujube witches'-broom phytoplasma, an obligate pathogen that cause jujube witches'-broom disease. *BMC Genomics* 19, 689. DOI: 10.1186/s12864-018-5075-1.

Wang, L., Liu, S., Gao, M., Wang, L., Wang, L. *et al.* (2022) The crosstalk of the salicylic acid and jasmonic acid signaling pathways contributed to different resistance to phytoplasma infection between the two genotypes in Chinese jujube. *Frontiers in Microbiology* 13, 800762. DOI: 10.3389/fmicb.2022.800762.

Xu, Y., Wang, C., Kong, D., Cao, M., Zhang, Q. *et al.* (2023) Identification of high tolerance to jujube witches' broom in Indian jujube (*Ziziphus mauritiana* Lam.) and mining differentially expressed genes related to the tolerance through transcriptome analysis. *Plants* 12, 2082. DOI: 10.3390/plants12112082.

Xue, C., Li, H., Liu, Z., Wang, L., Zhao, Y. *et al.* (2019) Genome-wide analysis of the WRKY gene family and their positive response to phytoplasma invasion in Chinese jujube. *BMC Genomics* 20, 464. DOI: 10.1186/s12864-019-5789-8.

Xue, C., Zhang, Y., Li, H., Liu, Z., Gao, W. *et al.* (2023) The genome of *Candidatus* Phytoplasma ziziphin provides insights into their biological characteristics. *BMC Biology* 23, 251. DOI: 10.1186/s12870-023-04243-6.

Xue, C., Zhang, L., Li, H., Liu, Z., Zhang, Y., Liu, M., and Zhao, J. (2024) The effector PHYL1$_{JWB}$ from *Candidatus* Phytoplasma ziziphi induces abnormal floral development by destablising flower development proteins. *Plant, Cell and Environment* 47, 4963–4976. DOI: 10.1111/pce.15079.

Yang, J., Shen, Z., Qu, P., Yang, R., Shao, A. *et al.* (2023a) Influence of jujube witches' broom (JWB) phytoplasma infection and oxytetracycline hydrochloride treatment on the gene expression profiling in jujube. *International Journal of Molecular Science* 24, 10313. DOI: 10.3390/ijms241210313.

Yang, M., Han, L., Zhang, S., Dai, L., Li, B. *et al.* (2023b) Insights into the evolution and spatial chromosome architecture of jujube from an updated gapless genome assembly. *Plant Communications* 4, 100662. DOI: 10.1016/j.xplc.2023.100662.

Yao, S. (2013) Past, present, and future of jujubes-Chinese dates in the United States. *HortScience* 48, 672–680. DOI: 10.21273/HORTSCI.48.6.672.

Ye, X., Wang, H., Chen, P., Fu, B., Zhang, M. *et al.* (2017) Combination of iTRAQ proteomics and RNA-seq transcriptomics reveals multiple levels of regulation in phytoplasma-infected *Ziziphus jujuba* Mill. *Horticulture Research* 4, 17080. DOI: 10.1038/hortres.2017.80.

Zhang, Y., Gao, W., Li, H., Wang, Y., Li, D. *et al.* (2020) Genome-wide analysis of the bZIP gene family in Chinese jujube (*Ziziphus jujuba* Mill.) *BMC Genomics* 21, 483. DOI: 10.1186/s12864-020-06890-7

Zhou, J., Ma, F., Yao, Y., Deng, M., Chen, M. *et al.* (2021) Jujube witches' broom phytoplasma effector SJP1 and SJP2 induce lateral bud outgrowth by repressing the ZjBRC1-controlled auxin efflux channel. *Plant, Cell and Environment* 44, 3257–3272. DOI: 10.1111/pce.14141.

16 Movement of Phytoplasmas and Disease Development in Plants

Carmine Marcone*
Department of Pharmacy, University of Salerno, Fisciano (Salerno), Italy

Abstract
Phytoplasmas colonize plants differently depending on the host's pathogen type, properties, and specific reactions. Knowledge of the movement of phytoplasmas within the host plant and their final distribution in various organs is usually essential for understanding the plant host–phytoplasma interactions. For most phytoplasma diseases for which the colonization behavior of the causal agent(s) has been elucidated through the use of several methods, the expression of symptoms depends on the presence and titer of the invading phytoplasma(s). However, in some diseases, because of the poor correlation between phytoplasma presence and symptom severity occurring in some parts of infected plants, a long-distance effect of phytoplasma infections may be involved.

Introduction

Phytoplasmas are a large group of plant-pathogenic, wall-less, obligate bacteria with a small genome size that ranges from 530 to 1350 kilobases (Marcone *et al.*, 1999). In diseased plants, phytoplasmas reside almost exclusively in the phloem sieve elements and are transmitted from plant to plant by phloem-feeding homopteran insects, mainly leafhoppers (Cicadellidae) and planthoppers (Fulgoromorpha) and less frequently psyllids (Psyllidae) (Weintraub and Beanland, 2006). Phytoplasmas can also access phloem sieve elements through the parasitic plant dodder (*Cuscuta* spp.). Once phytoplasmas have entered the phloem sieve elements, they spread systemically throughout the plant by passing through phloem sieve plate pores with the sieve tube flow (Fig. 16.1).

Occasionally, a few phloem parenchyma and companion cells adjacent to sieve tubes are also invaded (Seemüller *et al.*, 2002). However, it is unknown how phytoplasmas gain access to these cells. It seems impossible that phytoplasmas squeeze through plasmodesmata because these structures, which are responsible for the regulated exchange of compounds between sieve elements and companion cells, have a pore diameter of only 3–4 nm. Also, changes to the ultrastructure of plasmodesmata have not been observed in phytoplasma-infected plants (Christensen *et al.*, 2005). Another argument questioning the occurrence of phytoplasmas in parenchyma or companion cells is the marked physiological difference between sieve tube elements and parenchyma cells. Sieve elements are largely free from cell organelles when mature, thus forming a low-resistance pathway for assimilating and

*Corresponding author: cmarcone@unisa.it

 Phytoplasmas: Genomes, Plant Hosts and Vectors, 2nd Edition. (Ed. W. Wei)
DOI: 10.1079/9781800627031.0016

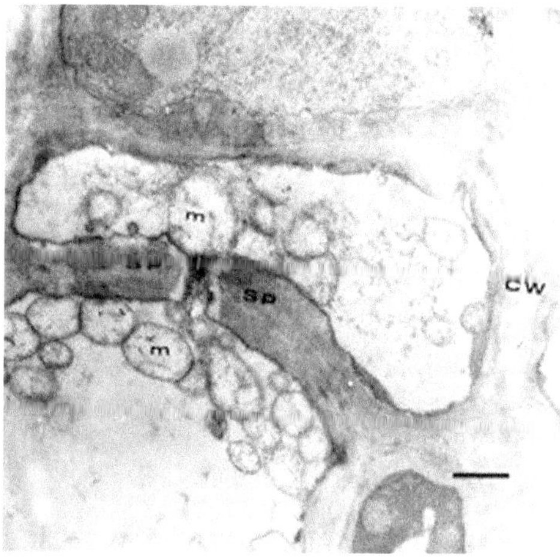

Fig. 16.1. Transmission electron micrograph showing phytoplasma bodies (m) passing through phloem sieve plate (SP) pores. Some phytoplasmas are attached to the inner surface of the sieve tube plasma membrane. CW, cell wall. The scale bar represents 700 nm.

facilitating the spread of phytoplasmas. The phloem sap is nutrient-rich and under high hydrostatic (turgor) pressure. The content varies depending on the plant species but always includes large amounts of carbohydrates. The most abundant is sucrose. Phloem parenchyma has a much lower osmotic pressure and a full complement of cell organelles. Companion cells also have a complete set of cell organelles. However, like sieve elements, they have a high sucrose content and osmotic pressure, and they may be a more appropriate environment than parenchyma cells for harboring phytoplasmas.

Phytoplasmas are associated with diseases in more than 1000 plant species worldwide. The list of plants known to harbor phytoplasmas is continually increasing, as is the number of taxonomically characterized phytoplasma strains (Lee *et al.*, 2000; Seemüller *et al.*, 2002; Bertaccini, 2022; Wei and Zhao, 2022). However, various aspects of the mechanisms by which phytoplasmas induce disease in plants, as well as the reasons for different reactions of the host plants to phytoplasma infections, are still poorly understood. Knowledge of the movement of phytoplasmas within the host plant and their final distribution in various organs is usually essential for understanding the plant host–phytoplasma interactions.

Disease Induction

Since phytoplasmas live and multiply in functional phloem sieve tube elements, the main effect of phytoplasma infections is the impairment of the sieve tube function. Numerous studies have provided evidence that inhibition of phloem transport occurs in phytoplasma-infected plants, which, in turn, leads to an accumulation of abnormal amounts of carbohydrates in source leaves (i.e. mature leaves) and a marked reduction of these compounds in sink organs (i.e. young leaves, flowers, growing fruits, and roots) (Braun and Sinclair, 1978; Kartte and Seemüller, 1991a; Lepka *et al.*, 1999; Guthrie *et al.*, 2001; Maust *et al.*, 2003; Wei *et al.*, 2022). Changes in photosynthate translocation, along with other impaired physiological functions, may account for symptoms observed in infected plants. These impairments include reduced photosynthesis, pigment content, stomatal conductance, transpiration, and root respiration, alterations in secondary metabolism, and disturbances in plant hormone balance, likely mediated by phloem dysfunction (León *et al.*, 1996; Lepka *et al.*, 1999; Tan and Whitlow, 2001; Maust *et al.*, 2003; Choi *et al.*, 2004; Endeshaw *et al.*, 2012; Vitale *et al.*, 2013; Gai *et al.*, 2014; Tan *et al.*, 2015; Zimmermann *et al.*, 2015; Wright *et al.*, 2021; Wei *et al.*, 2022).

Several plant host genes, which are differentially expressed upon phytoplasma infection, have been identified. These include genes involved in phytohormone synthesis and signaling, photosynthesis, carbohydrate and lipid metabolism, amino acid and sugar transport, secondary metabolite biosynthesis, cell wall biosynthesis and degradation, circadian rhythm and flowering time, morphological changes, and plant stress and/or defense response (Pracros *et al.*, 2006; Albertazzi *et al.*, 2009; Hren *et al.*, 2009a, 2009b; Cettul and Firrao, 2011; Himeno *et al.*, 2011; Landi and Romanazzi, 2011; Zamharir *et al.*, 2011; Ding *et al.*, 2013; Mou *et al.*, 2013; Wang *et al.*, 2018; Wei *et al.*, 2022; Wright *et al.*, 2023; Zhang *et al.*, 2024).

In addition, a considerable number of genes that are likely to play major roles in host–phytoplasma interactions are known from the complete genome sequences of the phytoplasmas that have been determined to date. Among these, there are genes encoding membrane proteins and effector (virulence) proteins (Bai *et al.*, 2009; Hoshi *et al.*, 2009; MacLean *et al.*, 2011; Sugio *et al.*, 2011a, 2011b; Kube *et al.*, 2012; Maejima *et al.*, 2014; Neriya *et al.*, 2014; Zimmermann *et al.*, 2015; Seemüller *et al.*, 2018; Wei *et al.*, 2019; Fernández *et al.*, 2024). There is an indication that effector proteins secreted by phytoplasmas are released directly inside phloem cells and are then transported through plasmodesmata to mesophyll and other cells of sink tissues, where interfere with host cell processes, leading to changes in plant physiology and development, and plant interactions with insect vectors (for reviews, see Hogenhout *et al.*, 2008; Tomkins *et al.*, 2018; Wang *et al.*, 2024).

Plants infected by phytoplasmas exhibit a wide range of specific and non-specific symptoms. Symptoms of affected plants may vary with the phytoplasma strain, host plant, stage of the disease, age of the plant at the time of infection, phytoplasma concentration in infected tissues, strain interactions, and environmental conditions (for reviews, see McCoy *et al.*, 1989; Lee *et al.*, 2000; Seemüller *et al.*, 2002; Marcone, 2014; Bertaccini, 2022). Specific symptoms include virescence, phyllody, big bud, flower proliferation, and other flower abnormalities, all resulting in sterility, witches'-brooms (Fig. 16.2), rosetting, internode elongation and etiolation, shortened internodes, enlarged stipules, off-season growth, and brown discoloration of phloem tissue. Less specific and non-specific symptoms that are most often common in woody plants include foliar yellowing and reddening, small leaves, leaf roll, leaf curl, vein clearing, vein enlargement, vein necrosis, premature fall coloration, premature defoliation, undersized fruits, poor terminal growth, sparse foliage, dieback, stunting of overall plant growth, and decline. In rare instances, phytoplasma-infected plants are fully non-symptomatic over their lifespan. Symptom remission, alternatively referred to as recovery, acquired tolerance, or resistance, has also been observed in a number of plant host–phytoplasma combinations. This phenomenon has been attributed to different factors, including differences in strain virulence, interference between strains (or cross-protection), shifts in phytoplasma populations, and accumulation of hydrogen peroxide, callose, and phloem proteins in affected tissues (Morvan *et al.*, 1986; Berges and Seemüller, 2002; Musetti *et al.*, 2010; Seemüller *et al.*, 2010, 2011b; Gambino *et al.*, 2013; Osler *et al.*, 2014).

Fig. 16.2. Witches'-broom: specific symptoms in *Spartium junceum* (Spanish broom) induced by spartium witches'-broom (SpaWB) disease.

Colonization Behavior of Phytoplasmas in Plants and its Relationship to Symptom Expression

Phytoplasmas differentially colonize plants, depending on the host's pathogen type, properties, and specific reactions. Movement, distribution, and multiplication of phytoplasmas in plants and relationship to symptom expression have been examined in both herbaceous and perennial woody hosts infected by taxonomically different phytoplasmas, using fluorescence and electron microscopy, bioimaging, histological and serological techniques, DNA hybridization, PCR, quantitative real-time PCR (qPCR) and digital PCR (dPCR) assays, and transmission experiments (Seemüller *et al.*, 1984a; Douglas, 1986; Kuske and Kirkpatrick, 1992; Lherminier *et al.*, 1994; Siddique *et al.*, 1998; Constable *et al.*, 2003; Christensen *et al.*, 2004; Wei *et al.*, 2004; Saracco *et al.*, 2006; Baric *et al.*, 2011; Lee *et al.*, 2012; Bahder *et al.*, 2019; Mazraie *et al.*, 2019; Carminati *et al.*, 2021; Wright *et al.*, 2021, 2022; Ivanauskas *et al.*, 2024).

Colonization in herbaceous hosts

Aster yellows

Work by Kuske and Kirkpatrick (1992) has shown that two strains of the aster yellows (AY) agent '*Candidatus* Phytoplasma asteris' – the severe (SAY) and dwarf (DAY) strains – had similar colonization patterns in the experimental host periwinkle (*Catharanthus roseus*), following graft-inoculation, over a 10-week period of observation. Both strains moved from the grafted shoots into ungrafted shoots that originated from the main stem below the grafted shoots, approximately 1 week after the graft unions had healed, and then spread systemically throughout the plant. Strain SAY colonized periwinkle more rapidly than strain DAY, and strain SAY was detected about 1 week earlier than the other. Moreover, periwinkle plants infected by SAY showed symptoms 1 week earlier than those infected by DAY. By 4 weeks after the plants were grafted, the titer of SAY in the various plant parts was higher than that of DAY, and the symptoms induced by SAY were more pronounced. However, the SAY titer declined after 4 weeks from inoculation, whereas the titer of DAY continued to increase, ultimately surpassing the SAY titer by the end of the observation period. Also, at the end of the observation period, symptoms in plants infected by DAY were less pronounced than those in plants infected by SAY. For both strains, titers were highest in actively growing meristematic regions, such as ungrafted shoots, with symptoms arising from the main stem below the grafted shoots and on side shoots that emerged from the grafted shoots, but were lowest in the roots. Distribution and titers of phytoplasma infections were clearly correlated to symptom expression in the aerial parts. A similar colonization pattern was observed in strawberry plants naturally infected by the clover phyllody phytoplasma, in which phytoplasma titers were highest in pedicels and receptacles with symptoms, followed by sepals, petals, and leaves. However, the clover phyllody agent was not detectable in strawberry roots (Clark *et al.*, 1983).

Wei *et al.* (2004) analyzed the dynamics of onion yellows (OY) phytoplasma in garland chrysanthemum (*Chrysanthemum coronarium*) using a localized inoculation technique and nested-PCR, qPCR, and immunohistochemical assays. Following localized inoculation of a leaf of garland chrysanthemum by the leafhopper vector *Macrosteles striifrons*, the OY agent spread sequentially from the inoculated leaf to the main stem, root, and upper leaves, and then the lower leaves. The OY phytoplasma systemically colonized the plant 21 days after inoculation, and its concentration increased approximately sixfold per week, from 14 to 28 days after inoculation. The phytoplasma concentration in the roots was higher than in the inoculated leaves. Within the aerial parts, the OY agent was most abundant in inoculated leaves and shoot apices, followed by the youngest leaves and lower leaves. Furthermore, at 28 days after inoculation, characteristic internal symptoms, such as phloem necrosis in the root and phloem hyperplasia in the stem, as well as external symptoms, such as stunted growth, were apparent. Similar data were obtained by Saracco *et al.* (2006), who investigated the colonization pattern of chrysanthemum yellows (CY) phytoplasma in daisy (*Chrysanthemum carinatum*). Following localized inoculation of either a basal or apical leaf by the leafhopper vector *M. quadripunctulatus*, the CY phytoplasma colonized and multiplied to high numbers in the roots and young, apical leaves rather than in the old, basal leaves. However, it was always more concentrated in the roots than aerial parts (Saracco *et al.*, 2006).

In India, in sesame (*Sesamum indicum*) plants affected by a phyllody disease associated with a 16SrI-B subgroup phytoplasma, phytoplasma titer varied greatly among the three different organs examined, the root, stem, and leaves (Singh *et al.*, 2018). The lowest titer occurred in the root, whereas the highest titer was observed in the leaves. Phytoplasma concentration was correlated to symptom severity during the early and mid stages of disease development. However, in the late stage, irrespective of symptom severity, a decrease in the phytoplasma titer occurred in each organ. Differences in phytoplasma titer between lower and upper leaves as well as between lower and upper parts of the stem throughout the disease development were also observed (Singh *et al.*, 2018).

Flavescence dorée and jujube witches'-broom

The movement and distribution of flavescence dorée (FD) phytoplasma within the experimental host *Vicia faba* have been described by Lherminier *et al.* (1994). This work has shown that, although the FD phytoplasma was inoculated into the aerial parts by the leafhopper *Euscelidius variegatus*, the FD agent was first detected in the roots, where it multiplied to a high extent prior to symptom development. Subsequently, the FD phytoplasma invaded the collar and axillary shoots, emerging about 20 days after inoculation from the basal part of the main stem and then moved acropetally along the main stem. As soon as symptoms became apparent, FD phytoplasma could be detected in all parts of the inoculated plant. However, the highest phytoplasma concentrations were recorded in the upper parts of basal axillary shoots, other than in the roots. In periwinkle plants graft-inoculated with the jujube witches'-broom (JWB) phytoplasma ('*Ca.* P. ziziphi'), the infection spread sequentially from the inoculated scion to the main stem and root. After accumulation in the root, the JWB agent gradually colonized the aerial parts from the root upwards. Systemic colonization of the plant was achieved by 82 days after inoculation (Lee *et al.*, 2012).

Potato purple top

Ivanauskas *et al.* (2024) reported on investigations on phytoplasma colonization and symptom development in tomato and periwinkle plants experimentally infected with potato purple top (PPT) phytoplasma. In tomato plants, big bud, cauliflower-like inflorescence, and witches'-broom symptoms developed at approximately 28, 45, and 60 days post-inoculation, respectively, while in periwinkle plants, virescence, phyllody, and witches'-broom occurred at approximately 30, 35, and 50 days after inoculation, respectively. In both host plants, the phytoplasma titer differed considerably at different stages of disease progression. It was higher in the early stages, characterized by big bud and phyllody symptoms, than in the late stage, which was characterized by witches'-broom. Notably, the highest phytoplasma titer in tomato was found in samples displaying big bud symptoms, whereas in periwinkle the highest titer was found in samples showing phyllody. Also, the phytoplasma titer was significantly higher in periwinkle than in tomato plants, indicating, differences in host susceptibility to the PPT agent.

Stolbur

Work by Carminati *et al.* (2021) showed that in graft-inoculated tomato plants, the stolbur agent '*Ca.* P. solani' moved first from the inoculation site to the main stem to reach sink tissues, such as the shoot apex and root, at 12 days post-inoculation. Subsequently, the invading phytoplasma spread to newly expanded leaves and advanced progressively to source tissues, such as mature leaves. Complete systemic colonization was achieved at 29 days post-inoculation. At this time, severe symptoms, such as big buds, were present. The phytoplasma titer was significantly higher in the root than in the aerial parts throughout the observation period. Mature basal leaves harbored the lowest phytoplasma titer in the early and mid stages of the disease progression. In contrast, all aerial parts, including mature basal leaves, showed a similar titer in advanced stages at 57 days post-inoculation.

X-disease

In celery (*Apium graveolens*) plants experimentally inoculated with the X-disease phytoplasma using the leafhopper vector *Colladonus montanus*, phytoplasma infections were first detected in the roots (Kirkpatrick, 1991). However, in plants with fully developed symptoms, the phytoplasma concentration was highest in the shoot apex, followed by the youngest, severely affected leaves. The X-disease phytoplasma was also present in much lower concentration in older, completely chlorotic leaves.

In all the colonization patterns mentioned, phytoplasma movement was directed towards actively growing regions of the plant with symptoms, which acted as metabolic sinks, to which nutrients were directed through the phloem conduit. In contrast, in free-branching-affected poinsettia plants, the infecting phytoplasma, the poinsettia branch-inducing (PoiBI) agent, which is a member of the X-disease phytoplasma group or 16SrIII group, preferentially colonized and accumulated in source leaves and, to a lesser extent, in petioles of source leaves and stems (Christensen *et al.*, 2004). PoiBI phytoplasma did not colonize or was present in very low numbers in sink organs such as roots and young leaves. Moreover, the extent of PoiBI phytoplasma invasion was correlated with symptom severity.

Colonization in woody hosts

For a number of deciduous woody hosts, it has been shown that phytoplasma colonization of the aerial parts of infected plants is subject to seasonal fluctuation.

Apple proliferation and pear decline

In apple (*Malus* spp.) and pear (*Pyrus* spp.) trees affected by apple proliferation (AP) and pear decline (PD) diseases, respectively, colonization patterns of the causal agents, '*Ca.* P. mali' and '*Ca.* P. pyri,' have been monitored over a period of several years with fluorescence microscopy using the DNA dye 4′-6-diamidino-2-phenylindole (DAPI) test and periodic transmission grafting (Schaper and Seemüller, 1982; Seemüller *et al.*, 1984a, 1984b). Since phytoplasmas depend on functional sieve tubes and because the sieve tubes in the above-ground parts of apple and pear trees degenerate in late fall and early winter, AP and PD phytoplasmas are, in almost all cases, eliminated in the aerial parts during winter. In contrast, they persist in the roots where intact sieve tubes are present throughout the year. Both pathogens may recolonize the aerial parts from the roots in spring when new phloem is being formed. This recolonization was studied by Schaper and Seemüller (1984). Over a 5-year period, from the middle of March to the middle of June, a narrow strip of bark was removed at regular intervals all around the trunks or scaffold limbs of orchard apple and pear trees that were colonized in the roots. The girdling interrupted the continuity of the phloem, which prevented the translocation of phytoplasmas above the girdling line. The study revealed that recolonization started as soon as newly differentiated sieve tubes were present. This point was usually reached in March. The upward movement of AP and PD phytoplasmas was relatively slow, reaching a velocity of 7.5–30 mm per day. However, the mechanism of this movement is still unknown. Recolonization usually occurs annually during the first few years of the disease. Later, the aerial parts may be only partly or weakly colonized or may not be colonized at all.

Because the expression of symptoms depends on the presence of the invading phytoplasmas, trees intensively colonized in the aerial parts develop AP-specific symptoms, whereas those only partially, weakly, or not colonized develop mild or no symptoms. Therefore, the variation in symptom expression over the years is related to the seasonal fluctuation of the phytoplasma population in the aerial parts of infected trees. The root system of infected trees remains colonized for the life of the trees. Baric *et al.* (2011) reported seasonal fluctuation of AP phytoplasma titer in AP-affected apple trees in northern Italy using qPCR assays. The average phytoplasma titer was higher in the root than in the aerial parts. Also, symptom severity was strictly related to the phytoplasma titer. AP phytoplasma was also detected in the aerial parts of non-symptomatic trees, although at low concentrations. However, the average phytoplasma titer in aerial parts of symptomatic trees was 14 times higher than that of non-symptomatic trees. This therefore shows that pathogen occurrence alone was not essential for the development of severe AP symptoms, but that there needed to be high numbers of pathogen in the aerial parts of affected trees for the development of severe AP symptoms, such as witches'-brooms and/or undersized fruits (Baric *et al.*, 2011).

The colonization behavior described is typical for trees grown on established rootstocks based on *Malus* × *domestica* (Schaper and Seemüller, 1982, 1984; Seemüller *et al.*, 1984a, 1984b; Carraro *et al.*, 2004). Extensive studies on AP resistance have shown that diseased apple trees grafted on experimental *Malus sieboldii*-based AP-resistant apomictic rootstocks either never developed symptoms or, only rarely, developed temporary, mild symptoms. Such trees were not colonized in the aerial parts and harbored extremely low phytoplasma numbers in the roots. Thus, the low starting concentration in the roots and the unsuitable host properties of *M. sieboldii*-based genotypes may impair the spread of the pathogen from the roots into the scion during recolonization of the stem in spring (Kartte and Seemüller, 1991b; Seemüller *et al.*, 2008, 2011a)

Similarly, pear trees on clonal quince (*Cydonia oblonga*) rootstocks are little affected by PD disease, owing to the poor host properties of quince. In this rootstock, the PD phytoplasma occurs in a very low titer, which has a negative effect on the recolonization of the aerial parts in spring, resulting in no or mild symptoms (Seemüller *et al.*, 1986). '*Ca.* P. mali' was also detected in the aerial parts of AP-affected apple trees during the dormant seasons (Baric *et al.*, 2011). However, according to previous findings, the pathogen in this case is not viable or transmissible (Seemüller *et al.*, 2011a). Also, in Spain, '*Ca.* P. pyri' was detected in aerial parts throughout the year by nested PCR and was transmitted by chip budding in January and February (Errea *et al.*, 2002; Garcia-Ciapa *et al.*, 2003). Climatic conditions occurring in Spain may allow overwintering of the PD agent in the aerial parts of affected trees (Seemüller *et al.*, 2011c).

X-disease

The X-disease agent '*Ca.* P. pruni,' which infects stone fruits in North America, following initial inoculation, moves basipetally towards the roots. However, it may be transported to a local sink near the inoculation site, depending on the strength of the sink pressure. After accumulation in the roots and lower trunk, the X-disease phytoplasma gradually colonizes the tree systemically from the basal parts upwards, accumulating in lower branches first, then aerial limbs (Wright *et al.*, 2022; Harper *et al.*, 2023). Complete systemic colonization of a tree can take several growing seasons and leads to an increased overall phytoplasma titer (Wright *et al.*, 2022). The pathogen overwinters in the roots and, to some extent, in aerial parts. The latter are then reinvaded in the following spring (for review, see Kirkpatrick, 1991).

Studies by Wright *et al.* (2022) showed that in X-disease-affected sweet cherry (*Prunus avium*) trees, the phytoplasma titer of woody stem (branch) tissues was consistently high throughout the growing season, mainly between bloom (April) and harvest (June/July), decreasing by only 1–2 times at postharvest (August) and predormancy (September) stages. This colonization behavior differed from that occurring in leaf tissues, which was characterized by a low titer at bloom that increased through the season, peaking at postharvest and predormancy, at which times the titer was 4–5 times higher than at bloom. '*Ca.* P. pruni' occurred at low titer in the peduncles at bloom stage and increased in titer by approximately three times throughout the growing season, reaching the highest concentration between pit hardening and harvest stages of fruit development (Kirkpatrick, 1991; Wright *et al.*, 2022).

A similar behavior was observed in developing fruits from shuck-fall (April/May), although titer peaked at pit hardening (May) and then decreased. This decrease may be due to phytoplasma-induced changes in source–sink relationships since developing fruits are strong sink tissues. Titer in fruits was lower than in peduncles throughout the season (Wright *et al.*, 2022). Symptom severity was correlated with '*Ca.* P. pruni' titer. In severely affected sweet cherry trees, the phytoplasma titer was 4–5 times higher than in those with milder symptoms (Wright *et al.*, 2021, 2022). Douglas (1986), who monitored the distribution of X-disease phytoplasma in diseased peach (*P. persica*) and chokecherry (*P. virginiana*) trees during two growing seasons by DAPI fluorescence, found a good correlation between symptom severity and extent of phytoplasma invasion. However, invasion occurred earlier and was more extensive in chokecherry trees than in peach trees. Also, the pathogen was more evenly distributed in chokecherry than in peach. A slow rate of colonization and symptom expression may account for the tolerance or resistance observed in some sweet cherry cultivars (Uyemoto *et al.*, 1991).

Since X-disease phytoplasma differentially colonizes plant tissues throughout the season, variations in transmission rates following graft inoculations have been observed in many instances for this pathogen. Work by Uyemoto and Luhn (2006) has shown that in sweet cherry the transmission rate was 10% in spring (June), whereas it reached 64% in summer (August), irrespective of the age of the shoots used as inoculum. Besides cherry sources, graft transmissions using scion buds from X-disease-affected peach trees, collected in June, yielded a high transmission rate (55%). Similar results were obtained by Suslow and Purcell (1982) and Kirkpatrick *et al.* (1990), who demonstrated that the ability of the leafhopper vector *Colladonus montanus* to acquire the X-disease phytoplasma from diseased sweet cherry trees correlated with the seasonal variation of the phytoplasma titer occurring in the leaves. The highest rates of infective *C. montanus* individuals were recorded in August–September. Rosenberg and Jones (1977) reported on the seasonal variations in the transmission of X-disease phytoplasma in peach and chokecherry trees following graft inoculations with stem scions. The highest transmission rates were obtained in June–August (82–100%) and May–June (100%) for peach and chokecherry, respectively. The pathogen was also transmitted during winter, although at a low percentage in peaches. A good correlation between symptom severity and phytoplasma titer in X-diseased affected peach was also reported by Wright *et al.* (2023) using qPCR assays.

European stone fruit yellows

The European stone fruit yellows (ESFY) agent '*Ca.* P. prunorum,' which is the only confirmed phytoplasma pathogen causing diseases of *Prunus* species in Europe, can persist in the aerial parts of diseased stone fruit trees during the dormant season, whereas it can be present at root level throughout the year (Seemüller *et al.*, 1998; Jarausch *et al.*, 1999). Furthermore, Jarausch *et al.* (1999) proved by tissue culture techniques that the ESFY pathogen colonizing the aerial parts during winter is viable. Therefore, it can be efficiently transmitted by grafting stem scions throughout the year (Seemüller *et al.*, 1998). In France, Jarausch *et al.* (1999) monitored the colonization pattern of the ESFY phytoplasma in apricot and Japanese and European plum trees using DAPI fluorescence and PCR assays. Although the pathogen was detected throughout the year in the aerial parts, increased colonization (titer) in diseased trees occurred during summer (July–September). Trees intensively colonized in the aerial parts usually develop characteristic symptoms (Marcone *et al.*, 2010).

Kison and Seemüller (2001) examined the response of many established and experimental *Prunus* rootstocks to ESFY phytoplasma infections using graft-inoculation, PCR assays, and DAPI fluorescence tests. They showed that phytoplasma infections occurred in all inoculated *Prunus* genotypes, irrespective of symptom expression. The colonization density was lower in little-affected or unaffected rootstocks than in susceptible rootstocks. There were also differences between the colonization of the roots and the aerial parts of rootstocks. Usually, phytoplasma numbers were higher in the roots than in the stems of the same genotype. In some rootstocks, infections were not found in the stem but were present in the roots. Colonization appeared to persist over the observation period of 5–8 years, even in the least affected rootstocks such as *P. domestica* stocks Achermann's, Brompton, and P 2175, and *P. cerasifera* stock Myrabi.

Australian papaya dieback and palm lethal yellowing

Working with Australian papaya dieback disease, Siddique *et al.* (1998) monitored, through PCR assays, the distribution of the presumed causal agent, '*Ca.* P. australiense,' within different parts of diseased papaya (*Carica papaya*) plants at progressive stages of disease development. In the early stages of the disease, the papaya-infecting phytoplasma proved to be present in expanding leaves, flowers, and upper part of the stem. As the disease progressed and symptoms became more pronounced, the pathogen occurred in the mid-lower part of the stem and even in the roots. However, no phytoplasma infections occurred in symptomatic mature leaves during all stages of disease development. Also, the pathogen occurred in very low numbers throughout the plant, and its concentration decreased as the disease progressed.

Similar distribution patterns have been observed in coconut palms (*Cocos nucifera*) affected by lethal yellowing (LY) disease. In this case, the colonizing phytoplasma occurs mainly in sink plant parts, such as the apical meristem, immature leaves, root meristem, and inflorescences, but is rarely detectable in mature leaves with symptoms (León *et al.*, 1996; Maust *et al.*, 2003). In Mexico, Oropeza *et al.* (2011) monitored the phytoplasma distribution in different parts of LY-affected coconut plants and symptom development using PCR assays. Stems showed the highest detection frequency, reaching 100% in all stages, including early stages when the infected plants were still without symptoms; mature leaves, however, did not harbor phytoplasma infections throughout development of the disease. Other parts, such as young leaves, inflorescences, stem apex, root apex, intermediate leaves, and roots without apex, had a detection frequency that varied among the different parts but showed the same behavior, characterized by an increase over time, except in the late stage of symptom development when it decreased. Accumulation in the stem in advance of symptom expression implies that this part of the plant is the site where the invading phytoplasma first moves from the leaves after initial inoculation by insect vectors and in which it multiplies before spreading to other parts, including roots (Oropeza *et al.*, 2011).

Córdova *et al.* (2014) reported on the detection and quantification of 16SrIV-A, -D, and -E subgroup phytoplasmas in different parts of LY-affected coconut palms in the Americas at symptom developmental stages of yellowing in lower and middle leaves, using qPCR assays. The highest concentration was found in the stem, followed in decreasing order by primary root apex, mature inflorescences at developmental stages 1 and 2, inflorescences at developmental stages 3–7, spear leaves (youngest emerged yet unopened leaves), flag leaves (first open leaves), and mature leaves. These results agree with those previously reported by Oropeza *et al.* (2011). Work on the occurrence and distribution of 16SrIV-D subgroup phytoplasma in different leaf tissues of common ornamental palm species (*Phoenix roebelinii*, *Sabatal palmetto*, and *Syagrus ramanzoffiana*) affected by the lethal bronzing disease (LBD) in Florida, at different stages of symptom development, using qPCR and dPCR assays, showed that phytoplasma infection was only detectable in spear leaves and not in other kind of leaves irrespective of symptom development (Bahder *et al.*, 2019). With the exception of *S. palmetto*, the phytoplasma titer was higher in spear leaves than in the trunk. Therefore, spear leaves may play a key role in the acquisition process of 16SrIV-D phytoplasma by insect vectors.

Work on the distribution and quantification of 16SrIV-D phytoplasma in the trunks of LBD-affected *Phoenix sylvestris* and *S. palmetto* trees in Florida, at different stages of decline, including early symptoms, moderate symptoms, and late symptoms, using qPCR assays, revealed that regardless of the stage of decline, the highest phytoplasma titer occurred at the base of the trunk and gradually decreased further up the trunk (Bahder *et al.*, 2020). Also, in LBD-affected *P. sylvestris* trees, the phytoplasma titer was higher at the early symptom stage than at the late symptom stage (Bahder *et al.*, 2020). In coconut palms affected by the Weligama coconut leaf wilt disease (WCLWD) in Sri Lanka, the highest detection rate of the associated phytoplasma, which is a member of the 16SrXI phytoplasma group, was recorded in bud leaves, followed by roots and young inflorescences. However, no phytoplasma infections were detectable in stems (De Silva *et al.*, 2023). The detection rate increased over time in the plant parts tested, except roots, reaching 100% in bud leaves after 1 year.

Grapevine yellows

The distribution and persistence of phytoplasmas in grapevine plants affected by the Australian grapevine yellows (AGY) disease have been studied over an observation period of several years using PCR assays (Constable *et al.*, 2003; Angelini *et al.*, 2018). AGY disease is known to be associated with three taxonomically different phytoplasmas: '*Ca.* P. australiense,' the tomato big bud (TBB) agent, and Buckland Valley grapevine yellows (BVGY) phytoplasma (Angelini *et al.*, 2018). Each year, the AGY phytoplasmas were detected in the shoots, cordons, trunks, and roots of diseased grapevine plants. Thus, they appear to infect Australian grapevines from year to year persistently. However, the detection frequency within the different plant organs differed considerably. In the roots, the detection frequency was extremely low compared with aerial parts throughout the observation period, thus indicating a poor colonization of the root system. The highest detection rates were recorded in symptomatic shoots during mid to late summer, before harvest (Constable *et al.*, 2003; Angelini *et al.*, 2018).

Also, there was a good correlation between detection frequency and symptom expression. In a few cases, phytoplasma infections were detected in non-symptomatic plants, which persisted over time (Angelini *et al.*, 2018). In FD-affected grapevine plants in Slovenia, the phytoplasma titer in aerial parts, including canes, leaves, tendrils, flowers, and berries, all collected from symptomatic shoots, showed an increasing trend throughout the growing season, reaching the highest level in berries in August

(Prezelj *et al.*, 2013). FD phytoplasmas were also detected in tissues from non-symptomatic shoots, although their concentration was considerably lower than in tissues from symptomatic shoots. Also, in tissues from non-symptomatic shoots, the titer decreased during the growing season (Prezelj *et al.*, 2013).

Alder yellows

Alder yellows (ALY) phytoplasma, which is known to affect several alder (*Alnus*) species in Europe (Marcone *et al.*, 2024), persists in the aerial parts of trees throughout the year. In this respect, the colonization differs significantly from that of AP and PD phytoplasmas. Of the different alder trees examined by Lederer and Seemüller (1991) using DAPI fluorescence, ALY infections proved to be detectable in all trees older than approximately 5 years, irrespective of symptom expression. About 80% of infected trees were non-symptomatic for ALY, whereas only 20% showed ALY symptoms. Usually, healthy-appearing trees were more heavily colonized than trees with symptoms. The phytoplasma population was always higher in petioles and young twigs than in several-year-old branches, trunks, and roots. Colonization was characterized by an uneven phytoplasma distribution in adjacent sieve tubes, with some tubes packed with phytoplasmas; in contrast, others contained a considerably lower population (Fig. 16.3). By

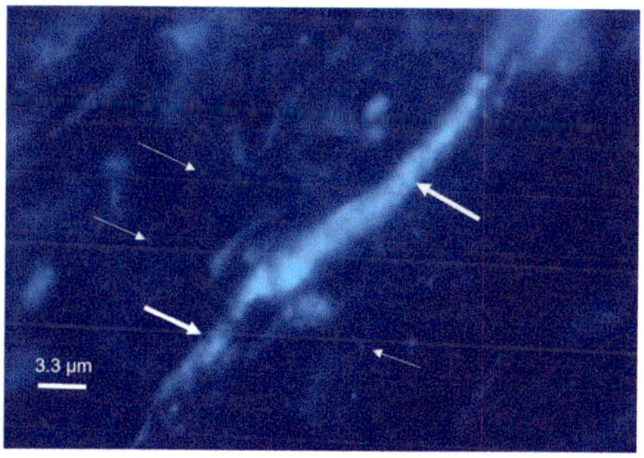

Fig. 16.3. Phytoplasma colonization in petioles of *Alnus glutinosa* (alder) infected by alder yellows (ALY), viewed under fluorescence microscopy using the DNA dye 4′-6-diamidino-2-phenylindole (DAPI fluorescence test). Phytoplasmas are visible within several sieve tubes, appearing as single particles (small arrows), small aggregates, or larger fluorescent areas (large arrows).

graft-inoculation of healthy alder seedlings with scion wood from differently affected and non-symptomatic trees and observation of disease development over a 5-year period, Berges and Seemüller (2002) provided evidence that the ALY phytoplasma is pathogenic to alder and may induce severe symptoms but that avirulent strains occurring within this taxon account for the latent infections that are widespread in Europe.

Other phytoplasma diseases of woody hosts

Similarly to AP and PD diseases, the degeneration of the phytoplasma population during winter in above-ground parts of trees, the overwintering in the root system, and the reinvasion of the stem in spring have been described for elm yellows (EY), mulberry dwarf, and paulownia witches'-broom diseases affecting their respective hosts (for references, see Schaper and Seemüller, 1984). Braun and Sinclair (1976) were the first to relate the conditions of the phloem to the survival of the causal agent. In winter, they found no functional sieve tubes in the stem of EY-affected elm (*Ulmus*) trees but did find intact sieve tubes in the roots. Thus, they concluded that the EY phytoplasma (*'Ca.* P. ulmi') overwinters in the roots and recolonizes the stem in spring.

Work by Mazraie *et al.* (2019) showed that after initial inoculation of lime (*Citrus aurantifolia*) seedlings, '*Ca.* P. aurantifolia' – the causal agent of lime witches'-broom (WBDL) disease – moved in both directions, downwards to the root and upwards to the apical leaves. The early accumulation of the pathogen in the apical leaves was not dependent on its occurrence in the root, and movement from the root to the aerial parts was relatively slow. Also, symptom severity was correlated with the phytoplasma titer. In the presence of pronounced witches'-brooms, which appeared approximately 10 months after inoculation, the phytoplasma titer was highest in the shoots and leaves of witches'-brooms, followed by the crown, root, and shoots and apical leaves of non-symptomatic branches. In field-grown WBDL-affected lime trees, which were still incubated, the phytoplasma titer in leaves was higher in August than in March and December (Mazraie *et al.*, 2019). This finding reflects the seasonal variations in the colonization pattern of the aerial parts of affected lime trees.

Strain Interference in Phytoplasma Colonization

Interference among strains of the same taxon has been described for several plant host–phytoplasma combinations during colonization. Studies conducted by Sinclair and Griffiths (2000) revealed that, in periwinkle plants that were co-inoculated with two strains of the ash yellows (AshY) phytoplasma that differed greatly in aggressiveness, the most aggressive strain appeared sooner and more frequently than the less aggressive strain in leaves located at a distance from the inoculation sites. Thus, aggressiveness was associated with a higher movement or multiplication rate of the most aggressive strain compared with the less aggressive one. However, when either strain was inoculated 11 weeks before the other into the same plant, only the initial strain could be detected after 12 weeks of incubation. Therefore, the initial strain or its effect on the host may have interfered with long-distance movement or multiplication of the second strain. Thus, Sinclair and Griffiths (2000) proposed a concept of pre-emptive dominance to explain the continuing dominance of the first strain that colonized a plant, regardless of its aggressiveness.

Interaction among three strains of the AY phytoplasma has been reported by Freitag (1964). When tobacco (*Nicotiana rustica* var. *humilis*) and plantain (*Plantago major*) plants were inoculated first with either DAY or SAY strains and 1 week later with the Tulelake AY (TLAY) strain, the plants mainly developed symptoms of the first strains. The results indicated a cross-protection reaction. The strain that was inoculated first nearly always predominated and prevented the development of symptoms of the challenging strain. Plants inoculated simultaneously with DAY and SAY strains usually developed symptoms of only one strain. However, some plants first showed symptoms of one strain and subsequently also symptoms of the other strain. In plantain, plants were inoculated first with the TLAY strain and then challenged by the DAY or SAY strain; the initial strain was displaced, and symptoms of the challenging strain developed. The most interesting results were obtained when tobacco plants were first inoculated with the TLAY strain and then, a week later, with either the DAY or the SAY strains. In these instances, the plants first developed pronounced vein-clearing symptoms indicative of TLAY strain infections and recovered completely. This was interpreted as a phenomenon of mutual suppression, which resulted in the neutralization of the two competing strains (Freitag, 1964).

In periwinkle plants, which were co-inoculated either simultaneously or sequentially with 16SrV-C and -D strains of the FD phytoplasma, the 16SrV-C strain suppressed over time the other strain, and at the end of the observation period only the 16SrV-C strain was detectable in roots and aerial parts of the inoculated plants. In contrast, in singly inoculated plants, both strains showed similar colonization patterns and symptom severity (Rossi *et al.*, 2020). It is conceivable that in nature, interaction among strains of the same taxon occurring within a given plant host may account for either the lack of symptom expression or mild symptom expression (Berges and Seemüller, 2002; Seemüller *et al.*, 2010, 2011b).

Conclusion

Phytoplasmas are obligate, bacterial plant pathogens that systemically colonize the plants by passing through phloem sieve plate pores. Since these prokaryotes live and multiply in functional phloem sieve tube elements, the main effect of phytoplasma infections is the impairment of sieve tube function. However, various aspects of the mechanisms by which phytoplasmas induce disease in plants and the reasons for different reactions of the host plants to phytoplasma infections are still poorly understood. Phytoplasmas colonize plants differently depending on the host's pathogen type, properties, and specific reactions. Knowledge of the movement of phytoplasmas within the host plant and their final distribution in various organs is usually essential for understanding the plant host–phytoplasma interactions. For most phytoplasma diseases for which the colonization behavior of the causal agent(s) has been elucidated through the use of several methods, the expression of symptoms depends on the presence and titer of the invading phytoplasma(s). However, in some diseases, because of the poor correlation between phytoplasma presence and symptom severity occurring in some parts of infected plants, a long-distance effect of phytoplasma infections may be involved. This is also supported by findings that effector proteins secreted by phytoplasmas into the cytoplasm of phloem sieve cells are then spread systemically to other cells of different tissues, which interfere with host cell processes, leading to changes in plant physiology and development. Also, interaction among strains of the same taxon occurring within a given plant host may alter the colonization behavior of the invading strains.

References

Albertazzi, G., Milc, J., Caffagni, A., Francia, E., Roncaglia, E. *et al.* (2009) Gene expression in grapevine cultivars in response to bois noir phytoplasma infection. *Plant Science* 176, 792–804.

Angelini, E., Constable, F., Duduk, B., Fiore, N., Quaglino, F., and Bertaccini, A. (2018) Grapevine phytoplasmas. In: Rao, G.P., Bertaccini, A., Fiore, N., and Liefting, L. (eds) *Phytoplasmas: Plant Pathogenic Bacteria - I. Characterization and Epidemiology of Phytoplasma-Associated Diseases*. Springer Nature, Singapore, pp. 123–151.

Bahder, B.W., Soto, N., Komondy, L., Mou, D.F., Humphries, A.R., and Helmick, E.E. (2019) Detection and quantification of the 16SrIV-D phytoplasma in leaf tissue of common ornamental palm species in Florida using qPCR and dPCR. *Plant Disease* 103, 1918–1922.

Bahder, B.W., Soto, N., Mou, D.F., Humphries, A.R., and Helmick, E.E. (2020) Quantification and distribution of the 16SrIV-D phytoplasma in the wild date palm, *Phoenix sylvestris*, at different stages of decline using quantitative PCR (qPCR) analysis. *Plant Disease* 104, 1328–1334.

Bai, X., Correa, V.R., Toruño, T.Y., Ammar, E.-D., Kamoun, S., and Hogenhout, S.A. (2009) AY-WB phytoplasma secretes a protein that targets plant cell nuclei. *Molecular Plant–Microbe Interactions* 22, 18–30.

Baric, S., Berger, J., Cainelli, C., Kerschbamer, C., Letschka, T., and Dalla Via, J. (2011) Seasonal colonisation of apple trees by '*Candidatus* Phytoplasma mali' revealed by a new quantitative TaqMan real-time PCR approach. *European Journal of Plant Pathology* 129, 455–467.

Berges, R. and Seemüller, E. (2002) Impact of phytoplasma infection of common alder (*Alnus glutinosa*) depends on strain virulence. *Forest Pathology* 32, 357–363.

Bertaccini, A. (2022) Plants and phytoplasmas: When bacteria modify plants. *Plants* 11, 1425.

Braun, E.J. and Sinclair, W.A. (1976) Histopathology of phloem necrosis in *Ulmus americana*. *Phytopathology* 66, 598–607.

Braun, E.J. and Sinclair, W.A. (1978) Translocation in phloem necrosis-diseased American elm seedlings. *Phytopathology* 68, 1733–1737.

Carminati, G., Brusa, V., Loschi, A., Ermacora, P., and Martini, M. (2021) Spatiotemporal and quantitative monitoring of the fate of '*Candidatus* Phytoplasma solani' in tomato plants infected by grafting. *Pathogens* 10, 811.

Carraro, L., Ermacora, P., Loi, N., and Osler, R. (2004) The recovery phenomenon in apple proliferation-infected trees. *Journal of Plant Pathology* 86, 141–146.

Cettul, E. and Firrao, G. (2011) Development of phytoplasma-induced flower symptoms in *Arabidopsis thaliana*. *Physiological and Molecular Plant Pathology* 76, 204–211.

Choi, Y.H., Tapias, E.C., Kim, H.K., Lefeber, A.W.M., Erkelens, C. *et al.* (2004) Metabolic discrimination of *Catharanthus roseus* leaves infected by phytoplasma using ^{1}H-NMR spectroscopy and multivariate data analysis. *Plant Physiology* 135, 2398–2410.

Christensen, N.M., Nicolaisen, M., Hansen, M., and Schulz, A. (2004) Distribution of phytoplasmas in infected plants as revealed by real-time PCR and biomaging. *Molecular Plant–Microbe Interactions* 17, 1175–1184.

Christensen, N.M., Axelsen, K.B., Nicolaisen, M., and Schulz, A. (2005) Phytoplasmas and their interactions with hosts. *Trends in Plant Science* 10, 526–535.

Clark, M.F., Barbara, D.J., and Davies, D.L. (1983) Production and characteristics of antisera to *Spiroplasma citri* and clover phyllody-associated antigens derived from plants. *Annals of Applied Biology* 103, 251–259.

Constable, F.E., Gibb, K.S., and Symons, R.H. (2003) Seasonal distribution of phytoplasmas in Australian grapevines. *Plant Pathology* 52, 267–276.

Córdova, I., Oropeza, C., Puch-Hau, C., Harrison, N., Collí-Rodríguez, A. *et al.* (2014) A real-time PCR assay for detection of coconut lethal yellowing phytoplasmas of group 16SrIV subgroups A, D and E found in the Americas. *Journal of Plant Pathology* 96, 343–352.

De Silva, P.R., Perera, C.N., Bahder, B.W., and Attanayake, R.N. (2023) Nested PCR-based rapid detection of phytoplasma leaf wilt disease of coconut in Sri Lanka and systemic movement of the pathogen. *Pathogens* 12, 294.

Ding, Y., Wu, W., Wei, W., Davis, R.E., Lee I.-M. *et al.* (2013) Potato purple top phytoplasma-induced disruption of gibberellin homeostasis in tomato plants. *Annals of Applied Biology* 162, 131–139.

Douglas, S.M. (1986) Detection of mycoplasmalike organisms in peach and chokecherry with X-disease by fluorescence microscopy. *Phytopathology* 76, 784–787.

Endeshaw, S.T., Murolo, S., Romanazzi, G., and Neri, D. (2012) Effects of Bois noir on carbon assimilation, transpiration, stomatal conductance of leaves and yield of grapevine (*Vitis vinifera*) cv. Chardonnay. *Physiologia Plantarum* 145, 286–295.

Errea, P., Aguelo, V., and Hormaza, J.I. (2002) Seasonal variations in detection and transmission of pear decline phytoplasma. *Journal of Phytopathology* 150, 439–443.

Fernández, F.D., Yan, X.-H., Kuo, C.-H., Marcone, C., and Conci, L.R. (2024) Improving the comprehension of pathogenicity and phylogeny in '*Candidatus* Phytoplasma meliae' through genome characterization. *Microorganisms* 12, 142.

Freitag, J.H. (1964) Interaction and mutual suppression among three strains of aster yellows virus. *Virology* 24, 401–413.

Gai, Y.-P., Han, X.-J., Li, Y.Q., Yuan, C.-Z., Mo, Y.-Y. *et al.* (2014) Metabolomic analysis reveals the potential metabolites and pathogenesis involved in mulberry yellow dwarf disease. *Plant, Cell and Environment* 37, 1474–1490.

Gambino, G., Boccacci, P., Margaria, P., Palmano, S., and Gribaudo, I. (2013) Hydrogen peroxide accumulation and transcriptional changes in grapevines recovered from Flavescence dorée disease. *Phytopathology* 103, 776–784.

Garcia-Chapa, M., Medina, V., Viruel, M.A., Lavina, A., and Batlle, A. (2003) Seasonal detection of pear decline phytoplasma by nested-PCR in different pear cultivars. *Plant Pathology* 52, 513–520.

Guthrie, J.N., Walsh, K.B., Scott, P.T., and Rasmussen, T.S. (2001) The phytopathology of Australian papaya dieback: A proposed role for the phytoplasma. *Physiological and Molecular Plant Pathology* 58, 23–30.

Harper, S.J., Northfield, T.D., Nottingham, L.R., DuPont, S.T., Thompson, A.A. *et al.* (2023) Recovery plan for X-disease in stone fruit caused by '*Candidatus* Phytoplasma pruni'. *Plant Health Progress* 24, 258–295.

Himeno, M., Neriya, Y., Minato, N., Miura, C., Sugawara, K. *et al.* (2011) Unique morphological changes in plant pathogenic phytoplasma-infected petunia flowers are related to transcriptional regulation of floral homeotic genes in an organ-specific manner. *Plant Journal* 67, 971–979.

Hogenhout, S.A., Oshima, K., Ammar, E.-D., Kakizawa, S., Kingdom, H.N., and Namba, S. (2008) Phytoplasmas: Bacteria that manipulate plants and insects. *Molecular Plant Pathology* 9, 403–423.

Hoshi, A., Oshima, K., Kakizawa, S., Ishii, Y., Ozeki, J. *et al.* (2009) A unique virulence factor for proliferation and dwarfism in plants identified from a phytopathogenic bacterium. *Proceedings of the National Academy of Sciences of the USA* 106, 6416–6421.

Hren, M., Nikolić, P., Rotter, A., Blejec, A., Terrier, N. *et al.* (2009a) 'Bois noir' phytoplasma induces significant reprogramming of the leaf transcriptome in the field grown grapevine. *BMC Genomics* 10, 460.

Hren, M., Ravnikar, M., Brzin, J., Ermacora, P., Carraro, L. *et al.* (2009b) Induced expression of sucrose synthase and alcohol dehydrogenase I genes in phytoplasma-infected grapevine plants grown in the field. *Plant Pathology* 58, 170–180.

Ivanauskas, A., Inaba, J., Zhao, Y., Bottner-Parker, K.D., and Wei, W. (2024) Differential symptomology, susceptibility, and titer dynamics manifested by phytoplasma-infected periwinkle and tomato plants. *Plants* 13, 787.

Jarausch, W., Lansac, M., and Dosba, F. (1999) Seasonal colonization pattern of European stone fruit yellows phytoplasmas in different *Prunus* species detected by specific PCR. *Journal of Phytopathology* 147, 47–54.

Kartte, S. and Seemüller, E. (1991a) Histopathology of apple proliferation in *Malus* taxa and hybrids of different susceptibility. *Journal of Phytopathology* 131, 149–160.

Kartte, S. and Seemüller, E. (1991b) Susceptibility of grafted *Malus* taxa and hybrids to apple proliferation disease. *Journal of Phytopathology* 131, 137–148.

Kirkpatrick, B.C. (1991) Mycoplasma-like organisms – plant and invertebrate pathogens. In: Balows, A., Trüper, H.G., Dworkin, M., Harder, W., and Schleifer K.-S. (eds) *The Prokaryotes*, Vol. III. Springe, New York, pp. 4050–4067.

Kirkpatrick, B.C., Fisher, G.A., Fraser, J.D., and Purcell, A.H. (1990) Epidemiological and phylogenetic studies on western X-disease mycoplasma-like organisms. *Zentralblatt für Bakteriologie* 20, 289–296.

Kison, H. and Seemüller, E. (2001) Differences in strain virulence of the European stone fruit yellows phytoplasma and susceptibility of stone fruit trees on various rootstocks to this pathogen. *Journal of Phytopathology* 149, 533–541.

Kube, M., Mitrovic, J., Duduk, B., Rabus, R., and Seemüller, E. (2012) Current view on phytoplasma genomes and encoded metabolism. *Scientific World Journal* 2012, article 185942.

Kuske, C.R. and Kirkpatrick, B.C. (1992) Distribution and multiplication of western aster yellows mycoplasmalike organisms in *Catharanthus roseus* as determined by DNA hybridization analysis. *Phytopathology* 82, 457–462.

Landi, L. and Romanazzi, G. (2011) Seasonal variation of defense-related gene expression in leaves from Bois noir affected and recovered grapevines. *Journal of Agricultural and Food Chemistry* 59, 6628–6637.

Lederer, W. and Seemüller, E. (1991) Occurence of mycoplasma-like organisms in diseased and non-symptomatic alder trees (*Alnus* spp.). *European Journal of Forest Pathology* 21, 90–96.

Lee, I.-M., Davis, R.E., and Gundersen-Rindal, D.E. (2000) Phytoplasma: Phytopathogenic mollicutes. *Annual Review of Microbiology* 54, 221–255.

Lee, S., Kim, C.-E., and Cha, B. (2012) Migration and distribution of graft-inoculated jujube witches'-broom phytoplasma within a *Cantharanthus roseus* plant. *Plant Pathology Journal* 28, 191–196.

Lherminier, J., Courtois, M., and Caudwell, A. (1994) Determination of the distribution and multiplication sites of flavescence dorée mycoplasma-like organisms in the host plant *Vicia faba* by ELISA and immunocytochemistry. *Physiological and Molecular Plant Pathology* 45, 125–138.

León, R., Santamaría, J.M., Alpizar, L., Escamilla, J.A., and Oropeza, C. (1996) Physiological and biochemical changes in shoots of coconut palms affected by lethal yellowing. *New Phytologist* 134, 227–234.

Lepka, P., Stitt, M., Moll, E., and Seemüller, E. (1999) Effect of phytoplasmal infection on concentration and translocation of carbohydrates and amino acids in periwinkle and tobacco. *Physiological and Molecular Plant Pathology* 55, 59–68.

Maejima, K., Iwai, R., Himeno, M., Komatsu, K., Kitazawa, Y. *et al.* (2014) Recognition of floral homeotic MADS domain transcription factors by a phytoplasmal effector, phyllogen, induces phyllody. *Plant Journal* 78, 541–554.

MacLean, A.M., Sugio, A., Makarova, O.V., Findlay, K.C., Grieve, V.M. *et al.* (2011) Phytoplasma effector SAP54 induces indeterminate leaf-like flower development in *Arabidopsis* plants. *Plant Physiology* 157, 831–841.

Marcone, C. (2014) Molecular biology and pathogenicity of phytoplasmas. *Annals of Applied Biology* 165, 199–221.

Marcone, C., Neimark, H., Ragozzino, A., Lauer, U., and Seemuller, E. (1999) Chromosome sizes of phytoplasmas composing major phylogenetic groups and subgroups. *Phytopathology* 89, 805–810.

Marcone, C., Jarausch, B., and Jarausch, W. (2010) '*Candidatus* Phytoplasma prunorum', the causal agent of European stone fruit yellows: an overview. *Journal of Plant Pathology* 92, 19–34.

Marcone, C., Pierro, R., and Palmieri, C. (2024) Occurrence, impact, and multilocus sequence analysis of alder yellows phytoplasma infecting common alder and Italian alder in southern Italy. *Microorganisms* 12, 1140.

Maust, B.E, Espadas, F., Talavera, C., Aguilar, M., Santamaría, J.M., and Oropeza, C. (2003) Changes in carbohydrate metabolism in coconut palms infected with the lethal yellowing phytoplasma. *Phytopathology* 93, 976–981.

Mazraie, M.A, Izadpanah, K., Hamzehzarghani, H., Salehi, M., and Faghihi, M.M. (2019) Spread and colonization pattern of '*Candidatus* Phytoplasma aurantifolia' in lime plants [*Citrus aurantifolia* (Christm.) Swingle] as revealed by real-time PCR assay. *Journal of Plant Pathology* 101, 629–637.

McCoy, R.E., Caudwell, A., Chang, C.J., Chen, T.A., Chiykowski, L.N. *et al.* (1989) Plant diseases associated with mycoplasma-like organisms. In: Whitcomb, R.F. and Tully, J.G. (eds) *The Mycoplasmas*, Vol. V. Academic Press, San Diego, CA, pp. 545–640.

Morvan, G., Arnoux, M., and Castelain, C. (1986) Prospective for the control of apricot chlorotic leaf roll a mycoplasma disease, by cross protection. *Acta Horticulturae* 193, 359–366.

Mou, H.-Q., Lu, J., Zhu, S.-F., Lin, C.-L., Tian, G.-Z., Xu, X., and Zhao, W.-J. (2013) Transcriptomic analysis of *Paulownia* infected by paulownia witches'-broom phytoplasma. *PloS One* 8, e77217.

Musetti, R., Paolacci, A., Ciaffi, M., Tanzarella, O.A., Polizzotto, R. *et al.* (2010) Phloem cytochemical modification and gene expression following the recovery of apple plants from apple proliferation disease. *Phytopathology* 100, 390–399.

Neriya, Y., Maejima, K., Nijo, T., Tomomitsu, T., Yusa, A. *et al.* (2014) Onion yellow phytoplasma P38 protein plays a role in adhesion to the hosts. *FEMS Microbiology Letters* 361, 115–122.

Oropeza, C., Cordova, I., Chumba, A., Narváez, M., Sáenz, M., Ashburner, R., and Harrison, N. (2011) Phytoplasma distribution in coconut palms affected by lethal yellowing disease. *Annals of Applied Biology* 159, 109–117.

Osler, R., Borselli, S., Ermacora, P., Loschi, A., Martini, M., Musetti, R., and Loi, N. (2014) Acquired tolerance in apricot plants that stably recovered from European stone fruit yellows. *Plant Disease* 98, 492–496.

Pracros, P., Renaudin, J., Eveillard, S., Mouras, A., and Hernould, M. (2006) Tomato flower abnormalities induced by stolbur phytoplasma infection are associated with changes of expression of floral development genes. *Molecular Plant–Microbe Interactions* 19, 62–68.

Prezelj, N., Nikolić, P., Gruden, K., Ravnikar, M., and Dermastia, M. (2013) Spatiotemporal distribution of flavescence dorée phytoplasma in grapevine. *Plant Pathology* 62, 760–766.

Rosenberger, D.A. and Jones A.L. (1977) Seasonal variation in infectivity of inoculum from X-diseased peach and chokecherry plants. *Plant Disease Reporter* 61, 1022–1024.

Rossi, M., Vallino, M., Galetto, L., and Marzachì, C. (2020) Competitive exclusion of flavescence dorée phytoplasma strains in *Catharanthus roseus* plants. *Plants* 9, 1594.

Saracco, P., Bosco, D., Veratti, F., and Marzachì, C. (2006) Quantification over time of chrysanthemum yellows phytoplasma (16Sr-I) in leaves and roots of the host plant *Crysanthemum carinatum* (Schousboe) following inoculation with its insect vector. *Physiological and Molecular Plant Pathology* 67, 212–219.

Schaper, U. and Seemüller, E. (1982) Condition of the phloem and persistence of mycoplasmalike organisms associated with apple proliferation and pear decline. *Phytopathology* 72, 736–742.

Schaper, U. and Seemüller, E. (1984) Recolonization of the stem of apple proliferation and pear decline-diseased trees by the causal organisms in spring. *Zeitschrift für Pflanzenkrankheiten und Pflanzenschutz* 91, 608–613.

Seemüller, E., Kunze, L., and Schaper, U. (1984a) Colonization behavior of MLO, and symptom expression of proliferation-diseased apple trees and decline-diseased pear trees over a period of several years. *Zeitschrift für Pflanzenkrankheiten und Pflanzenschutz* 91, 525–532.

Seemüller, E., Schaper, U., and Zimbelmann, F. (1984b) Seasonal variation in the colonization patterns of mycoplasmalike organisms associated with apple proliferation and pear decline. *Zeitschrift für Pflanzenkrankheiten und Pflanzenschutz* 91, 371–382.

Seemüller, E., Schaper, U., and Kunze, L. (1986) Effect of pear decline on pear trees on 'Quince A' and *Pyrus communis* seedling rootstocks. *Zeitschrift für Pflanzenkrankheiten und Pflanzenschutz* 93, 44–50.

Seemüller, E., Stolz, H., and Kison, H. (1998) Persistence of the European stone fruit yellows phytoplasma in aerial parts of *Prunus* taxa during the dormant season. *Journal of Phytopathology* 146, 407–410.

Seemüller, E., Garnier, M., and Schneider, B. (2002) Mycoplasmas of plants and insects. In: Razin, S. and Hermann, R. (eds) *Molecular Biology and Pathogenicity of Mycoplasmas*. Kluwer Academic/Plenum Publishers, New York, pp. 91–115.

Seemüller, E., Moll, E., and Schneider, B. (2008) Apple proliferation resistance of *Malus sieboldii*-based rootstocks in comparison to rootstocks derived from other *Malus* species. *European Journal of Plant Pathology* 121, 109–119.

Semüller, E., Kiss, E., Sule, S., and Schneider, B. (2010) Multiple infection of apple trees by distinct strains of '*Candidatus* Phytoplasma mali' and its pathological relevance. *Phytopathology* 100, 863–870.

Seemüller, E., Carraro, L., Jarausch, W., and Schneider, B. (2011a) Apple proliferation phytoplasma. In: Hadidi, A., Barba, M., Candresse, T., and Jelkmann, W. (eds) *Virus and Virus-Like Diseases of Pome and Stone Fruits*. American Phytopathological Society (APS Press), St Paul, MN, pp. 67–73.

Seemüller, E., Kampmann, M., Kiss, E., and Schneider, B. (2011b) *HflB* gene-based phytopathogenic classification of '*Candidatus* Phytoplasma mali' strains and evidence that strain composition determines virulence in multiply infected apple trees. *Molecular Plant–Microbe Interactions* 24, 1258–1266.

Seemüller, E., Schneider, B., and Jarausch, B. (2011c) Pear decline phytoplasma. In: Hadidi, A., Barba, M., Candresse, T., and Jelkmann, W. (eds) *Virus and Virus-Like Diseases of Pome and Stone Fruits*. American Phytopathological Society (APS Press), St Paul, MN, pp. 77–84.

Seemüller, E., Zikeli, K., Furch, A.C.U., Wensing, A., and Jelkmann, W. (2018) Virulence of '*Candidatus* Phytoplasma mali' strains is closely linked to conserved substitutions in AAA+ ATPase AP460 and their supposed effect on enzyme function. *European Journal of Plant Pathology* 150, 701–711.

Siddique, A.B.M., Guthrie, J.N., Walsh, K.B., White, D.T., and Scott, P.T. (1998) Histopathology and within-plant distribution of the phytoplasma associated with Australian papaya dieback. *Plant Disease* 82, 1112–1120.

Sinclair, W.A. and Griffiths, H.M. (2000) Variation in aggressiveness of ash yellows phytoplasmas. *Plant Disease* 84, 282–288.

Singh, V., Kumar, S., and Lakhanpaul, S. (2018) Differential distribution of phytoplasma during phyllody progression in sesame (*Sesamum indicum* L.) under field conditions – An important consideration for effective sampling of diseased tissue. *Crop Protection* 110, 200–201.

Sugio, A., Kingdom, H.N., MacLean, A.M., Grieve, V.M., and Hogenhout, S.A. (2011a) Phytoplasma protein effector SAP11 enhances insect vector reproduction by manipulating plant development and defense hormone biosynthesis. *Proceedings of the National Academy of Sciences of the USA* 108, 4252–4257.

Sugio, A., MacLean, A.M., Kingdom, H.N., Grieve, V.M., Manimekalai, R., and Hogenhout, S.A. (2011b) Diverse targets of phytoplasma effectors: From plant development to defense against insects. *Annual Review of Phytopathology* 49, 175–195.

Suslow, K.G. and Purcell, A.H. (1982) Seasonal transmission of X-disease agent from cherry by leafhopper *Colladonus montanus*. *Plant Disease* 66, 28–30.

Tan, P.Y. and Whitlow, T. (2001) Physiological responses of *Catharanthus roseus* (periwinkle) to ash yellows phytoplasmal infection. *New Phytologist* 150, 757–769.

Tan, Y., Wei, H.R., Wang, J.W., Zong, X.J., Zhu, D.Z., and Liu, Q.Z. (2015) Phytoplasmas change the source-sink relationship of field-grown sweet cherry by disturbing leaf function. *Physiological and Molecular Plant Pathology* 92, 22–27.

Tomkins, M., Kliot, A., Marée, A.F.M., and Hogenhout, S.A. (2018) A multi-layered mechanistic modelling approach to understand how effector genes extend beyond phytoplasma to modulate plant hosts, insect vectors and the environment. *Current Opinion in Plant Biology* 44, 39–48.

Uyemoto, J.K. and Luhn, C.F. (2006) In-season variations in transmission of cherry X-phytoplasma and implication in certification programs. *Journal of Plant Pathology* 88, 317–320.

Uyemoto, J.K., Kirkpatrick, B.C., and Cummins, J.N. (1991) Susceptibility of selected cherry clones and related species to Western X-disease. *HortScience* 26, 1510–1511.

Vitale, M., Chitarra, W., Galetto, L., Bosco, D., Marzachì, C. *et al.* (2013) Flavescence dorée phytoplasma deregulates stomatal control of photosynthesis in *Vitis vinifera*. *Annals of Applied Biology* 162, 335–346.

Wang, H., Ye, X., Li, J., Tan, B., Chen, P. *et al.* (2018) Transcriptome profiling analysis revealed co-regulation of multiple pathways in jujube during infection by '*Candidatus* Phytoplasma ziziphi'. *Gene* 665, 82–95.

Wang, R., Bai, B., Li, D., Wang, J., Huang, W., Wu, Y., and Zhao, L. (2024) Phytoplasma: A plant pathogen that cannot be ignored in agricultural production – Research progress and outlook. *Molecular Plant Pathology* 25, e13437.

Wei, W. and Zhao, Y. (2022) Phytoplasma taxonomy: Nomenclature, classification, and identification. *Biology* 11, 1119.

Wei, W., Kakizawa, S., Suzuki, S., Jung, H.-Y., Nishigawa, H. *et al.* (2004) In planta dynamic analysis of onion yellows phytoplasma using localized inoculation by insect transmission. *Phytopathology* 94, 244–250.

Wei, W., Davis, R.E., Bauchan, G.R., and Zhao, Y. (2019) New symptoms identified in phytoplasma-infected plants reveal extra stages of pathogen-induced meristem fate-derailment. *Molecular Plant–Microbe Interactions* 32, 1314–1323.

Wei, W., Inaba, J., Zhao, Y., Mowery, J.D., and Hammond, R. (2022) Phytoplasma infection blocks starch breakdown and triggers chloroplast degradation, leading to premature leaf senescence, sucrose reallocation, and spatiotemporal redistribution of phytohormones. *International Journal of Molecular Sciences* 23, 1810.

Weintraub, P.G. and Beanland, L. (2006) Insect vectors of phytoplasmas. *Annual Review of Entomology* 51, 91–111.

Wright, A.A., Shires, M.K., Beaver, C., Bishop, G., DuPont, S.T., Naranjo, R., and Harper, S. (2021) Effect of '*Candidatus* Phytoplasma pruni' infection on sweet cherry fruit. *Phytopathology* 111, 2195–2202.

Wright, A.A., Shires, M., Molnar, C., Bishop, G., Johnson, A., Frias, C., and Harper, S.J. (2022) Titer and distribution of '*Candidatus* Phytoplasma pruni' in *Prunus avium*. *Phytopathology* 112, 1406–1412.

Wright, A.A., Molnar, C., Shires, M.K., Bishop, G., and Harper, S.J. (2023) Physiological and transcriptomic analysis of '*Candidatus* Phytoplasma pruni' infection in *Prunus persica*. *PhytoFrontiers™* 3, 323–336.

Zamharir, M.G., Mardi, M., Alavi, S.M., Hasanzadeh, N., Nekouei, M.K. *et al.* (2011) Identification of genes differentially expressed during interaction of Mexican lime tree infected with "*Candidatus* Phytoplasma aurantifolia". *BMC Microbiology* 11, 1.

Zhang, Y., Qiao, Z., Li, J., and Bertaccini, A. (2024) Paulownia witches' broom disease: A comprehensive review. *Microorganisms* 12, 885.

Zimmermann, M.R., Schneider, B., Mithöfer, A., Reichelt, M., Semüller, E., and Furch, A.C.U. (2015) Implications of *Candidatus* Phytoplasma mali infection on phloem function of apple trees. *Journal of Endocytobiosis and Cell Research* 26, 67–75.

17 Understanding Mixed Infections of Phytoplasmas with Diverse Plant Pathogens

Shao-Shuai Yu*
Chinese Academy of Tropical Agricultural Sciences, Wenchang, China

Abstract

Under natural conditions, plants can be simultaneously infected by phytoplasmas from different 16Sr groups or by various plant pathogens, including '*Candidatus* Liberibacter' species, viruses, spiroplasmas, '*Ca.* Phloemobacter' species, '*Ca.* Arsenophonus' species, and fungi. Mixed infections in plants caused by phytoplasmas and diverse pathogens lead to challenges and complexities in diagnosing and controlling the associated plant diseases. This chapter describes and summarizes the current research on mixed infections of phytoplasmas with various plant pathogens. The difficulties in diagnosing, monitoring, preventing, and controlling diseases related to mixed infections are analyzed and discussed. Strategies for monitoring and preventing plant diseases associated with mixed infections of phytoplasmas and different pathogens are suggested to reduce the agricultural impact of co-infection and ensure the health and safety of related germplasm resources.

Introduction

Phytoplasmas, previously known as mycoplasma-like organisms (MLOs), are a large group of cell wall-less, pleomorphic, and phloem limited pathogenic prokaryotes that cannot be cultured *in vitro* (Doi *et al.*, 1967; Yu *et al.*, 2016; Hemmati *et al.*, 2021). Under natural conditions, prokaryotic pathogens can infect nearly 1000 plant species and cause serious economic losses around the world (Yu *et al.*, 2016). The pathogens were originally discovered by Doi *et al.* in 1967. In 1994, the term "phytoplasma" was introduced at the 10th Congress of the International Organization of Mycoplasmology, replacing the term MLO. Phytoplasmas were proposed to be within the novel genus '*Candidatus* Phytoplasma' in 2004 (IRPCM, 2004). Additionally, the description guidelines of '*Ca.* Phytoplasma' species were revised in 2022 (Bertaccini *et al.*, 2022). Phytoplasmas are transmitted by insects, parasitic plants, and grafting, manipulating the affected host plants and insects by secreting pathogenic effectors (Sugio *et al.*, 2011; Bertaccini *et al.*, 2014; Mori *et al.*, 2020; Ramos *et al.*, 2020; Wei *et al.*, 2021). After being infected by phytoplasmas, host plants undergo a series of related pathological changes, such as the spatio-temporal redistribution of phytohormones and degradation of chloroplasts. These changes lead to related plant diseases characterized by abnormal symptoms, such as witches'-broom, virescence, phyllody, cluster top, internode shortening, yellowing of leaflets, dwarfing, crinkling, decline

*Corresponding author: yushaoshuai@catas.cn

 Phytoplasmas: Genomes, Plant Hosts and Vectors, 2nd Edition. (Ed. W. Wei)
DOI: 10.1079/9781800627031.0017

in plant growth, and even death (Sugio *et al.*, 2011; Bertaccini *et al.*, 2014; Ahmed *et al.*, 2022; Wei *et al.*, 2022).

In the fields of medicine, aquaculture, and agriculture, mixed infections caused by different pathogens lead to difficulties and complexities in the diagnosis and control of diseases in humans, animals, and plants (Kohda *et al.*, 2021; Paul *et al.*, 2022; Xu *et al.*, 2022). The structure and functions of human skin are influenced by the colonization of multiple bacterial species and their interspecific interactions, particularly between bacteria such as *Staphylococcus epidermidis* and *S. aureus* (Byrd *et al.*, 2017; Chen *et al.*, 2018; Kohda *et al.*, 2021). Severe physiological distress in fish in aquaculture can be caused by co-infection with bacteria, viruses, and other parasites (Manivannan *et al.*, 2002; Swaminathan *et al.*, 2016; Thamizhvanan *et al.*, 2019; Sahoo *et al.*, 2020; Paul *et al.*, 2022). In agriculture, co-infection with pathogens exacerbates plant disease conditions, creating challenges for monitoring and early warning, as well as prevention and control management of the related diseases (Galvão *et al.*, 2021; Luis-Pantoja *et al.*, 2021; Salehi *et al.*, 2022; Xu *et al.*, 2022; Yu *et al.*, 2022b; Behrmann *et al.*, 2023; Lee and Han, 2023; Mall and Vishwakarma, 2023).

Under natural conditions, plants can be simultaneously infected by phytoplasmas belonging to different 16Sr groups or by phytoplasmas along with other pathogens such as '*Ca.* Liberibacter' species, viruses, and spiroplasmas, as shown in Table 17.1. Phytoplasmas, which cannot be cultured *in vitro*, are cell wall-less, phloem-limited, prokaryotic pathogens. Mixed infections with strains belonging to different 16Sr taxonomic groups or with other pathogens tend to occur frequently (IRPCM, 2004; Teixeira *et al.*, 2008; Chen *et al.*, 2009; Lee *et al.*, 2009; Sun *et al.*, 2013; Arratia-Castro *et al.*, 2014; Bertaccini *et al.*, 2014; Galvão *et al.*, 2021; Luis-Pantoja *et al.*, 2021; Salehi *et al.*, 2022; Yu *et al.*, 2022b; Behrmann *et al.*, 2023; Duduk *et al.*, 2023; Lee and Han, 2023). Co-infections of phytoplasmas belonging to different 16Sr groups, as well as interactions among various pathogens, exacerbate the development of diseases associated with phytoplasmas, '*Ca.* Liberibacter' species, viruses, and others, thereby increasing the complexity of detecting and monitoring related diseases. Therefore, it is crucial to clarify the types and characteristics of the pathogens involved in mixed infections for the diagnosis, early warning, prevention, and control of related diseases.

Mixed Infections of Phytoplasmas with '*Ca.* Liberibacter' Species

'*Ca.* Phytoplasma' and '*Ca.* Liberibacter' species are phloem-limited and unculturable bacteria that are transmitted by insects with piercing–sucking mouthparts (Doi *et al.*, 1967; Liefting *et al.*, 2008; Satta *et al.*, 2016; Yu *et al.*, 2016; Arce-Leal *et al.*, 2020; Sumner-Kalkun *et al.*, 2020; Yang *et al.*, 2020; Bao *et al.*, 2021; Quiroga *et al.*, 2022). '*Ca.* Liberibacter' species are associated with economically important plant diseases, such as huanglongbing (HLB) in citrus plants and zebra chip disease in potato tubers (Liefting *et al.*, 2008; Munyaneza *et al.*, 2010; EPPO, 2012; Satta *et al.*, 2016; Arce-Leal *et al.*, 2020; Sumner-Kalkun *et al.*, 2020; Yang *et al.*, 2020; Bao *et al.*, 2021; Quiroga *et al.*, 2022). There are reports that plants can be mixed-infected with phytoplasmas and '*Ca.* Liberibacter' species (Teixeira *et al.*, 2008; Chen *et al.*, 2009; Arratia-Castro *et al.*, 2014; Satta *et al.*, 2016; Alizadeh *et al.*, 2017; Yu *et al.*, 2022b), as shown in Table 17.1. HLB is among the most destructive diseases affecting citrus plants, including *Citrus maxima*, *C. sinensis*, *C. aurantifolia*, and *Citrus* × *limon* (Teixeira *et al.*, 2008; Chen *et al.*, 2009; Arce-Leal *et al.*, 2020; Yang *et al.*, 2020). This disease impacts citrus production in China and around the world (Arce-Leal *et al.*, 2020; Li *et al.*, 2020; Bao *et al.*, 2021; Dorado *et al.*, 2021; Nan *et al.*, 2021; Quiroga *et al.*, 2022). Based on presumptive geographic origins (Asian, African, and American continents, respectively) and 16S rRNA gene-based molecular classification, three distinct species, '*Ca.* Liberibacter asiaticus,' '*Ca.* Liberibacter africanus,' and '*Ca.* Liberibacter americanus,' have been reported to be associated with citrus HLB (Arce-Leal *et al.*, 2020; Yang *et al.*, 2020; Bao *et al.*, 2021; Quiroga *et al.*, 2022). *C. maxima* is an important fruit tree in Hainan Province, a tropical island in China. The plants exhibiting disease symptoms, including yellowing, crinkled leaves, and mottled leaves, were found on the island, causing serious economic losses and affecting the local citrus industry.

Table 17.1. Information on representative mixed infections of phytoplasmas affiliated with different 16Sr groups or with other pathogens.

Pathogens in mixed infection						
Kinds	Name	Phytoplasma	Hosts	Symptoms	Study location	Reference
Phytoplasma	16SrV	16SrI	Jujube	Witches'-broom	Korea	Lee *et al.*, 2009
	16SrV	16SrI	Jujube	Witches'-broom	Henan, China	Sun *et al.*, 2013
	16SrXXXII	16SrI	*Elaeocarpus sylvestris*	Decline	Jeju Island, Korea	Lee & Han, 2023
'*Ca*. Liberibacter' spp.	'*Ca*. Liberibacter asiaticus'	16SrII-V	*Citrus maxima*	Huanglongbing including yellowing, crinkled leaves, mottled leaves	Hainan, China	Yu *et al.*, 2022b
	'*Ca*. Liberibacter asiaticus'	16SrII	*Citrus sinensis*	Huanglongbing	Iran	Alizadeh *et al.*, 2017
	'*Ca*. Liberibacter asiaticus'	16SrI	Citrus plants	Huanglongbing	Guangdong, China	Chen *et al.*, 2009
	'*Ca*. Liberibacter asiaticus'	16SrI	Citrus plants	Huanglongbing	Mexico	Arratia-Castro *et al.*, 2014
	'*Ca*. Liberibacter asiaticus'	16SrIX	Citrus plants	Huanglongbing	Brazil	Teixeira *et al.*, 2008
	'*Ca*. Liberibacter asiaticus'	16SrI, 16SrII, 16SrIV, 16SrV, 16SrVII, 16SrX, 16SrXI, 16SrXII	*Citrus latifolia*, *C. paradisi*, *C. sinensis*, *C. reticulate* × *C. sinensis*	Huanglongbing	Caribbean	Luis-Pantoja *et al.*, 2021
	'*Ca*. Liberibacter solanacearum'	16SrI	Carrot	Leaf curling with yellowing, twisting of petioles, stunted growth of shoots and roots	Spain	Satta *et al.*, 2016

Continued

Table 17.1. Continued

Pathogens in mixed infection						
Kinds	Name	Phytoplasma	Hosts	Symptoms	Study location	Reference
Virus	Cucumber mosaic virus	16SrII	*Solanum melongena*	Mosaic, mottling, small leaves	India	Abirami *et al.*, 2022
	Begomovirus	16SrVI	*Solanum melongena*	Mosaic, small leaves	India	Venkataravanappa *et al.*, 2018
	Potato virus X, potato virus Y	16SrVI	*Solanum melongena*	Small leaves, leaf chlorosis and malformations with mosaic mottling	India	Kumar *et al.*, 2016
	Ageratum enation virus	16SrI	*Capsicum annuum*	Yellowing, leaf curling, small leaves	India	Tiwari *et al.*, 2022
	Banana bunchy top virus, banana streak MY virus	16SrI-D	Banana plants	Bunchy top, small leaves, leaf chlorosis, necrosis and stunted growth	India	Mitra *et al.*, 2022
	Sugarcane yellow leaf virus	16SrXI	Sugarcane	Yellow, grassy shoot	India	Nithya *et al.*, 2020
	Sugarcane yellow leaf virus	16SrI	Sugarcane	Yellowing	Egypt	El Sayed *et al.*, 2016
	Sugarcane yellow leaf virus	16SrI-A	Sugarcane	Yellow discoloration of the midrib	Cuba	Arocha *et al.*, 1999
	Papaya ringspot virus	16SrII	*Carica papaya*	Mosaic, yellowing, crinkling	Cuba	Arocha *et al.*, 2009
	Mungbean yellow mosaic India virus	16SrI	*Glycine max*	Small leaf, yellowing, mosaic, witches'-broom	India	Kumar *et al.*, 2015
	Tomato yellow leaf curl virus, potato virus Y	16SrI, 16SrV, 16SrXII	Tomato	Witches'-broom, stunting, yellowing	Mauritius	Gungoosingh-Bunwaree *et al.*, 2013
	Blueberry ringspot virus	16SrI-E	Blueberry	Yellowing, witches'-broom	Canada	Arocha Rosete *et al.*, 2019

Spiroplasma	Spiroplasma	Phytoplasma	*Dalbulus maidis*	ND	Brazil	Oliveira *et al*., 2015
	Spiroplasma	16SrIV	Palm species	Yellowing	Mexico	Lebsky *et al*., 2019
	Spiroplasma	Phytoplasma	Corn (maize)	Stunting	Brazil	Galvão *et al*., 2021
	Spiroplasma citri	16SrII-C, 16SrII-D, 16SrVI-A, 16SrIX-C	Sesame	Phyllody	Iran	Salehi *et al*., 2022
'*Ca.* Phloemobacter' species	'*Ca.* Phloemobacter fragariae'	Stolbur phytoplasma	Sugar beet	"Basses richesses" syndrome	Europe	Gatineau *et al*., 2002
	'*Ca.* Phloemobacter fragariae'	Stolbur phytoplasma	Strawberry	Marginal chlorosis	France	Danet *et al*., 2003
'*Ca.* Arsenophonus' species	'*Ca.* Arsenophonus phytopathogenicus'	'*Ca.* Phytoplasma solani'	Sugar beet	"Basses richesses" syndrome	Germany	Duduk *et al*., 2023
	'*Ca.* Arsenophonus phytopathogenicus'	'*Ca.* Phytoplasma solani'	Beet, potato plants	"Basses richesses" syndrome	Germany	Behrmann *et al*., 2023
	'*Ca.* Arsenophonus phytopathogenicus'	'*Ca.* Phytoplasma solani'	*Pentastiridius leporinus*	ND	Germany	Behrmann *et al*., 2023
Fungus	*Macrophomina phaseolina*	'*Ca.* Phytoplasma solani'	Sugar beet	Rubbery taproot, charcoal root rot	Central Europe	Duduk *et al*., 2023

ND, not determined.

The isease symptoms resemble those of citrus HLB. Several plant diseases associated with phytoplasmas have been reported previously in the same regions (Yu *et al.*, 2021a, 2021b, 2021c, 2021d, 2022a, 2022b). Therefore, plants exhibiting these disease symptoms were suspected of having been infected with phytoplasma pathogens and '*Ca.* Liberibacter asiaticus.'

Yu *et al.* (2022b) investigated the causal agents associated with disease symptoms in *C. maxima*, including yellowing and leaf crinkling. Their study identified two distinct phytoplasma strains, which belong to the 16SrII and 16SrXXXII groups, in addition to '*Ca.* Liberibacter asiaticus' present in symptomatic plants. Sequence analysis confirmed that these pathogens were absent in healthy plants, indicating a strong association with the observed disease symptoms. Among 54 symptomatic samples, 22.2% tested positive for 16SrII phytoplasma, 3.7% for 16SrXXXII, and 11.1% for '*Ca.* Liberibacter asiaticus.' Additionally, 7.4% of the samples exhibited mixed infections involving both phytoplasma and '*Ca.* Liberibacter asiaticus.' These findings highlight the complexity of *C. maxima* disease etiology and suggest that mixed infections may contribute to more severe or variable symptoms. The study underscores the importance of considering multiple pathogens when diagnosing and managing citrus diseases.

Yu *et al.* (2022b) further examined the genetic diversity of phytoplasmas and '*Ca.* Liberibacter asiaticus' infecting symptomatic *C. maxima* plants in China. Their analysis revealed that phytoplasma strains belonging to two ribosomal subgroups (16SrII-V and 16SrXXXII-D) were closely related to previously reported strains infecting various plant hosts in Asia. These included associations with species such as *Tephrosia purpurea*, *Melochia corchorifolia*, and *Trema tomentosa*, with sequence identities reaching 100%. Similarly, the '*Ca.* Liberibacter asiaticus' strains found in *C. maxima* shared complete sequence homology with reference strains (CP010804) from GenBank, indicating genetic stability across geographic regions. Phylogenetic analysis confirmed that the 16SrII-V and 16SrXXXII-D phytoplasma strains clustered with related phytoplasmas from multiple plant hosts across South-east Asia, while the '*Ca.* Liberibacter asiaticus' strains formed a well-supported, distinct clade consistent with global reference sequences. These results reinforce the view that *C. maxima* is susceptible to genetically diverse strains of phytoplasmas, which can co-occur with '*Ca.* Liberibacter asiaticus' in the same host.

The occurrence of mixed infections in citrus is not unique to China. Similar associations have been reported in *C. sinensis* in Iran (Alizadeh *et al.*, 2017), as well as in citrus species in Brazil, Mexico, and the Caribbean (Teixeira *et al.*, 2008; Chen *et al.*, 2009; Arratia-Castro *et al.*, 2014; Luis-Pantoja *et al.*, 2021). In these cases, '*Ca.* Liberibacter asiaticus' has been found co-infecting citrus alongside phytoplasmas from multiple ribosomal groups, including 16SrI, 16SrIV, and 16SrIX, further confirming the widespread nature of such interactions.

The findings from Yu *et al.* (2022b) contribute to a growing body of evidence showing that mixed infections involving phytoplasmas and '*Ca.* Liberibacter' species are a significant factor in citrus decline. The presence of diverse and coexisting pathogens complicates both diagnosis and management, particularly in tropical regions where vector populations are active year-round. Comprehensive pathogen characterization is therefore crucial for effective epidemic surveillance and control strategies (details are provided later in this chapter).

Additionally, the broader relevance of '*Ca.* Liberibacter' species as emerging plant pathogens is supported by other studies reporting '*Ca.* Liberibacter solanacearum,' a close relative of '*Ca.* Liberibacter asiaticus,' as a causal agent in solanaceous and apiaceous crops in Europe and New Zealand (Liefting *et al.*, 2008; Munyaneza *et al.*, 2010; Alfaro-Fernández *et al.*, 2012; EPPO, 2012; Satta *et al.*, 2016; Sumner-Kalkun *et al.*, 2020). In these cases, too, mixed infections involving phytoplasmas and '*Ca.* Liberibacter' species have been associated with overlapping symptoms, reinforcing the global importance of integrating molecular diagnostics into plant disease management frameworks.

Mixed Infection of Phytoplasmas Belonging to Different 16Sr Groups

Plants can be co-infected by phytoplasmas belonging to different 16Sr groups (Table 17.1). Based on 16S rRNA gene fragments amplified using primers P1/P7 and R16F2n/R16R2, jujube trees showing symptoms of typical witches'-broom like yellowing, small leaves, short internodes, and proliferation of shoots were identified as being co-infected by 16SrI and 16SrV phytoplasmas in Korea (Lee *et al.*, 2009). In 2013, a mixed infection involving 16SrI and 16SrV phytoplasma groups was identified in a single jujube tree in Henan Province, China, based on the 16S rRNA gene sequence analysis (Sun *et al.*, 2013). Lee and Han (2023) reported the presence of *Elaeocarpus sylvestris* trees with decline diseases in Jeju Island, South Korea that were co-infected by phytoplasmas belonging to 16SrI and 16SrXXXII groups based on conserved sequence fragments of 16S rRNA, *rp* operon, *tuf*, and *secA* genes.

This ability of phytoplasmas to infect multiple plant hosts facilitates their spread across different species. In China, both paulownia and jujube are widely distributed plants and susceptible to witches'-broom diseases caused by 16SrI and 16SrV phytoplasma groups, respectively (Wang *et al.*, 2010; Sun *et al.*, 2013). Mixed infections of 16SrI and 16SrV phytoplasma groups in jujube trees suggest that 16SrI phytoplasmas may be transmitted by insect vectors, such as leafhoppers, between paulownia and jujube trees (Wang *et al.*, 2010; Sun *et al.*, 2013).

Mixed Infection of Phytoplasmas with Viruses

Viruses are pathogens that cause diseases in plants, animals, and bacteria. Plant diseases caused by viruses pose a significant threat to global agricultural production. Under natural conditions, plants frequently encounter multiple viruses and pathogens, resulting in common mixed infections (Singhal *et al.*, 2021; Xu *et al.*, 2022). Cassava mosaic disease is associated with a mixed infection of African cassava mosaic virus (ACMV) and the Uganda strain of East African cassava mosaic virus (EACMV) (Legg and Fauquet, 2004). Potato rugose mosaic disease is caused by the synergistic combination of potato virus X (PVX) and potato virus Y (PVY), and crinkle disease by PVX along with potato virus A (PVA) (Jones and Barbetti, 2012). Tomato streak disease is caused by a mixed infection of PVX and tobacco mosaic virus (TMV) (Jones and Barbetti, 2012). Corn (maize) can be co-infected by corn chlorotic mottle virus and PVY, leading to the devastating disease known as maize lethal necrosis (Wangai *et al.*, 2012; Redinbaugh and Stewart, 2018).

In addition to viral co-infection, viruses can also infect plants along with other pathogens like phytoplasma, causing severe losses of affected plants (Kumar *et al.*, 2016; Venkataravanappa *et al.*, 2018; Abirami *et al.*, 2022; Mitra *et al.*, 2022; Tiwari *et al.*, 2022; Mall and Vishwakarma, 2023), as listed in Table 17.1. In India, disease symptoms of mosaic, mottling, and small leaves in *Solanum melongena* were associated with mixed infections, including cucumber mosaic virus (CMV) with 16SrII phytoplasma (Abirami *et al.*, 2022), begomovirus with 16SrVI phytoplasma (Venkataravanappa *et al.*, 2018), as well as PVX, PVY, and 16SrVI phytoplasma (Kumar *et al.*, 2016). In sugarcane, the disease symptom of yellowing was caused by sugarcane yellow leaf virus (ScYLV) along with phytoplasmas belonging to the 16SrI or 16SrXI group (Arocha *et al.*, 1999; El Sayed *et al.*, 2016; Nithya *et al.*, 2020). Similarly, *Carica papaya* can be co-infected by the papaya ringspot virus (PRSV) and phytoplasmas affiliated with the 16SrII group in Cuba (Arocha *et al.*, 2009). Banana plants can be co-infected by banana bunchy top virus (BBTV), banana streak MY virus (BSMYV), along with phytoplasmas belonging to the 16SrI-D subgroup (Mitra *et al.*, 2022). Mixed infection also affects *Capsicum annuum*, with ageratum enation virus and the 16SrI phytoplasma, resulting in plant diseases characterized by symptoms such as yellowing, leaf curling, and small leaves (Tiwari *et al.*, 2022). The disease symptoms of plants caused by phytoplasmas, or viruses, are very similar and difficult to distinguish, highlighting the importance of determining whether a plant disease results from a single pathogen or a mixed infection for effective disease management.

Mixed Infection and Interaction of Phytoplasmas with Other Pathogens

Spiroplasmas, belonging to the class Mollicutes, are cell wall-deficient helical bacteria associated with corn stunt diseases (Davis and Worley, 1973; Schmitt *et al.*, 1984; Yu *et al.*, 2009;

Harne *et al.*, 2020; Galvão *et al.*, 2021). Many studies demonstrated that spiroplasma can also be detected in plants with symptoms associated with phytoplasma (Oliveira *et al.*, 2015; Lebsky *et al.*, 2019; Galvão *et al.*, 2021; Salehi *et al.*, 2022), as shown in Table 17.1. For example, Oliveira *et al.* (2015) found that phytoplasma and spiroplasma could be simultaneously transmitted by *Dalbulus maidis* leafhopper to corn plants. In palm species infected with the lethal yellowing phytoplasma belonging to the 16SrIV group in Mexico, spiroplasmas were detected by scanning electron microscopy and nested PCR (Gundersen and Lee, 1996; Lebsky *et al.*, 2019). Galvão *et al.* (2021) also reported that corn stunting was associated with mixed infection of phytoplasma and spiroplasma in Brazil. Based on molecular analyses with specific PCR primers, mixed infection of *Spiroplasma citri* and phytoplasmas belonging to 16SrII-C, 16SrII-D, 16SrVI-A, and 16SrIX-C subgroups was identified in sesame plants showing phyllody disease symptoms in Iran (Salehi *et al.*, 2022).

In addition to co-infection with '*Ca.* Liberibacter' species, viruses, and spiroplasmas, phytoplasmas can also cause plant disease symptoms similar to those caused by the phloem-restricted and unculturable pathogens of '*Ca.* Phloemobacter fragariae' and '*Ca.* Arsenophonus phytopathogenicus' (Table 17.1). The "basses richesses" syndrome of sugar beet, first described in 1991 in France, is associated with a stolbur phytoplasma and '*Ca.* Phloemobacter fragariae' (Gatineau *et al.*, 2002). Marginal chlorosis of strawberries, which has influenced strawberry production in France for more than 35 years, was associated with a color phytoplasma and '*Ca.* Phloemobacter fragariae' (Danet *et al.*, 2003). In 2023, two phloem-restricted bacterial pathogens, the stolbur phytoplasma, '*Ca.* Phytoplasma solani' and the γ-proteobacterium '*Ca.* Arsenophonus phytopathogenicus,' were found in sugar beet and potato crops in Germany (Behrmann *et al.*, 2023; Duduk *et al.*, 2023). Although the symptoms of plant diseases caused by the phytoplasma 'Ca. Phloemobacter fragariae' or other pathogens are particularly similar, the pathogenic consequences of these pathogens are relatively different. For example, phytoplasma has been associated with cell necrosis and cell wall lignification, while '*Ca.* Phloemobacter fragariae' has also been linked to these same abnormal symptoms as well as deposits of phenolic compounds in the lumen of phloem cells (Gatineau *et al.*, 2002).

Phytoplasma can not only co-infect plants with prokaryotic pathogens, but also can co-infect plants with eukaryotic pathogens like fungus. Sugar beet can be co-infected by phytoplasmas and fungal pathogens (Duduk *et al.*, 2023). When infected by '*Ca.* Phytoplasma solani' and *Macrophomina phaseolina*, sugar beets develop a rubbery taproot and charcoal root rot, respectively. Plants affected by rubbery taproot become more susceptible to charcoal root rot, which increases crop losses associated with phytoplasmas (Duduk *et al.*, 2023). This situation illustrates that mixed infections involving different 16Sr groups or with other different prokaryotic and eukaryotic pathogens require varied diagnostic methods, prevention, and control strategies.

Diversity of the Pathogens and Their Influences on Mixed Infections

Within certain geographic regions with diverse plant pathogens, mixed infections are common. For example, in Hainan Province, China, many plant pathogens, including phytoplasmas from different 16Sr groups and '*Ca.* Liberibacter asiaticus,' have been identified in plants exhibiting similar and distinct abnormal symptoms (Che *et al.*, 2024). in Hainan, periwinkles (*Catharanthus roseus*) infected with 16SrI-B phytoplasma show little leaf, phyllody, and virescence, whereas periwinkles infected with 16SrV group phytoplasma display yellowing symptoms. Similarly, *Melochia corchorifolia* shows phyllody when infected with 16SrI-B phytoplasma and witches'-broom when infected with 16SrII-A phytoplasma. Other examples include *Carica papaya* infected by 16SrI-AP, 16SrI-B, and 16SrII-U phytoplasma, showing malformation, yellowing, and little leaf, respectively. *Praxelis clematidea* infected with 16SrII-A and16SrII-V phytoplasmas shows witches'-broom and phyllody, respectively. However, rubber trees (*Hevea brasiliensis*), whether infected by 16SrI or 16srII group phytoplasmas, singly or mixed, showed stem fasciation (Yang *et al.* 2016). The same symptoms in the same plant species might be caused by infection by two different phytoplasmas. In China, sugarcane white leaf disease is one of the most severe diseases that threaten sugarcane production in Yunnan and Hainan. However, the pathogens that cause the disease differ: 16SrII in Hainan and 16SrXI in Yunnan (Li *et al.*, 2014). Additionally, in *Citrus maxima*, symptoms such as yellowing and mottled leaves have been associated with mixed infections by 16SrII-V phytoplasma, 16SrXXXII-D phytoplasma, and '*Ca.* Liberibacter asiaticus' (Yu *et al.*, 2022b).

The same plant can be infected by phytoplasmas belonging to different 16Sr groups, whereas some plants can be infected by different types of plant pathogens (Yu *et al.*, 2022b, 2024; Che *et al.*, 2024). Although there have been no reports of co-infection in some plants, the diversity of host species suggests a potential risk of co-infection, especially in areas with diverse pathogen reservoirs, including weeds and wild plants. These reservoir plants play an important role in the epidemiology and emergence of phytoplasma diseases, therefore identification of weeds that may transmit phytoplasmas is vital for understanding pathosystems and improving disease management (Duduk *et al.*, 2018; Fernández *et al.*, 2020; Banzato *et al.*, 2021). A phytoplasma was detected in *Richardia scabra*, a weed found in a sugarcane field affected by white leaf disease. The similarity of 16S rRNA between sugarcane white leaf phytoplasma strain SCWL-DZCM and *R. scabra* phyllody phytoplasma strain RiSP-WZS was 99.7% (Che *et al.*, 2024). Furthermore, the phytoplasma strains infecting cassava identified in China, Vietnam, and Thailand shared >99.3% homology; and the phytoplasma strains infecting *Piper nigrum* shared 99.8% homology with the phytoplasma strains infecting *Carica papaya*, *Melia azedarach*, *Capsicum annuum*, *Pericampylus glaucus*, *Catharanthus roseus*, and *Malvastrum coromandelianum*, which were all identified on Hainan Island (Che *et al.*, 2024). The phytoplasma strains might be transmitted among the plants that can be infected by pathogens sharing comparatively higher levels of homology (Che *et al.*, 2024).

In the same geographic region, comparatively higher-level diversity of plant pathogens and their plant hosts leads to overlapping plant hosts infected by different pathogens, resulting in the occurrence of mixed infection (Yu *et al.*, 2024). For example, 16SrXXXII group phytoplasma are mainly distributed in Asian regions. Phytoplasma strains belonging to the 16SrXXXII-A, 16SrXXXII-B, and 16SrXXXII-C subgroups are only distributed in the South-east Asian region, including Malaysia (Nejat *et al.*, 2009, 2013). The phytoplasma strains belonging to the 16SrXXXII-D, 16SrXXXII-E, and 16SrXXXII-F subgroups are only distributed in the East Asian region, including China, Japan, and South Korea (Iwabuchi *et al.*, 2018; Wang *et al.*, 2021; Yu *et al.*, 2021b, 2022b; Lee *et al.*, 2022; Lee and Han, 2023; Zhu *et al.*, 2023). Nejat *et al.* (2009, 2013) identified and proposed the novel taxon '*Ca.* Phytoplasma malaysianum' belonging to the 16SrXXXII group including 16SrXXXII-A, 16SrXXXII-B, and 16SrXXXII-C subgroups in Malaysia; Yu *et al.* (2021b, 2024) identified and proposed 16SrXXXII-D and 16SrXXXII-F subgroups in 16SrXXXII group in China; and Lee *et al.* (2022; Lee and Han, 2023) identified and proposed 16SrXXXII-E subgroup in the 16SrXXXII group in South Korea. The phytoplasmas belonging to the 16SrXXXII group can infect plants causing plant diseases such as *Trema tomentosa* witches'-broom, *Citrus maxima* yellow leaf, *Camptotheca acuminate* witches'-broom, *Elaeocarpus sylvestris* decline, and *Areca catechu* palm yellow leaf disease in China (Wang *et al.*, 2021; Yu *et al.*, 2021b, 2022b; Zhu *et al.*, 2023); periwinkle virescence and phyllody, and yellow dwarf coconut and oil palm in Malaysia (Nejat *et al.*, 2009, 2013); *Elaeocarpus sylvestris* decline in South Korea (Lee *et al.*, 2022; Lee and Han, 2023) as well as *E. zollingeri* yellows in Japan (Iwabuchi *et al.*, 2018). Co-infection of phytoplasmas belonging to 16SrXXXII and the other 16Sr groups have been reported (Lee and Han, 2023). For example, *E. sylvestris* trees with decline diseases were mixed infected by phytoplasmas belonging to 16SrXXXII and 16SrI groups in Jeju Island of South Korea (Lee and Han, 2023). Currently, apart from the co-infection of plants with 16SrXXXII and 16SrI phytoplasma, there is no evidence of co-infection of 16SrXXXII phytoplasma with phytoplasmas belonging to the other 16Sr groups in plants. However, the diversity of phytoplasma and their plant hosts in Asian regions is abundant. In particular, there are some plants that are co-hosts of phytoplasmas affiliated with different 16Sr groups within a limited geographic region (Yu *et al.*, 2024). Therefore, further attention to the mixed infection of these pathogens is still needed. Given the potential for mixed infections of phytoplasmas from different 16Sr groups or with other pathogens, it is necessary to conduct simultaneous inspection and quarantine of host plants (such as *Areca catechu* and *C. maxima*) and their related products during transportation across regions. This approach will help prevent the spread of phytoplasma and other pathogens as well as their related plant diseases.

Monitoring and Control of the Plant Diseases Associated with Mixed Infections

A plant can be simultaneously infected by different pathogens. Therefore, in disease diagnosis and pathogen detection, testing should include all the potential pathogens a host may carry, rather than focusing on a single pathogen. For example, HLB diseases on citrus crops were associated with '*Ca.* Liberibacter' species; however, citrus crops showing symptoms of HLB diseases can also carry phytoplasmas. Therefore, both '*Ca.* Liberibacter' species and phytoplasma should be looked for in symptomatic citrus crops. Citrus crops and their products should undergo comprehensive testing for both pathogens during transportation to prevent the spread and the epidemiology of their related diseases.

The interactions among pathogens within a co-infected host remain to be elucidated. As demonstrated in previous research (Yu *et al.*, 2022b), phytoplasmas belonging to the 16SrII-V subgroup and '*Ca.* Liberibacter asiaticus' strains were simultaneously detected in *C. maxima* with HLB disease symptoms. However, the relationships among the different 16Sr group phytoplasmas or between the phytoplasma and '*Ca.* Liberibacter asiaticus' within the diseased plant remain unknown. Therefore, to better understand the related diseases, it is essential to clarify the diversity of pathogens, their main and secondary effects, and their roles in disease symptom development. Specifically, identifying interaction patterns between coexisting pathogens – whether between phytoplasmas or between phytoplasmas and other pathogens – could reveal disease occurrence. This would help develop targeted control strategies and improve their control effect, thereby reducing the efficiency of related diseases.

In a mixed infection system, the diversity of pathogens leads to a complexity of pathogenic factors and disease mechanisms, which poses difficulties for the prevention and control management of related diseases. In terms of prevention and control of the disease associated with mixed infections of phytoplasmas and other pathogens, a single prevention and control method on its own is not enough. Instead, a comprehensive approach combining multiple measures is essential. For plants susceptible to multiple pathogens such as phytoplasma, '*Ca.* Liberibacter' species, and viruses, it is crucial to consider the possibility of co-infections during the breeding process of disease-resistant varieties, especially in natural screening environments. Usually, the breeding effort targets a single pathogen. However, this approach may fail when co-infections occur, as mixed infections can interfere with resistance screening. Disease-resistant varieties are fundamental for protecting plants from infections, so targeted screening for resistance against multiple pathogens – including phytoplasmas from different 16Sr groups and other pathogens such as '*Ca.* Liberibacter' species and viruses – is necessary to enhance the durability and effectiveness of these resistant varieties.

Conclusion

Plants can be co-infected by phytoplasmas from various 16Sr groups and other pathogens, such as '*Ca.* Liberibacter asiaticus,' '*Ca.* Liberibacter solanacearum,' viruses, spiroplasma, '*Ca.* Phloemobacter fragariae,' '*Ca.* Arsenophonus phytopathogenicus,' and fungi. These infections cause symptoms such as witches'-broom, yellowing, virescence, phyllody, small leaves, crinkled leaves, mottled leaves, and dwarfing. The genetic diversity of these pathogens and their complex interactions in the mixed infection system pose great difficulties for diagnosing, monitoring, preventing, and controlling related diseases. In the future, comprehensive pathogen detection will be crucial for the accurate diagnosis and monitoring of diseases associated with mixed infections. In terms of prevention and control of related diseases, it is essential to understand the interactions between different pathogens, integrate targeted technologies, and select disease-resistant varieties for multiple pathogens. Implementing prevention and control management measures will help reduce the agricultural impact of co-infection and ensure the health and safety of the related germplasm resources.

References

Abirami, R., Manoranjitham, S.K., Mohankumar, S., and Karthikeyan, G. (2022) Preponderance of mixed infection of Cucumber mosaic virus and '*Candidatus* Phytoplasma australasia' on brinjal in India. *Microbial Pathogenesis* 169, 105596.

Ahmed, E.A., Farrag, A.A., Kheder, A.A., and Shaaban, A. (2022) Effect of phytoplasma associated with sesame phyllody on ultrastructural modification, physio-biochemical traits, productivity and oil quality. *Plants* 11, 477.

Alfaro-Fernández, A., Cebrián, M.C., Villaescusa, F.J., Hermoso de Mendoza, A., Ferrándiz, J.C., Sanjuán, S., and Font, M.I. (2012) First report of '*Candidatus* Liberibacter solanacearum' in carrot in mainland Spain. *Plant Disease* 96, 582.

Alizadeh, H., Quaglino, F., Azadvar, M., Kumar, S., Alizadeh, A. *et al.* (2017) First report of a new citrus decline disease (CDD) in association with double and single infection by '*Candidatus* Liberibacter asiaticus' and '*Candidatus* Phytoplasma aurantifolia' related strains in Iran. *Plant Disease* 101, 2145.

Arce-Leal, A.P., Bautista, R., Rodriguez-Negrete, E.A., Manzanilla-Ramirez, M.A., Velazquez-Monreal, J.J. *et al.* (2020) Gene expression profile of Mexican Lime (*Citrus aurantifolia*) trees in response to Huanglongbing disease caused by *Candidatus* Liberibacter asiaticus. *Microorganisms* 8, 528.

Arocha, Y., Gonzalez, L., Peralta, E.L., and Jones, P. (1999) First report of virus and phytoplasma pathogens associated with yellow leaf syndrome of sugarcane in Cuba. *Plant Disease* 83, 1177.

Arocha, Y., Piñol, B., Acosta, K., Almeida, R., Devonshire, J. *et al.* (2009) Detection of phytoplasma and potyvirus pathogens in papaya (*Carica papaya* L.) affected with bunchy top symptom (BTS) in eastern Cuba. *Crop Protection* 28, 640–646.

Arocha Rosete, A., Lambert, L., Joly-Seguin, V. Michelutti, R., Schilder, A., and Bertaccini, A. (2019) Surveys reveal a complex association of phytoplasma and viruses with the blueberry stunt disease in Canadian blueberry farms. *Annals of Applied Biology* 174, 142–152.

Arratia-Castro, A.A., Santos-Cervantes, M.E., Fernández-Herrera, E., Chávez-Medina, J.A., Flores-Zamora, G.L. *et al.* (2014) Occurrence of '*Candidatus* Phytoplasma asteris' in citrus showing Huanglongbing symptoms in Mexico. *Crop Protection* 62, 144–151.

Banzato, T.C., Ferreira, J., and Bedendo, I.P. (2021) Field mustard (*Brassica rapa*) an invasive weed species in cauliflower fields is a host of multiple phytoplasmas. *Australasian Plant Pathology* 50, 403–405.

Bao, M., Zhong, Z., Chen, J., and Deng, X. (2021) Investigation of citrus HLB symptom variations associated with "*Candidatus* Liberibacter asiaticus" strains harboring different phages in Southern China. *Agronomy* 11, 2262.

Behrmann, S.C., Rinklef, A., Lang, C., Vilcinskas, A., and Lee, K.Z. (2023) Potato (*Solanum tuberosum*) as a new host for *Pentastiridius leporinus* (Hemiptera: Cixiidae) and *Candidatus* Arsenophonus Phytopathogenicus. *Insects* 14, 281.

Bertaccini, A., Duduk, B., Paltrinieri, S., and Contaldo, N. (2014) Phytoplasmas and phytoplasma diseases: A severe threat to agriculture. *American Journal of Plant Sciences* 5, 1763–1788.

Bertaccini, A., Arocha-Rosete, Y., Contaldo, N., Duduk, B., Fiore, N. *et al.* (2022) Revision of the '*Candidatus* Phytoplasma' species description guidelines. *International Journal of Systematic and Evolutionary Microbiology* 72, 005353.

Byrd, A., Deming, C., Cassidy, S., Harrison, O., Ng, W. *et al.* (2017) *Staphylococcus aureus* and *Staphylococcus epidermidis* strain diversity underlying pediatric atopic dermatitis. *Science Translational Medicine* 9, eaal4651.

Che, H.Y., Yu, S.S., Chen, W., Zheng, W.H., Cao, X.R., and Luo, D.Q. (2024) Molecular identification and characterization of novel taxonomic subgroups and new host plants in 16SrI and 16SrII group phytoplasmas and their evolutionary diversity on Hainan Island, China. *Plant Disease* 108, 1703–1718.

Chen, J., Pu, X., Deng, X., Liu, S., Li, H., and Civerolo, E. (2009) A phytoplasma related to '*Candidatus* Phytoplasma asteri' detected in citrus showing Huanglongbing (yellow shoot disease) symptoms in Guangdong, P.R. China. *Phytopathology* 99, 236–242.

Chen, Y., Fischbach, M., and Belkaid, Y. (2018) Skin microbiota-host interactions. *Nature* 553, 427–436.

Danet, J.L., Foissac, X., Zreik, L., Salar, P., Verdin, E., Nourrisseau, J.G., and Garnier, M. (2003) '*Candidatus* Phlomobacter fragariae' is the prevalent agent of marginal chlorosis of strawberry in French production fields and is transmitted by the planthopper *Cixius wagneri* (China). *Phytopathology* 93, 644–649.

Davis, R.E. and Worley, J.F. (1973) Spiroplasma: Motile, helical microorganism associated with corn stunt disease. *Phytopathology* 63, 403–408.

Doi, Y., Teranaka, M., Yora, K., and Asuyama, H. (1967) Mycoplasma-or PLT group-like microorganisms found in the phloem elements of plants infected with mulberry dwarf, potato witches' broom, aster yellows, or paulownia witches'-broom. *Japanese Journal of Phytopathology* 33, 259–266.

Dorado, C., Bowman, K.D., Cameron, R.G., Manthey J.A., Bai, J., and Ferguson, K.L. (2021) Steam explosion (STEX) of *Citrus* × *Poncirus* hybrids with exceptional tolerance to *Candidatus* Liberibacter asiaticus (CLas) as useful sources of volatiles and other commercial products. *Biology* 10, 1285.

Duduk, B., Stepanović, J., Yadav, A., and Rao, G.P. (2018) Phytoplasmas in weeds and wild plants. In: Rao, G.P., Bertaccini, A., Fiore, N., and Liefting, L.W. (eds) *Phytoplasmas: Plant Pathogenic Bacteria I – Characterisation and Epidemiology of Phytoplasma-Associated Diseases*. Springer, Singapore, pp. 313–345.

Duduk, B., Kosovac, A., Stepanovic, J., Rekanovic, E., Curcic, Z. *et al.* (2023) Phytoplasma, proteobacterium and fungus in single and mixed infections of sugar beet in central Europe. *Phytopathogenic Mollicutes* 13, 97–98.

El Sayed, A.I., Soufi, Z., Wahdan, K.M., and Komor, E. (2016) Detection and characterization of phytoplasma and sugarcane yellow leaf virus associated with leaf yellowing of sugarcane. *Journal of Phytopathology* 164, 217–225.

EPPO (European and Mediterranean Plant Protection Organization) (2012) First report of '*Candidatus* Liberibacter solanacearum' on carrots and celery in Spain, in association with *Bactericera trigonica*. *EPPO Reporting Service – Pests and Diseases* 6, article 2012/121.

Fernández, F.D., Galdeano, E., and Conci, L.R. (2020) Phytoplasmas diversity and identification of new aster yellows subgroup (16SrI) associated with weed species in Argentina. *International Journal of Systematic and Evolutionary Microbiology* 70, 35–43.

Galvão, S.R., Sabato, E.O., and Bedendo, I.P. (2021) Occurrence and distribution of single or mixed infection of phytoplasma and spiroplasma causing corn stunting in Brazil. *Tropical Plant Pathology* 46, 152–155.

Gatineau, F., Jacob, N., Vautrin, S., Larrue, J., Lherminier, J., Richard-Molard, M., and Boudon-Padieu, E. (2002) Association with the syndrome "basses richesses" of sugar beet of a phytoplasma and a bacterium-like organism transmitted by a *Pentastiridius* sp. *Phytopathology* 92, 384–392.

Gundersen, D.E. and Lee, I.M. (1996) Ultrasensitive detection of phytoplasmas by nested-PCR assays using two universal primer pairs. *Phytopathologia Mediterranea* 35, 114–151.

Gungoosingh-Bunwaree, A., Contaldo, N., Mejia, J.E., Paltrinieri, S., Benimadhu, S.P., and Bertaccini, A. (2013) Phytoplasma and virus diseases on tomato in Mauritius. *Australasian Journal of Plant Pathology* 42, 313–320.

Harne, S., Gayathri, P., and Béven, L. (2020) Exploring spiroplasma biology: Opportunities and challenges. *Frontiers in Microbiology* 11, 589279.

Hemmati, C., Nikooei, M., Al-Subhi, A.M., and Al-Sadi, A.M. (2021) History and current status of phytoplasma diseases in the middle east. *Biology* 10, 226.

IRPCM (International Research Programme on Comparative Mycoplasmology) (2004) '*Candidatus* Phytoplasma', a taxon for the wall-less, non-helical prokaryotes that colonize plant phloem and insects. *International Journal of Systematic and Evolutionary Microbiology* 54, 1243–1255.

Iwabuchi, N., Endo, A., Kameyama, N., Satoh, M., Miyazaki, A. *et al.* (2018) First report of "*Candidatus* Phytoplasma malaysianum" associated with *Elaeocarpus* yellows of *Elaeocarpus zollingeri*. *Journal of General Plant Pathology* 84, 160–164.

Jones, R.A.C. and Barbetti, M.J. (2012) Influence of climate change on plant disease infections and epidemics caused by viruses and bacteria. *CABI Reviews* 7, 1–32.

Kohda, K., Li, X., Soga, N., Nagura, R., Duerna, T. *et al.* (2021) An in vitro mixed infection model with commensal and pathogenic staphylococci for the exploration of interspecific interactions and their impacts on skin physiology. *Frontiers in Cellular and Infection Microbiology* 11, 712360.

Kumar, M., Katiyar, A., Madhupriya, and Rao, G.P. (2016) First report of association of potato virus X and potato virus Y and '*Candidatus* Phytoplasma trifolii' in brinjal in India. *Virus Disease* 27, 207–208.

Kumar, S., Sharma, P., Sharma, S., and Rao, G.P. (2015) Mixed infection and natural spread of 'Candidatus Phytoplasma asteris' and mungbean yellow mosaic India virus affecting soya bean crop in India. *Journal of Phytopathology* 163, 395–406.

Lebsky, V., Hernandez, J., Barraza, A., Oropeza, C., Narvaez, M. *et al.* (2019) Ultrastructural analysis of spiroplasmas detected in palm species infected with the lethal yellowing phytoplasma from Yucatan and Baja California Sur, Mexico. *Phytopathogenic Mollicutes* 9, 155–156.

Lee, G.W. and Han, S.S. (2023) Molecular detection of phytoplasmas of the 16SrI and 16SrXXXII groups in *Elaeocarpus sylvestris* trees with decline disease in Jeju Island, South Korea. *Plant Pathology Journal* 39, 149–157.

Lee, S., Han, S., and Cha, B. (2009) Mixed infection of 16S rDNA I and V groups of phytoplasma in a single jujube tree. *Plant Pathology Journal* 25, 21–25.

Lee, G.W., Han, T.W., Lee, S.K., and Han, S.S. (2022) *Candidatus* phytoplasma malaysianum (16SrXXXII) associated with *Elaeocarpus sylvestris* decline in South Korea. *Forest Science and Technology* 18, 7–13.

Legg, J.P. and Fauquet, C.M. (2004) Cassava mosaic geminiviruses in Africa. *Plant Molecular Biology* 56, 31–39

Li, B., Wang, S., Zhang, Y., and Qiu, D. (2020) Acid soil improvement enhances disease tolerance in citrus infected by *Candidatus* Liberibacter asiaticus. *International Journal of Molecular Sciences* 21, 3614.

Li, Y., Piao, C.G., Tian, G.Z., Liu, Z.X., Guo, M.W., Lin, C.L., and Wang, X.Z. (2014) Multilocus sequences confirm the close genetic relationship of four phytoplasmas of peanut witches'-broom group 16SrII-A. *Journal of Basic Microbiology* 54, 818–827.

Liefting, L.W., Egusquiza, Z.C., Clover, G.R.G., and Anderson, J.A.D. (2008) A new '*Candidatus* Liberibacter' species in *Solanum tuberosum* in New Zealand. *Plant Disease* 92, 1474.

Luis-Pantoja, M., Paredes-Tomás, C., Uneau, Y., Myrie, W., Morillon, R. *et al.* (2021) Identification of '*Candidatus* Phytoplasma' species in "huanglongbing" infected citrus orchards in the Caribbean. *European Journal of Plant Pathology* 160, 185–198.

Mall, S. and Vishwakarma, R. (2023) An update status of coinfection of phytoplasmas with other pathogens in plants. *Phytopathogenic Mollicutes* 13, 151–162.

Manivannan, S., Otta, S.K., Karunasagar, I., and Karunasagar, I. (2002) Multiple viral infection in *Penaeus monodon* shrimp postlarvae in an Indian hatchery. *Diseases of Aquatic Organisms* 48, 233–236.

Mitra, S., Debnath, P., Rai, R., Srivastava, N., Rao, G.P., and Baranwal, V.K. (2022) Identification of '*Ca.* Phytoplasma asteris', banana bunchy top virus and banana streak MY virus associated with Champa and Sabri banana cultivars in Tripura, a north eastern state of India. *European Journal of Plant Pathology* 163, 907–922.

Mori, N., Cargnus, E., Martini, M., and Pavan, F. (2020) Relationships between *Hyalesthes obsoletus*, its herbaceous hosts and bois noir epidemiology in northern Italian vineyards. *Insects* 11, 606.

Munyaneza, J.E., Fisher, T.W., Sengoda, V.G., Garczynski, S.F., Nissinen, A., and Lemmetty, A. (2010) First report of '*Candidatus* Liberibacter solanacearum' in carrots in Europe. *Plant Disease* 94, 639.

Nan, J., Zhang, S., Zhan, P., and Jiang, L. (2021) Discovery of novel GMPS inhibitors of *Candidatus* Liberibacter asiaticus by structure based design and enzyme kinetic. *Biology* 10, 594.

Nejat, N., Sijam, K., Abdullah, S.N.A., Vadamalai, G., and Dickinson, M. (2009) Phytoplasmas associated with disease of coconut in Malaysia: Phylogenetic groups and host plant species. *Plant Pathology* 58, 1152–1160.

Nejat, N., Vadamalai, G., Davis, R.E., Harrison, N.A., Sijam, K. *et al.* (2013) '*Candidatus* Phytoplasma malaysianum', a novel taxon associated with virescence and phyllody of Madagascar periwinkle (*Catharanthus roseus*). *International Journal of Systematic and Evolutionary Microbiology* 63, 540–548.

Nithya, K., Parameswari, B., and Viswanathan, R. (2020) Mixed infection of sugarcane yellow leaf virus and grassy shoot phytoplasma in yellow leaf affected Indian sugarcane cultivars. *Plant Pathology Journal* 36, 364–377.

Oliveira, E., Landau, E.C., and Sousa, S.M. (2015) Simultaneous transmission of phytoplasma and spiroplasma by *Dalbulus maidis* leafhopper and symptoms of infected maize. *Phytopathogenic Mollicutes* 5, S99.

Paul, A., Baral, B., Parida, S., Naik, D.N., and Sahoo, P.K. (2022) Mixed infections in tropical freshwater fish culture systems: A potential emerging threat for successful aquaculture. *Aquaculture Asia Magazine* 26, 20–24.

Quiroga, N., Gamboa, C., Medina, G., Contaldo, N., Torres, F. *et al.* (2022) Survey for '*Candidatus* Liberibacter' and '*Candidatus* Phytoplasma' in Citrus in Chile. *Pathogens* 11, 48.

Ramos, A., Esteves, M.B., Cortes, M.T.B., and Lopes, J.R.S. (2020) Maize bushy stunt phytoplasma favors its spread by changing host preference of the insect vector. *Insects* 11, 600.

Redinbaugh, M.G. and Stewart, L.R. (2018) Maize lethal necrosis: An emerging, synergistic viral disease. *Annual Review of Virology* 5, 301–322.

Sahoo, P.K., Paul, A., Sahoo, M.K., Pattanayak, S., Rajesh Kumar, P., and Das, B.K. (2020) Incidences of infectious diseases in freshwater aquaculture farms of eastern India: A passive surveillance based study from 2014–2018. *Journal of Aquaculture Research and Development* 11, 1–5.

Salehi, M., Faghihi, M.M., Salehi, E., and Bertaccini, A. (2022) Occurrence of single and mixed infection of *Spiroplasma citri* and phytoplasmas in sesame plants in Iran. *Australasian Plant Pathology* 51, 13–26.

Satta, E., Ramirez, A.S., Paltrinieri, S., Contaldo, N., Benito, P., Poveda, J.B., and Bertaccini, A. (2016) Simultaneous detection of mixed '*Candidatus* Phytoplasma asteris' and '*Ca*. Liberibacter solanacearum' infection in carrot. *Phytopathologia Mediterranea* 55, 401–409.

Schmitt, U., Petzold, H., and Marwitz, R. (1984) An in situ freeze-fracture study of *Spiroplasma citri* and the corn stunt Spiroplasma. *Phytopathology* 111, 297–304.

Singhal, P., Nabi, S.U., Yadav, M.K., and Dubey, A. (2021) Mixed infection of plant viruses: diagnostics, interactions and impact on host. *Journal of Plant Diseases and Protection* 128, 353–368.

Sugio, A., Maclean, A.M., Kingdom, H.N., Grieve, V.M., Manimekalai, R., and Hogenhout, S.A. (2011) Diverse targets of phytoplasma effectors: From plant development to defense against insects. *Annual Review of Phytopathology* 49, 175–195.

Sumner-Kalkun, J.C., Highet, F., Arnsdorf, Y.M., Back, E., Carnegie, M. *et al.* (2020) '*Candidatus* Liberibacter solanacearum' distribution and diversity in Scotland and the characterisation of novel haplotypes from *Craspedolepta* spp. (Psylloidea: Aphalaridae). *Scientific Reports* 10, 16567.

Sun, X.C., Mou, H.Q., Li, T.T., Tian, Q., and Zhao, W.J. (2013) Mixed infection of two groups (16SrI & V) of phytoplasmas in a single jujube tree in China. *Journal of Phytopathology* 161, 661–665.

Swaminathan, T.R., Kumar, R., Dharmaratnam, A., Basheer, V.S., Sood, N. *et al.* (2016) Emergence of carp edema virus in cultured ornamental koi carp, *Cyprinus carpio* koi, in India. *Journal of General Virology* 97, 3392–3399.

Teixeira, D.D.C., Wulff, N.A., Martins, E., Elliot, K., Bassanezi, R.B. *et al.* (2008) A phytoplasma closely related to the pigeon pea witches'-broom phytoplasma (16SrIX) is associated with citrus huanglongbing symptoms in the state of São Paulo, Brazil. *Phytopathology* 98, 977–984.

Thamizhvanan, S., Sivakumar, S., Santhosh Kumar, S., Vinoth Kumar, D., Suryakodi, S. *et al.* (2019) Multiple infections caused by white spot syndrome virus and *Enterocytozoon hepatopenaei* in pond-reared *Penaeus vannamei* in India and multiplex PCR for their simultaneous detection. *Journal of Fish Disease* 42, 447–454.

Tiwari, N.N., Jain, R.K., Prajapati, M.R., Singh, J., Srivastava, S., Tiwari, A.K., and Marcone, C. (2022) Evidence of mixed infection of phytoplasma and begomovirus associated with *Withania somnifera* and *Capsicum annum* plants from Uttar Pradesh, India. *Archives of Phytopathology and Plant Protection* 55, 2146–2157.

Venkataravanappa, V., Prasanna, H.C., Lakshminarayana, C.N., and Reddy, M.K. (2018) Molecular detection and characterization of phytoplasma in association with begomovirus in eggplant. *Acta Virologica* 62, 246–258.

Wang, J., Tian, G.Z., Xu, Q.C., Liu Y.G., Gao, R., Li, X.D., and Zhu, X.P. (2010) Molecular detection of phytoplasma strains from several plants around diseased paulownia infected with paulownia witches'-broom phytoplasma. *Scientia Agricultura Sinica* 43, 304–312.

Wang, Z., Wang, W., Yuan, E., Su, F., Mao, Q., Wan, Q., and Cai, H. (2021) Identification of *Camptotheca acuminate* witches-broom phytoplasma and *Taq*Man-based fluorescence qPCR method for detection. *Acta Phytopathologica Sinica* 51, 429–440.

Wangai, A.W., Redinbaugh, M.G., Kinyua, Z.M., Miano, D.W., Leley, P.K. *et al.* (2012) First report of maize chlorotic mottle virus and maize lethal necrosis in Kenya. *Plant Disease* 96, 1582.

Wei, W., Trivellone, V., Dietrich, C.H., Zhao, Y., Bottner-Parker, K.D., and Ivanauskas, A. (2021) Identification of phytoplasmas representing multiple new genetic lineages from phloem-feeding leafhoppers highlights the diversity of phytoplasmas and their potential vectors. *Pathogens* 10, 352.

Wei, W., Inaba, J., Zhao, Y., Mowery, J.D., and Hammond, R. (2022) Phytoplasma infection blocks starch breakdown and triggers chloroplast degradation, leading to premature leaf senescence, sucrose reallocation, and spatiotemporal redistribution of phytohormones. *International Journal of Molecular Sciences* 23, 1810.

Xu, Y., Ghanim, M., and Liu, Y. (2022) Editorial: Mixed infections of plant viruses in nature and the impact on agriculture. *Frontiers in Microbiology* 13, 922607.

Yang, Y., Jiang, L., Che, H.Y., Cao, X.R., and Luo, D.Q. (2016) Phytoplasma in association with rubber tree (*Hevea brasiliensis*) stem fasciation in China. *Plant Disease* 100, 2520.

Yang, Y., Li, D., Duan, Y., and Jiang, L. (2020) Investigation and distribution of the citrus Huanglongbing and identification of Huanglongbing Liberibacter species in Hainan. *Plant Quarantine* 34, 43–47.

Yu, H., Liu, S., Ruan, K., Chen, Y., and Wang, Z. (2009) Advances in classification and biodiversity of spiroplasmas-a review. *Acta Microbiologica Sinica* 49, 567–572.

Yu, S.S., Xu, Q.C., Lin, C.L., Wang, S.J., and Tian, G.Z. (2016) Genetic diversity of phytoplasmas: research status and prospects. *Biodiversity Science* 24, 205–215.

Yu, S.S., Song, W.W., and Qin, W.Q. (2021a) Molecular detection and phylogenetic relationship of areca palm yellow leaf phytoplasma in Hainan, China. *Chinese Journal of Tropical Crops* 42, 3066–3072.

Yu, S.S., Tang, Q.H., Wu, Y., Lin, M.X., Zhao, R.L., Song, W.W., and Qin, W.Q. (2021b) First report of 16SrXXXII group related phytoplasma associated with *Trema tomentosa* witches'-broom disease in China. *Plant Disease* 105, 1191.

Yu, S.S., Zhao, R.L., Lin, M.X., Wu, Y., Chen, S.G. *et al.* (2021c) Occurrence of phytoplasma belongs to 16SrII group associated with witches'-broom symptoms in *Emilia sonchifolia* in Hainan Island of China. *Plant Disease* 105, 4151.

Yu, S.S., Zhao, R.L., Lin, M.X., Wu, Y., Chen, S.G. *et al.* (2021d) *Waltheria indica* is a new host of phytoplasma belongs to 16SrI-B subgroup associated with virescence symptoms in China. *Plant Disease* 105, 2012.

Yu, S.S., Wu, Y., and Song, W.W. (2022a) Occurrence of a 16SrII-V subgroup phytoplasma associated with witches'-broom disease in *Melochia corchorifolia* in China. *Plant Disease* 106, 754.

Yu, S.S., Zhu, A.N., Song, W.W., and Yan, W. (2022b) Molecular identification and characterization of two groups of phytoplasma and *Candidatus* Liberibacter asiaticus in single or mixed infection of *Citrus maxima* on Hainan Island of China. *Biology* 11, 869.

Yu, S.S., Zhu, A.N., Che, H.Y., and Song, W.W. (2024) Molecular identification of '*Candidatus* Phytoplasma malaysianum'-related strains associated with *Areca catechu* palm yellow leaf disease and phylogenetic diversity of the phytoplasmas within 16SrXXXII group. *Plant Disease* 108,1331–1343.

Zhu, A., Yu, S., Su, L., Liu, L., Song, W., and Yan, W. (2023) Molecular detection and genetic variation of phytoplasmas from eight plants in garden of areca with yellow leaf disease in China. *Chinese Journal of Tropical Crops* 44, 1190–1202.

Index

Note: page numbers in **bold** refer to figures and tables.